Lecture Notes in Computer Science 963

Edited by G. Goos, J. Hartmanis and J. van Leeuwen

Advisory Board: W. Brauer   D. Gries   J. Stoer

Springer-Verlag Berlin Heidelberg GmbH

Don Coppersmith (Ed.)

Advances in Cryptology – CRYPTO '95

15th Annual International Cryptology Conference
Santa Barbara, California, USA, August 27-31, 1995
Proceedings

 Springer

Series Editors

Gerhard Goos, Universität Karlsruhe, Germany

Juris Hartmanis, Cornell University, NY, USA

Jan van Leeuwen, Utrecht University, The Netherlands

Volume Editor

Don Coppersmith
IBM T.J. Watson Research Center, Mathematical Sciences 32-256
P.O.Box 218, Yorktown Heights, NY 10598, USA

Cataloging-in-Publication data applied for

Die Deutsche Bibliothek - CIP-Einheitsaufnahme

Advances in cryptology : proceedings / CRYPTO '95, 15th
Annual International Cryptology Conference, Santa Barbara,
California, USA, August 27 - 31, 1995. Don Coppersmith (ed.).

(Lecture notes in computer science ; 963)
ISBN 3-540-60221-6
NE: Coppersmith, Don [Hrsg.]; CRYPTO <15, 1995, Santa Barbara,
Calif.>; GT

CR Subject Classification (1991): E.3-4, G.2.1, D.4.6, F2.1-2, C.2, J.1

1991 Mathematics Subject Classification: 94A60, 11T71, 11Yxx, 68P20,
68Q20, 68Q25

ISBN 978-3-540-60221-7   ISBN 978-3-540-44750-4 (eBook)

DOI 10.1007/978-3-540-44750-4

© Springer-Verlag Berlin Heidelberg 1995

Originally published by Springer-Verlag Berlin Heidelberg New York in 1995

Typesetting: Camera-ready by author
SPIN 10486614   06/3142 – 5 4 3 2 1 0   Printed on acid-free paper

PREFACE

The Crypto '95 conference was sponsored by the International Association for Cryptologic Research (IACR), in cooperation with the IEEE Computer Society Technical Committee on Security and Privacy, and the Computer Science Department of the University of California, Santa Barbara. It took place at the University of California, Santa Barbara, from August 27-31, 1995. This was the fifteenth annual Crypto conference; all have been held at UCSB. For the second time, proceedings were available at the conference. The General Chair, Stafford Tavares, was responsible for local organization and registration.

The Program Committee considered 151 papers and selected 36 for presentation. There were also two invited talks. Robert Morris, Sr. gave a talk on "Ways of Losing Information," which included some non-cryptographic means of leaking secrets that are often overlooked by cryptographers. The second talk, "Cryptography - Myths and Realities," was given by Adi Shamir, this year's IACR Distinguished Lecturer. Shamir is the second person to receive this honor, the first having been Gus Simmons at Crypto '94.

These proceedings contain revised versions of the 36 contributed talks. Each paper was sent to at least three members of the program committee for comments. Revisions were not checked on their scientific aspects. Some authors will write final versions of their papers for publication in refereed journals. Of course, the authors bear full responsibility for the contents of their papers.

I am very grateful to the members of the Program Committee for their hard work and the difficult task of selecting one quarter of the submitted papers. Following recent traditions, the submissions were anonymous; and each program committee member could be the author of at most one accepted paper.

We thank the following referees and external experts for their help on various papers: Philippe Béguin, Mihir Bellare, Charles Bennett, Gilles Brassard, Florent Chabaud, Chris Charnes, Yair Frankel, Atsushi Fujioka, Thomas Hardjono, Philippe Hoogvorst, Nobuyuki Imoto, Toshiya Itoh, Sushil Jajodia, Lars Knudsen, Paul Kocher, Mitsuru Matsui, Tsutomu Matsumoto, David M'Raihi, Yi Mu, Rafail Ostrovsky, Eiji Okamoto, Tatsuaki Okamoto, David Pointcheval, Rei Safavi-Naini, Kouichi Sakurai, Jennifer Seberry, Hiroki Shizuya, Dan Simon, Othmar Staffelbach, Jacques Stern, Moti Yung and Xian-Mo Zhang. I apologize for any omissions.

I thank Baruch Schieber and Prabhakar Raghavan for help with software and LaTeX; Barbara White and Peg Cargiulo for secretarial help; and Yvo Desmedt, Jimmy Upton and Peter Landrock for advice on the mechanics.

Finally, thanks go to all who submitted papers for Crypto '95. The success of the conference depends on the quality of its submissions. I am also thankful for all the authors, who cooperated by delivering their final copy to me in a timely fashion for the proceedings.

Don Coppersmith
Program Chair, Crypto '95
IBM Research Division, Yorktown Heights, New York, USA
June, 1995

CRYPTO '95

University of California, Santa Barbara
August 27-31, 1995

Sponsored by the

International Association for Cryptologic Research

in cooperation with the

*IEEE Computer Society Technical Committee
on Security and Privacy*

and the

*Computer Science Department,
University of California, Santa Barbara*

General Chair

Stafford Tavares, Queen's University, Canada

Program Chair

Don Coppersmith, IBM T.J. Watson Research Center, USA

Program Committee

Ross Anderson — Cambridge University, UK
Ernest Brickell — Sandia National Laboratories, USA
Hugo Krawczyk — IBM T.J. Watson Research Center, USA
Susan Langford — Stanford University, USA
Kevin McCurley — Sandia National Laboratories, USA
Willi Meier — HTL Brugg-Windisch, Switzerland
Moni Naor — Weizmann Institute of Science, Israel
Andrew Odlyzko — AT&T Bell Laboratories, USA
Kazuo Ohta — NTT Laboratories, Japan
Josef Pieprzyk — University of Wollongong, Australia
Jean-Jacques Quisquater — UCL-MathRIZK, Belgium
Alan Sherman — Univ. of Maryland Baltimore County, USA
Scott Vanstone — University of Waterloo, Canada
Serge Vaudenay — Ecole Normale Supérieure, France

CONTENTS

MAC and Hash

Number Theory I

Oblivious Transfer

Cryptanalysis I

Key Escrow

Protocols

Cryptanalysis II

Zero Knowledge, Interactive Protocols

Secret Sharing

Number Theory II

Secret Sharing II

Everything Else

MDx-MAC and Building Fast MACs from Hash Functions

Bart Preneel[1]* Paul C. van Oorschot[2]

[1] Katholieke Universiteit Leuven, Dept. Electrical Engineering-ESAT,
Kardinaal Mercierlaan 94, B-3001 Heverlee, Belgium
bart.preneel@esat.kuleuven.ac.be
[2] Bell-Northern Research, P.O. Box 3511 Station C,
Ottawa, Ontario, K1Y 4H7, Canada
paulv@bnr.ca

Abstract. We consider the security of message authentication code
(MAC) algorithms, and the construction of MACs from fast hash functions. A new forgery attack applicable to all iterated MAC algorithms is
described, the first known such attack requiring fewer operations than exhaustive key search. Existing methods for constructing MACs from hash
functions, including the secret prefix, secret suffix, and envelope methods, are shown to be unsatisfactory. Motivated by the absence of a secure,
fast MAC algorithm not based on encryption, a new generic construction
(*MDx*-MAC) is proposed for transforming any secure hash function of
the MD4-family into a secure MAC of equal or smaller bitlength and
comparable speed.

1 Introduction

Hash functions play a fundamental role in modern cryptography. One main application is their use in conjunction with digital signature schemes; another is
in conventional techniques for message authentication. In the latter, it is preferable that a hash function take as a distinct secondary input a secret key. Such
hash functions, commonly known as *message authentication codes* (MACs), have
received widespread use in practice for data integrity and data origin authentication, e.g. in banking applications (see [7, 17]).

Compared to the extensive work on the design and analysis of hash functions,
little attention has been given to the design of efficient MACs [22] (although see
[2]). One apparent reason is that the first proposals for MAC algorithms were
quickly turned into standards and proved adequate in practice. The first constructions are based on the Cipher Block Chaining (CBC) and Cipher FeedBack
(CFB) modes of a block cipher [14, 15]. Most standards and applications use
the CBC mode (CBC-MAC); theoretical support for this construction was given
recently in [1]. Another proposal dating back to 1984 is the Message Authenticator Algorithm (MAA) [5, 6, 14], for which no significant weaknesses have

---

* N.F.W.O. postdoctoral researcher, sponsored by the National Fund for Scientific
Research (Belgium).

D. Coppersmith (Ed.): Advances in Cryptology - CRYPTO '95, LNCS 963, pp. 1-14, 1995.
© Springer-Verlag Berlin Heidelberg 1995

previously been identified. MAA is a current ISO standard, and is relatively fast in software (about 40% slower than MD5). Its main disadvantage is that the result, being 32 bits, is considered unacceptably short for many applications. Recent research on authentication codes has resulted in very fast, scalable, and information theoretically secure constructions [16, 19, 28], which require relatively short keys. The disadvantage is that a different key must be used for every message. If this is not acceptable, one can generate the key using a cryptographically strong pseudo-random string generator, but the resulting scheme is then (at most) computationally secure.

Around 1991, Rivest proposed two very fast hash functions, namely MD4 [24] and MD5 [25]. Later RIPEMD [23] and SHA [12] were introduced by other research groups. In software, these hash functions may be as much as one order of magnitude faster than DES. Several factors have motivated their adoption as the basis for MAC algorithms: system designers quickly realized that MACs based on these outperform other available options; the additional implementation effort required to use these as MACs is very small; and the fact that such MACs do not involve encryption algorithms has favorable export implications. Because of these factors, MAC constructions based on these hash functions were adopted in Kerberos [20] and SNMP [13] and are proposed for IPSEC [18].

In this paper, a new general attack is proposed which applies to all iterated MACs, including MAA and CBC-MAC. It is a birthday attack on known text-MAC pairs, which with a few additional chosen text-MAC pairs, allows MAC forgery. An extension of the attack is also given. The best previous general attack on MAC algorithms was an exhaustive search for the key. The new attack requires a number of known text-MAC pairs which is $O(2^{n/2})$, where n is the bitlength of the internal memory (chaining variable) of the MAC algorithm.

We then analyze three proposals for MAC algorithms based on hash functions: prepending a secret key to the message input of the hash function (secret prefix method), appending a secret key to the input (secret suffix method), and combining both operations (envelope method). For the secret prefix and secret suffix method, a systematic analysis is given which generalizes the known attacks. For the envelope method, the new general attack applies and illustrates that the proof outline in [26] for the security of this method is incorrect. Our conclusion is that these approaches do not achieve the security level suggested by the size of the parameters. Moreover, some variants of these methods are susceptible to more serious attacks.

In addition to the concerns this raises about these constructions for MAC algorithms from hash functions, the new attack calls into question the strength of MAA and CBC-MAC. This creates a need for a new fast MAC algorithm offering security substantially better than those which succumb to attacks requiring on the order of 2^{32} known or chosen text-MAC pairs.

Motivated by the above reasons, and the lack of an acceptable, non-encryption based fast MAC algorithm providing more than 32-bit results, we propose a new generic construction (MDx-MAC), suitable for application to MD4-family hash functions including MD5, RIPEMD, and SHA, and yielding comparable throughput. The security of the proposed construction is also examined.

The remainder of this paper is organized as follows. §2 reviews the definitions of a hash function and a MAC. §3 discusses the new general attack on MACs. §4 analyzes three previous proposals for constructing a MAC based on a secure hash function. §5 describes and examines the new MAC construction. §6 concludes the paper.

2 Definitions and Background

Hash functions are functions that map bitstrings of arbitrary finite length into strings of fixed length. Given h and an input x, computing $h(x)$ must be easy. First we give definitions of hash functions which do not involve secret parameters. A *one-way hash function* must satisfy the following properties:

- **preimage resistance**: it is computationally infeasible to find any input which hashes to any pre-specified output.
- **second preimage resistance**: it is computationally infeasible to find any second input which has the same output as any specified input.

For an *ideal* one-way hash function with an m-bit result, finding a preimage or a second preimage requires $O(2^m)$ operations. A *collision resistant hash function* is a one-way hash function that satisfies an additional condition:

- **collision resistance**: it is computationally infeasible to find a collision, i.e. two distinct inputs that hash to the same result.

For an *ideal* collision resistant hash function with an m-bit result, no attack finding a collision betters a birthday or square root attack of $O(2^{m/2})$ operations.

A MAC is a hash function with a secondary input, the secret key K. Given h, an input x, and the secret key K, computing $h(x)$ must be easy. (Note K here is assumed to be an implicit parameter of $h(x)$.) The strongest condition one may impose on a MAC is that for someone who does not know the secret key, it be computationally infeasible to perform an *existential forgery*, i.e. to find an arbitrary message and its corresponding MAC. This should be contrasted to a *selective forgery*, where an opponent can determine the MAC for a message of his choice. For a practical attack, one often requires that the forgery is *verifiable*, which means that the MAC is correct with probability close to 1. Here we assume that the opponent is capable of performing a *chosen text attack*, i.e. may obtain MACs corresponding to a number of messages of his choice. We allow in fact a stronger notion, namely an *adaptive* chosen text attack, in which his requests may depend on the outcome of previous requests. To be meaningful, a forgery must be for a message different than any for which a MAC was previously obtained.

For an *ideal* MAC, any method to find the key is as expensive as an exhaustive search of $O(2^k)$ operations for a k-bit key. The number of text-MAC pairs required for verification of such an attack is k/m. An opponent who has identified the correct key can compute the MAC for any message (i.e. key recovery allows selective forgery). If the opponent knows no text-MAC pairs, or if $m < k$, his best strategy may be to simply guess the MAC corresponding to a chosen message; the probability of success is $1/2^m$. The disadvantage of a guessing attack is that it is not verifiable. A further desirable property of an ideal MAC is that finding a second preimage should require $O(2^m)$ known text-MAC pairs. In some settings (e.g. multi-destination electronic mail [21]) it may be desirable that this requires $O(2^m)$ off-line MAC computations even for someone who knows the key.

Most hash functions, including MACs, are designed as iterative processes which hash inputs of arbitrary length by processing successive fixed-size b-bit blocks of the input. The input x is divided into t blocks x_1 through x_t. If the total length is not a multiple of b, the input is padded using an unambiguous padding rule. The hash function h can be described as follows:

$$H_0 = IV; \qquad H_i = f(H_{i-1}, x_i), 1 \leq i \leq t \qquad h(x) = H_t.$$

Here f is the *compression function* of h, and H_i is the *chaining variable* between stage $i - 1$ and stage i with bitlength n ($n \geq m$).

In the case of a MAC, one often applies an output transformation g to H_t to obtain the hash-result, i.e. $h(x) = g(H_t)$. For example, in the CBC-MAC as specified in [14, 15] the output transformation g consists of selecting the leftmost m bits. The secret key can be introduced in the IV, in the compression function f, and in the output transformation g.

3 A New General Attack on MAC Algorithms

We describe a new attack which is applicable to all iterated MACs. The parameters depend only on the bitsize n of the chaining variables and on the bitsize m of the hash-result. We first make no assumptions about the texts being hashed, and further below give an optimization in the case that texts have a common sequence of s trailing blocks.

To facilitate Proposition 2 below, we first make two simple definitions followed by a lemma. Consider the pair (x, x') with $h(x) = g(H_t)$ and $h(x') = g(H_t')$, where g is the output transformation as defined above. Given a collision $h(x) = h(x')$, it may have arisen in one of two ways. An *internal* collision is said to have occurred if $H_t = H_t'$. An *external* collision is said to have occurred if $H_t \neq H_t'$ but $g(H_t) = g(H_t')$.

Lemma 1. *An internal collision for an iterated MAC algorithm can be used to obtain a verifiable MAC forgery with a chosen text attack requiring only one requested MAC.*

Proof: For an internal collision (x, x'), note $h(x \| y) = h(x' \| y)$ for any single block y. Thus requesting a MAC for the single chosen text $x \| y$, permits forgery – the MAC for $x' \| y$ is the same (here $\|$ denotes concatenation). ■

Note that this observation has been made independently by others including H. Krawczyk (e.g. see [18]). It follows that a security requirement for MACs is that it be infeasible for an adversary to find internal collisions (cf. collision resistance for hash functions).

Proposition 2. *Let h be an iterated MAC with n-bit chaining variable and m-bit result. An internal collision for h can be found using u known text-MAC pairs and v chosen texts. The expected values for u and v are as follows: $u = \sqrt{2} \cdot 2^{n/2}$ and[1] $v = 0$ if the output transformation g is a permutation; otherwise, v is approximately $2 \cdot 2^{n-m} + 2 \lceil \frac{n}{m} \rceil$.*

Proof: If the number of known texts is $r = \sqrt{2} \cdot 2^{n/2}$, a single internal collision is expected by the birthday paradox (note $\binom{r}{2}/2^n \approx 1$). If g is a permutation (e.g. the identity mapping), all collisions are internal and the results follows by Lemma 1. If g behaves as a random function[2], $\binom{r}{2}/2^m \approx r^2/2^{m+1} = 2^{n-m}$ external collisions are expected and additional work is required – for a verifiable forgery – to distinguish the internal collision from the external collisions. (Note Lemma 1 requires internal collisions.) This may be done by appending a string y to both elements of each collision pair and checking whether the corresponding MACs are equal. This requires $2(1 + 2^{n-m})$ chosen text-MAC requests. For an internal collision both results will always be equal, while for an external collision this will be so with probability[3] $1/2^m$. Discard collision pairs corresponding to unequal MACs. The expected number of remaining collision pairs after this stage is 2^{n-2m} external plus one internal (but these cannot yet be distinguished). If the (total) number of remaining collision pairs is 2 or more (e.g. $n - 2m > 0$), further external collisions must be discarded by appending a different y, and continuing in this manner until only a single collision remains; with high probability this is an internal collision. This may require a small number of additional chosen texts and a total number $2 \cdot 2^{n-m} \cdot 2^m/(2^m - 1) + 2 \lceil \frac{n}{m} \rceil$. ∎

Note that creating t MAC forgeries by this method requires one internal collision and t chosen-text MAC requests. The cost of this one internal collision is given by Proposition 2.

The attack outlined in the proof of Proposition 2 yields an internal collision (x, x'). If x and x' have a common sequence of s trailing blocks and if the compression function f is a permutation (for fixed x_i), the collision must occur at H_{t-s}, i.e. just before the common blocks. After deleting the s common blocks in x and x', one still has an internal collision. In this case the attack can be enhanced since this provides additional freedom in the choice of the forged text by Lemma 1. In particular, if x and x' have the same length one can obtain a forgery on a text of that length. As a significant consequence, *in this case the attack cannot be precluded by prepending the length of the input before the MAC calculation or by fixing the length of the input.*

If all the texts in the known text-MAC pairs of Proposition 2 have a common sequence of s trailing blocks, and if the compression function behaves as a random

---

[1] For this choice of u the probability of the attack failing is $1/e$; however, such failure can be detected in which case selecting additional known text-MAC pairs is required. By doubling the number of known text-MAC pairs the probability of failure decreases to $1/e^4$. The same probabilities hold for the value u given in Proposition 4.

[2] This is formalized in the full paper. The effective random mapping is from n bits to m bits; if the image is smaller, the collision probability increases.

[3] If y has i blocks, the probability increases by a factor i (see Lemma 3 with $r = 2$).

mapping, fewer known and chosen texts are required. In order to prove this we use a generalization of the birthday attack (proof to be given in the full paper):

Lemma 3. *Let h be an iterated MAC with n-bit chaining variable, a compression function f which behaves like a random function (for fixed x_i), and an output transformation g that is a permutation. Consider a set of $r \geq 2$ distinct messages which have the last s blocks in common, with $r^2 s = O(2^n)$. The probability that the set contains at least two messages that collide under h is approximately*

$$1 - \exp\left(-\frac{r^2(s+1)}{2^{n+1}}\right).$$

∎

(Note this reduces to the well known birthday attack for $s = 0$.) This lemma can be extended easily to the case where g is a random function. It yields an optimization of Proposition 2 as follows (proof to be given in the full paper):

Proposition 4. *Let h be an iterated MAC with n-bit chaining variable, m-bit result, a compression function f which behaves like a random function (for fixed x_i), and output transformation g. An internal collision for h can be found using u known text-MAC pairs, where each text has the same substring of $s \geq 0$ trailing blocks, and v chosen texts. The expected values for u and v are as follows: $u = \sqrt{2/(s+1)} \cdot 2^{n/2}$; $v = 0$ if g is a permutation or $s + 1 \geq 2^{n-m+6}$ (the expected number of external collisions is sufficiently small), and otherwise v is approximately*

$$2 \cdot \frac{2^{n-m}}{(s+1)} + 2 \left\lceil \frac{n - \log_2(s+1)}{m} \right\rceil.$$

∎

If we have an internal collision with $s \geq 1$, the probability that it occurs before the last w blocks equals $1 - w/(s+1)$. This event can be checked with a small number of additional chosen texts. Again the attack still works if one appends an arbitrary block y after the internal collision rather than at the end. This means that an attacker can replace or delete $w \leq s$ trailing blocks, and that *the attack is applicable even if the input is of fixed length or if the length is prepended to the input* (cf. [1]). A non-verifiable version of the attack requires $\sqrt{2/(s+1)} \cdot 2^{n/2}$ known texts and only a single chosen text, with success probability approximately $1/(2^{n-m}/(s+1) + 1)$.

Applying Proposition 4 to MAA, $n = 64$, $m = 32$, and $s \geq 2$ (since two key-dependent blocks are always appended). For $s = 2$, the attack of Lemma 1 requires $0.82 \cdot 2^{32}$ known text-MAC pairs and $0.67 \cdot 2^{32}$ chosen text-MAC pairs. For $s = 2^{16} + 2$ (corresponding to a fixed but arbitrary 256 Kbyte trailing block), $2^{24.5}$ known texts and 131 071 chosen texts are required. Note that the designer of MAA realized that its compression function not being a bijection might lead to weaknesses, motivating a special mode in [14] for messages longer than 1024 bytes. However, it turns out that the above attack is applicable to this mode as well. This is the first attack on MAA (that we are aware of) which is more efficient than an exhaustive key search or guessing the MAC.

For CBC-MAC with $m = n = 64$, Proposition 2 requires $2^{32.5}$ known text-MAC pairs and one chosen text; for $m = 32$, about 2^{33} additional chosen texts are required. The attack of Proposition 4 fails for CBC-MAC and CFB-MAC with maximal feedback, since for these the compression function is bijective on the chaining variable for fixed x_i (e.g. $H_i = f(H_{i-1}, x_i) = E_K(H_{i-1} \oplus x_i)$, where $E_K(H_{i-1})$ denotes the encryption of H_{i-1} with key K). However, it does apply to CFB-MAC with feedback shorter than one block and to RIPE-MAC [23]. In §4 we discuss other specific schemes to which the attacks apply. Proposition 4 answers in part an open question arising in the discussion of CBC-MAC [1] on whether a bijective compression function (for fixed input x_i) allows stronger security claims. Note that Lemma 3 is of independent interest for parallelizing a collision search when the constraint is the number of hash function evaluations rather than the number of evaluations of the round function.

4 Three Previous MACs Based on Hash Functions

In this section we discuss the security of three proposals to construct a MAC based on a hash function: the secret prefix, secret suffix, and envelope methods.

4.1 The Secret Prefix Method

The *secret prefix* method consists of prepending a secret key K_1 to the message x before the hashing operation: $\text{MAC}(x) = h(K_1\|x)$ for h an unkeyed hash function. If the key consists of a complete block, this corresponds to a hash function with a secret IV. This method was suggested for MD4 independently by Tsudik [26] and by the Internet Security and Privacy Working Group for use in the Simple Network Management Protocol (SNMP) [13]. In the 1980s this was already proposed for at least two other schemes (for example [3]). As has been pointed out in several papers, this MAC is insecure: a single text-MAC pair contains information essentially equivalent to the secret key, independent of its size. An attacker may append any blocks to the message and update the MAC accordingly, using the old MAC as the initial chaining variable for the update. The messages for which an attacker can compute the MAC are restricted to those having known texts as prefix, but this is only a very weak restriction. The appending attack may be precluded if only a subset of the hash output bits are used as the MAC (e.g. $m = n/2$ as for MD2.5 below), or by prepending the length of the message before hashing [26]. However, relying on a prepended length for security appears to make additional demands on the properties of the hash function and the attack discussed following Proposition 4 still applies for $s \geq 1$. Indeed, the compression function in MD4-based hash functions is of the form $H_i = E_{x_i}(H_{i-1}) + H_{i-1}$, which behaves as a random function (for fixed x_i); the addition here is modulo 2^{32}.

A variation of the prefix method with MD5 is used in Kerberos V5, under the name MD2.5 [20]. The 128-bit key K_1 is derived from a 56-bit DES key K by using DES as a keystream generator in Output Feedback Mode (OFB) with

$IV = 0$. The MAC consists of the leftmost 64 bits of the 128-bit hash-result. While the expansion does not thwart an exhaustive search for the DES key, it appears to provide some benefit. Moreover, revealing only 64 bits of the hash-result makes it hard to append one or more blocks and update the MAC. However, the attack of Proposition 4 still applies if $s \geq 1$ and if we have an internal collision before the last block ($w \geq 1$). Also, it remains conceivable that one could append carefully chosen blocks at the end in such a way that the new MAC depends only on the 64 known bits, implying that choosing $m < n$ imposes additional conditions on the hash function beyond those for which it was designed or has yet been analyzed. An unfortunate additional drawback is that, while one advantage of an MD5-based MAC over a DES-CBC MAC is avoidance of block ciphers and associated possible export issues, DES is nonetheless required for key expansion in MD2.5. Another concern is that a 56-bit key does not offer sufficient protection against an exhaustive key search [29].

4.2   The Secret Suffix Method

A second proposal is to append a secret key K_2 to the message: $\mathrm{MAC}(x) = h(x\|K_2)$. This approach, proposed for SNMP [13], is called the *secret suffix* method in [26]. A concern with this method is that an off-line collision attack on the hash function may be used to obtain an internal collision; therefore by a birthday attack, finding a pair (x, x') such that $h(x) = h(x')$ requires about $O(2^{n/2})$ off-line operations. (Note that the candidates for the collision search can be chosen from a controlled set.) Lemma 1 may then be applied. Moreover, this method is weak if an (off-line) second preimage attack on the underlying hash function is feasible – given one known text-MAC pair, a second hash function preimage (for that text) allows an existential MAC forgery. If t text-MAC pairs are known, finding a MAC second preimage requires $2^n/t$ rather than 2^n off-line trials; here, if the length of the message is not appended, t is the total number of blocks rather than the number of messages. Finally, it should be noted that an attacker can remove the feedforward in the last iteration, since the the chaining variable entering this iteration can be computed using x only.

4.3   The Envelope Method

The *envelope method* [26] combines the prefix and suffix methods. One prepends a secret key K_1 and appends a secret key K_2 to the message input: $\mathrm{MAC}(x) = h(K_1\|x\|K_2)$. It is claimed in [26] (along with a sketch of proof) that a divide and conquer attack against K_1 and K_2 is not possible, and that breaking this method requires exhaustive search for a key of $k_1 + k_2$ bits (with $k_i = |K_i|$ and $k_1 = n$, the size of the chaining variable). We now show this statement is false.

The approach suggested in §3 can be used in a divide and conquer key recovery attack on K_1 and K_2. Once an attacker finds an internal collision for the chaining variables, he can perform an exhaustive search for K_1, eliminating all trial key values which do not yield a collision before the last block (i.e. an internal collision). This requires 2^{k_1} off-line operations. Slightly more than k_1/n internal collisions are required to determine K_1 uniquely [22]; if $k_1 = n$, two collisions are certainly sufficient. After this, the envelope method is effectively reduced to the secret suffix method. Once K_1 is determined, an exhaustive search can be used to find K_2. This disproves the claim of [26]. Note that choosing $K_1 \neq K_2$ does not offer nearly as much additional security as one might think relative to $K_1 = K_2$ (which provides k_1 bits of security from exhaustive search); an attack on the former however does require a large number of known text-MAC pairs.

The envelope method used with MD4-based hash functions is also subject to the forgery as discussed in §4.1; the MAC forgery possible by Proposition 4 applies with $m = n = 128$ (regardless of k_i) and $s \geq 1$ (assume the last block consists of K_2 only). For $s = 2^{16}$, $2^{56.5}$ known text-MAC pairs are required and one chosen text. The result is that the security of this scheme is significantly less than that suggested by the key size $k_1 + k_2$.

Three MD5-based MAC proposals for the IPSEC working group are made in [18]: one is the envelope method with $K_1 = K_2$ and $k_1 = 128$ (K_1 is padded to a complete block), the other two are $\text{MAC}(x) = h(K_1 \| h(K_2 \| x))$ and $\text{MAC}(x) = h(K_1 \| h(K_1 \| x))$. It is suggested that the best known attack on these schemes requires 2^{64} chosen messages; however, Proposition 4 shows that $2^{56.5}$ known text-MAC pairs are sufficient (if $s = 2^{16}$). Also, the second scheme is vulnerable to the divide and conquer attack described above.

4.4 Summary of Results on the Three Previous Proposals

The weaknesses of the three existing proposals discussed above are summarized in Table 1. Storage requirements (e.g. for known pairs) have been omitted, as well as the potential improvements due to common trailing blocks as discussed in §3. The tabulated values, corresponding to the best known attacks, give *upper* bounds on the security of these constructions. Depending on the parameters, finding a second preimage may be easier by first obtaining the key with an exhaustive search; this type of attack is not noted in the table.

If the underlying hash function is collision resistant (implying n is sufficiently large), the figures in Table 1 (aside from the secret prefix method without additional precautions) indicate that the corresponding attacks are only *certificational* – breaking these schemes is easier than breaking an ideal MAC with the same parameters, but the attacks are still not feasible in practice. In particular, the number of known or chosen texts required is much smaller than one would expect, and known texts can be replaced by off-line computations. It is however clear from Table 1 that if the hash function is only a one-way hash function (with n typically between 64 and 80 bits), then both the suffix and envelope methods are vulnerable as well. Also, it follows that in case of the envelope method k_1 must not be too small.

Even if keys are chosen sufficiently large that these attacks are computationally infeasible, one should keep in mind the attacks are independent of possible weaknesses of the hash function. More sophisticated attacks might be found which exploit such weaknesses.

Table 1. Security of 3 proposals to build n-bit MACs ($n = m$) from hash functions. "#MAC" is the number of known text-MAC pairs; "C" the number of chosen texts; "#opn" the number of off-line compression function operations required for best known attacks; t is the number of messages (or blocks) available to an attacker; k, k_1, k_2 are key bitlengths.

	ideal MAC (k)		secret prefix (k_1)		secret suffix (k_2)		envelope ($k_1 + k_2$)	
	#MAC	#opn	#MAC	#opn	#MAC	#opn	#MAC	#opn
key recovery	$\lceil \frac{k}{n} \rceil$	2^k	1	0‡	$\lceil \frac{k_2}{n} \rceil$	2^{k_2}	$\lceil \frac{k_1+k_2}{n} \rceil$ $2^{n/2}$	$2^{k_1+k_2}$ $2^{k_1} + 2^{k_2}$
MAC forgery	$\lceil \frac{k}{n} \rceil$	2^k	1	1	1C	$2^{n/2}$	5C$+2^{n/2}$ $2^{n/2}$	0 2^{k_1}†
2nd preim.	2^n	0	t	$2^n/t$	t	$2^n/t$	2^n	0

†This attack reduces the envelope method to the secret suffix method only.
‡Information essentially equivalent to the secret key is known.

5 A New MAC Construction: MDx-MAC

From the previous section it may be concluded that extreme care must be exercised in constructing a MAC from a hash function. With this in mind, we propose a new construction, with the following design goals:

1. The secret key should be involved at the beginning, at the end, and in every iteration of the hash function (cf. MAA [5, 6, 14]).
2. The deviation from the original hash function should be minimal (to minimize implementation effort and maximize on confidence previously gained).
3. The performance should be close to that of the hash function.
4. The additional memory requirements should be minimized (keeping smart card implementations in mind).
5. The approach should be generic, i.e. should apply to any hash function based on the same principles as MD4.

The new construction converts the hash function MDx into the MAC algorithm MDx-MAC with a key K up to 128 bits in length. Here MDx may be any of MD5, RIPEMD, SHA, or similar algorithms. (We omit MD4 itself from recommendation because of weaknesses identified in [8] and [27].)

MDx-MAC uses three 16-byte constants T_0, T_1, T_2, which define three further 96-byte constants U_0, U_1, U_2 (see below). The first run-time step is key expansion. If K is shorter than 128 bits, concatenate K to itself a sufficient number of times, and select the leftmost 128 bits. Let $\overline{MDx}$ denote algorithm MDx with both padding and appended length omitted. The 16-byte secret key K is expanded to three 16-byte (or for SHA, 20-byte) subkeys K_0, K_1, and K_2:

for $i := 0$ *to* 2 $K_i := \overline{MDx}(K \,\|\, U_i \,\|\, K)$

The leftmost 16 bytes of the derived key K_1 are split into four 32-bit substrings denoted $K_1[i]$ $(0 \le i \le 3)$. Also, only the leftmost 16 bytes of K_2 are retained. Note for SHA, $\overline{MDx}$ produces a 20-byte output versus 16 bytes for e.g. MD5.

MDx-MAC is then obtained from MDx with modifications as follows:

1. The initial value IV of MDx is replaced by K_0.
2. $K_1[i \bmod 4]$ is added mod 2^{32} to the constants which are used in round i of each iteration of MDx.[4]
3. Following the block containing the padding and appended length as defined by MDx (i.e. the last block after normal post-processing), an additional complete 64-byte block is appended which has the following form:

$$K_2 \,\|\, K_2 \oplus T_0 \,\|\, K_2 \oplus T_1 \,\|\, K_2 \oplus T_2$$

4. The MAC result is the leftmost m bits of the hash value. In view of the attack of Proposition 4, $m = n/2$ is recommended for most applications.

The interpretation of strings as integers is defined to match that used in MDx.

The computational overhead of the MAC construction is 6 block operations for the key expansion (2 for each K_i); a single block operation is required for each 64 bytes of message. The additional storage requirements are 16 bytes for K, 48 bytes for the T_i, and 16 bytes for K_2; K_0 may be computed when required, and K_1 may be added immediately to the constants. Software implementations indicate MD5-MAC is 5-20% slower than MD5 (one factor is that the modified constants must be read from memory): on a 33 MHz 80486, MD5-MAC runs at 11.3 Mbit/s, while MD5 achieves 14.3 Mbit/s; on a HP 715–80, MD5-MAC runs at 43.6 Mbit/s compared to 46.9 Mbit/s for MD5. For RIPEMD and SHA, the performance difference is smaller since the modified constants can be stored in a register. Below we give the hex constants T_i and three test vectors $(x,$ MD5-MAC$(x))$ for hex key $K =$ 00112233445566778899aabbccddeeff:

```
T0:   97 ef 45 ac 29 0f 43 cd 45 7e 1b 55 1c 80 11 34
T1:   b1 77 ce 96 2e 72 8e 7c 5f 5a ab 0a 36 43 be 18
T2:   9d 21 b4 21 bc 87 b9 4d a2 9d 27 bd c7 5b d7 c3
("",                         1f1ef2375cc0e0844f98e7e811a34da8)
("abc",                      e8013c11f7209d1328c0caa04fd012a6)
("abcdefghijklmnopqrstuvwxyz", 9172867eb60017884c6fa8cc88ebe7c9)
```

The 16-byte constants T_i and 96-byte constants U_i are defined as follows. The definition of T_i involves the 62-byte constant R. $=$"ab...yzAB...YZ01...89" and 2-byte constants S_0, S_1, S_2, where S_i is the 16-bit string formed by repeating twice the 8-bit hexadecimal representation of i (e.g. $S_1 = 3131$).

for $i := 0$ *to* 2 $T_i := \overline{MDx}(S_i \,\|\, R)$ (leftmost 16 bytes of)
for $i := 0$ *to* 2 $U_i := T_i \,\|\, T_{i+1} \,\|\, T_{i+2} \,\|\, T_i \,\|\, T_{i+1} \,\|\, T_{i+2}$

---

[4] For RIPEMD the rounds in the two independent iterations are numbered 0-2 and 3-5 respectively. For MD5 and SHA, the rounds are 0-3.

where the subscripts in T_i are taken modulo 3. The constants T_i and U_i are fixed for all time, given MDx.

The idea of constructing the T_i in this manner is to obtain "random" (in the sense of un-contrived) bit strings which are easy to compute if an implementation of MDx is available (as opposed to the constants $\sqrt{2}$ and $\sqrt{3}$ in MD4, and the sine constants in MD5). The U_i are defined in terms of the T_i, so that it suffices to store 48 bytes to define the three 96-byte strings U_i. The key expansion makes use of two compression functions in order to preclude recovering K from any of the K_i. (If only a single iteration were used, a cryptanalyst could remove the feedforward in MDx, which would reduce the strength.) The three derived keys K_0, K_1, and K_2 are computed from K by applying a one-way function, implying that even if two of the three values are known, it is computationally infeasible to compute the third. Also, the relation between these derived keys is hard to predict. While the mapping from K to K_i is not bijective, the expected reduction in entropy for each K_i is negligible.

The role of K_0 and K_2 here is similar to that of the secret keys K_1 and K_2 in the envelope method, but with the difference that a divide-and-conquer attack now provides no advantage. The intention is that the use of K_1 in MDx-MAC provides additional protection over the envelope method in the case that weaknesses of the hash function become known. (Use of the iteration-invariant K_1 is not very strong by itself: for MD5 and RIPEMD this is almost equivalent to an offset in each message block.) Finally, an exhaustive search for each of the K_i is as hard as an exhaustive search for K. An advantage of the overall approach is that it minimizes the difference between MDx and MDx-MAC, reducing the probability of introducing new weaknesses.

If, as recommended, the bitsize of the chaining variable is equal to twice that of the MAC (i.e. $n = 2m$), a forgery attack on the new scheme requires $O(2^m/(s+1))$ chosen text-MAC pairs and $O(2^m/\sqrt{(s+1)})$ known texts (Proposition 4), and thus MDx-MAC is better than the envelope method as given in [26] for which $m = n$. In addition, MDx-MAC does not succumb to the divide and conquer key search attack. Moreover, the key size of 128 bits gives a better idea of the actual strength of MDx-MAC than the 128+512=640 bits previously proposed for the envelope method. We believe the new scheme would also be stronger against attacks exploiting the internal structure of MDx. Based on the above remarks, we state the following conjecture:

Conjecture 1 *If* MDx *is a secure hash function of the MD4-family, the best attacks on* MDx-MAC *are exhaustive (with respect to key search), and the attack of Proposition 4 (with respect to forgery).*

Although it was shown in [9] that the compression function of MD5 is not collision resistant, we do not expect this results in a weakness of MD5-MAC.

For SHA, a more natural construction might be to use a 160-bit key K, construct subkeys K_i of the same length, and make appropriate modifications. However, a 128-bit key appears adequate against all plausible key search attacks, and as defined (with $n = 2m = 160$), the forgery attack of Proposition 2 against SHA-MAC requires about 2^{80} chosen text-MAC pairs and 2^{80} known texts. Note that the corresponding numbers of known and chosen texts for MD5-MAC and RIPEMD-MAC are about 2^{64}.

6 Concluding Remarks

The new forgery attack on iterated MACs requires $O(2^{n/2})$ known text-MAC pairs and $O(2^{n-m})$ chosen texts, where m is the bitlength of the hash-result and n is that of the chaining variable. A naive non-verifiable attack always succeeds with probability 2^{-k} by guessing the k-bit key and computing the MAC, or 2^{-m} by guessing the MAC. These attack scenarios differ, but nonetheless suggest using $n = 2m = k$.

The new attack may pose a serious threat to certain applications of CBC-MAC (e.g. when $n = m = 64$). Its implications for the security of MAA are also serious. There are disadvantages with MAC functions such as DES CBC-MAC built from block ciphers (including speed and exportability). The analysis of existing proposals indicates that one must exercise care in designing MACs based on hash functions. The new MAC construction addresses all of these concerns and appears to be the first such proposal based on existing hash functions. It differs from previous such proposals in that it involves a secret key in each iteration of the hash function and yet is sufficiently generic to apply to any function of the MD4-family. As with all new proposals, we caution that it would be imprudent to employ the new MAC algorithm in practice prior to adequate peer review.

Acknowledgements

We would like to thank Antoon Bosselaers and Roland Lockhart for efficient independent software implementations confirming the test vectors in §5.

References

1. M. Bellare, J. Kilian, P. Rogaway, "The security of cipher block chaining," *Proc. Crypto'94, LNCS 839*, Springer-Verlag, 1994, pp. 341–358.
2. M. Bellare, R. Guérin, P. Rogaway, "XOR MACs: new methods for message authentication using block ciphers," *Proc. Crypto'95* (this volume).
3. F. Cohen, "A cryptographic checksum for integrity protection," *Computers & Security*, Vol. 6, No. 5, 1987, pp. 505–510.
4. I.B. Damgård, "A design principle for hash functions," *Proc. Crypto'89, LNCS 435*, Springer-Verlag, 1990, pp. 416–427.
5. D. Davies, "A message authenticator algorithm suitable for a mainframe computer," *Proc. Crypto'84, LNCS 196*, Springer-Verlag, 1985, pp. 393–400.
6. D. Davies, D.O. Clayden, "The message authenticator algorithm (MAA) and its implementation," *NPL Report DITC 109/88*, Feb. 1988.
7. D. Davies, W. Price, *Security for Computer Networks*, 2nd ed., Wiley, 1989.
8. B. den Boer, A. Bosselaers, "An attack on the last two rounds of MD4," *Proc. Crypto'91, LNCS 576*, Springer-Verlag, 1992, pp. 194–203.

9. B. den Boer, A. Bosselaers, "Collisions for the compression function of MD5," *Proc. Eurocrypt'93, LNCS 765*, Springer-Verlag, 1994, pp. 293–304.

10. FIPS 46, *Data encryption standard*, NBS, U.S. Department of Commerce, Washington D.C., Jan. 1977.

11. FIPS 81, *DES modes of operation*, NBS, US Department of Commerce, Washington D.C., Dec. 1980.

12. FIPS 180-1, *Secure hash standard*, NIST, US Department of Commerce, Washington D.C., April 1995.

13. J.M. Galvin, K. McCloghrie, J.R. Davin, "Secure management of SNMP networks," *Integrated Network Management, II*, North Holland, 1991, pp. 703–714.

14. ISO 8731:1987, *Banking – approved algorithms for message authentication, Part 1, DEA*, IS 8731-1, *Part 2, Message Authentication Algorithm (MAA)*, IS 8731-2.

15. ISO/IEC 9797:1993, *Information technology - Data cryptographic techniques - Data integrity mechanisms using a cryptographic check function employing a block cipher algorithm*,.

16. T. Johansson, G. Kabatianskii, B. Smeets, "On the relation between A-codes and codes correcting independent errors," *Proc. Eurocrypt'93, LNCS 765*, Springer-Verlag, 1994, pp. 1–11.

17. R.R. Jueneman, S.M. Matyas, C.H. Meyer, "Message authentication with Manipulation Detection Codes," *Proc. 1983 IEEE Symposium on Security and Privacy*, IEEE Computer Society Press, 1983, pp. 33–54.

18. B. Kaliski, M. Robshaw, "Message authentication with MD5," *CryptoBytes (RSA Laboratories Technical Newsletter)*, Vol. 1, No. 1, Spring 1995, pp. 5–8.

19. H. Krawczyk, "LFSR-based hashing and authentication," *Proc. Crypto'94, LNCS 839*, Springer-Verlag, 1994, pp. 129–139.

20. J. Linn, "The Kerberos Version 5 GSS-API Mechanism," *Internet Draft*, Feb. 1995.

21. C. Mitchell, M. Walker, "Solutions to the multidestination secure electronic mail problem," *Computers & Security*, Vol. 7, No. 5, 1988, pp. 483–488.

22. B. Preneel, *Cryptographic Hash Functions*, Kluwer Academic Publishers, 1995 (to appear).

23. RIPE, *Race Integrity Primitives Evaluation (RIPE-RACE 1040): Final Report*, LNCS, Springer-Verlag, 1995 (to appear).

24. R.L. Rivest, "The MD4 message digest algorithm," *Proc. Crypto'90, LNCS 537*, Springer-Verlag, 1991, pp. 303–311.

25. R.L. Rivest, "The MD5 message-digest algorithm," *Request for Comments (RFC) 1321*, Internet Activities Board, Internet Privacy Task Force, April 1992.

26. G. Tsudik, "Message authentication with one-way hash functions," *ACM Computer Communications Review*, Vol. 22, No. 5, 1992, pp. 29–38.

27. S. Vaudenay, "On the need for multipermutations: cryptanalysis of MD4 and SAFER," *Fast Software Encryption, LNCS*, Springer-Verlag, 1995 (to appear).

28. M.N. Wegman, J.L. Carter, "New hash functions and their use in authentication and set equality," *J. Computer Sys. Sciences*, Vol. 22, No. 3, 1981, pp. 265–279.

29. M.J. Wiener, "Efficient DES key search," *Technical Report TR-244*, School of Computer Science, Carleton University, Ottawa, Canada, May 1994. Presented at the rump session of Crypto'93.

XOR MACs: New Methods for Message Authentication Using Finite Pseudorandom Functions

Mihir Bellare[1] and Roch Guérin[1] and Phillip Rogaway[2]

[1] IBM T.J. Watson Research Center, PO Box 704, Yorktown Heights, NY 10598,
USA. e-mail: {mihir, guerin}@watson.ibm.com
[2] Dept. of Computer Science, Eng. II Bldg., University of California, Davis,
CA 95616, USA. e-mail: rogaway@cs.ucdavis.edu

Abstract. We describe a new approach for authenticating a message
using a finite pseudorandom function (PRF). Our "XOR MACs" have
several nice features, including parallelisability, incrementality, and provable security. The finite PRF can be "instantiated" via DES (yielding
an alternative to the CBC MAC), via the compression function of MD5
(yielding an alternative to various "keyed MD5" constructions), or in
a variety of other ways. The proven security is quantitative, expressing
the adversary's inability to forge in terms of her (presumed) inability to
break the underlying finite PRF. This is backed by attacks showing the
analysis is tight. Our proofs exploit linear algebraic techniques.

1 Introduction

A message authentication scheme enables two parties sharing a key a to authenticate their transmissions. This is one of the most widely used cryptographic
primitives, and it may become even more so: as security concerns grow, it is
reasonable to anticipate that virtually every transmitted message (or packet)
will use cryptographic means to ensure authenticity. (For example, the ubiquitous use of message authentication is already being contemplated for the next
generation of Internet Protocols.)

Message authentication is usually accomplished by including with each transmitted message M a short string, called its "message authentication code"
(MAC) or "signature," computed as a function of M and the shared key a.
The most prevalent MAC is the "cipher block chaining message authentication code" (CBC MAC) specified in the International Standard ISO 9797 [ISO]
and the U.S. Standard ANSI X9.9 [X9.9]. In recent years another type of MAC
has started to become prevalent: these are constructed by somehow "keying" a
cryptographic hash, as in $\text{MAC}_a(x) = \text{MD5}(x.a)$ (see, for example, [Ts]).

The goal of the present work is to provide new methods which have certain
efficiency and security advantages. We call our methods "XOR schemes." They
are simple to describe and implement. They use as their underlying primitive any
finite pseudorandom function (PRF). In particular, a finite PRF can be defined

D. Coppersmith (Ed.): Advances in Cryptology - CRYPTO '95, LNCS 963, pp. 15-28, 1995.

from a block cipher (e.g. DES) or from the compression function of a crypto-graphic hash (e.g., MD5) yielding concrete alternatives to the above mentioned MACs.

What is an XOR MAC? At the highest level, the computation of an XOR MAC consists of three steps: (1) encode the message M as a collection of blocks (each block will depend on a small number of bits from the message); (2) apply the finite PRF to each of the blocks, thus creating a collection of PRF images (the MAC key a is the index for all of these PRF computations); and (3) XOR the set of PRF images together, building the MAC out of the result. Different ways of implement the encoding step (and different choices of the finite PRF) yield different XOR MACs. (Obviously not all encodings will result in secure MACs. We specify several simple ones which do, and also specify general conditions to determine which encodings work.)

This paper specifies, for every finite PRF family F and every value of a block size b, two XOR MACs—a stateless (and probabilistic) one called XMACR$_{F,b}$, and a stateful (and deterministic) one called XMACC$_{F,b}$. (In a stateful MAC the signer maintains information, in our case a counter, which he updates each time a message is signed.) The schemes are described concretely in Section 2, as are their main efficiency advantages, namely parallelizability and incrementality.

Security of our schemes. Our XOR schemes are *proven* secure– we show that if the F is a "secure" finite PRF family then the MAC schemes based on it are also "secure." In formalizing this, security of a finite PRF family means indistinguishability from a family of random functions in the sense of [GGM], while security of a MAC means it resists chosen message attack. To make these results meaningful for practice, the security in both cases is made quantitative: we measure the success probabilities as a function of the resources (time and chosen message queries) available to the adversary, and specify exact reductions, enabling the protocol designer to compute, given some assumed security on the finite PRF, how many queries an XOR MAC based on it will withstand. This type of security analysis for a MAC, starting from a finite PRF, begins with [BKR].

Our XOR schemes are so simple that it is tempting to think one can easily find attacks. This is why we stress the importance of the proofs of security which show that no attacks short of breaking the underlying PRF will succeed.

An advantage of quantified security is that it allows one to compare the securities of different MACs based on the same finite PRF family. (Note that a concrete finite PRF family F, eg. a block cipher like DES, may possess strengths which are not reflected in the model of F being a finite PRF family, and these other strengths are potentially relevant in determining the strength of a MAC based on the block cipher. In making security determinations and comparisons we are treating the underlying primitives, eg. DES, as being known to only possess the properties which have been formally modeled, here the property of being a finite PRF.) We will see that our counter based MAC is "more secure" than our randomized one, and that both are "more secure" than the CBC MAC. In particular, the success probability of the adversary in the XOR schemes is independent of the lengths of the messages in her chosen message attack (as long as they stay below a certain specified but very large length) while the attacks of [Kr, PV] show that the success probability of the adversary in the CBC scheme grows as a linear function of the message length. See Section 6.

We also describe the best attacks we know on the XOR schemes. They use birthday attacks (collisions) and indicate that the analysis from our proofs is tight.

2 The Schemes and Their Properties — Concretely

We begin by presenting concrete instantiations of our two main schemes using DES. (But we stress this is just an example. Other instantiations are possible, using other block ciphers, or even methods such as MD5, as discussed later.) We let $l = 64$ and $L = 48$. For any 56-bit key a and l-bit plaintext x we let $F_a(x)$ be the first L bits of $DES_a(x)$. (We stress that F_a outputs only 48 bits, and not the full 64-bit DES output. We have truncated the output because DES is a pseudorandom *permutation*, while what we want is a pseudorandom function.) Sender and receiver share a 56-bit DES key a which specifies F_a.

Message formatting and notation. We assume the length $|M|$ of M is a multiple of 32 bits, which can easily be achieved by a suitable padding. (For example, append a one and then append enough zeros to bring the length to a multiple of 32 bits.) The message is then viewed as a sequence of 32-bit blocks, $M = M[1] \ldots M[n]$ with $|M[i]| = 32$ for $i = 1, \ldots, n$. We assume that the number n of blocks is less than 2^{31} —equivalently $|M| \leq 32 * 2^{31} = 2^{36}$ bits— which would not normally be a significant restriction in practice.

Let $\langle i \rangle$ denote the binary representation of block index $i \in \{1, \ldots, n\}$ as a string of exactly 31 bits. (This is why we assumed the bound on n.) Let $\alpha . \beta$ denote the concatenation of strings α and β. We give two schemes:

Scheme XMACR. The first scheme is called the randomized XOR scheme, XMACR. To authenticate the message $M = M[1] \ldots M[n]$ do the following:

- Pick at random a 63-bit string r, hereafter called the *seed*
- Set $z = F_a(0.r) \oplus F_a(1.\langle 1 \rangle . M[1]) \oplus F_a(1.\langle 2 \rangle . M[2]) \oplus \cdots \oplus F_a(1.\langle n \rangle . M[n])$
- Set the MAC of M to the pair $\mu = (r, z)$.

Thus the sender will transmit (M, μ). The receiver, receiving (M', μ'), where $\mu' = (r', z')$, computes $z = F_a(0 . r') \oplus F_a(1 . \langle 1 \rangle . M'[1]) \oplus F_a(1 . \langle 2 \rangle . M'[2]) \oplus \cdots \oplus F_a(1 . \langle n \rangle . M'[n])$. The receiver accepts M' if and only if $z = z'$.

We stress that new coins are flipped to determine the seed each time the sender wants to authenticate a message, and also that the seed is included in the signature.

Scheme XMACC. The second scheme is called the counter-based XOR scheme. Here it is required that the sender maintain a 63-bit counter C which is initially 0 and is incremented for each message. (Thus at most 2^{63} messages can be signed.) To authenticate message $M = M[1] \ldots M[n]$ do the following:

- Increment the counter C by 1

- Set $z = F_a(0.C) \oplus F_a(1.\langle 1 \rangle.M[1]) \oplus F_a(1.\langle 2 \rangle.M[2]) \oplus \cdots \oplus F_a(1.\langle n \rangle.M[n])$
- Set the MAC of M to the pair $\mu = (C, z)$.

Thus the sender will transmit (M, μ). The receiver, receiving (M', μ') where $\mu' = (C', z')$, computes $F_a(0.C') \oplus F_a(1.\langle 1 \rangle.M'[1]) \oplus F_a(1.\langle 2 \rangle.M'[2]) \oplus \cdots \oplus F_a(1.\langle n \rangle.M'[n])$ and accepts iff this value equals z'. Note the counter is included in the signature. Also the receiver maintains no state.

Stateful schemes are not necessarily "worse" than stateless ones; programmatically, a "static" variable is easy, but a good approximation to randomness is hard. We now discuss properties of XMACR and XMACC.

Parallelizability. The DES computations on different blocks can be made in *parallel*. In general, the throughput of an XOR MAC can be doubled by doubling the amount (and not speed) of the underlying hardware. An environment where this is crucial is message authentication over high speed networks (where packets will flow over optical links at rates of 1–10 GBit/second). In that setting one cannot realistically use the CBC MAC because of its sequential nature; an XOR scheme is a more appropriate choice. Note that even in the software setting parallelizability can be relevant: with an adequate degree of parallelism, multiple machine pipelines can all be kept busy doing useful work.

Incrementality. An XOR MAC is *incremental* [BGG1] with respect to block replacement. Suppose $M[i]$ is modified to a new 32-bit value m. Then, for a long message M, one can update the MAC much quicker than it would take to re-compute it. Let's illustrate for XMACR. Let $\mu = (r, z)$ be a MAC of M and let M' denote M with block i replaced by m. To compute a MAC for M', pick r' at random and let $z' = z \oplus F_a(0.r) \oplus F_a(0.r') \oplus F_a(1.\langle i \rangle.M[i]) \oplus F_a(1.\langle i \rangle.m)$. Then (r', z') is a MAC for M'. Extensions of this scheme to support insertion and deletion of blocks (not just replacement) appear in [BGG2].

Out-of-order verification. Tag verification can proceed even if message blocks arrive out of order. Here it is only necessary that the each block be accompanied by its index. With other mechanisms MAC verification cannot proceed before the first block has been received, for example. Out-of-order MAC verification is useful since networks always have some degree of packet loss and re-transmission.

DES computations. The number of DES computations is twice that of the CBC MAC. (The overhead can be reduced as discussed in Section 4 by increasing the block size, currently set to 32, at the cost of reducing the maximum allowable message length.) So, in software, the above schemes are slower than the CBC MAC. But an XOR MAC based on DES is interesting for hardware efficiency, particularly for high-speed networks, or in settings where the incrementality compensates for the slower from-scratch MACing time. For a software-efficient XOR MAC use the MD5-based instantiation discussed later.

MD5-based instantiation. A software-efficient XOR MAC would start not with DES but with a software-efficient PRF. For example, from the compression function of a cryptographic hash function, say $md : \{0,1\}^{640} \to \{0,1\}^{128}$, one can define a finite PRF, say $F_a(x)$ equals the first 64 bits of $md(x.a)$, where

$|x| = 560$ and $|a| = 80$. Using 48-bits for the block index, we would get a MAC which uses one application of md for every 512 bits of message. This is as efficient as proposals like $MD5(x \cdot a)$ or $MD5(a \cdot x \cdot a)$ which are currently being considered for the Internet, and has the advantages of parallelizability and incrementality.

Security. Observe that including the block indices in the argument to F_a is necessary— if they are omitted, permuting the message blocks would leave the MAC unchanged. One can also see that the block containing the random string r (resp. counter) of XMACR (resp. XMACC) cannot be omitted. In other words, the scheme in which the MAC is set to $F_a(1 \cdot \langle 1 \rangle \cdot M[1]) \oplus F_a(1 \cdot \langle 2 \rangle \cdot M[2]) \oplus \cdots \oplus F_a(1 \cdot \langle n \rangle \cdot M[n])$ is easily broken—e.g., set $M_1 = A \cdot B$, $M_2 = A' \cdot B$, $M_3 = A \cdot B'$ and $M_4 = A' \cdot B'$, and note that the MACs of M_1, M_2, M_3 sum to give the MAC of M_4.

The idea behind the nonces is to prevent the attacker from forming new MACs via linear combinations of old ones. This is in fact the only attack short of breaking the PRF. This is not obvious, of course; indeed it is far from clear why XMACR and XMACC should be secure. That is why we have our proofs.

3 Definitions

In order to prove the security we first need to provide appropriate definitions of security for block ciphers and message authentication schemes.

We model block ciphers as finite pseudorandom functions in the manner of [BKR]. The underlying notion is the pseudorandom function notion of [GGM], appropriately tailored to take into account the fact that block ciphers have fixed input and output lengths that can't be treated asymptotically. This approach builds on a suggestion of [LuRa] that DES be viewed as a "pseudorandom in practice" function family.

Denote by $|x|$ the length of a string x. If $i \in \{1, \ldots, 2^n\}$ is an integer then we denote by $\langle i \rangle_n$ the natural binary encoding of i as an n-bit string. (Thus the $\langle \cdot \rangle$ of Section 2 is $\langle \cdot \rangle_{31}$ in our current notation.) If S is a set (resp. probability space) then $x \xleftarrow{R} S$ denotes the operation of selecting an element uniformly at random from S (resp. at random according to the distribution specified by S).

Finite pseudorandom function families. A *function family* is a set of functions, and an associated set of strings called keys. Each key names a function in the family according to some fixed convention, and the function corresponding to key a is denoted F_a. (Note that two keys can name the same function.) To pick a function f at random from a family F means to pick a key a uniformly at random and let $f = F_a$; we write $f \xleftarrow{R} F$ for this operation. For example DES is a function family where the set of keys is the set of all 56-bit strings.

A family F has input length l and output length L if each $f \in F$ maps $\{0,1\}^l$ to $\{0,1\}^L$. (Eg. $l = L = 64$ for DES.) It has key length κ if the associated set of keys is the set of all strings of length κ. The family of random functions with input length l and output length L is the family R of all functions mapping $\{0,1\}^l$ to $\{0,1\}^L$. The key of a function f in this family is the string which describes its truth table. Note this is a very large family, consisting of 2^{L2^l} functions.

A finite function family F is "pseudorandom" if the input-output behavior of F_a "looks random" to someone who doesn't know the key a. This is formalized via the notion of statistical tests [GGM]. Formally, such a test is an oracle algorithm A. Let F, G be finite function families. The advantage of A in distinguishing F from G is defined by

$$\mathsf{Adv}_A(F, G) \;=\; \Pr_{g \overset{R}{\leftarrow} F} [\, A^g = 1 \,] - \Pr_{g \overset{R}{\leftarrow} G} [\, A^g = 1 \,].$$

The probability is over the indicated random choice of g and the coin tosses of A.

Let family F have input length l and output length L, and let R be the family of random functions with the same parameters. To discuss security quantitatively, we say that statistical test A $[t, q, \epsilon]$-breaks F if A runs in at most t steps, makes at most q oracle queries, and achieves $\mathsf{Adv}_A(F, R) \geq \epsilon$. (The running time here is measured in a standard RAM model of computation.) In informal discussion, the finite function family F is said to be $[t, q, \epsilon]$-pseudorandom if there is no statistical test that $[t, q, \epsilon]$-breaks F. (To be fully formal one ought to consider also other parameters such as the "code size".) In other words, in time t and given q examples one cannot distinguish a random member of F from a random function with advantage more than ϵ.

Notice that the key size of the finite PRF family F does not need to be explicitly specified in the definition of security: its influence is captured in that it influences the values of t, q, ϵ for which the F is $[t, q, \epsilon]$-pseudorandom.

Message authentication. We provide formal definitions of schemes and their security in the exact security setting. We begin with stateless schemes, in which no counters or other state information need be maintained. Then we briefly indicate how the definitions should be updated to take account of state.

A *(stateless) message authentication scheme* consists of a *signing* algorithm Sig and a *verifying* algorithm Vf. The signing algorithm may be probabilistic; the verifying one typically is not. Associated to the scheme are parameters κ and L_{sig} describing the key length and signature length, respectively. On input a κ-bit key a and a message M, algorithm Sig outputs an L_{sig}-bit string μ called the *signature*, or MAC, of M. On input a κ-bit key a, a message M and an L_{sig}-bit string μ, algorithm Vf outputs a bit, with 1 standing for accept and 0 for reject. We ask for a basic validity condition, namely that authentic signatures are accepted with probability one.

An adversary for a message authentication scheme is allowed a chosen message attack, and declared successful if, following this attack, she produces a forgery. Formally the adversary is a probabilistic algorithm E which is given oracle access to the signer and verifier—more precisely, to $\mathsf{Sig}(a, \cdot)$ and $\mathsf{Vf}(a, \cdot, \cdot)$ for a random but hidden choice of a. E can request a signature of a message of her choice; to do this, she writes M on a special query tape. She can also ask the verifier to verify that μ is a valid signature for M; to do this she writes (M, μ) on a special query tape. E's *attack* on the scheme is described by the following experiment:

(1) A random string a of length κ is selected as the shared secret. A random string r_E is selected as the coin tosses of E. E now starts computing.

(2) Suppose E makes a signing query M. Then the oracle computes a signature $\mu \overset{R}{\leftarrow} \mathsf{Sig}(a, M)$ and returns it to E. (Since Sig may be probabilistic, this step requires making the necessary underlying choice of a random string for Sig, anew for each signing query.)

(3) Suppose E makes a verify query (M, μ). The oracle computes the decision $d = \mathsf{Vf}(a, M, \mu)$ and returns it to E.

The adversary is allowed an adaptive chosen message attack, as in the notion of [GMR], but we also allow verify queries because, unlike the setting in digital signatures, E cannot compute the verify predicate on her own (since the verify algorithm is not public). Note that E does not see a nor the coin tosses of Sig.

We say that E's attack on $\mathcal{M}$ is a (q_s, q_v)-attack if during the course of the attack she makes no more than q_s signing queries and no more than q_v verify queries. A (q_s, q_v)-attack is a (t, q_s, q_v)-attack if, in addition, E runs for no more than t steps, in the RAM model of computation we fixed above. It is useful to say that a verify query (M, μ) is *known-authentic* if a signing query M was made prior to this verify query and the signature returned was μ. Note validity implies that known-authentic verify queries are accepted. We thus assume of any adversary E that she never makes any known-authentic queries.

The outcome of running the protocol in the presence of an adversary is used to define security. We say that E is *successful* if she makes a verify query (M, μ) which is accepted but which is not known-authentic.[3] (The verify query (M, μ) in question is called a *forgery*, and the definition reflects the notion of existential forgery [GMR].) We say that E $[q_s, q_v, \epsilon]$-breaks the scheme if her attack is a (q_s, q_v, ϵ)-attack and her probability of success is at least ϵ. We say she $[t, q_s, q_v, \epsilon]$-breaks the scheme if her attack is a (t, q_s, q_v, ϵ)-attack and her probability of success is at least ϵ. In informal discussion we'll say the scheme is $[t, q_s, q_v, \epsilon]$-unforgeable if there is no adversary who can $[t, q_s, q_v, \epsilon]$-break it. (To be fully formal we would have to consider also other parameters like the "code size.")

In a stateful message authentication scheme the signer maintains state across consecutive signing requests. (For example, in our counter-based scheme the signer maintains a message counter.) In such a case the signing algorithm can be thought of as taking an additional input —the "current" state C_s of the signer— and returning an additional output —the signer's next state. We must modify the experiment describing E's attack: in Step (1) we also have that C_s is initialized to a value specified by the scheme; and in Step (2) we compute $(\mu, C_s') \overset{R}{\leftarrow} \mathsf{Sig}(a, M, C_s)$, then return μ to the adversary and replace C_s by C_s'. Note the adversary doesn't see the revised state (though in the stateful scheme

---

[3] This is slightly stronger than the more standard definition in which one would only ask that the message M was not a previous signing query. We make this stronger requirement because we achieve it and because it is useful in contexts like entity authentication.

of this paper this wouldn't matter). Also note that we allow the signer a state, but not the verifier.

4 The Randomized XOR Scheme

We first present the general scheme, of which the scheme XMACR in Section 2 is a special case, and then proceed to the security analysis.

SPECIFICATION. Let F be a family of functions with key length κ, input length l, and output length L. We fix in addition a parameter $b \leq l - 1$ which will be the block size. We will assume that any message M to be authenticated has length at most $|M| \leq b2^{l-b-1}$. By standard padding arguments we may assume wlog that the message length is a multiple of b. We then regard M as a sequence of b-bit blocks. The number of blocks is denoted $\|M\|_b$, and with b understood the i-th block is denoted $M[i]$, for $i = 1, \ldots, \|M\|_b$. Let $r \in \{0,1\}^{l-1}$, and let $a \in \{0,1\}^{\kappa}$ be the shared key. We define $\text{tag}_{F,b}(a, M, r)$ by

$$F_a(0 . r) \oplus F_a(1 . \langle 1 \rangle_{l-b-1} . M[1]) \oplus \cdots \oplus F_a(1 . \langle n \rangle_{l-b-1} . M[n]) . \tag{1}$$

We use this function in both the randomized and the counter-based schemes. We call r the *seed*. The (stateless) message authentication scheme is:

function $\text{SigR}_{F,b}(a, M)$	function $\text{VfR}_{F,b}(a, M', (r', z'))$
$r \xleftarrow{R} \{0,1\}^{l-1}$;	$z \leftarrow \text{tag}_{F,b}(a, M', r')$
$z \leftarrow \text{tag}_{F,b}(a, M, r)$	if $z = z'$ then return 1
return (r, z)	else return 0

We call $\text{XMACR}_{F,b}$ the randomized XOR scheme based on function family F and using block size b. The validity condition is easy to verify. Note that the XMACR scheme of Section 2 is, in the current terminology, $\text{XMACR}_{F,32}$ with F being the family specified by $F_a(\cdot) = $ first 48 bits of the output of $\text{DES}_a(\cdot)$.

TRADING EFFICIENCY FOR MESSAGE LENGTH. Note that choosing different values of b will tradeoff the number of F_a computations with the allowable length of messages that can be signed. Namely, the scheme calls for $1 + \|M\|_b = 1 + (|M|/b)$ evaluations of F_a and allows $|M|$ to be $b2^{l-b-1}$ so that increasing b reduces the number of F_a evaluations at the cost of restricting the scheme to shorter messages. For example, the XMACR scheme of Section 2, with $b = 32$, currently has twice the DES operations of the CBC MAC, and allows $|M|$ up to 2^{36}. But we could set $b = 48$ and have only 33% more DES operations than the CBC MAC, now with $|M| \leq 48 * 2^{15} = 3 * 2^{19}$.

SECURITY: INFORMATION THEORETIC CASE. Begin by thinking of F as ideal (i.e., truly random). Namely, we consider $\text{XMACR}_{R,b}$, which we call the information theoretic case. The following theorem provides an absolute bound on the success of the adversary in terms of the number of sign and verify queries she makes.

Theorem 1. *Let R be the family of random functions with input length l and output length L, let b be at most $l - 1$, and let E be any adversary making a (q_s, q_v)-attack on XMACR$_{R,b}$. Then the probability that E is successful is at most $\delta_R \stackrel{\text{def}}{=} 2q_s^2 \cdot 2^{-l} + q_v \cdot 2^{-L}$.*

Proof. Due to page limits we provide a very brief sketch; for a full proof see [BGR]. The proof has two parts: first we relate the security of the scheme to the probability that a certain matrix has full rank; then we bound this probability.

Since E is computationally unbounded she is wlog deterministic. The probabilistic choices in E's attack on the scheme are thus the initial choice a of key (naming a random function $R_a \in R$), and the choices of seeds made by the signer in the course of signing. Let M_i denote the random variable whose value is the i-th message whose signature E requests. Let R_i be the random seed chosen by the signer to sign M_i and let $Z_i = \text{tag}_{R,b}(a, M_i, R_i)$ denote its tag, $i = 1, \ldots, q_s$. Let Distinct be the event that $R_1, \ldots, R_{q_s}$ are all distinct and Succ the event that E is successful. By birthday bounds we can show that $\Pr[\neg\text{Distinct}] \leq q_s^2 \cdot 2^{l-1}$. Now we want to show that $\Pr[\text{Succ} \mid \text{Distinct}] \leq q_v \cdot 2^{-L}$ whence the theorem follows.

Fix a particular sequence of messages $M_1, \ldots, M_{q_s}$, a particular choice $r_1, \ldots, r_{q_s} \in \{0, 1\}^{l-1}$ of *distinct* seeds and a particular choice $z_1, \ldots, z_{q_s}$ of L-bit strings, for which $\Pr[M_i = M_i$ and $R_i = r_i$ and $Z_i = z_i$ for $i = 1, \ldots, q_s] > 0$. We let

$$\Pr_1[\cdot] = \Pr[\cdot \mid M_i = M_i \text{ and } R_i = r_i \text{ and } Z_i = z_i \text{ for } i = 1, \ldots, q_s]$$

denote the indicated conditional probability measure. (The probability is effectively over only the random choice of the shared key a, since everything else is fixed.) Fix a message M_{q_s+1} distinct from $M_1, \ldots, M_{q_s}$, a seed $r_{q_s+1} \in \{0, 1\}^{l-1}$ and an L-bit string z_{q_s+1}. These are intended to stand for a possible forgery $(M_{q_s+1}, (r_{q_s+1}, z_{q_s+1}))$. (Notice that although M_{q_s+1} is distinct from previous messages, r_{q_s+1} is not assumed distinct from previous seeds– indeed, since the adversary may choose it, we cannot make such an assumption.) Below we will show that

$$\Pr_1[\text{tag}_{R,b}(a, M_{q_s+1}, r_{q_s+1}) = z_{q_s+1}] \leq 2^{-L} . \tag{2}$$

Given this, standard conditioning arguments can then be used to show that $\Pr[\text{Succ} \mid \text{Distinct}] \leq q_v \cdot 2^{-L}$. In what follows, we make the wlog assumption that E first makes its q_s signing queries, and then makes its q_v verify queries.

Recall $M_i[j] \in \{0, 1\}^b$ denotes the j-th block of M_i. We define a $q_s + 1$ by 2^l matrix B over $GF(2)$. Its rows are indexed $1, \ldots, q_s + 1$ and its columns are indexed by the l-bit strings in lexicographic order. The entry in row i, column x is denoted $B[i, x]$, and is defined as follows. First consider the case where the first bit of x is 0, so that $x = 0 . y$. Then we set $B[i, x] = 1$ if $y = r_i$ and 0 otherwise. Now suppose the first bit of x is 1, and write it as $x = 1 . \langle j \rangle_{l-b-1} . y$, where $|y| = b$. Then we set $B[i, x] = 1$ if $M_i[j] = y$ and 0 otherwise. (In particular, $B[i, x] = 0$ if $j > \|M_i\|_b$.) Note the matrix is not a random variable—it is fixed given that $M_1, \ldots, M_{q_s+1}$ and $r_1, \ldots, r_{q_s+1}$ are fixed.

LEMMA. The matrix B has full rank.

PROOF. Transform B by row and column operations until it has a $q_s + 1$ by $q_s + 1$ identity matrix in its upper left corner. At any time, the left half of B means the first 2^{l-1} columns and the right half means the rest. Initially the left (resp. right) half consists of those columns whose index has first bit 0 (resp. 1). Since $r_1, \ldots, r_{q_s}$ are distinct, we can permute columns until the first q_s rows of the left half of B consist of a q_s by q_s identity matrix followed by a q_s by $2^{l-1} - q_s$ matrix of zeroes. We now consider two cases. First if r_{q_s+1} is distinct from $r_1, \ldots, r_{q_s}$ then a single column swap suffices. The case when $r_{q_s+1} = r_\alpha$ for some $\alpha \in \{1, \ldots, q_s\}$ uses that M is by assumption different from M_α and is more complex, requiring a few operations. $\square$

Equation (2) is established via standard relations of linear to probabilistic independence that have been used in several places (eg. [ABI, BeRo]). $\square$

Note the bound is independent of b: the latter figures only in our assumption that any query M made by E above satisfies $\|M\|_b \leq 2^{l-b-1}$. We stress that δ_R grows with the *square* of the number of signing queries: a "birthday" type behavior. Attacks we present later will show that this analysis and behavior is essentially the "best possible."

SECURITY: COMPUTATIONAL CASE. We now assume we are given a family F which is not truly random, but $[t', q', \epsilon']$-pseudorandom. In that case, how secure is XMACR$_{F,b}$? This is what the following tells us. It is the result of more direct interest in practice (although Theorem 1 is in some ways more basic). The constant c below depends only on details of the computational model. The proof is not too hard and can be found in [BGR].

Theorem 2. *There is an oracle machine U and a constant c such that the following is true. Let F be a family of functions with input length l and output length L and let b be at most $l - 1$. Let E be an adversary who $[t, q_s, q_v, \epsilon]$-breaks XMACR$_{F,b}$ and suppose any message M in a query of E has a number $\|M\|_b$ of blocks which is bounded by n. Let $\delta_R = 2q_s^2 \cdot 2^{-l} + q_v \cdot 2^{-L}$. Then U^E $[t', q', \epsilon']$-breaks F, where*

$$t' = t + c(l + L)q' \quad ; \quad q' = (q_s + q_v) \cdot (n + 1) \quad ; \quad \epsilon' = \epsilon - \delta_R .$$

In other words if F is $[t', q', \epsilon']$-pseudorandom (the values t', q', ϵ' depending on the key size and cryptanalytic strength of the finite PRF F) then XMACR$_{F,b}$ is $[t, q_s, q_v, \epsilon]$-unforgeable, where $t = t' - c(l + L)q'$, $q_s + q_v = q'/(n + 1)$ and $\epsilon = \epsilon' + \delta_R$. Thus a success probability of δ_R for the adversary is unavoidable, even if the PRF is "ideal;" beyond that, the success of the adversary is bounded in terms of the parameters of the block cipher.

ATTACKS. We present the best attacks we know. Since we think of F as pseudorandom, we will do the attack assuming it is in fact random; that is, we look at XMACR$_{R,b}$ where R is the family of random functions with input length l and output length L. Given q_s, q_v we specify a particular adversary E who makes q_s sign queries and q_v verify queries, and then outputs a forgery with probability

$\epsilon = \Omega(\delta_R)$, where $\delta_R = 2q_s^2 \cdot 2^{-l} + q_v \cdot 2^{-L}$. The attack is based on birthday attacks, and finds enough collisions that linearity can be exploited.

Proposition 3. *Let R be the family of random functions with input length l and output length L, and let b be at most $l - 1$. Then there is a constant $c > 0$ such that for any q_s, q_v satisfying $q_s^2 \le 2^l$ and $q_v \le 2^L$, there is an adversary E who $[t, q_s, q_v, \epsilon]$-breaks $\mathrm{XMACR}_{R,b}$, where*

$$t = c(l + L)(q_s + q_v) \quad ; \quad \epsilon = \max\left\{ \left(1 - \frac{1}{e}\right) \cdot \frac{q_s^2 - 3q_s}{4 \cdot 2^l}, \frac{q_v}{2^L} \right\} .$$

Proof. We provide a very brief sketch; for a full proof see [BGR]. Given distinct b-bit strings A', B' we show how to forge the signature of the message $M_4 = A' . B'$. E chooses a b-bit string $A \notin \{A', B'\}$ and a b bit string $B \notin \{A', B', A\}$. She sets $M_1 = A . B$, $M_2 = A' . B$ and $M_3 = A . B'$. She sets $q = \lfloor (q_s - 1)/2 \rfloor$. Now she mounts the following attack—

(1) For $i = 1, 2$, she makes the signing query M_i a total of q times. Let $(r_{i,j}, z_{i,j})$ denote the answers, $i = 1, 2$ and $j = 1, \dots, q$.

(2) She makes the signing query M_3. Let (r, z_3) denote the answer.

Notice that the total number of signing queries made is $2q + 1 \le q_s$. Now let Coll be the event that there exist j_1, j_2 such that $r_{1,j_1} = r_{2,j_2}$. Then:

(3) If Coll is true then E sets $\mu = (r, z)$ where $z = z_{1,j_1} \oplus z_{2,j_2} \oplus z_3$. She then makes the verify query (M_4, μ).

(4) Else, she picks a random $r' \in \{0, 1\}^{l-1}$ and lets $z_1', \dots, z_{q_v}'$ be distinct L-bit strings, for example the q_v lexicographically least L-bit strings. She makes the q_v verify queries $(M_4, (r', z_j'))$ for $j = 1, \dots, q_v$.

Note the number of verify queries made is at most q_v. For the analysis first check that if Coll is true then E is successful in forgery; otherwise via step (4) she is successful with probability $q_v \cdot 2^{-L}$. To conclude we derive some birthday lower bounds to show that $\Pr[\mathrm{Coll}]$ is at least $(1 - e^{-1}) \cdot (q_s^2 - 3q_s)/(4 \cdot 2^l)$. $\square$

The above indicates our analysis in Theorem 1 is tight up to small constant factors. Thus we have been able to give a pretty complete picture of the security. Namely, because of Theorem 1, the attack in the proof of Proposition 3 represents the best possible one (up to constant factors) short of breaking the PRF. We remark that the proof shows that the adversary forges the signature of essentially any message of her choice. This makes the attack all the more relevant.

5   The Counter-Based XOR Scheme

Here we present another scheme, one which allows the signer to maintain state in the form of a counter. The gain is greater security: the success probability of the adversary in the analogue of Theorem 1 does not depend on the number q_s of signing queries at all (as long as the latter is bounded by a certain exponential function of l)!

SPECIFICATION. Let F, l, L, κ be as before, and $a \in \{0,1\}^\kappa$ the shared key. The idea is to use the the same tagging function as above, but use the current counter value as the seed. Formally the scheme $\text{XMACC}_{F,b}$ is specified by functions $\text{SigC}_{F,b}, \text{VfC}_{F,b}$. The signing function depends on a counter C maintained by the signer; it is initially 0 and then incremented by the signing function itself. (In Section 2 we loosely identified C with its 63 bit representation. Now we are more precise, viewing it as an integer and writing $\langle C \rangle_{l-1}$ for the corresponding string.) The verifying function has no state. Below $\text{tag}_{F,b}$ is the function specified in Equation (1) of Section 4.

$$
\begin{array}{l|l}
\textbf{function } \text{SigC}_{F,b}(a, M, C) & \textbf{function } \text{VfC}_{F,b}(a, M', (C', z')) \\
\quad C \leftarrow C + 1 \,; & \quad z \leftarrow \text{tag}_{F,b}(a, M', \langle C' \rangle_{l-1}) \\
\quad z \leftarrow \text{tag}_{F,b}(a, M, \langle C \rangle_{l-1}) & \quad \textbf{if } z = z' \textbf{ then return } 1 \\
\quad \textbf{return } ((C, z),\ C) & \quad \textbf{else return } 0
\end{array}
$$

We call $\text{XMACC}_{F,b}$ the counter-based XOR scheme based on function family F and using block size b. The validity of the counter-based XOR scheme is easy to verify. Note that the XMACC scheme of Section 2 is, in the current terminology, $\text{XMACC}_{F,32}$ with F being the family specified by $F_a(\cdot) = $ first 48 bits of the output of $\text{DES}_a(\cdot)$.

As before the length of any message whose signature the adversary requests is assumed bounded by $b2^{l-b-1}$. But also we will now assume that the total number of signing requests is bounded by 2^{l-1}. That is, we require that C not exceed 2^{l-1}. (Typically, this is not a significant restriction.) These assumptions are made in the theorems that follow.

SECURITY: INFORMATION THEORETIC CASE. There is a dramatic increase in security: the success probability of the adversary now depends only on the number q_v of its verify queries, rather than this plus $2q_s^2 \cdot 2^{-l}$. The proof is like that of Theorem 1 and can be found in [BGR].

Theorem 4. *Let R be the family of random functions with input length l and output length L, let b be at most $l - 1$, and let E be any adversary making a (q_s, q_v)-attack on $\text{XMACC}_{R,b}$, where $q_s < 2^{l-1}$. Then the probability that E is successful is at most $\delta_C \stackrel{\text{def}}{=} q_v \cdot 2^{-L}$.*

To see concretely what this improvement means, think of $F_a = $ first 48 bits of DES_a, where we have $l = 64$ and $L = 48$. If $q_s = 2^{20}$ and $q_v = 1$, then the success probability is a marginal 2^{-23} in the randomized scheme, but it is 2^{-48} in the counter-based one.

SECURITY: COMPUTATIONAL CASE. We get a corresponding improvement. The proof is like that of Theorem 2 and can be found in [BGR].

Theorem 5. *There is an oracle machine U and a constant c such that the following is true. Let F be a family of functions with input length l and output length L and let b be at most $l - 1$. For $q_s < 2^{l-1}$ let E be an adversary who*

$[t, q_s, q_v, \epsilon]$-*breaks* XMACC$_{F,b}$, *and suppose any message* M *in a query of* E *has a number* $\|M\|_b$ *of blocks which is bounded by* n. *Let* $\delta_C = q_v \cdot 2^{-L}$. *Then* U^E $[t', q', \epsilon']$-*breaks* F, *where*

$$ t' = t + c(l + L)q' \quad ; \quad q' = (q_s + q_v) \cdot (n + 1) \quad ; \quad \epsilon' = \epsilon - \delta_C . $$

Again, what this means is that if F is $[t, q', \epsilon']$-pseudorandom then XMACC$_{F,b}$ is $[t, q_s, q_v, \epsilon]$-unforgeable, where $t = t' - c(l + L)q'$, $q_s + q_v = q'/(n + 1)$, and, most importantly, $\epsilon = \epsilon' + \delta_C$.

ATTACKS. The best attack is just to guess signatures! Furthermore one does not expect better than the following (short of breaking the PRF) by virtue of the above theorems.

Proposition 6. *Let* R *be the family of random functions with input length* l *and output length* L, *and let* b *be at most* $l - 1$. *There is a constant* $c > 0$ *such that for any* $q_v \leq 2^L$, *there is an adversary* E *who* $[t, 0, q_v, \epsilon]$-*breaks* XMACC$_{R,b}$, *where* $t = c(l + L)q_v$ *and* $\epsilon = q_v \cdot 2^{-L}$

6 Comparison with the CBC MAC

We compare the security of our schemes to that of the CBC MAC. First, let us recall that scheme. Let F be a family of functions with input and output length l. A message $M = M[1] \ldots M[n]$ is viewed as a sequence of l-bit blocks. The (full) CBC scheme is specified by the following:

function SigCBC$_{F,n}(a, M[1] \ldots M[n])$	function VfCBC$_{F,n}(a, M', \mu')$
$y_0 \leftarrow 0^l$	$\mu \leftarrow$ SigCBC$_{F,n}(a, M')$
for $i \leftarrow 1$ to n do $y_i \leftarrow F_a(y_{i-1} \oplus M[i])$	if $\mu = \mu'$ then return 1
return y_n	else return 0

The scheme is denoted CBC-MAC$_{F,n}$. We will consider the information theoretic case. The following was proved in [BKR]. Let R be the family of random functions of input and output length l, and let E be any adversary. Then the probability that E $[q_s, q_v, \epsilon]$-breaks CBC-MAC$_{R,n}$ is at most $\delta_{CBC} \overset{\text{def}}{=} 3(n^2 + 1) \cdot (q_s + q_v)^2 \cdot 2^{-l}$. To compare this to our schemes set $L = l$ in Theorems 1 and 4. Clearly, δ_R is smaller than δ_{CBC}, and δ_C is considerably smaller than δ_{CBC}; in particular, δ_R and δ_C don't depend on n while δ_{CBC} does, a significant difference.

Yet this by itself is not proof that our schemes are more secure, because it may by that the analysis of [BKR] is not tight. In fact, however, there are attacks (lower bounds) which indicate that the best improvement one could hope for in their analysis would be that $\delta_{CBC} = \Omega(nq_s^2 + q_v)2^{-l}$. This result is due independently to Krawczyk [Kr] and Preneel and Van Oorschot [PV]—what they show is an attack on the CBC MAC which succeeds in forging the signature of a new message with probability $\Omega(nq_s^2) \cdot 2^{-l}$, after having made q_s signing queries on n-block messages. Thus the dependence on n in δ_{CBC} is unavoidable.

We comment that the CBC-MAC$_{F,n}$ is only secure for fixed n; the scheme must be modified to accommodate n's of varying length. In contrast, both XMACR$_{F,b}$ and XMACC$_{F,b}$ operate on inputs of varying lengths (with the security bounds given by our theorems).

Acknowledgments

We thank Mike Luby for pointing out an oversight in a previous version of the proof of Theorem 1. We thank Hugo Krawczyk for a great deal of useful advice and information.

References

[ABI] N. ALON, L. BABAI AND A. ITAI. A fast and simple randomized parallel algorithm for the maximal independent set problem. *J. of Algorithms*, Vol.7, 567-583, 1986.

[BGR] M. BELLARE, R. GUÉRIN AND P. ROGAWAY. XOR MACs: New methods for message authentication using finite pseudorandom functions. Available from the authors or out of http://www.cs.ucdavis.edu/~rogaway/

[BKR] M. BELLARE, J. KILIAN AND P. ROGAWAY. On the security of cipher block chaining. *Advances in Cryptology – Crypto 94 Proceedings*.

[BGG1] M. BELLARE, O. GOLDREICH AND S. GOLDWASSER. Incremental cryptography: The case of hashing and signing. *Advances in Cryptology – Crypto 94 Proceedings*.

[BGG2] M. BELLARE, O. GOLDREICH AND S. GOLDWASSER. Incremental cryptography and application to virus protection. *Proceedings of the Twenty Seventh Annual Symposium on the Theory of Computing*, ACM, 1995.

[BeRo] B. BERGER AND J. ROMPEL, "Simulating $(\log^c n)$-wise independence in NC," *Proceedings of the Thirtieth Annual Symposium on the Foundations of Computer Science*, IEEE, 1989.

[GGM] O. GOLDREICH, S. GOLDWASSER, AND S. MICALI. How to construct random functions. *Journal of the ACM*, Vol. 33, No. 4, 210–217, 1986.

[GMR] S. GOLDWASSER, S. MICALI, AND R. RIVEST. A digital signature scheme secure against adaptive chosen-message attacks. *SIAM Journal of Computing*, 17(2):281–308, April 1988.

[ISO] ISO/IEC 9797. Data cryptographic techniques – Data integrity mechanism using a cryptographic check function employing a block cipher algorithm, 1989.

[Kr] H. KRAWCZYK. Personal communication, September 1994.

[LuRa] M. LUBY AND C. RACKOFF, "How to construct pseudorandom permutations from pseudorandom functions," *SIAM J. Comput*, Vol. 17, No. 2, April 1988.

[PV] B. PRENEEL AND P. VAN OORSCHOT. A new generic attack on message authentication codes. *Advances in Cryptology – Crypto 95 Proceedings*.

[Ri] R. RIVEST, "The MD5 message digest algorithm." IETF RFC-1321, 1992.

[Ts] G. TSUDIK, "Message authentication with one-way hash functions." *Proceedings of Infocom 92*, IEEE Press, 1992.

[X9.9] ANSI X9.9, American National Standard for Financial Institution Message Authentication (Wholesale), American Bankers Association, 1981. Revised 1986.

Bucket Hashing and its Application to Fast Message Authentication

Phillip Rogaway

Department of Computer Science, University of California,
Davis, CA 95616, USA. rogaway@cs.ucdavis.edu

Abstract. We introduce a new technique for generating a message authentication code (MAC). At its center is a simple metaphor: to (non-cryptographically) hash a string x, cast each of its words into a small number of *buckets*; xor the contents of each bucket; then collect up all the buckets' contents. Used in the context of Wegman–Carter authentication, this style of hash function provides the fastest known approach to software message authentication.

1 Introduction

MESSAGE AUTHENTICATION. Message authentication is one of the most common cryptographic aims. The setting is that two parties, a signer S and verifier V, share a (short, random, secret) key, a. When S wants to send V a message, x, S computes for it a *message authentication code* (MAC), $\mu \leftarrow \mathsf{MAC}_a(x)$, and S sends V the pair (x, μ). On receipt of (x', μ'), verifier V checks that $\mathsf{MACV}_a(x', \mu') = 1$.

To describe the security of a message authentication scheme an adversary is given an oracle for $\mathsf{MAC}_a(\cdot)$. Following [11], she is declared *successful* if she outputs an (x^*, t^*) such that $\mathsf{MACV}_a(x^*, t^*) = 1$ but x^* was never asked of the $\mathsf{MAC}_a(\cdot)$ oracle. For a scheme to be "good," reasonable adversaries should rarely succeed.

SOFTWARE-EFFICIENT MACs. In the current computing environment it is often necessary to compute MACs frequently and over strings which are commonly hundreds to thousands of bytes long. Despite this, there will usually be no special-purpose hardware to help out: MAC generation and verification will need to be done in software on a conventional workstation or personal computer. So to reduce the impact of message authentication on the machine's overall performance, and to facilitate more pervasive use of message authentication, we need substantially faster techniques. That is what this paper provides.

TWO APPROACHES TO MESSAGE AUTHENTICATION. The fastest software MACs in common use today are exemplified by $\mathsf{MAC}_a(x) = h(x\|a)$, with h a (software-efficient) cryptographic hash function, such as $h =$ MD5. Such methods are described in [22]. A scheme like this might seem to be about as software-efficient as one might realistically hope for: after all, we are computing one of the fastest types of cryptographic primitives over a string nearly identical in length to

D. Coppersmith (Ed.): Advances in Cryptology - CRYPTO '95, LNCS 963, pp. 29-42, 1995.

that which we want to authenticate. But it is well-known that this reasoning is specious: in particular, Wegman and Carter [23] showed back in 1981 that we do not have to "cryptographically" transform the entire string x.

In the Wegman–Carter approach communicating parties S and V share a secret key a which specifies both a random pad p and a hash function h drawn randomly from a strongly universal$_2$ family of hash functions $\mathcal{H}$. (Recall that $\mathcal{H}$ is strongly universal$_2$ if for all $x_0 \neq x_1$, the random variable $h(x_0) \parallel h(x_1)$ is uniformly distributed.) To authenticate a message x, the sender transmits $h(x)$ xor-ed with the next piece of the pad p. The thing to notice is that x is transformed first by a non-cryptographic operation (universal hashing) and only then is it subjected to a cryptographic operation (encryption)—now applied to a much shorter string.

As it turns out, to make a good MAC you don't need to start from a strongly universal$_2$ family. Carter and Wegman [7] also introduced the notion of an almost universal$_2$ family, $\mathcal{H}$. This must satisfy the weaker condition that $\Pr_{h \in \mathcal{H}}[h(x_0) \neq h(x_1)]$ is small for all $x_0 \neq x_1$. As observed by Stinson [19], an almost universal$_2$ family can easily be turned into an almost strongly universal$_2$ family (which can, in turn, be used to authenticate ones messages). In this manner the problem of finding an efficient MAC has effectively been reduced to that of finding an efficient almost universal$_2$ family of hash functions.

OUR CONTRIBUTION. This paper provides a novel almost universal$_2$ family of hash functions. We call our hash family *bucket hashing*. It is distinguished by its member functions being extremely fast to compute—as few as 6 elementary machine instructions per word (independent of word size) for the version of bucket hashing we concentrate on in this paper. Putting such a family of hash functions to work in the framework of known constructions gives rise to the most efficient software MACs now known. For example, we estimate that a MAC so constructed can authenticate (reasonably long) messages in about 10–15 instructions per 32-bit word. For comparison, authenticating messages using an MD5-based technique requires some 40–50 instructions per word [21].

A bucket hash MAC has advantages in addition to speed. Bucket hashing is a *linear* function —it is a special case of matrix multiplication over GF(2)— and this linearity yields many pleasant characteristics for a bucket hash MAC. In particular, bucket hashing is *parallelizable,* since each word of the hash is just the xor of certain words of the message. Bucket hashing is *incremental* in the sense of [2] with respect to both append and substitute operations. Finally, the only processor instructions a bucket hash needs are word-aligned load, store, and xor; thus a bucket hash MAC is essentially endian-indifferent.

One might worry that the linearity of bucket hashing might give rise to some "weakness" in a MAC which exploits it. But it does not. A bucket hash MAC, like any MAC which follows the Wegman–Carter paradigm, enjoys the assurance advantages of provable security. Moreover, this provable security is achieved under extremely "tight" reductions, so that an adversary who can successfully break the MAC can break the underlying cryptographic primitive (e.g., DES) with essentially identical efficacy. In contrast, a scheme like $\mathsf{MAC}_a(x) = h(x \| a)$

is not known to be provably secure under *any* standard assumption on h.

RELATED WORK. The general theory of unconditional authentication was developed by Simmons; see [18] for a survey. As we have already explained, the universal-hash-and-then-encrypt paradigm is due to Wegman and Carter [23]. The idea springs from their highly influential [7].

In Wegman–Carter authentication the size of the hash family corresponds to the number of bits of shared key—one reason to find smaller families of universal hash functions than those of [7, 23]. Stinson does this in [19], and also gives general results on the construction of universal hash functions. We exploit some of these ideas. Subsequent improvements (rooted in coding theory) came from Bierbrauer, Johansson, Kabatianskii and Smeets [6], and Gemmell and Naor [9].

The above work concentrates on unconditionally-secure authentication. Brassard [5] first connects the Wegman–Carter approach to the complexity-theoretic case. The complexity-theoretic notion for a secure MAC is a straightforward adaptation of the definition of a digital signature due to Goldwasser, Micali and Rivest [11]. Their notion of an adaptive chosen message attack is equally at home for defining an unconditionally-secure MAC. Thus we view work like ours as making statements about unconditionally-secure authentication which give rise to corresponding statements and concrete schemes in the complexity-theoretic tradition. To make this translation we regard a finite pseudorandom function (PRF) as the most appropriate tool. Bellare, Kilian and Rogaway [3] were the first to formalize such objects, investigate their usage in the construction of efficient MACs, and suggest them as a desirable starting point for practical, provably-good constructions. Finite PRFs are a refinement of the PRF notion of Goldreich, Goldwasser and Micali [10] to take account of the fixed lengths of inputs and outputs in the efficient primitives of cryptographic practice.

Zobrist [25] gives a hashing technique which predates [7] and which, in implementation, essentially coincides with the scheme $\mathcal{H}_M$ described in Section 2 and due to [7]. Arnold and Coppersmith [1] give an interesting hashing technique which allows one to map a set of keys k_i into a set of corresponding values v_i using a table only slightly bigger than $\sum_i v_i$. The proof of our main technical result is somewhat reminiscent of the analysis in [1].

Lai, Rueppel and Woolven [14], Taylor [20], and Krawczyk [12] have all been interested in computationally efficient MACs. The last two works basically follow the Wegman–Carter paradigm. In particular, Krawczyk obtains efficient message authentication codes from hash families which resemble traditional cyclic redundancy codes (CRCs), and matrix multiplication using Toeplitz matrices. Though originally intended for hardware, these techniques are fast in software, too. We recall Krawczyk's CRC-like hash in Section 2.

2 Preliminaries

This section provides background drawn from Carter and Wegman [7, 23], Stinson [19], and Krawczyk [12]. The only new material is the (simple) scheme $\mathcal{H}_N$ and the statement of Theorem 7. Proofs are omitted.

A *family of hash functions* is a finite multiset $\mathcal{H}$ of string-valued functions, each $h \in \mathcal{H}$ having the same nonempty domain $A \subseteq \{0,1\}^*$ and range $B \subseteq \{0,1\}^*$.

Definition 1. [7] A family of hash functions $\mathcal{H} = \{h : A \to B\}$ is ϵ-**almost universal$_2$**, written ϵ-AU$_2$, if for all $x_0, x_1 \in A$, $x_0 \neq x_1$, $\mathrm{Pr}_{h \in \mathcal{H}}[h(x_0) = h(x_1)] \leq \epsilon$. Family $\mathcal{H}$ ϵ-**almost XOR universal$_2$**, written ϵ-AXU$_2$, if for all $x_0, x_1 \in A$, $y \in B$, $x_0 \neq x_1$, $\mathop{\mathrm{Pr}}_{h \in \mathcal{H}}[h(x_0) \oplus h(x_1) = y] \leq \epsilon$.

The value of $\epsilon^* = \max_{x_0 \neq x_1}\{\mathrm{Pr}_h[h(x_0) = h(x_1)]\}$ is called the *collision probability*. For us, the principle measures of the worth of an AU$_2$ hash family are how small is ϵ^* and how fast can one compute its functions.

To make a fast MAC we will want to "glue together" various universal hash families. The following are our the basic methods for doing this.

First we need a way to make the domain of a hash family bigger. Let $\mathcal{H} = \{h : \{0,1\}^a \to \{0,1\}^b\}$. By $\mathcal{H}^m = \{h : \{0,1\}^{am} \to \{0,1\}^{bm}\}$ we denote the family of hash functions whose elements are the same as in $\mathcal{H}$ but where $h(x_1 x_2 \cdots x_m)$, for $|x_i| = a$, is defined by $h(x_1) \parallel h(x_2) \parallel \cdots \parallel H(x_m)$.

Proposition 2. [19] If $\mathcal{H}$ is ϵ-AU$_2$ then $\mathcal{H}^m$ is ϵ-AU$_2$.

Sometimes one needs a way to make the collision probability smaller. Let $\mathcal{H}_1 = \{h : \{0,1\}^a \to \{0,1\}^{b_1}\}$ and $\mathcal{H}_2 = \{h : \{0,1\}^a \to \{0,1\}^{b_2}\}$ be families of hash functions. By $\mathcal{H}_1 \& \mathcal{H}_2 = \{h : \{0,1\}^a \to \{0,1\}^{b_1+b_2}\}$ we mean the family of hash functions whose elements are pairs of functions in $\mathcal{H}_1$ and $\mathcal{H}_2$ and where $(h_1, h_2)(x)$ is defined as $h_1(x) \parallel h_2(x)$.

Proposition 3. If $\mathcal{H}_1$ is ϵ_1-AU$_2$ and $\mathcal{H}_2$ is ϵ_2-AU$_2$ then $\mathcal{H}_1 \& \mathcal{H}_2$ is $\epsilon_1 \epsilon_2$-AU$_2$.

Next is a way to make the image of a hash function shorter. Let $\mathcal{H}_1 = \{h : \{0,1\}^a \to \{0,1\}^b\}$ and $\mathcal{H}_2 = \{h : \{0,1\}^b \to \{0,1\}^c\}$ be families of hash functions. Then by $\mathcal{H}_2 \circ \mathcal{H}_1 = \{h : \{0,1\}^a \to \{0,1\}^c\}$ we mean the family of hash function whose elements are pairs of functions in $\mathcal{H}_1$ and $\mathcal{H}_2$, and where $(h_1, h_2)(x)$ is defined as $h_2(h_1(x))$.

Proposition 4. [19] If $\mathcal{H}_1$ is ϵ_1-AU$_2$ and $\mathcal{H}_2$ is ϵ_2-AU$_2$ then $\mathcal{H}_2 \circ \mathcal{H}_1$ is $(\epsilon_1 + \epsilon_2)$-AU$_2$.

Composition can also be used to turn an AU$_2$ family into AXU$_2$ family:

Proposition 5. [19] Suppose $\mathcal{H}_1 = \{h : A \to B\}$ is ϵ_1-AU$_2$, and $\mathcal{H}_2 = \{h : B \to C\}$ is ϵ_2-AXU$_2$. Then $\mathcal{H}_2 \circ \mathcal{H}_1 = \{h : A \to C\}$ is $(\epsilon_1 + \epsilon_2)$-AXU$_2$.

Now given a family of hash functions $\mathcal{H} = \{A \to \{0,1\}^b\}$ we can construct from it a MAC. In the scheme we denote WC[$\mathcal{H}$], the signer S and verifier V share a random element $h \in \mathcal{H}$, as well as an infinite random string $p = p_0 p_1 p_2 \cdots$, where $|p_i| = b$. Together, h and p comprise the shared secret. The signer maintains a counter, cnt, which is initially 0. We let the MAC of x under key (h, p) be given by $(\text{cnt}, p_{\text{cnt}} \oplus h(x))$. The scheme is stateful: after the MAC of x is computed, the number cnt is incremented. The following theorem says that it is impossible (regardless of time, number of queries, or amount of MACed text) to forge with probability exceeding the collision probability (see Appendix B for definitions):

Theorem 6. [23, 12] Let $\mathcal{H}$ be ϵ-AXU$_2$ and suppose E (t, q, μ, δ)-breaks WC[$\mathcal{H}$]. Then $\delta \leq \epsilon$.

A natural complexity-theoretic variant is to use, instead of the random pad p, an index $a \in \{0, 1\}^\kappa$ into a finite PRF F. The signer S maintains a counter cnt $\in \{0, 1\}^l$. Function F_a maps l-bit strings to b-bit ones. Now S and V share a random $a \in \{0, 1\}^k$ and a random $h \in \mathcal{H}$. The cntth MAC of x under key (h, a) is given by $(\text{cnt}, F_a(\langle\text{cnt}\rangle_l) \oplus h(x))$. At most 2^l messages may be MACed before the key must be changed. We call the scheme just described WC[$\mathcal{H}, F$]. Its security is described by the following:

Theorem 7. Let $\mathcal{H} = \{h : A \rightarrow \{0, 1\}^b\}$ be an ϵ-AXU$_2$ family of hash functions, let $F : \{0, 1\}^\kappa \times \{0, 1\}^l \rightarrow \{0, 1\}^b$ be a finite PRF, and let E be an adversary which (t, q, μ, δ)-breaks WC[$\mathcal{H}, F$]. Suppose one can in time $T_\mathcal{H}$ compute a representation of a random element $h \in \mathcal{H}$, and from this representation one can compute h-values on q strings totalling μ bits in $T_h(q, \mu)$ time. Then there is an algorithm D which $(t + \Delta t, q + 1, \delta - \epsilon)$-breaks F, where $\Delta t = O(T_h(q, \mu) + T_\mathcal{H} + ql + qb)$.

The value of Δt would usually be insignificant compared to t.

In the above two theorems the forging probability is independent of the number of queries (q) and the length of the queried messages (μ). This is a significant advantage compared with constructions based on the iterated application of a finite PRF.

We emphasize that the signer is stateful in both WC[$\mathcal{H}$] and WC[$\mathcal{H}, F$]. This improves these schemes' security (compared with using a random value) at little practical cost. Note that the verifier is not stateful, as we have chosen a definition of security (see Appendix B) which does not address "replay attacks."

We end this section with some constructions for software-efficient hash families. In the first $A = \{0, 1\}^a$ is the strings we want to hash and $B = \{0, 1\}^b$ is the space we want to hash our strings into. An element of $\mathcal{H}_M[a, b]$ is described by a $b \times a$ binary matrix. If $h \in \mathcal{H}_M$ is such a matrix we define $h(x)$ as the product hx of matrix h and column vector x.

Proposition 8. [7] $\mathcal{H}_M[a, b]$ is 2^{-b}-AXU$_2$.

We can modify this family to trade ϵ-AXU$_2$ for ϵ-AU$_2$ and some speed. Let $A = \{0, 1\}^{a+b}$ be the strings we want to hash and let $B = \{0, 1\}^b$ be the space we want to hash into. Each $h \in \mathcal{H}_N[a+b, b]$ is described by a $b \times a$ binary matrix. If $h \in \mathcal{H}_N[a+b, b]$ is such matrix, we define $h(x_1 e_0)$, where $|x_1| = a$ and $|x_0| = b$, by $hx_1 \oplus x_0$—the product of matrix h and column vector x_1, xor-ed with x_0.

Proposition 9. $\mathcal{H}_N[a+b, b]$ is $2^{-b} - \text{AU}_2$.

Here's a final construction, this one from [12]. An element of the hash family $\mathcal{H}_K[a, b] = \{h : \{0,1\}^a \to \{0,1\}^b\}$ is described by an irreducible polynomial over GF[2] of degree b. Given such a polynomial h, the value of $h(M)$, where $M \in \{0,1\}^a$, is the coefficients of $M(X)X^b \bmod h(X)$, where X is a formal variable and $M(X)$ is the polynomial over X whose coefficients are given by M.

Theorem 10. [12] $\mathcal{H}_K[a, b]$ is $\frac{a+b}{2^b-1}$-AXU$_2$.

3 Bucket Hashing

In software, hash families such as $\mathcal{H}_N$, $\mathcal{H}_M$ and $\mathcal{H}_K$ are all reasonably efficient. Still, for reasonably long message, we can do quite a bit better.

3.1 Defining $\mathcal{H}_B[w, n, N]$

Fix a "word size" $w \geq 1$. For some particular $n \geq 1$ and $N \geq 3$ we will be hashing from $A = \{0,1\}^{wn}$ to $B = \{0,1\}^{wN}$. We call N the "number of buckets" for reasons soon to be clear. As a typical example, take $w = 32$, $n = 1024$, and $N = 140$. We require that $\binom{N}{3} \geq n$.

Each $h \in \mathcal{H}_B[w, n, N]$ is specified by a length-n list of cardinality-3 subsets of $[0..N-1]$. We denote this list by $h = (h_0, \ldots, h_{n-1})$. We denote the three elements of h_i by $h_i = \{h_i^1, h_i^2, h_i^3\}$.

Choosing a random h from $\mathcal{H}_B[w, n, N]$ means choosing a random length-n list of three-element subsets of $[0..N-1]$ subject to the constraint that no two of these sets are the same. That is, for all $i \neq j$, we demand that $h_i \neq h_j$.

Let $h = (h_0, \ldots, h_{n-1}) \in \mathcal{H}_B[w, n, N]$ be as above. Then $h(x)$ is defined by the following algorithm. Let $x = x_0 \cdots x_{n-1} \in A$, with each $|x_i| = w$. First, initialize y_j to 0^w for each $j \in [0..N-1]$. Then, for each $i \in [0..n-1]$ and $k \in h_i$, replace y_k by $y_k \oplus x_i$. When done, set $h(x) = y_0 \| y_1 \| \cdots \| y_{N-1}$. In pseudocode we have:

> **for** $j \leftarrow 0$ to $N-1$ **do** $y_j \leftarrow 0^w$
> **for** $i \leftarrow 0$ to $n-1$ **do**
> $\quad y_{h_i^1} \leftarrow y_{h_i^1} \oplus x_i$
> $\quad y_{h_i^2} \leftarrow y_{h_i^2} \oplus x_i$
> $\quad y_{h_i^3} \leftarrow y_{h_i^3} \oplus x_i$
> **return** $y_0 \| y_1 \| \cdots \| y_{N-1}$

The computation of a $h(x)$ can be envisioned as follows. We have N buckets, each initially empty. The first word of x is thrown into the three buckets specified by h_0. The second word of x is thrown into the three buckets specified by h_1. And so on, with the last word of x being thrown into the three buckets specified by h_{n-1}. Our N buckets now contain a total of $3n$ words. Compute the xor of the words in each of the buckets (with the xor of no words being defined as 0^w). The hash of x, $h(x)$, is the concatenation of the final contents of the buckets.

3.2 Collision probability of $\mathcal{H}_B[w, n, N]$

The following theorem shows that $\mathcal{H}_B[w, n, N]$ is $\Theta(N^{-6})$-good. For large N and n, where $N \ll n$, the bound approaches $3348 N^{-6}$.

Theorem 11. [Bucket hash bound] Assume $w \geq 1$, $N \geq 20$ and $n \leq \binom{N}{3}$. Then $\mathcal{H}_B[w, n, N]$ is $(\lambda_{n,N} \cdot \alpha(N))$-AU$_2$, where $\lambda_{n,N} = \dfrac{1}{1 - n/\binom{N}{3}}$ and $\alpha(n) =$

$$\frac{720(N-3)(N-4)(N-5) + 1944(N-3)(N-4)^2 + 648(N-2)(N-3)^2 + 36N(N-1)(N-2)}{N^3(N-1)^3(N-2)^3}.$$

The proof (given in Appendix A) involves a tedious calculation on a Markov chain. For intuition, it is not hard to guess that $\max_{x_0 \neq x_1} \Pr[h(x_0) = h(x_1)]$ is achieved on strings x_0 and x_1 which differ in exactly four (appropriately selected) words. Bounding the collision probability for this case gives the formula of above. Most of the effort is showing that x_0 and x_1 differing by four words really *is* the case which which maximizes the collision probability.

By way of example, suppose we use $\mathcal{H}_B[w, n, N]$ to hash $n \in \{256, 1024, 4096\}$ words down to $N \in \{20, 40, \cdots, 200\}$ words. From Theorem 11, upper bounds on the probability that distinct but equal length strings collide are given by:

	20	40	60	80	100	120	140	160	180	200
256	$2^{-14.01}$	$2^{-20.25}$	$2^{-23.76}$	$2^{-26.24}$	$2^{-28.17}$	$2^{-29.75}$	$2^{-31.08}$	$2^{-32.23}$	$2^{-33.25}$	$2^{-34.16}$
1024	$2^{-11.08}$	$2^{-20.13}$	$2^{-23.73}$	$2^{-26.23}$	$2^{-28.16}$	$2^{-29.74}$	$2^{-31.08}$	$2^{-32.23}$	$2^{-33.25}$	$2^{-34.16}$
4096		$2^{-19.51}$	$2^{-23.59}$	$2^{-26.17}$	$2^{-28.14}$	$2^{-29.73}$	$2^{-31.07}$	$2^{-32.22}$	$2^{-33.24}$	$2^{-34.16}$

The first entry of the last row is missing because it does not satisfy the condition of Theorem 11: there are not enough distinct triples of 20 buckets to accommodate 4096 words.

4 The Efficiency of Universal Hash Methods

To make a practical MAC we want a fast-to-compute ϵ-AU$_2$ hash family $\mathcal{H} = \{h : \{0,1\}^{\leq a} \to \{0,1\}^b\}$ where, for example, $a \approx 2^{30}$, $b \approx 64$, and $\epsilon \approx 2^{-30}$. This section compares the efficiency of various universal hash families useful to construct such an $\mathcal{H}$. Efficiency comparisons are given in a very concrete way: we count the machine instructions per word of x to compute $h(x)$. We assume a contemporary 32-bit architecture. Though instruction counting is only a crude predictor of speed, an analysis like this is still the best implementation-independent way to get a feel for our methods' efficiency.

EFFICIENCY OF $\mathcal{H}_B$. From Section 3.2 and the above preamble it is apparent that we need more buckets than can be accommodated by a typical machine's register set. So there are two natural strategies to hash $x = x_0 \ldots x_{n-1}$:

- **Method 1** – Process words $x_0, \ldots, x_{n-1}$. We can read each x_i from memory and then, three times: (1) read the value y_j of some bucket j from memory; (2) compute $x_i \oplus y_j$; (3) write the result back to bucket j. Total instruction count is 10 instructions per word.

- **Method 2** – Fill buckets $y_0, \ldots, y_{N-1}$. We can xor together all words that should wind up in bucket 0; then xor all words that go in bucket 1; and so forth, for each of the N buckets. A total of $3n$ reads into $x_0, \ldots, x_{n-1}$ will be needed; plus a total of $3n - N$ xor operations; and (possibly) another N writes back to memory. Total instruction count is 6 instructions per word.

Achieving the stated instruction counts requires the use of self-modifying code ("sm code"); above, it is implicitly assumed that the representation of $h \in \mathcal{H}_B$ is the piece of executable code which computes it. In implementation, this can be tricky. If we must spend the time to load the bucket location (Method 1) or word location (Method 2) out of memory ("kiss code"), these loads comprise extra overhead. For Method 2, a refusal to use self-modifying code will further increase the instruction count because of overhead to control the looping: it is key-dependent how many words fall into a given bucket.

method	implementation	table size (Bytes)	$\approx$ instrs/wd
$\mathcal{H}_B[32, n, N]$	Method 1 (sm code)	code – no table	10
$\mathcal{H}_B[32, n, N]$	Method 2 (sm code)	code – no table	6
$\mathcal{H}_B[32, n, N]$	Method 1 (kiss code)	$3n$ or $12n$	13
$\mathcal{H}_B[32, n, N]$	Method 2 (kiss code)	$6n$ or $12n$	9+

For Method 2, n should be at most a few thousand to retain reasonable cache performance.

EFFICIENCY OF $\mathcal{H}_N$ AND $\mathcal{H}_M$. The software-efficient implementation of $\mathcal{H}_N$ relies on a table of pre-computed inner products. There is a tradeoff between the size of this table and the number of instructions. Consider $\mathcal{H}_N[64, 32]$ and assume we partition words as 8-bit bytes to look up inner products in a pre-computed table TN. The table size is then 4×2^8 words, or 4 KBytes. To hash the two-word string $x_0 x_1 x_2 x_3 \; x_4 x_5 x_6 x_7$ to the one-word string $y_0 y_1 y_2 y_3$ we must isolate the four bytes x_0, x_1, x_2, x_3; lookup in TN the entries these bytes index; then xor what we get out of the table. We need 3–5 instructions per table lookup, plus an extra couple of instructions to read $x_4 x_5 x_6 x_7$ and xor it with what we have already. Other examples:

method	cmprsn	coll prob	table size (Bytes)	$\approx$ instrs/wd
$\mathcal{H}_2 = \mathcal{H}_N[2 \times 32, 32]$	2	2^{-32}	4 K	7–11
$\mathcal{H}_N[8 \times 32, 32]$	8	2^{-32}	28 K	7–11
$\mathcal{H}_2 \circ \mathcal{H}_2 \circ \mathcal{H}_2$	8	$2^{-30.4}$	12 K	13.5–19.5

Overall compression is poor unless the table size is quite large, leading one towards multiple applications of concatenation and composition (using Propositions 2 and 4) as in the third row of the table.

The instruction count and table size of $\mathcal{H}_M$ are worse than $\mathcal{H}_N$—for example, 12–20 instructions with an 8 KByte table for $\mathcal{H}_M[2 \times 32, 32]$.

EFFICIENCY OF $\mathcal{H}_K[n, 64]$. The collision probability of $\mathcal{H}_K[n, 32]$ is inadequate for the stated goal, and there is no apparent way to implement $\mathcal{H}_K[n, w]$ for

$w \in [33..64]$ any faster than implementing $\mathcal{H}_K[n, 64]$. The software-efficient implementation of $\mathcal{H}_K[n, 64]$ relies on a table TK of pre-computed quotients. Let us assume that we index into TK in 8-bit bytes. Computing $h \in \mathcal{H}_K[n, 64]$ on $x_0 \cdots x_{n-1}$, where $|x_i| = 8$, could be done as in

$\qquad$ crc $\leftarrow 0^{64}$
$\qquad$ **for** $i = 0$ **to** $n/8 - 1$ **do**
$\qquad\qquad$ index $\leftarrow x_i \oplus ($crc & **0xFF**$);$ crc $\leftarrow$ crc $\rangle\rangle$ 8; crc $\leftarrow$ crc $\oplus$ TK[index]
$\qquad$ **return** crc

which, coded in the natural manner on a 32-bit machine gives, gives 44–48 instrs/wd with a 2 KByte table.

5 Towards a Fully-Specified Scheme

AN EXAMPLE. This section provides an illustrative example of a concrete MAC based on the ideas presented so far. This is only a "toy" example; doing a good job at specifying a software-optimized bucket hash MAC will involve further design, experimental, and theoretical work.

To keep things simple, first suppose we want to MAC strings which are always 4096 bytes (1024 words). Let $F : \{0,1\}^\kappa \times \{0,1\}^{64} \to \{0,1\}^{64}$ be a (candidate) finite PRF. Here's how to MAC the cnt^{th} message $x \in \{0,1\}^{32 \times 1024}$. Recall the $\mathcal{H}^i$ notation from Section 2.

1. Hash x with $h_1 \in \mathcal{H}_B[32, 1024, 144]$ to get a 144-word result, $y_1 \leftarrow h_1(x)$.
2. Hash y_1 with $h_2 \in (\mathcal{H}_N[64, 32])^{72}$ to get a 72-word result, $y_2 \leftarrow h_2(y_1)$.
3. Hash y_2 with $h_3 \in (\mathcal{H}_N[64, 32])^{36}$ to get a 36-word result, $y_3 \leftarrow h_3(y_2)$.
4. Hash y_3 with $h_4 \in (\mathcal{H}_N[64, 32])^{18}$ to get an 18-word result, $y_4 \leftarrow h_4(y_3)$.
5. Hash y_4 with $h_5 \in \mathcal{H}_K[576, 64]$ to get a 2-word result, $y_5 \leftarrow h_5(y_4)$.
6. Compute $p = F_a(\langle \text{cnt} \rangle_{64})$.
7. Return $(\text{cnt}, p \oplus y_5)$.

Conceptually, the MAC key is $(a, h_1, h_2, h_3, h_4, h_5)$; in practice, it would be the seed used by a pseudorandom generator to generate such a tuple.

Let us estimate the speed of the above scheme, TOY-MAC. Assume the most pessimistic instruction counts in Section 4, and assume 200 instructions for the computation of F_a. With bucket hashing achieved by Method 2 (sm code), we get $6 + 11(144/1024) + 11(72/1024) + 11(36/1024) + 48(18/1024) + 200(2/1024) \approx 10.3$ instructions per word. The tables add up to 14 KBytes and the total collision probability (using Propositions 2,4,5,9; Theorems 10,11) is $\leq 2^{-31.32} + 3 \times 2^{-32} + 2^{-53} \approx 2^{-29.8}$. For comparison, recall that authenticating messages using an MD5-based technique requires some 40–50 instructions per word [21].

Notice that the "cryptographic" contribution (Step 6) of TOY-MAC takes just $200(2/1024) \approx 0.4$ instructions, which is less than 5% of the total work. In a Wegman-Carter MAC one is afforded the luxury of conservative (slow) cryptography even in an aggressive (fast) MAC, since one arranges that the time complexity for the MAC is dominated by the non-cryptographic work.

Schemes simpler than TOY-MAC still perform well. Replacing steps (1)–(5) by $y_4 \leftarrow h_5(h_1(x))$ gives $6 + (144/1024)48 + (2/1024)200 \approx 13.2$ instructions/word.

Alternatively, since MD5 is about the same speed as $\mathcal{H}_K$, a similar estimate would hold for a scheme like $MAC_{a,cnt}(x) = \langle cnt, MD5(a \cdot cnt \cdot |x| \cdot h_1(x)) \rangle$.

Our instruction counts are attractive even if we do not use self-modifying code: naive implementations would take fewer than 20 instrs/wd.

SHORT STRINGS. What if the strings we are MACing has lengths which are a constant substantially less than 4096 bytes? Using the layers of hash functions chosen for TOY-MAC makes no sense if $|x|$ is too short. In general, the hash method which will be fastest for a string depends strongly on its length. For a software-optimized MAC to be most useful it should be as fast as possible for each input length. We thus suggest designing a Wegman–Carter MAC using multiple AU_2-hash functions, choosing the best sequence of these to apply to x based on $|x|$. While this may seem complex, the user of a MAC typically cares about performance and security, not definitional simplicity.

While instruction counts for fast AU_2-hashing are best (per word) when the strings being hashed are long, even strings of a few words benefit from having their lengths reduced (e.g. by $\mathcal{H}_N$) before being cryptographically acted on.

LONG STRINGS. Modifying TOY-MAC to deal with long strings is easy: for example, if x is $1024k$ words then one can replace Steps 1–4 by $y_4 \leftarrow (h_4 \circ h_3 \circ h_2 \circ h_1)^k$.

LENGTH VARIABILITY. TOY-MAC is only correct when the strings to MAC have some fixed length. The reason is that $\mathcal{H}_B$ is ϵ-AU_2 only when restricted to strings of a fixed length. For example, $h(x) = h(x0^w)$ for any $h \in \mathcal{H}_B$. Fortunately, eliminating the fixed-length restriction is easy. One solution is to include $|x|$ in the scope of $F_a(\cdot)$, which ensures that $WC[\mathcal{H}, F]$ is secure across variable-length strings as long as $\mathcal{H}$ is ϵ-AXU_2 on equal-length ones. The bound in Theorem 7 is unchanged. A second approach is to use constructions like $h_f(\langle |x| \rangle \cdot h_0(x))$. Neither approach requires $|x|$ be known before x is hashed.

6 Extensions and Directions

Generalizing $\mathcal{H}_B$ we call by "bucket hashing" any scheme in which the hash function h is a given by a list $(h_0, \ldots, h_{n-1})$ of "small" subsets of $[0..N-1]$ and the hash of $x = x_0 \cdots x_{n-1}$, where $|x_i| = w$, is:

    for $j \leftarrow 0$ **to** $N-1$ **do** $y_j \leftarrow 0^w$
    for $i \leftarrow 0$ **to** $n-1$ **do**
        for each $k \in h_i$ **do**
            $y_k \leftarrow y_k \oplus x_i$
    return $y_0 \| y_1 \| \cdots \| y_{N-1}$

In the general case the distribution on h-values is arbitrary. So $\mathcal{H}_B[w, n, N]$ is the special case in which we use the uniform distribution on three-elements subsets of distinct triples in $[0..N-1]$.

One could imagine many alternate distributions, some of which will give rise to faster-to-compute hash functions or better bounds on the collision probability. As an example, suppose $\mathcal{H}$ is given by choosing $h_0, h_1, \ldots$, in sequence, where h_i is a random triple of $[0..N-1]$ subject to the constraint that among

$\{h_0, \ldots, h_i\}$ there are no two *and no four* of the h_j's such that the multiset $\cup h_j$ has an even number of each point $0, 1, \ldots, N-1$. (Dropping the italicized words we would recover the definition of $\mathcal{H}_B[w, n, N]$.) This new family of hash functions, $\mathcal{H}'_B[w, n, N]$, may have substantially smaller collision probability than $\mathcal{H}_B[w, n, N]$, allowing one to choose a smaller value of N.

THE BUCKET HASH SCHEME OF A GRAPH. Hash family $\mathcal{H}_B$ would have been more efficient had each word gone into two buckets instead of three. One way to specify such a scheme is with a graph G whose N vertices comprise the N buckets and whose m edges $[0..m-1]$ indicate the pairs of buckets into which a word may fall. A random hash function from the family is given by a random permutation π on $[0..m-1]$. To hash a string $x_0 \ldots x_{n-1}$ using π, where $|x_i| = w$ and $n \leq m$, each word x_i is dropped into the two buckets at the endpoints of edge $\pi(i)$. As before, we xor the contents of each bucket and output their concatenation in some canonical order. We call the above scheme $\mathcal{H}_G[w, n, N]$.

Finding a "good" $\mathcal{H}_G[w, n, N]$ amounts to finding a graph G on a small number of vertices N, a large number of edges m, and such that for all $1 \leq k \leq n \leq m$, if k distinct edges are selected at random from G, then the probability that their union (with multiplicities) comprises a union of cycles is at most some small number ϵ

One interesting set of graphs in this regard are the (d, g)-*cages* (see [4]). A (d, g)-cage is a smallest d-regular graph whose shortest cycle has g edges. These graphs have been explicitly constructed for various values of (d, g). Though (d, g)-cages are rather large (for even g they have at least $(2(d-1)^{g/2} - 2)/(d-2)$ nodes) some (d, g)-cages may give rise to useful hash families. The following is an example.

For $d - 1$ a prime power, let $C[d, 6]$ be the $(d, 6)$-cage. This is the the point-line incidence graph of the projective plane of order $d - 1$. I conjecture that $\mathcal{H}_{C[d,6]}[w, d^3 - d^2 + d - 1, 2d^2 - 2d + 2]$ is ϵ-AU$_2$ for $\epsilon = \binom{d^2 - d + 1}{3} / \binom{d(d^2 - d + 1)}{6}$. Assuming this, $\mathcal{H}_{C[10,6]}[w, 909, 182]$ achieves compression ≈ 5, collision probability $2^{-32.572}$, and Method 2 cost of 4 (sm code) or 6 (kiss code) instrs/wd.

OPEN QUESTIONS. The generalized notion of bucket hashing amounts to saying that hashing is achieved for each bit position $1 \ldots w$ by matrix multiplication with a sparse Boolean matrix H. Expressing the method in this generality raises questions like the following: for all distributions $\mathcal{D}[N, n]$ of binary $N \times n$ matrices H having k ones per column (e.g., $k = 2, 3, 4$) for which is $\max_{x \in \{0,1\}^n - \{0^n\}} \Pr_{H \in \mathcal{D}[N,n]}[Hx = 0^N]$ minimized? What if we also demand that each pair of rows have the same number of ones? (This is true in the matrix of $C[d, g]$, and having this property eliminates the Method 2 (kiss code) overhead mentioned in Section 4.) What if we demand density $\leq \rho$ but make no further restriction on H? Answers will lead to still faster bucket hash MACs.

Acknowledgments

The author thanks Mihir Bellare, Don Coppersmith, Hugo Krawczyk, David Zuckerman, and an anonymous reviewer for their comments and suggestions.

References

1. R. ARNOLD AND D. COPPERSMITH, "An alternative to perfect hashing." IBM RC 10332 (1984).

2. M. BELLARE, O. GOLDREICH AND S. GOLDWASSER. "Incremental cryptography: The case of hashing and signing." *Advances in Cryptology - CRYPTO '94 Proceedings*, Springer-Verlag (1994).

3. M. BELLARE, J. KILIAN AND P. ROGAWAY, "The security of cipher block chaining." *Advances in Cryptology - CRYPTO '94 Proceedings*, 341-358 (1994).

4. J. BONDY AND U. MURTY, *Graph theory with Applications*. North Holland (1976).

5. G. BRASSARD, "On computationally secure authentication tags requiring short secret shared keys." *Advances in Cryptology - CRYPTO '82 Proceedings*, 79-86 (1983).

6. J. BIERBRAUER, T. JOHANSSON, G. KABATIANSKII AND B. SMEETS, "On families of hash functions via geometric codes and concatenation." *Advances in Cryptology -CRYPTO '93 Proceedings*, Springer-Verlag, 331-342 (1994).

7. L. CARTER AND M. WEGMAN, "Universal hash functions," *J. of Computer and System Sciences* 18, 143-154 (1979).

8. Y. DESMEDT, "Unconditionally secure authentication schemes and practical and theoretical consequences." *Advances in Cryptology - CRYPTO '85 Proceedings*, Springer-Verlag, 42-45 (1985).

9. P. GEMMELL AND M. NAOR, "Codes for interactive authentication." *Advances in Cryptology - CRYPTO '93 Proceedings*, Springer-Verlag, 355-367 (1994).

10. O. GOLDREICH, S. GOLDWASSER AND S. MICALI, "How to construct random functions." *Journal of the ACM*, Vol. 33, No. 4, 210-217 (1986).

11. S. GOLDWASSER, S. MICALI AND R. RIVEST, "A digital signature scheme secure against adaptive chosen-message attacks," *SIAM Journal of Computing*, 17(2):281-308, April 1988.

12. H. KRAWCZYK, "LFSR-based hashing and authentication." *Advances in Cryptology - CRYPTO '94 Proceedings*, Springer-Verlag, 129-139 (1994).

13. M. LUBY AND C. RACKOFF, "How to construct pseudorandom permutations from pseudorandom functions," *SIAM J. Comput*, Vol. 17, No. 2, April 1988.

14. X. LAI, R. RUEPPEL AND J. WOOLLVEN, "A fast cryptographic checksum algorithm based on stream ciphers." *Advances in Cryptology, Proceedings of AUSCRYPT 92.* Springer-Verlag (1992).

15. P. PEARSON, "Fast hashing of variable-length text strings." *Communications of the ACM*, 33 (6), 677-680 (1990).

16. R. RIVEST, "The MD5 message digest algorithm." IETF RFC-1321 (1992).

17. P. ROGAWAY, "Bucket hashing and its application to fast message authentication." (Full version of this paper.) Available from the author or out of http://www.cs.ucdavis.edu/~rogaway/

18. G. SIMMONS, "A survey of information authentication." In *Contemporary Cryptography, The Science of Information Integrity*, G. Simmons, editor. IEEE Press, New York (1992).

19. D. STINSON, "Universal hashing and authentication codes." *Designs, Codes and Cryptography*, vol. 4, 369-380 (1994). Earlier version in *Advances in Cryptology - CRYPTO '91 Proceedings*, Springer-Verlag, 74-85 (1991).

20. R. TAYLOR, "An integrity check value algorithm for stream ciphers." *Advances in Cryptology - CRYPTO '93 Proceedings*, Springer-Verlag, 40-48 (1994).

21. J. Touch, "Performance analysis of MD 5." Manuscript, February 1995.
22. G. Tsudik, "Message authentication with one-way hash functions." *Proceedings of Infocom 92*, IEEE Press (1992)
23. M. Wegman and L. Carter, "New hash functions and their use in authentication and set equality." *J. of Computer and System Sciences* 22, 265–279 (1981).
24. D. Wheeler, "A bulk data encryption algorithm." *Fast Software Encryption, Cambridge Security Workshop, 1993 Proceedings*, R. Anderson, ed., 127–134. Lecture Notes in Computer Science, vol. 809, Springer-Verlag (1994).
25. A. Zobrist, "A new hashing method with applications for game playing." University of Wisconsin, Dept. of Computer Science, TR #88 (April 1970).

A   Proof of Theorem 11

Due to page limits we give only the briefest sketch. The complete proof is in [17].

We argue first that wlog we may assume $w = 1$. Then we observe that $h(x)$ is a product Hx over GF[2] for an $N \times n$ matrix H and so h is linear and the collision probability is $\epsilon^* = \max_{x \in \{0,1\}^n - \{0^n\}} \Pr_{h \in \mathcal{H}_B}[h(x) = 0^N]$. This depends only on the number of ones in x. Thus our problem has been reduced to deciding which string $x_t = 1^t 0^{n-t}$, $t \geq 1$, maximizes $\epsilon_t^* = \Pr_{h \in \mathcal{H}_B}[h(x_t) = 0^N]$. It is easy to see that it's not any odd-indexed x_o. Furthermore, $\epsilon_2^* = 0$ because of our exclusion of $h_i = h_j$ for $i \neq j$. So $\epsilon^* = \max\{\epsilon_4^*, \epsilon_6^*, \epsilon_8^*, \ldots\}$.

Upperbounding ϵ^* is facilitated by looking not at $\mathcal{H}_B$, where no $h_i = h_j$ for $i \neq j$, but at the related hash family $\mathcal{H}'_B$ which removes this constraint. Let ϵ'_t be defined by $\Pr_{h \in \mathcal{H}'_B}[h(x_t) = 0^N]$ and let $\epsilon' = \max_{t=4,6,8,\ldots}\{\epsilon'_t\}$.

To analyze $\mathcal{H}'_B$ we construct an $(N+1)$-state Markov chain $\overline{\mathcal{M}}$. States are numbered $0, 1, \ldots, N$. The e-th state models that e buckets have an even number of ones while $N - e$ buckets have an odd number of ones. So N is the start state. Let $\mathcal{N} = N(N-1)(N-2)$. For $3 \leq i \leq N-3$ chain $\mathcal{M}$ has transition probabilities $p_{i,j}$ given by $p_{i,i-3} = i(i-1)(i-2)/\mathcal{N}$, $p_{i,i-1} = 3i(i-1)(N-i)/\mathcal{N}$, $p_{i,i+1} = 3i(N-i)(N-i-1)/\mathcal{N}$ and $p_{i,i+3} = (N-i)(N-i-1)(N-i-2)/\mathcal{N}$, for all $p_{i,j}$ where both i and j are in $[1..N]$ and they differ by ± 1 or ± 3. For $0 \leq i \leq 2$ and $N-2 \leq i \leq N$ we have $p_{0,3} = p_{N,N-3} = 1$, $p_{1,2} = p_{N-1,N-2} = 3/N$, $p_{1,4} = p_{N-1,N-4} = N-3/N$, $p_{2,1} = p_{N-2,N-1} = 6/N(N-1)$, $p_{2,3} = p_{N-2,N-3} = 6(N-3)/N(N-1)$, and $p_{2,5} = p_{N-2,N-5} = (N-3)(N-4)/N(N-1)$. For all other transition probabilities $p_{i,j}$ we have $p_{i,j} = 0$. A simple calculation now shows that $\epsilon'_4 = \alpha(N)$. Using that $N \geq 20$ we get that $\alpha(N) \geq 1997/\mathcal{N}^2$ and so the following proves that $\epsilon' = \epsilon'_4$.

Lemma 12. Suppose $N \geq 20$. If $t \geq 6$ then $\epsilon'_t \leq 1788/\mathcal{N}^2$.

To prove this we bound selected probability masses $\pi'_i(t)$ for $i \in [1..N]$ in $\mathcal{M}'$ by considering the 7-state process $\overline{\mathcal{M}}$ which we get by collapsing a certain $N - 6$ set of states of $\mathcal{M}'$. Process $\overline{\mathcal{M}}$ has states which we shall call N, $N-1$, $N-2$, $N-3$, $N-4$, $N-6$, and R. The first six represent the corresponding states in $\mathcal{M}'$ while state R represents the remaining states, combined. One can verify that the nonzero transition probabilities $\bar{p}_{i,j}$ of $\overline{\mathcal{M}}$ are bounded by $\bar{p}_{N,N-3} = 1$, $\bar{p}_{N-1,N-4} \leq 1$, $\bar{p}_{N-1,N-2} = 3/N$, $\bar{p}_{N-2,R} \leq 1$, $\bar{p}_{N-2,N-3} = 6(N-2)(N-3)/\mathcal{N}$,

$\bar{p}_{N-2,N-1} = 6(N-2)/\mathcal{N}$, $\bar{p}_{N-3,N} = 6/\mathcal{N}$, $\bar{p}_{N-3,N-6} \leq 1$, $\bar{p}_{N-3,N-4} = 9(N-3)(N-4)/\mathcal{N}$, $\bar{p}_{N-3,N-2} = 18(N-3)/\mathcal{N}$, $\bar{p}_{N-4,R} \leq 1$, $\bar{p}_{N-4,N-3} = 36(N-4)/\mathcal{N}$, $\bar{p}_{N-4,N-1} = 24/\mathcal{N}$, $\bar{p}_{N-6,N-3} = 120/\mathcal{N}$, $\bar{p}_{N-6,R} \leq 1$, $\bar{p}_{R,R} \leq 1$, $\bar{p}_{R,N-6} \leq 15(N-5)(N-6)/\mathcal{N}$, $\bar{p}_{R,N-4} \leq 60(N-5)/\mathcal{N}$ $\bar{p}_{R,N-2} = 60/\mathcal{N}$, We show that $\bar{\pi}_N(6) \leq 1788/\mathcal{N}^2$ by induction on t. The basis and induction are obtained by calculations using the above bounds.

We conclude the theorem by showing that $\gamma_{n,N}$ adequately compensates for the error we have induced by examining $\mathcal{H}'_B$ instead of $\mathcal{H}_B$. This follows from a lemma asserting that if p is the probability that a random set of n distinct random triples of elements drawn from $[0..N-1]$ contains no repeated triple, then $\epsilon^* \leq \epsilon'/(1-p)$.

B   Definitions of Security

SECURITY OF A MAC. We define (deterministic, counter-based) message authentication schemes. In this case a MAC scheme $\mathcal{M}$ specifies a set Messages $= \{0,1\}^{\leq L}$ that can be authenticated; a finite set Keys $\subseteq \{0,1\}^*$ of keys; a set Tags $\subseteq \{0,1\}^*$; a number MAX $= 2^r$ which is the number of messages that can be authenticated; and a pair of functions (MAC, MACV) where MAC : Messages $\times$ Keys $\times [0..\text{MAX}-1] \to$ Tags, and MACV : Messages $\times$ Keys $\times$ Tags $\to \{0,1\}$. We write $\text{MAC}_a^{\text{cnt}}(x)$ for $\text{MAC}(x,a,\text{cnt})$, $\text{MACV}_a(x,t)$ for $\text{MACV}(x,a,t)$, and $\text{MAC}_a(x)$ for $\text{MAC}_a^0(x)$, We demand that for any $x \in$ Messages, $a \in$ Keys, and cnt $\in [0..\text{MAX}-1]$, $\text{MACV}_a(x,\text{MAC}_a^{\text{cnt}}(x)) = 1$.

Let $\mathcal{M}$ be a message authentication scheme. A *MAC oracle* $\text{MAC}_a(\cdot)$ for $\mathcal{M}$ behaves as follows: it answers its first query $x_0 \in$ Messages with the string $\text{MAC}_a^0(x_0)$; it answers its second query $x_1 \in$ Messages with the string $\text{MAC}_a^1(x_1)$; and so forth.

An *adversary* E for a message authentication scheme $\mathcal{M}$ is an algorithm equipped with a MAC oracle. Adversary E is said to *succeed* on a particular execution having MAC oracle $\text{MAC}_a(\cdot)$ if E outputs a string (x^*, t^*) where $\text{MACV}_a(x^*, t^*) = 1$ yet E made no earlier query of x^*.

Adversary E is said to be $[t,q,m,\epsilon]$-*break* $\mathcal{M}$, where $0 \leq q <$ MAX and $0 \leq \epsilon \leq 1$, if E makes at most q queries of its oracle, asks its oracle a total of m bits, runs in at most t time, and $\epsilon \geq \Pr\left[a \leftarrow \text{Keys} : E^{\text{MAC}_a(\cdot)} \text{ succeeds}\right]$.

SECURITY OF A FINITE PRF. A finite PRF is a map $F : \{0,1\}^\kappa \times \{0,1\}^l \to \{0,1\}^L$. We write $F_a(x)$ in place of $F(a,x)$. We let $R_{l,L}$ be the set of all functions mapping $\{0,1\}^l$ to $\{0,1\}^L$. Following [10], a *distinguisher* is an oracle algorithm D. We say that D $[t,q,\epsilon]$-breaks F if D runs in at most t steps, makes at most q oracle queries, and $\Pr_{a \leftarrow \{0,1\}^k}[D^{F_a(\cdot)} = 1] - \Pr_{g \leftarrow R_{l,L}}[D^g = 1] \geq \epsilon$. Complexity is measured in a standard RAM model of computation, with oracle queries counting as one step.

Fast Key Exchange with Elliptic Curve Systems

Richard Schroeppel   Hilarie Orman   Sean O'Malley   Oliver Spatscheck *
{rcs, ho, sean, spatsch}@cs.arizona.edu
Department of Computer Science, University of Arizona

Abstract. The Diffie-Hellman key exchange algorithm can be implemented using the group of points on an elliptic curve over the field $\mathbb{F}_{2^n}$. A software version of this using $n = 155$ can be optimized to achieve computation rates that are slightly faster than non-elliptic curve versions with a similar level of security. The fast computation of reciprocals in $\mathbb{F}_{2^n}$ is the key to the highly efficient implementation described here.

1   Introduction

The Diffie-Hellman key exchange algorithm [11] is a method for initiating an encrypted conversation between two previously unintroduced parties. It relies on exponentiation in a large group, and the software implementation of the group operation is usually computationally intensive. The algorithm has been proposed as an Internet standard [15], and the benefit of an efficient implementation would be that it could be widely deployed across a variety of platforms, greatly enhancing the security of the Internet by solving the problem of key exchange for millions of host machines.

The Diffie-Hellman algorithm was implemented several years ago as part of the Sun SecureRPC system used by Sun Microsystems, and the implementation used numbers of a size that was determined in [21] to be attackable using a method described in [9]. This work indicated that instead of using a 192 bit modulus, which could be "cracked" in only about 3 months of effort (including software development), system designers should use at least a 512 bit modulus. Informal conversations with people associated with developing the Sun SecureRPC system indicated that they did not wish to increase the size of the numbers, in part because of the amount of time needed for the computation. The extra time results because of the number of large-number arithmetic operations that must be carried out.

In our work with implementations of cryptographic protocols, we developed a simple version of the Diffie-Hellman protocol in what has been termed *transient* mode, where the two parties each select a random exponent e and exchange values of g^e, where g is a group element. If party A selects $e = a$ and party B selects $e = b$, then each party can compute g^{ab}, but no eavesdropper can do so. In our implementation, we used the ring $\mathbb{Z}_p$, with p a 512 bit prime, a size that should resist attacks with hardware resources known today. The protocol took from 2 to 10 seconds on a variety of modern and popular hardware platforms.

---

* This work was supported by the National Computer Security Center, contract MDA904-92-C-5151.

D. Coppersmith (Ed.): Advances in Cryptology - CRYPTO '95, LNCS 963, pp. 43-56, 1995.

This speed is unpalatable for machines that need to participate in many keyed conversations with a large set of peers. We would estimate that no busy machine should devote more than .1% of its cycles to key exchange, and this limits even a very fast machine ($\leq$ 200 MHz) to fewer than 20 key exchanges per hour.

This work motivated our research into faster software implementations of the basic operations behind the protocol. Elliptic curve systems, first suggested by Victor Miller [25] and Neal Koblitz [19], were a natural choice because they are (insofar as is known today) immune to the index calculus attack. This means that smaller numbers can be used to achieve the same degree of security for the Diffie-Hellman algorithm as the 512 bit version described above. It is interesting to note that the numbers in our implementation are even smaller than the Sun RPC version. In addition to this basic savings in computation cost, there are several software optimization techniques that result in a significantly faster algorithm.

An additional advantage is that, as computers get faster, the size of the numbers needed to achieve a particular level of security grows much more slowly for elliptic curve systems when compared to methods that use ordinary integers.

The elliptic curve method uses a different group operation than multiplication of integers mod p. Instead, the operation is over the group of points on an elliptic curve, and the operation is arithmetically more complicated. The size of the group used in our implementation is approximately 2^{155}. The group operation is implemented using numbers from the Galois field $\mathbb{F}_{2^{155}}$. Our initial implementation of this was more than twice as fast as the implementation using integers modulo a 512 bit prime, and there was obvious room for improvement. For the DH key exchange algorithm, a properly chosen elliptic curve over $\mathbb{F}_{2^{155}}$ offers somewhat more security than does working modulo a 512 bit prime.

The improvements described here are how to efficiently compute the field operations in $\mathbb{F}_{2^{155}}$, especially reciprocals, and a minor improvement in the formula for doubling an elliptic curve point. The most important contributor to the success of the algorithm is the fast reciprocal routine.

2 Overview of the Method

We include here brief descriptions of the field and elliptic curve manipulations; this material is from draft document [24]. See Silverman [32] for a general introduction to elliptic curves; Menezes [23] provides a cookbook approach and an introduction to the cryptographic methods. Other good references are [1, 2, 3, 5].

For our purposes, an elliptic curve E is a set of points (x, y) with coordinates x and y lying in the field $\mathbb{F}_{2^{155}}$ and satisfying a certain cubic equation. The points (x, y) form a commutative group under "addition"; the rule for "addition" involves several field operations, including computing a reciprocal; the formulas are in section 3.1.

The elliptic curve analog of the Diffie-Hellman key exchange method uses an elliptic curve E, and a point $P = (x_0, y_0)$ which generates the whole addition group of E. The base field, curve equation, and starting point are all system-wide public parameters.

When user A wants to start a conversation, he chooses a secret integer multiplier K_A in the range $[2, order(E) - 2]$ and computes $K_A P$ by iterating the addition of P using a "double and add" scheme. He then sends the x and y coordinates of the point $K_A P$ to user B. User B selects his own secret multiplier K_B and computes and sends to user A the point $K_B P$. Each user can then compute the point $(K_A K_B)P$. Some bits are selected from the coordinates to become the secret session key for their conversation. Insofar as is known, there is no effective method for recovering $(K_A K_B)P$ by eavesdropping on this conversation, other than solving the discrete logarithm problem. The discrete logarithm problem is hard for elliptic curves, because the index calculus attack that is so effective modulo p does not work.

The elliptic curve operations require addition, multiplication, squaring, and inversion in the underlying field. The number of applications of each operation depends on the exact details of the implementation; in all implementations the inversion operation is by far the most expensive (by a factor of 5 to 20 over multiplication).

3 Working with an Elliptic Curve

3.1 Adding and Doubling Points

The two elliptic curve operations that are most relevant to the complexity of multiplying a group element by a constant are the Add and Double operations. They are presented slightly modified from their presentation in [23]. The elliptic curve $E_{a,b}$ is the set of all solutions (x, y) to the equation $y^2 + xy = x^3 + ax^2 + b$. Here a and b are constants from the field $\mathbb{F}_{2^{155}}$; b must be nonzero. The variables x and y are also elements of $\mathbb{F}_{2^{155}}$. A solution (x, y) is called a *point* of the curve. An extra point O is used as the group identity. We use $(0,0)$ to represent O. (Because $b \neq 0$, $(0,0)$ is never a solution of our equation.)

(i) Adding two points (x_1, y_1) and (x_2, y_2):
   If either point is O, return the other point as the sum.
   If $x_1 = x_2$: When $y_1 \neq y_2$, return O; otherwise use the Doubling rule.
   If $x_1 \neq x_2$, then $(x_1, y_1) + (x_2, y_2) = (x_3, y_3)$, where

$$x_3 = \lambda^2 + \lambda + x_1 + x_2 + a, \quad y_3 = \lambda(x_1 + x_3) + x_3 + y_1, \quad \text{and} \quad \lambda = \frac{y_1 + y_2}{x_1 + x_2}.$$

(ii) Doubling a point (x, y):
   If $x = 0$, then $2(x, y) = O$. Otherwise, $2(x, y) = (x_2, y_2)$, where

$$x_2 = \lambda^2 + \lambda + a, \quad y_2 = x^2 + (\lambda + 1) x_2, \quad \text{and} \quad \lambda = \left(x + \frac{y}{x} \right).$$

x_2 as given in [23] uses an extra multiplication. Our formula can be further improved to remove one multiplication [31].

(iii) Negating a point (x, y): $\qquad -1(x, y) = (x, x + y)$.

From these formulas, we can determine the number of field operations required for each kind of elliptic curve operation. An Addition step usually requires eight additions, two multiplications, one squaring, three reductions modulo the field polynomial $T(u)$, and one inversion. A Doubling step usually requires four additions, two multiplications, two squarings, four reductions mod $T(u)$, and one inversion. A Negation step requires one addition. The important contributors to the running time are the multiplications and inversions.

3.2 Choosing the Curve

The constant a in the elliptic curve equation can be chosen to simplify the operations of doubling a point and of adding two points. We use $a = 0$, which eliminates one addition from the formula for the x coordinate in both operations.

The order of the group is roughly 2^{155}, but the exact size depends on the choice of a and b. There is a complicated algorithm due to Schoof [30], with improvements by Atkin, Elkies, Morain, and Couveignes [10] for determining the group order. For maximum security, the order should have as large a prime factor as possible. In our equation, with $a = 0$, the best possible order is $4p$, with p a prime near 2^{153} [20]. (If $a \neq 0$, the order can have the form $2p$ with p near 2^{154}, giving a bit of extra security.) Lay and Zimmer [22] give a method for creating a curve with a given order, but we are reluctant to use their scheme because it produces curves closely related to *rational* curves with an extra structural property called *complex multiplication*. We don't know of any way to "crack" such curves, but it seems prudent to avoid this extra structure.

A curve is selected by choosing small values for (x, y) and computing b from the equation $y^2 + xy = x^3 + b$. The curve order is computed with Schoof's algorithm, and tested to see if it is of the form $4p$. For curves based on $\mathbb{F}_{2^{155}}$, a few hundred tries may be necessary.

The best known methods for computing elliptic curve discrete logarithms take time proportional to the square root of the largest prime factor of the group order [27, 28, 14]. In our case, the largest prime factor will be about 2^{153}, so finding discrete logarithms will take about $2^{76.5} \approx 10^{23}$ operations.

3.3 Computing a Multiple of a Point

The number of additions and doublings necessary for computing nP (where P is a point on the curve and n is an integer) is an important factor in the speed of the key exchange algorithm. We initially implemented the straightforward double-and-add approach based on the binary expansion of n. For a random 155 bit multiplier n, computing nP requires 154 doubling steps and an average of 77 addition steps. The number of doubling steps is roughly fixed, but it is possible to reduce the number of addition steps. This problem is studied in a large literature on *addition chains* [18, 7, 8, 29]. We implemented two well-known easy speedups that apply to randomly chosen multipliers.

If the starting point P is available ahead of time, preparation of tables of multiples of P is useful [8]. This is the situation for the first two of the four point multiplications in key exchange where both parties begin with the system-wide generator $P = (x_0, y_0)$. The precomputed table consists of $16^K P$ for $K = 0, \ldots, 38$. Using radix 16 and the digits ± 1, a typical 155 bit multiplier n requires about 42 point additions (and no doublings) to compute nP.

When the starting point P is not known ahead of time, as for the final two key exchange point multiplications, a different speedup is available. This is a blend of the *m-ary* method ([18] p. 404) with Booth's algorithm [6]. In this case, the computed table is the odd multiples from P to $15P$. Then we proceed as in the usual double-and-add method, scanning the bit representation of the multiplier n from the high end. Using the table of small multiples of P, we can do several doubling steps before an addition is necessary. If we require an even multiple of P, such as $12P$, we instead use an odd multiple $(3P)$, introduced a couple of steps earlier in the doubling process. Because the subtraction of two curve points is no costlier than addition, we have the option of subtracting a small multiple of P when convenient. On the average, the number of addition/subtraction steps needed is $\frac{1}{6}$ the size of the exponent. For a random 155 bit exponent, we will use about 152 doubling steps and 32 addition/subtractions (including the cost of preparing the table).

For DH key exchange, the first two of the point multiplications are with a known starting point, and the last two are with new points. The last two point multiplications can take place in parallel. The total time for the key exchange is $2(42A) + (152D + 32A) = 152D + 116A$. The double-and-add approach would use $3(154D + 77A) = 462D + 231A$, nearly three times as many operations.

4 Field Operations in $\mathbb{F}_{2^{155}}$

4.1 Representation of the Field Elements

We represent field elements as bitstrings of length 155. For a 64 bit processor, this is only 3 words; the brevity of the representation means that much of the computation can be done in hardware registers.

Let $k[u]$ be the ring of polynomials over $\mathbb{F}_2$. We will work in the extension field of the trinomial $T(u) = u^{155} + u^{62} + 1$. The extension is a field because the polynomial is irreducible over $\mathbb{F}_2$. The field elements are members of $k[u]$ modulo the field polynomial $T(u)$, with coefficients drawn from the set $0, 1$. Each polynomial in $k[u]$ can be reduced to a remainder of degree at most 154.

The irreducible trinomial $T(u)$ has a structure that makes it a pleasant choice for representing the field. In $\mathbb{F}_2$, there are only two irreducible polynomials of this degree. The fact that the middle term, u^{62}, has an exponent that is roughly half of the field degree is important to the optimizations for calculating modular reductions (as described in section 4.3) and to the division by large powers of u, (as explained in section 4.4).

4.2 Addition and Multiplication

Field elements (for a prime power field) are added and multiplied as follows:

– *Field addition:* $(a_{n-1} \cdots a_1 a_0) + (b_{n-1} \cdots b_1 b_0) = (c_{n-1} \cdots c_1 c_0)$, where $c_i = a_i + b_i$ in the field $\mathbf{F}_2$. That is, field addition is performed componentwise.

– *Field multiplication:* $(a_{n-1} \cdots a_1 a_0) \cdot (b_{n-1} \cdots b_1 b_0) = (r_{n-1} \cdots r_1 r_0)$, where the polynomial $(r_{n-1} u^{n-1} + \cdots + r_1 u + r_0)$ is the remainder when the polynomial $(a_{n-1} u^{n-1} + \cdots + a_1 u + a_0) \cdot (b_{n-1} u^{n-1} + \cdots + b_1 u + b_0)$ is divided by $T(u)$ over $\mathbf{F}_2$.

The addition algorithm for field elements is trivial: the two blocks of bits are simply combined with the bitwise *xor* operation. Because our field has characteristic 2, subtraction is the same as addition, and negation is the identity operation.

Multiplication of field elements uses the same shift-and-add algorithm as is used for multiplication of integers, except that the "add" is replaced with "xor". This has the virtue that the operation can no longer generate carries, simplifying the implementation. We experimented with several different ways of organizing the multiplication routines and found that different architectures had different optimal routines. (Our timings are done with the optimal routines for each architecture.)

The Squaring Operation Squaring a polynomial in a modulo 2 field is a *linear* operation. In the formula for squaring a binomial, $(a + b)^2 = a^2 + 2ab + b^2$, the cross-term vanishes modulo 2 and the square reduces to $a^2 + b^2$. Consequently, we can square a sum by squaring the individual terms. For example, $(u^3 + u + 1)^2 = u^6 + u^2 + 1$.

In terms of bitstrings, to square a polynomial, we spread it out by interleaving a 0 bit between each polynomial bit. For example, $u^3 + u + 1$ is represented as **1011**, and the square is **1000101**. This can be done quickly using table lookup to convert each byte to its 15 bit square. The squared polynomial is then reduced modulo $T(u)$. Squaring is so much faster than regular multiplication that it can be ignored in rough comparisons of the timings.

4.3   Modular Reduction

The field elements are polynomials with coefficients in the ring $\mathbf{Z}_2$. After each multiplication or squaring, the result must be reduced modulo $T(u) = u^{155} + u^{62} + 1$. The product of two polynomials of degree 154 produces a polynomial of degree 308. The product is represented as 10 words on a 32 bit architecture, or 5 words on a 64 bit architecture.

A hand tailored reduction method, specific for $T(u)$, takes advantage of the degree of the middle term to minimize the number of operations required. First note that

$$u^{155} \equiv u^{62} + 1, \quad \text{and} \quad u^n \equiv u^{n-93} + u^{n-155} \bmod T(u).$$

Assume the polynomial to be reduced is

$$P(u) = a_{308} u^{308} + \cdots + a_2 u^2 + a_1 u + a_0.$$

As many as 93 of the leading terms of $P(u)$ can be reduced modulo $T(u)$ by replacing each non-zero term by its congruent two-term expression, i.e. $au^n \equiv au^{n-93} + au^{n-155} \bmod T(u)$. We can think of this as zeroing out the upper 93 bits of the 309 bits of the expression (subtracting each term) and adding in the representation of each original term right-shifted by 93 (i.e., multiplied by u^{-93}) and also right-shifted by 155 (i.e., multiplied by u^{-155}):

$$P(u) \equiv P(u)_{215-0} + P(u)_{308-216}(u^{-93} + u^{-155}) \bmod T(u)$$

where $P(u)_{j-k} = \sum_{i=j}^{k} a_i u^i$ is the portion of $P(u)$ from degrees j through k. This yields a length 216 partial result. This reduction can be repeated to make the degree less than 155.

In practice, we work one computer word at a time, lowering the degree by either 32 or 64, proceeding from the high order terms (bits) to the low. The results are accumulated into the original expression, i.e. the bitstring representing $P(u)$ is the operand for each shift and xor operation.

The benefit of using a trinomial as the modulus is that each word only needs to be xored into two places for the accumulation operation. Having the middle term of relatively low degree is beneficial because the accumulation operation with a high-order word does not affect that word, so that each reduction step reduces the degree by a full word. (If the middle term were u^{150} instead of u^{62}, we would only shorten our dividend by 5 bits each time instead of 32, and we would have to do the reduction operation multiple times.)

Some other recommended trinomials are $u^{127} + u^{63} + 1$, $u^{140} + u^{65} + 1$, $u^{172} + u^{81} + 1$, $u^{185} + u^{69} + 1$, $u^{191} + u^{71} + 1$, $u^{217} + u^{64} + 1$, $u^{223} + u^{91} + 1$, and $u^{255} + u^{82} + 1$. Irreducible trinomials are somewhat sparse: for the degrees from $100 - 199$, 43 have no irreducible trinomial. If one needs to work with a field of a specific degree, and the field has no good trinomial, a pentanomial (at least) is required. (In fields of characteristic 2, polynomials with an even number of terms are always divisible by $u + 1$.)

4.4 Computing Reciprocals

The rules for doubling an elliptic curve point, and for adding two elliptic curve points, involve computing a reciprocal, either $1/x$ or $1/(x_1 + x_2)$ (see section 3.1). Multiplicative inversion of elements in a field is usually so slow that people have gone to great lengths to avoid it. Menezes [23] (p. 90) and Beth and Schaefer [5] discuss projective schemes, which use about nine multiplications per elliptic curve step, but use very few reciprocals. We report here on a relatively fast algorithm for field inversion, which allows direct use of the simple formulas for operating on elliptic curve points. Our field inversion time is about three multiplication times, a substantial improvement over [23]. [2]

For the field we are working in, the problem to be solved is

---

[2] Menezes' inversion scheme computes $1/A(u)$ as $A(u)^{2^{155}-2} \bmod T(u)$. This can be done with 10 multiplications and 154 squarings [16].

Given a non-zero polynomial $A(u)$ of degree less than or equal to 154, find the (unique) polynomial $B(u)$ of degree less than or equal to 154 such that

$$A(u)B(u) \equiv 1 \bmod u^{155} + u^{62} + 1.$$

The problem has a simple, but relatively slow, recursive solution, exactly analogous to the related algorithm for integers. We have developed an algorithm that is considerably faster. It borrows ideas from Berlekamp [4] and from the low-end GCD algorithm of Roland Silver, John Terzian, and J. Stein (described in Knuth [18] p. 297). Our Almost Inverse algorithm computes $B(u)$ and k such that

$$AB \equiv u^k \bmod M, \quad deg(B) < deg(M), \quad \text{and} \quad k < 2deg(M),$$

where $deg(B)$ denotes the polynomial degree of B. After executing the algorithm, we will need to divide B by u^k $(\bmod M)$ to get the true reciprocal of A.

The pseudo-code for the algorithm is given below. The computer implementation relies on a few representational items:

- Multiplication of a polynomial by u is a left-shift by 1 bit.
- Division of a polynomial by u is a right-shift by 1 bit.
- A polynomial is **even** if its least significant bit, the coefficient of u^0 (the constant term), is 0. Otherwise it is **odd**.

The algorithm will work whenever $A(u)$ and $M(u)$ are relatively prime, $A(u) \neq 0$, $M(u)$ is odd, and $deg(M) > 0$. In our application, $M(u)$ is the irreducible trinomial $T(u)$, so the algorithm works for any nonzero $A(u)$ with degree less than $T(u)$.

The Almost Inverse Algorithm

```
        Initialize integer k=0, and polynomials B=1,C=0,F=A,G=M.

loop:   While F is even, do F=F/u, C=C*u, k=k+1.
        If F = 1, then return B,k.
        If deg(F) < deg(G), then exchange F,G and exchange B,C.
        F=F+G, B=B+C.
        Goto loop.
```

We improved the performance of this raw algorithm considerably with the following programming tricks:

- The operations on the polynomials B, C, F, G are made into inline, loop-unrolled code within the inversion routine. This is a crucial optimization, resulting in a factor of 3 reduction in the overall running time.
- Instead of using small arrays for B, C, F, G, use separate named variables $B0, B1, \ldots, G4$ etc. to hold the individual words of the polynomials. Assign as many of these as possible to registers.
- F is even at the bottom of the loop, so the "Goto" can skip over the test for the "While". This non-structured jump into the body of a "While" loop saves about 10% of the time.

- Instead of exchanging F, G and B, C, make two copies of the code, one with the names exchanged. Whenever an exchange would be called for, instead jump to the other copy.
- During the execution of the code, the lengths of the variables F, G shrink, while B, C grow. Detect when variables' lengths cross a word boundary, and switch to a copy of the code which knows the exact number of words required to hold the variables. This optimization makes the code much larger, because either 25 (for a 32 bit architecture) or 9 (for a 64 bit architecture) copies are required. Fortunately the code still fits within the DEC Alpha on-chip cache.

The following additional optimization is possible:

- Because F, G shrink while B, C expand, some of the variables representing the high-order terms can share a machine register. This would be useful on register-poor machines.

Dividing out u^k To find the reciprocal of $A(u)$, we need to divide $B(u)$ by u^k, working mod $T(u)$. The typical value of k is 260, although k can be as large as 309. The strategy is to divide B successively by u^w, where w is the number of bits in the wordsize of the computer, and finish up with a final division by a smaller power of u.

The operation of dividing by u^w is broken into two parts. First, a suitably chosen multiple of T is added to B, so as to zero out the w low order bits of B. The new B can have degree as large as $154 + w$. Second, the new B is right-shifted by w bits, effectively dividing it by u^w. Since the low order bits are 0, the division is exact; and the right-shift reduces the degree to (at most) 154.

The "suitably chosen multiple of T" is just T times the low order 32 (or 64) bits of B. For the 32 bit SPARC, using the notation of section 4.3,

$$B \equiv B + B_{31-0}(u^{155} + u^{62} + 1) \equiv B_{154-32} + B_{31-0}(u^{155} + u^{62}) \bmod T(u).$$

The second term is computed by left-shifting the low-order 32 bits of B by 62 bits and 155 bits and is xored directly into B. The zeroing operation has a complication on the Alpha, where we work with 64 bits at a time: After the shift-and-two-xors step, there are two possibly unzeroed bits, B_{63-62}. An additional shift-and-two-xors step is performed with this twenty-five cent field to clear it.

The same logic, modified for the smaller shift size, is used for the final division by a less-than-wordsize power of u.

5 Timings

The timings on two platforms are presented in Figure 1. The Sun SPARC IPC is a 25 MHz RISC architecture, with a 32 bit word size. The DEC Alpha 3000 is a 175 MHz RISC architecture, with a 64 bit word size. If everything is just right (all the data in registers, etc.), the SPARC machine can execute 25 million instructions per second, the Alpha 175 million. The Alpha has an 8K byte on-chip instruction cache; assuring that the critical field operations are loaded into

Field and Curve Operations	SPARC IPC	Alpha
155 bit add	3.5	.22
155 x 155 bit multiply	112.6	7.10
32 x 32 bit multiply	8.0	
64 x 64 bit multiply		1.65
155 bit square	8.1	.49
Modular reduction, 310 bits to 155 bits	3.8	.15
Reciprocal, including divide by u^k ($k = 261$)	280.1	25.21
Double an elliptic curve point	481.0	40.46
Add two elliptic curve points	550.8	42.05
Multiply known elliptic curve point	23 msec	1.8 msec
Multiply new elliptic curve point	92 msec	7.8 msec
Key exchange times		
Elliptic curve DH key exchange (155 bits)	137 msec	11.5 msec
Ordinary integer DH key exchange (512 bits)	2670 msec	185 msec
... with four tricks (estimate)	523 msec	34.8 msec
... no tricks, 128 bit key (estimate)	670 msec	46 msec
Best DH key exchange, estimates for 128 bit multiplier/exponent		
Elliptic curve over $\mathbb{F}_{2^{155}}$ (estimate)	114 msec	9.6 msec
Modular arithmetic, 512 bit prime (estimate)	150 msec	10.0 msec

Fig. 1. Times for various field and elliptic curve operations. Unlabeled times are in microseconds.

the cache without conflict is crucial to achieving the results reported here. The SPARC instruction cache is 64K bytes, and the code fits easily.

We made a few measurements on other architectures. The Intel 486 (66MHz) and the DEC MIPS (25MHz) are both within 10% of the SPARC times. Both machines have a 32 bit word size.

Assumptions for Timing Estimates Our modular arithmetic implementation of Diffie-Hellman key exchange uses the GNU bignum package GMP [13]. This package includes assembly language routines for the low-level primitives.

For a fair comparison with ordinary modular key exchange, we assume that all reasonable optimizations are made to the arithmetic. Many of the tricks that we have used in our elliptic curve program have analogs in ordinary arithmetic. For comparison purposes, we assume that the four tricks described below are added to GMP.

- Special case code for squaring a number is used to reduce the time to 60% of a multiplication.
- The modulus is chosen to be a prime P of the form $2^{512} - K$ with small K, reducing the time for a modular reduction to 10% of a multiplication.
- When the base g to be exponentiated is known ahead of time, we assume that tables of $g^{\pm 32^K} \pmod{P}$ are prepared in advance. For a random 512 bit

exponent, this will reduce the required number of modular multiplications to 114 on the average, and eliminate all squarings.

- When the base B is not known in advance, we assume the *m-ary* method [18], using the odd powers $B^1, \ldots, B^{31}(\text{mod } P)$. For a random 512 bit exponent, this will require an average 508 modular squarings and 100 modular multiplications.

Under these assumptions, the estimated times for modular arithmetic DHKX are given in the table above. These times are compared with actual measured times for the implemented elliptic curve routines.

Phil Karn has suggested speeding up the key exchange by limiting the exponent size to 128 bits, in Photuris [15]. Therefore, we also provide estimates for both methods of key exchange on the assumption that the exponents/multipliers are reduced to 128 bits. In this case, the elliptic curve method has only a slightly better running time than the modular arithmetic method.

5.1 Planning Ahead

A major advantage of elliptic curves is that they scale well with increasing computer power. In the discrete logarithm problem for modular arithmetic, changing the modulus from a 512 bit prime to a 1024 bit prime multiplies the cracking effort by a factor of 8 million.[3] In the elliptic curve case, forcing a corresponding increase in cracking effort requires adding only 46 bits to the size of the field. This would raise our field size from 155 bits to 201 bits.

Now compare the changes in encryption effort for this increased security. In the ordinary arithmetic case, changing from a 512 bit modulus to 1024 bits will make the basic operation of modular multiplication take 3 times as long if the Karatsuba method [17] is used. The corresponding elliptic curve times increase by only $(201/155)^2 = 1.7$. (Both methods would also require that exponent/multiplier sizes increase by 46 bits, so each method incurs an additional penalty of about $174/128 = 1.4$.)

This advantage of elliptic curves is even more pronounced if long-term security is required: each additional bit added to the field size increases the cracking work by a factor of 1.4. In the modular arithmetic case, adding a bit to a 1024 bit prime only multiplies the required work by 1.026. To double the cracking effort requires increasing the the field size by 2 bits in the elliptic curve case, versus adding 27 bits to the length of the modulus for modular arithmetic.

6   Other Applications

ElGamal Encryption The elliptic curve improvements also make ElGamal style encryption and signatures more attractive. The total effort of signing and checking a signature is less with elliptic curve methods than with RSA.

---

[3] The time required to find a discrete logarithm mod p is estimated with the formula $\exp(1.93 \sqrt[3]{\log p (\log \log p)^2})$ [21].

Cryptographic Operation	SPARC IPC	Alpha
Elliptic curve ElGamal encryption (155 bits)	116 msec	9.8 msec
Elliptic curve ElGamal decryption (155 bits)	94 msec	8.0 msec
Elliptic curve ElGamal signature (estimate)	24 msec	1.8 msec
Elliptic curve ElGamal check signature (estimate)	220 msec	17.8 msec

Fig. 2. Times for elliptic curve versions of ElGamal cryptographic operations.

The ElGamal encryption method [12, 23] can be implemented using these elliptic curve routines. The method uses a "semi-static" Diffie-Hellman key exchange: there is a public elliptic curve with generator $P = (x_0, y_0)$, and participants choose secret multipliers m_p and contribute $P_p = m_p P$ to a public key phone book. When party A wishes to communicate with party B, he selects a random multiplier k and computes kP and kP_B. The latter quantity is used to create a key for encrypting the message. The quantity kP is attached to the ciphertext. Party B can recover the key by computing $m_B kP$. Timings of the encryption and decryption operations are presented in Figure 2, along with estimates for the related signature operations.

Other Finite Fields The new reciprocal algorithm is useful for doing arithmetic in other finite fields. Because it makes inversion less costly, it will be worthwhile to reanalyze other formulas for operations with elliptic curves. The reciprocal algorithm can also be used, with slight modifications, to compute reciprocals in ordinary integer modular arithmetic. (The algorithm is most efficient with moduli of the form $2^A - 2^B - 1$, but will work reasonably with $2^A - k2^B - 1$ for 32 bit k. For generic odd moduli, Peter Montgomery's trick [26] is useful for dividing by the required power of 2.) Another benefit may be in ordinary modular exponentiation, as used in RSA and many other schemes: if a reciprocal costs only a few multiplications, addition-subtraction chains can be used to compute powers; this allows shorter chains which more than recoups the investment in computing the reciprocal.

7   Conclusions

We have shown that the software implementation of the Diffie-Hellman algorithm can be done more efficiently using elliptic curve systems over $\mathbf{F}_{2^n}$ than using integers modulo p. Assuming that no equivalent to the discrete logarithm attack exists for an elliptic curve, smaller number representations of the group elements can be used, and the software becomes quadratically faster. RISC machines with 64 bit wide words show excellent performance.

Our implementation's major speed advantage over previous implementations derives from its use of an efficient procedure for computing reciprocals in $\mathbf{F}_{2^n}$.

If network protocols were to rely on this method for establishing DES key pairs between hosts, four times as many connections could be made as compared with our baseline modulo p implementation. Even if the mod p arithmetic is improved as outlined in section 5, the elliptic curve method remains very competitive. As computing power increases, and the search capabilities of opponents improve accordingly, it is cheaper to improve the security of elliptic curve methods than to improve the security of mod p methods.

Key exchange nonetheless remains an expensive operation ... over 100 times as expensive as computing the MD5 one-way hash function, for example.

8 Acknowledgments

We thank R. W. Gosper for using MACSYMA to compute a table of factorizations of trinomials, and Alfred Menezes for providing us with reference [24].

References

1. G. AGNEW, T. BETH, R. MULLIN AND S. VANSTONE, "Arithmetic Operations in $GF(2^m)$", *Journal of Cryptology*, 6 (1993), 3-13.
2. G. AGNEW, R. MULLIN AND S. VANSTONE, "An Implementation of Elliptic Curve Cryptosystems over $F_{2^{155}}$", *IEEE Journal on Selected Areas in Communications*, 11 (1993), 804-813.
3. G. AGNEW, R. MULLIN, I. ONYSZCHUK AND S. VANSTONE, "An Implementation for a Fast Public-Key Cryptosystem", *Journal of Cryptology*, 3 (1991), 63-79.
4. ELWYN BERLEKAMP, *Algebraic Coding Theory*, McGraw-Hill, 1968, p.41.
5. T. BETH AND F. SCHAEFER, "Non Supersingular Elliptic Curves for Public Key Cryptosystems", *Advances in Cryptology – EUROCRYPT '91*, Lecture Notes in Computer Science, 547 (1991), Springer-Verlag, 316-327.
6. A. D. BOOTH, "A Signed Binary Multiplication Technique", *Q. J. Mech. Appl. Math.* 4 (1951), 236-240.
7. J. BOS AND M. COSTER, "Addition Chain Heuristics", *Advances in Cryptology – CRYPTO '89*, Lecture Notes in Computer Science, 435 (1990), Springer-Verlag, 400-407.
8. E. BRICKELL, D. GORDON, K. MCCURLEY, AND D. WILSON, "Fast Exponentiation with Precomputation (Extended Abstract)", *Advances in Cryptology – EUROCRYPT '92*, Lecture Notes in Computer Science, 658 (1993), Springer-Verlag, 200-207.
9. D. COPPERSMITH, A. ODLYZKO, AND R. SCHROEPPEL, "Discrete Logarithms in $GF[p]$", *Algorithmica*, 1 (1986), 1-15.
10. JEAN-MARC COUVEIGNES AND FRANÇOIS MORAIN *Algorithmic Number Theory: First International Symposium*, Lecture Notes in Computer Science, 877 (1994), Springer-Verlag, 43-58.
11. WHITFIELD DIFFIE AND M. E. HELLMAN, "New Directions in Cryptography", *IEEE Transactions on Information Theory*, IT-22, n. 6, Nov. 1976, pp 644-654
12. T. ELGAMAL, "A Public Key Cryptosystem and a Signature Scheme Based on Discrete Logarithms", *IEEE Trans. on Information Theory*, 31 (1985), 469-472.

13. TORBJORN GRANLUND, GMP, the GNU bignum package, version 1.3.2a, July 1994. ftp://prep.ai.mit.edu/pub/gnu/gmp-1.3.2.tar.gz

14. GREG HARPER, ALFRED MENEZES, AND SCOTT VANSTONE "Public-Key Cryptosystems with Very Small Key Lengths", *Advances in Cryptology – EUROCRYPT '92*, Lecture Notes in Computer Science, **658** (1993), Springer-Verlag, 163-173.

15. The Internet Engineering Task Force Working Group on Security for IPv4; drafts on key management available via FTP from the archives at ds.internic.net ; internet-drafts/draft-karn-photuris-00.txt

16. T. ITOH, O. TEECHI, AND S. TSUJII, "A Fast Algorithm for Computing Multiplicative Inverses in $GF(2^t)$ Using Normal Bases" (in Japanese), *J. Society for Electronic Communications (Japan)*, **44** (1986), 31-36.

17. A. KARATSUBA, *Doklady Akademiia Nauk SSSR* **145** (1962), 293-294.

18. DONALD E. KNUTH, *Seminumerical Algorithms*, The Art of Computer Programming, **2** Addison Wesley 1969

19. NEAL KOBLITZ, "Elliptic Curve Cryptosystems", *Mathematics of Computation*, **48** n. 177 (1987), 203-209.

20. NEAL KOBLITZ, "Constructing Elliptic Curve Cryptosystems in Characteristic 2", *Advances in Cryptology – CRYPTO '90 Proceedings*, Lecture Notes in Computer Science, **537** (1991), Springer-Verlag, 156-167.

21. B. LA MACCHIA AND A. ODLYZKO, "Computation of Discrete Logarithms in Prime Fields", *Designs, Codes and Cryptography*, **1** (1991), p. 47-62.

22. G. LAY AND H. ZIMMER, "Constructing Elliptic Curves with Given Group Order over Large Finite Fields", *Algorithmic Number Theory: First International Symposium*, Lecture Notes in Computer Science, **877** (1994), Springer-Verlag, 250-263.

23. ALFRED J. MENEZES, *Elliptic Curve Public Key Cryptosystems*, Kluwer Academic Publishers, 1993.

24. ALFRED J. MENEZES, MINGHUA QU, AND SCOTT A. VANSTONE, "Standard for RSA, Diffie-Hellman and Related Public Key Cryptography", Working Draft of IEEE P1363 Standard, April 24, 1995.

25. VICTOR S. MILLER, "Use of Elliptic Curves in Cryptography", *Advances in Cryptology – CRYPTO '85 Proceedings*, Lecture Notes in Computer Science, **218** (1986), Springer-Verlag, 417-426.

26. PETER L. MONTGOMERY, "Modular Multiplication without Trial Division", *Mathematics of Computation*, **44** (1985), 519-521.

27. P. VAN OORSCHOT AND M. WIENER, "Parallel Collision Search with Application to Hash Functions and Discrete Logarithms", 2nd ACM Conference on Computer and Communications Security, Fairfax, Virginia, November 4, 1994.

28. J. POLLARD, "Monte Carlo Methods for Index Computation mod p", *Mathematics of Computation*, **32** (1978), 918-924.

29. JÖRG SAUERBREY AND ANDREAS DIETEL "Resource Requirements for the Application of Addition Chains in Modulo Exponentiation", *Advances in Cryptology – EUROCRYPT '92*, Lecture Notes in Computer Science, **658** (1993), Springer-Verlag, 174-182.

30. R. SCHOOF, "Elliptic Curves Over Finite Fields and the Computation of Square Roots mod p", *Mathematics of Computation*, **44** (1985), 483-494.

31. RICH SCHROEPPEL, HILARIE ORMAN, SEAN O'MALLEY, AND OLIVER SPATSCHECK, "Fast Key Exchange with Elliptic Curve Systems", Univ. of Ariz. Comp. Sci. Tech. Report 95-03 (1995).

32. J. H. SILVERMAN, *The Arithmetic of Elliptic Curves*, Springer Graduate Texts in Mathematics **106** (1992).

Fast Server-Aided RSA Signatures
Secure Against Active Attacks

Philippe Béguin[1,*] and Jean-Jacques Quisquater[2]

[1] Laboratoire d'Informatique **
Ecole Normale Supérieure
45 Rue d'Ulm, F–75 230 Paris Cédex 05, France
e-mail: Philippe.Beguin@ens.fr

[2] Laboratoire DICE
Université Catholique de Louvain
Place du Levant 3, B–1 348 Louvain-la-Neuve, Belgium
e-mail: Quisquater@dice.ucl.ac.be

Abstract. Small units like chip cards have the possibility of computing, storing and protecting data. Today such chip cards have limited computing power, then some cryptoprotocols are too slow. Some new chip cards with secure fast coprocessors are coming but are not very reliable at the moment and a little bit expensive for some applications. In banking applications there are few servers (ATM) relative to many small units: it is a better strategy to put the computing power into few large servers than into the not-very-often used cards.
A possible solution is to use the computing power of the (insecure) server to help the chip card. But it remains an open question whether it is possible to accelerate significantly RSA signatures using an insecure server with the possibility of active attacks: that is, when the server returns false values to get some part of secret from the card.
In this paper, we propose a new efficient protocol for accelerating RSA signatures, resistant against all known active and passive attacks. This protocol does not use expensive precomputations; the computation done by the card, the used RAM and the data transfers between the card and the server are small. With current chip cards it is thus possible to implement efficiently this protocol.

1 Introduction

Small devices, like chip cards or smart cards are easy to carry and have the possibility of computing, storing and protecting data. Unfortunately, today such

---

* Part of this work was done while the author was visiting the Laboratoire de Microélectronique, Université Catholique de Louvain, Belgium.
** Supported by the Centre National de la Recherche Scientifique URA 1327.

D. Coppersmith (Ed.): Advances in Cryptology - CRYPTO '95, LNCS 963, pp. 57-69, 1995.
© Springer-Verlag Berlin Heidelberg 1995

chip cards have limited computing power and some protocols are not executed in an efficient way: for example, public-key cryptographic protocols. Some new chip cards with fast and secure coprocessors are coming but are not reliable at the moment (due to problems of auxiliary memory) and are in some cases too expensive. Anyway, it is useful to have many cheap secure cards and to put the expensive part into one or few insecure servers.

A possible solution is to use an auxiliary unit such as a banking terminal, a card reader, ... in order to help the chip card. In this paper we shall use the words *card* for the main unit and *server* for the auxiliary unit. From a theoretical point of view it is interesting to study how to share the computing power of two parties with some security constraints. This paper is a new efficient contribution in this important field.

If the server is secure and will not leak the secrets, it is possible to imagine a secure link between the card and the server: the card sends the secret values to be used to the server; the server computes the result and sends it using again the secure link. The interesting (real life) case is working with an insecure server. This server may then be under the influence of an opponent trying to obtain the secrets of the card or to cheat with a false result. The conclusion of this short analysis is that the card must protect its secrets and verify the computations received from the server. Let us remark that many proposed protocols ([11],[16],[10]) did not have a strong verification step and then were broken ([2],[13],[9]).

There exist two kinds of attacks against such protocols: classical searching ones are called *passive attacks*; specific ones where the server returns false values to get some information from the card, are called *active attacks*.

Such protocols were first studied by Matsumoto, Kato and Imai [11] and by Quisquater and De Soete [14] for accelerating RSA signatures [15]. Next Pfitzmann and Waidner [13] proposed some passive attacks against all protocols presented in [11]. And Anderson [2] proposed a very efficient active attack against one of the two protocols presented in [11], where a dishonest server could obtain the secret key, by using only one false signature. But these attacks could be very easily defeated if the card checks the correctness of the computed signatures and by increasing the parameters used in [11]. On the other hand Quisquater and De Soete's protocol is provably secure against passive attacks but it is very less efficient than Matsumoto, Kato and Imai's ones. Then Yen and Laih [16] and Matsumoto, Imai, Laih and Yen [10] presented an improvement of protocols presented in [11] secure against passive attacks: unfortunately they need expensive precomputations and thus, they are not very efficient. Next Kawamura and Shimbo [6] proposed four protocols provably secure against passive attacks: the two first are not very efficient and the two last are more efficient but they need expensive precomputations. So, all previously known protocols secure against passive attacks are not efficient or need expensive precomputations. Moreover, *absolutely none* of these protocols (provably or not provably secure against passive attacks) are secure against active attacks presented in [13]. As it is said in [6], it was an open question whether it is possible or not to construct a secure protocol against active attacks.

Burns and Mitchell [5] construct improvements of the two protocols presented in [11] (RSA-S1 and RSA-S2) secure against active attacks. Unfortunately, the first one (RSA-S1) is inefficient: the card has to do at least 188 modular multiplications for each signature. Otherwise, the second one (RSA-S2) is much efficient but it ignores Pfitzmann and Waidner's attacks [13] and thus is not secure.

Lim and Lee [9] developed other protocols using precomputations, based on the two-phase protocols due to Matsumoto, Imai, Laih and Yen [10].

Otherwise, Béguin and Quisquater [3] give the first method for accelerating significantly DSS signatures [12] provably secure against both passive and active attacks.

But it remains an open question whether it is possible to accelerate significantly (without expensive precomputations) RSA signatures using an insecure server, in a secure way against both passive and active attacks. In this paper, we propose a new efficient protocol resistant against all known passive and active attacks, including those presented in [13]. Moreover, we will show that our protocol is secure against more specific passive attacks. This protocol does not use expensive precomputations; the computation done by the card, the used RAM and the data transfers between the card and the server are small. Then, it is possible to implement efficiently this protocol with current chip cards.

We begin by giving some preliminaries, then we outline the protocol of Brickell, Gordon, McCurley and Wilson [4], which we will use. Then we describe our protocol and study its security. Finally, we expose the performances of this protocol.

2 Preliminaries

We denote by n the public modulus of RSA ($n = p \cdot q$), by s the secret signature key and by v the public verification key such that $s \cdot v = 1 \bmod \phi(n)$ with $\phi(n) = (p-1)(q-1)$. The card receives the message M and wants to compute, using the server, the signature of M: i.e. $S = M^s \bmod n$.

For a number a, we denote by $l(a)$ the number of bits of a, i.e. $l(a) = \lfloor \log_2 a \rfloor + 1$, and for a set F, we denote by $\#F$ the cardinality of F.

Let $k = l(n)$ and let $t = \max(l(p), l(q)) - 1$. In this paper, we will study the acceleration of the RSA signatures with 512 bit numbers ($k = 512$) and with 768 bit numbers ($k = 768$). We denote by modular multiplication the multiplication of two k bit numbers modulo a k bit number. In this paper, the computations done by the card will be measured in terms of modular multiplications. Hence we consider that the computation of $a \times b$ is about the equivalent of $\frac{1}{2} \times \frac{l(a)}{k} \times \frac{l(b)}{k}$ modular multiplications and the computation of $a \bmod b$ is about the equivalent of $\frac{\theta}{2} \times \frac{l(a)}{2k} \times \frac{l(b)}{k}$ modular multiplications. By θ we denote a constant near 1 varying with the implementation of the operation modulo. Here we use the practical value of 1.25 for θ. We suppose that the implementation of multiplications modulo is classical and does not use clever tricks or the Montgomery's method: such methods need a specific study.

3 Fast Exponentiation with Precomputation

Several protocols are known to compute exponentiation with precomputation [4], [8]... We present a protocol due to Brickell, Gordon, McCurley and Wilson [4]. The goal of this protocol is to compute a^x using some precomputations.

If $x = \sum_{i=0}^{m-1} a_i x_i$ with $0 \le a_i \le h$, and if a^{x_i} is known for each i, then the algorithm [BGMW] for computing a^x is the following one:

$$B \leftarrow \prod_{a_i = h} a^{x_i}$$
$$A \leftarrow B$$
for $d = h - 1$ to 1 by -1
$$B \leftarrow B \times \prod_{a_i = d} a^{x_i}$$
$$A \leftarrow A \times B$$
return(A)

In our protocol, the x_i will be known by the server which computes a^{x_i}, but x must be kept secret. Then, the card must use constant time to obtain a^x; otherwise, by observing the time used to compute a^x, an opponent could obtain some information about x. A solution is an algorithm with a constant number of multiplications, using, if necessary, the simulation (same time, no operation) of some multiplications.

During the computations, the card uses one of the two following methods.

Case 1. The card stores $a^{x_0}, \ldots, a^{x_{m-1}}$, next computes A. Let $k_d = \#\{a_i : a_i = d\}$; the number of multiplications done by the card during the algorithm is $(k_h - 1) + \sum_{d=1}^{h-1} (k_d + 1) = h - 2 + \sum_{d=1}^{h} k_d \le h - 2 + m$. Some simulation of multiplications is needed; thus the card computes exactly the equivalent of $h - 2 + m$ multiplications.

Case 2. Let $c_d = \prod_{a_i = d} a^{x_i}$. The card computes $c_1, \ldots, c_h$ while receiving the a^{x_i}'s. First $c_i = 1$ for each i, thus after receiving the first value there is no multiplication to do. But after this first step, in order to avoid to give any information to the server (or to any opponent), if the received number has to be multiplied by 1, the card needs to simulate a complete multiplication. Thus the card needs exactly the equivalent of $m - 1$ multiplications to obtain $c_1, \ldots, c_h$. Next the card needs $2(h - 1)$ multiplications to obtain A. Hence the card computes exactly the equivalent of $2h - 3 + m$ multiplications.

4 The RSA Signature

Using the Extended Euclidean Algorithm, the values w_p, w_q can be computed such that $w_p + w_q = 1$, $w_p \bmod p = 0$, $w_q \bmod q = 0$ and $0 \leq |w_p|, |w_q| \leq n$. Thus, if $y_p \equiv y \bmod p$ and $y_q \equiv y \bmod q$, we have $y \equiv y_p w_q + y_q w_p \bmod n$. The protocol is the following.

1. The card receives M to sign.
2. The card chooses randomly $a_0, a_1, \ldots, a_{m-1}$ s. t. $a_i \in \{0, \ldots, h\}$,
$$x_0, \ldots, x_{m-1} \text{ s. t. } l(x_i) \leq t - \log_2(m \cdot h) - 2.$$
3. The card computes $s_1 = \sum_{i=0}^{m-1} a_i x_i$.
4. The card sends M, n and $x_0, \ldots, x_{m-1}$ to the server.
5. The server returns $z_0, \ldots, z_{m-1}$ such that $z_i = M^{x_i} \bmod n$.
6. The card computes $z_p = \prod_{i=0}^{m-1} z_i^{a_i} \bmod p$ and $z_q = \prod_{i=0}^{m-1} z_i^{a_i} \bmod q$ using the algorithm [BGMW] already described.
7. The card computes $s_2 = s - s_1$,
   represents s_2 under the form $\sigma_p = s_2 \bmod (p-1) + \varrho_p(p-1)$,
   $$\sigma_q = s_2 \bmod (q-1) + \varrho_q(q-1),$$
   where ϱ_p is a random number of $\{0, \ldots, q-2\}$,
   ϱ_q is a random number of $\{0, \ldots, p-2\}$.
8. The card sends to the server σ_p, σ_q.
9. The server computes and sends to the card $y_p = M^{\sigma_p} \bmod n$
   $$y_q = M^{\sigma_q} \bmod n.$$
10. The card computes $S_p = y_p \times z_p \bmod p$ and $S_q = y_q \times z_q \bmod q$.
11. Now the card computes $S = w_q \cdot S_p + w_p \cdot S_q \bmod n$.
12. The card verifies $S^v \bmod n = M$.
13. If during the step 12, the verification is correct, then the card transmits S.

5 Security

5.1 An Exhaustive Search

A possible attack is to make an exhaustive search over s_1. An attacking server knowing M, S, y_p, computes $\gcd(n, S - y_p M^{\sum_{i=0}^{m-1} a_i x_i} \bmod n)$ for all a_i such that $0 \leq a_i \leq h$.

When $\sum_{i=0}^{m-1} a_i x_i = s_1$, this server obtains $\gcd(n, S - y_p M^{\sum_{i=0}^{m-1} a_i x_i} \bmod n) = p$ with very high probability. The complexity of this attack is $(h+1)^m$.

We have $\sigma_p = (s - s_1) \bmod (p-1) + \varrho_p(p-1)$. Hence $v\sigma_p = (1 - vs_1) \bmod (p-1) + v\varrho_p(p-1)$. But $(p-1) \bmod 2 = 0$, then $\sum_{i=0}^{m-1} (vx_i \bmod 2)(a_i \bmod 2) = 1 + v\sigma_p \bmod 2$ is known. Hence, using this equation, the complexity of the exhaustive search becomes $(h+1)^m/2$.

We suppose that the other factors of $p - 1$ and $q - 1$ are unknown (which is a normal supposition for RSA signatures), hence it is not possible to improve this search.

5.2 A Pseudo-Exhaustive Search

This attack is similar to the Pfitzmann and Waidner's attack [13] against the protocols of Matsumoto, Kato and Imai [11].

For all $A = \{a_0, \ldots, a_{\lceil m/2 \rceil - 1}\}$ such that $0 \le a_i \le h$, the attacking server computes

$$y_A = M^{\sum_{i=0}^{\lceil m/2 \rceil - 1} a_i x_i} \bmod n.$$

For all $B = \{b_{\lceil m/2 \rceil}, \ldots, b_{m-1}\}$ such that $0 \le b_i \le h$, the server computes

$$z_B^* = S \cdot y_p^{-1} \cdot z_B^{-1} \bmod n \qquad \text{with} \qquad z_B = M^{\sum_{i=\lceil m/2 \rceil}^{m-1} b_i x_i} \bmod n.$$

We have

$$s_1 = \sum_{i=0}^{m-1} a_i x_i = \sum_{i=0}^{\lceil m/2 \rceil - 1} a_i x_i + \sum_{j=\lceil m/2 \rceil}^{m-1} a_j x_j,$$

with $0 \le a_i, a_j \le h$. Let $A = \{a_0, \ldots, a_{\lceil m/2 \rceil - 1}\}$ and $B = \{a_{\lceil m/2 \rceil}, \ldots, a_{m-1}\}$, then $M^{s_1} \bmod n = y_A \cdot z_B \bmod n$. Thus

$$z_B^* \bmod p = (S \cdot y_p^{-1} \cdot z_B^{-1} \bmod n) \bmod p = S \cdot y_p^{-1} \cdot M^{-s_1} \cdot y_A \bmod p = y_A \bmod p.$$

Hence, with very high probability $\gcd(y_A - z_B^*, n) = p$.

Let $X = (h+1)^{\lceil m/2 \rceil}$, we denote by $y_1, \ldots, y_X$ and $z_1^*, \ldots, z_X^*$ respectively all the possible values of y_A and z_B^*.

We here describe two ways to perform this attack.

Way a. For all i and j, the attacking server computes $\gcd(y_i - z_j^*, n)$. The complexity of this attack is

$$X^2 = \left((h+1)^{\lceil m/2 \rceil} \right)^2.$$

Way b. Define the polynomial

$$P(z) = \prod_{j=1}^{X} \left(z - z_j^* \right) \bmod n.$$

For all $i \in \{1, \ldots, X\}$, the attacking server computes $P(y_i) \bmod n$. Using FFT algorithms described in [1] (chapter 8), we can theoretically evaluate the polynomial P in X points in $X(\log X)^2$ steps. Hence the complexity of this attack becomes

$$(h+1)^{\lceil m/2 \rceil} \left(\log \left((h+1)^{\lceil m/2 \rceil} \right) \right)^2.$$

The second way needs a lot of cache and RAM memory. Theoretically this way b's attack needs $X(\log X)^2$ steps, but practically the needed hundred Gigabytes will be stored on "slow" discs instead of fast access-memory. So, the practical complexity of the second way is quite the same as the complexity of the first one. Hence, the parameters used in section 6.2 to counter way b's attack will be more secure than we need.

5.3 An Attack Using the LLL Algorithm [7]

We consider that the card computes several signatures. We denote by $s_1^{[i]}$ the value s_1 used during the ith signature. Let us suppose that $l(s_1^{[i]}) \leq \frac{t}{\alpha} - 2$ for all i with $\alpha > 1$. We suppose without lost of generality $l(p) = t + 1$.
We have $\sigma_p^{[i]} = s - s_1^{[i]} + \varrho_p^{[i]}(p-1)$, then $\sigma_p^{[i]} - \sigma_p^{[i+1]} = s_1^{[i+1]} - s_1^{[i]} + (\varrho_p^{[i]} - \varrho_p^{[i+1]})(p-1)$. Let us use the following notations $\sigma_i = \sigma_p^{[i]} - \sigma_p^{[i+1]}$, $r_i = s_1^{[i+1]} - s_1^{[i]}$ and $\varrho_i = \varrho_p^{[i]} - \varrho_p^{[i+1]}$; we have $\sigma_i = r_i + \varrho_i(p-1)$. The server knows σ_i for all i, but r_i, ϱ_i and $p-1$ are unknown. We have $l(\sigma_i) = k$, $|\varrho_i| \leq q-1$, and $|r_i| \leq 2^{\frac{t}{\alpha}-2}$.
Consider the following matrix where l is a non negative integer:

$$A = \begin{pmatrix} \sigma_2 & \sigma_3 & \cdots & \sigma_{l+1} & \varepsilon \\ \sigma_1 & 0 & \cdots & 0 & 0 \\ 0 & \sigma_1 & 0 & \cdots & 0 \\ \vdots & \ddots & \ddots & \ddots & \vdots \\ 0 & \cdots & 0 & \sigma_1 & 0 \end{pmatrix}$$

where ε is a random number of size $\leq \frac{t}{\alpha} - 1$.
Consider the left multiplication by the vector $Y_0 = (\varrho_1, -\varrho_2, \ldots, -\varrho_{l+1})$, we obtain the vector $(\varrho_1\sigma_2 - \varrho_2\sigma_1, \varrho_1\sigma_3 - \varrho_3\sigma_1, \ldots, \varrho_1\sigma_{l+1} - \varrho_{l+1}\sigma_1, \varrho_1\varepsilon)$.
But $\varrho_1\sigma_i - \varrho_i\sigma_1 = \varrho_1 r_i - \varrho_i r_1$, then $|\varrho_1\sigma_i - \varrho_i\sigma_1| \leq q2^{\frac{t}{\alpha}-1}$ and $|\varrho_1\varepsilon| \leq q2^{\frac{t}{\alpha}-1}$.
Let $Y = (y_1, -y_2, \ldots, -y_{l+1})$ be a non null vector. Then

$$(y_1, -y_2 \ldots, -y_{l+1}) A = (y_1\sigma_2 - y_2\sigma_1, \ldots, y_1\sigma_{l+1} - y_{l+1}\sigma_1, y_1\varepsilon).$$

If y_1 is much larger than q then the last element will have a too large size. So consider all values of y_1 smaller than $4q$.
We have $y_1\sigma_i = \lfloor \frac{y_1\sigma_i}{\sigma_1} \rfloor \sigma_1 + r$, then $y_1\sigma_i - y_i\sigma_1 = \left(\lfloor \frac{y_1\sigma_i}{\sigma_1} \rfloor - y_i \right) \sigma_1 + r$. The LLL algorithm will minimize $y_1\sigma_i - y_i\sigma_1$ for all i, then $y_i = \lceil \frac{y_1\sigma_i}{\sigma_1} \rfloor$. Hence the ith coordinate of $Y \times A$ will be $\sigma_i y_1 \bmod \sigma_1$ if $\sigma_i y_1 \bmod \sigma_1 \leq \frac{\sigma_1}{2}$ and $-(\sigma_1 - \sigma_i y_1 \bmod \sigma_1)$ if $\sigma_i y_1 \bmod \sigma_1 \geq \frac{\sigma_1}{2}$.
Consider the following probability:

$$\mathrm{Pr}_{y_1 < 4q} \left(\sigma_i y_1 \bmod \sigma_1 < q2^{\frac{t}{\alpha}-1} \text{ or } (\sigma_1 - \sigma_i y_1) \bmod \sigma_1 < q2^{\frac{t}{\alpha}-1} \right) \simeq 2\frac{q2^{\frac{t}{\alpha}-1}}{\sigma_1}$$
$$\simeq \frac{q2^{\frac{t}{\alpha}}}{pq}$$
$$= \frac{2^{\frac{t}{\alpha}}}{p},$$

then

$$\mathrm{Pr}_{y_1 < 4q} \left(\sigma_i y_1 \bmod \sigma_1 < q2^{\frac{t}{\alpha}} \text{ or } (\sigma_1 - \sigma_i y_1) \bmod \sigma_1 < q2^{\frac{t}{\alpha}} \ \forall i \right) \simeq \left(2^{\frac{t}{\alpha}}/p \right)^l.$$

Then the average number of integers $y_1 < 4q$ such that $\sigma_i y_1 \bmod \sigma_1 < q2^{\frac{t}{\alpha}}$ or $(\sigma_1 - \sigma_i y_1) \bmod \sigma_1 < q2^{\frac{t}{\alpha}}$ for all i is about $4q \left(2^{\frac{t}{\alpha}}/p \right)^l$.

We have $l(p) = t + 1$ then $p \geq 2^t$. Hence, if $\alpha > 1$, there exists l such that $4q\left(2^{\frac{1}{\alpha}}/p\right)^l \ll 1$. We have proven that there exists a vector Y_0 such that $y_1\sigma_i - y_i\sigma_1 \leq q2^{\frac{1}{\alpha}}$ for all i: the vector $Y_0 = (\varrho_1, -\varrho_2, \ldots, -\varrho_{l+1})$. The probability that there exists another such a vector Y is very small; then the LLL algorithm will find the vector Y_0.

But in our protocol, we have $s_1 = \sum_{i=0}^{m-1} a_i x_i$, with $l(x_i) \leq t - \log_2(mh) - 2$ and $0 \leq a_i \leq h$, then $0 \leq s_1 \leq mh2^{t-\log_2(mh)-2}$. Hence $l(s_1) \leq t - 2$. Then, in our protocol $\alpha = 1$, and we cannot find such an l, hence this attack is ineffective.

5.4 Modification of the Previous Attack

In our protocol $l(s_1^{[i]}) \leq t - 2$ for all i; then $l(s_1^{[i]}) \leq t - \beta - 2$, with probability $1/2^\beta$. Then to find an $s_1^{[i]}$ such that $l(s_1^{[i]}) \leq t - \beta - 2$, we must consider 2^β different $\sigma_p^{[i]}$. But in the previous attack, we must obtain $l + 1$ different $\sigma_p^{[i]}$ such that $l(s_1^{[i]}) \leq t/\alpha - 2$; here $\alpha = \frac{t}{t-\beta}$. Hence we must obtain $(l + 1)2^\beta$ different $\sigma_p^{[i]}$; next, we must find the $l + 1$ values $\sigma_p^{[i]}$ such that $l(s_1^{[i]}) \leq t/\alpha - 2$: there are $\binom{(l+1)2^\beta}{l+1}$ possibilities, and for each we must perform a LLL reduction. Then the total number of LLL reductions done is $\binom{(l+1)2^\beta}{l+1}$.

Consider for simplicity $l(p) = l(q) = t + 1 = k/2$. Then the constraint $q\left(2^{\frac{1}{\alpha}}/p\right)^l \ll 1$ will become $t + tl(1/\alpha - 1) \ll 0$, thus $1 + l(1/\alpha - 1) < 0$. Hence $l > \alpha/(\alpha - 1) = t/\beta$. Then, the workload w of this attack satisfies

$$w > w(\beta) = \binom{\frac{t}{\beta}2^\beta}{\frac{t}{\beta}}.$$

With RSA-512 ($k = 512$) the minimum is obtained when $\beta = 129$: $w(129) = 2^{130}$. And with RSA-768 ($k = 768$) the minimum is obtained when $\beta = 193$: $w(193) = 2^{194}$. Recall that we must perform $w(\beta)$ LLL reductions, then these attacks are ineffective.

5.5 Active Attacks

If the server cheats to obtain some information, the card will detect that. Then the card will not reveal the false value of S. Hence an active attack using only one false signature like the Anderson's one [2] is impossible.

Moreover, since s_1 and the x_i's are different for each signature, active attacks in several rounds like Pfitzmann and Waidner's ones [13] are impossible. The main difference between our protocol and all previously known ones ([11], [14], [10], [6], [16]...) is the use of random values which prevents our protocol from active attacks using several signatures.

6 Performances

We suppose $v = 3$. We also suppose for evaluating the computations done by the card that $l(p) = k/2$ and $l(q) = k/2$. We use the results explained in section 2; all values used here are clearly more precise than needed, but in order to overcome the addition of error factors, we will approximate the results only at the end of the analysis.

6.1 Computations done by the Card

The card must multiply a $k/2$ bit number (say y) by a k bit number (say z) modulo a $k/2$ bit number (say p). The best way is to compute $z \bmod p$, then multiply the two $k/2$ bit numbers and then take the result modulo p. With this way the computation is $\frac{\theta}{2} \cdot \frac{k}{2k} \cdot \frac{k/2}{k} + \frac{1}{2} \cdot \frac{k/2}{k} \cdot \frac{k/2}{k} + \frac{\theta}{2} \cdot \frac{k}{2k} \cdot \frac{k/2}{k} = \frac{7}{16}$ modular multiplication.

The card must also multiply two $k/2$ bit numbers modulo a $k/2$ bit number which takes $\frac{1}{2} \cdot \frac{k/2}{k} \cdot \frac{k/2}{k} + \frac{\theta}{2} \cdot \frac{k}{2k} \cdot \frac{k/2}{k} = \frac{9}{32}$ modular multiplication.

To obtain z_p and z_q:

case 1 The card must do two modulo $(k, k/2)$ to obtain $z_0 \bmod p$ and $z_0 \bmod q$, then $m - 1$ multiplications $(k/2, k)$ modulo p and modulo q, and $h - 1$ multiplications $(k/2, k/2)$ modulo p and modulo q to obtain z_p and z_q. Hence the card must do $\frac{5}{16} + (m - 1) \times \frac{7}{8} + (h - 1) \times \frac{9}{16}$ modular multiplications.

case 2 The card must do two modulo $(k, k/2)$ to obtain $z_0 \bmod p$ and $z_0 \bmod q$, then $m - 1$ multiplications $(k/2, k)$ modulo p and modulo q to obtain $c_1, \ldots c_h$ modulo p and q, and $2h - 2$ multiplications $(k/2, k/2)$ modulo p and modulo q to obtain z_p and z_q. Hence the card must do $\frac{5}{16} + (m - 1) \times \frac{7}{8} + (2h - 2) \times \frac{9}{16}$ modular multiplications.

The card must do $\frac{1}{4}$ modular multiplication to obtain σ_p and σ_q: we suppose that $s \bmod (p-1)$ and $s \bmod (q-1)$ are stored in memory; next it must do $2 \times \frac{7}{16}$ modular multiplication to obtain S_p and S_q, and $2 \cdot \frac{1}{2} \cdot \frac{k/2}{k} \cdot \frac{k}{k} + \frac{\theta}{2} \cdot \frac{3k/2}{2k} \cdot \frac{k}{k} = \frac{31}{32}$ to obtain S. Finally the card needs two modular multiplications to verify S.

Hence the total number of modular multiplications done by the card is

case 1 $\dfrac{5}{16} + (m - 1) \times \dfrac{7}{8} + (h - 1) \times \dfrac{9}{16} + \dfrac{67}{32} + 2,$

case 2 $\dfrac{5}{16} + (m - 1) \times \dfrac{7}{8} + (2h - 2) \times \dfrac{9}{16} + \dfrac{67}{32} + 2.$

6.2 Choice of the Parameters

In order to leave the attacks studied in section 5 ineffective, that is of complexity at most 2^{64}, and to minimize the number of modular multiplications done by the card, we take

way a
   case 1 $h = 10$, $m = 19$,
   case 2 $h = 7$, $m = 22$.
way b
   case 1 $h = 17$, $m = 25$,
   case 2 $h = 11$, $m = 29$.

6.3   The Needed Memory of the Card

We will evaluate the amount of memory the card needs to achieve the above protocol. We separate this memory into RAM and EEPROM.

In EEPROM are put all fixed values $s, s \bmod (p-1), s \bmod (q-1), v, n, p, q, w_p, w_q$, that is $6k + 1$ bytes, and some other values.

case 1 M and $z_0, \ldots, z_{m-1}$ are stored in the EEPROM that is $(m+1) \times k$ bits. The total number of bits written into the EEPROM for each signature is $(m+1) \times k$, and each bit of memory is used only one time for each signature.

case 2 M and $c_{1,p}, \ldots, c_{h,p}, c_{1,q}, \ldots, c_{h,q}$ are stored in the EEPROM, that is $(h+1) \times k$ bits. When the card receives a new z_i, it must compute the new corresponding $c_{i,p}$ and $c_{i,q}$ and write these new values into the EEPROM. Then the total number of bits written into the EEPROM is $(m+1) \times k$, and each bit of memory is used at most $\frac{m}{h}$ times for each signature.

Hence the need EEPROM is

case 1 $(m+7) \times k + 1$ bits,
case 2 $(h+7) \times k + 1$ bits.

We consider now the RAM. In step 2., the card chooses a_0 and x_0 and computes $s_1 = a_0 x_0$, then sends x_0 to the server. Then it chooses a_1, x_1 and computes $s_1 := s_1 + a_1 x_1$ ... The card keeps in mind $s_1, a_0, \ldots, a_{m-1}, x_i$. Hence it is easy to see that the maximum needed RAM for the card is in step 6. In step 6., the card must store

case 1 $s_1, a_0, \ldots, a_{m-1}, A, B, z_i, n$: the needed RAM is $4k + t - 2 + m \log_2(h+1)$ bits.

case 2 $s_1, a_0, \ldots, a_{m-1}, A_p, B_p, A_q, B_q, z_i, p, q$. To obtain the new A_p, B_p, the card must compute and store $z_i \bmod p$, next to obtain the new A_q, B_q, it must compute and store $z_i \bmod q$ which can be put instead of $z_i \bmod p$. Then the needed RAM is $t - 2 + m \log_2(h+1) + 4k + t + 1$ bits.

6.4   The Data Transfers

In the two cases, the card must send to the server $M, n, x_0, \ldots, x_{m-1}, \sigma_p, \sigma_q$, and the server must send to the card $z_0, \ldots, z_{m-1}, y_p, y_q$. Then the data transfers are $(6 + m) \times k + m \times (t - \log_2(mh) - 2)$ bits.

6.5 Computations done by the Server

All computations done by the server are the evaluations of $M^x \bmod n$ for several $x \le N$. It will do it by using the protocol BGMW [4] already describe with $x_i = b^i$, $m = \lceil \log_b N \rceil$ and $h = b - 1$ for an appropriate b. Then knowing $M^{b^i} \bmod n$ for all $i = 0, \dots, m - 1$, the number of modular multiplications done by the server is in the average $\frac{b-1}{b} \lceil \log_b N \rceil + b - 3$. In our protocol, the server must do this for $N = 2^{\frac{k}{2}}$ (step 5.) and for $N = 2^k$ (step 9.).

RSA 512: $k = 512$

The server computes and stores $M^{16^i} \bmod n$ for $i = 0, \dots, 63$ and $M^{32^i} \bmod n$ for $i = 0, \dots, 102$ which needs 510 modular squares. Then in the step 5. for each modular exponentiation, the server must do in the average 73 modular multiplications using $b = 16$, and in the step 9. for each modular exponentiation, in the average 128.8 modular multiplications using $b = 32$. Then the total number of modular multiplications done by the server is $510 + 73 \times m + 2 \times 128.8$.

RSA 768: $k = 768$

The server computes and stores $M^{16^i} \bmod n$ for $i = 0, \dots, 95$ and $M^{32^i} \bmod n$ for $i = 0, \dots, 153$ which needs 765 modular squares. Then in the step 5. for each modular exponentiation, the server must do in the average 103 modular multiplications using $b = 16$, and in the step 9. for each modular exponentiation, in the average 178.2 modular multiplications using $b = 32$. Then the total number of modular multiplications done by the server is $765 + 103 \times m + 2 \times 178.2$.

6.6 Results

The following tables give for the two cases, the number of modular multiplications done by the card and by the server, the needed RAM, EEPROM (in bytes), the number of bytes exchanged between the card and the server and the factor of acceleration by using our protocol. We also give the total number of bytes written into the EEPROM during the protocol and the maximal average times we must write in each bytes. We take for t the value $k/2 - 1$.

In the first table, we give results for the RSA 512 bits, and in the second for the RSA 768 bits. We recall that, using the Chinese Remainder Theorem, the card can compute a RSA signature in 260 modular multiplications when $k = 512$ and in 388 when $k = 768$.

Let us notice that today it is possible to write into the EEPROM in parallel with other computations (without any penalty of time) and to use a data transfer of 100 kbits/s.

	way a		way b	
	case 1	case 2	case 1	case 2
multiplications (card)	25	30	35	40
without server	260			
factor of acceleration	10.4	8.7	7.4	6.5
EEPROM (bytes)	1665	897	2049	1153
write (bytes)	1280	1472	1664	1920
number of writing	1	3.1	1	2.6
RAM	296	328	301	333
data transfers (bytes)	2183	2468	2748	3127
multiplications (server)	2155	2374	2593	2885

Table 1. RSA–512

	way a		way b	
	case 1	case 2	case 1	case 2
multiplications (card)	25	30	35	40
without server	388			
factor of acceleration	15.5	12.9	11.1	9.7
EEPROM (bytes)	2497	1345	3073	1729
write (bytes)	1920	2208	2496	2880
number of writing	1	3.1	1	2.6
RAM	440	488	445	493
data transfers (bytes)	3287	3716	4140	4711
multiplications (server)	3078	3387	3696	4108

Table 2. RSA–768

7 Conclusion

We have presented a new efficient protocol for accelerating RSA signatures using
an insecure and fast server. This protocol is resistant against all known active
and passive attacks. It does not use expensive precomputations; the computation
done by the card, the needed RAM and the data transfers between the card and
the server are small. Then, it is possible to implement efficiently this protocol
with current chip cards.

It remains an open question: the existence of efficient protocols (without the
use of precomputation) for accelerating RSA signatures *provably* secure against
passive and active attacks.

References

1. Aho, A. V., Hopcroft, J. E., Ullman, J. D.: The design and analysis of computer algorithms. Addison-Wesley, Reading, Mass. (1974).
2. Anderson, R. J.: Attack on server-assisted authentication protocols. Electronic Letters (1992) p. 1473.
3. Béguin, P., Quisquater, J.-J.: Secure acceleration of DSS signatures using insecure server. In Proceedings of Asiacrypt '94 (To appear).
4. Brickell, E., Gordon, D. M., McCurley, K. S., Wilson, D.: Fast exponentiation with precomputation. In Advances in Cryptology – Proceedings of Eurocrypt '92 (1993) Lecture Notes in Computer Science vol. 658 Springer-Verlag pp. 200–207.
5. Burns, J., Mitchell, C. J.: Parameter selection for server-aided RSA computation schemes. IEEE Transactions on computers 43 (1994) pp. 163–174.
6. Kawamura, S., Shimbo, A.: Fast server-aided secret computation protocols for modular exponentiation. IEEE Journal on selected areas communications 11 (1993).
7. Lenstra, A. K., Lenstra, H. W., Lovász, L.: Factoring polynomials with rational coefficients. Math. Ann. 261 (1982) pp. 515–534.
8. Lim, C. H., Lee, P. J.: More flexible exponentiation with precomputation. In Advances in Cryptology – Proceedings of Crypto '94 (1994) vol. Lecture Notes in Computer Science 839 Springer-Verlag pp. 95–107.
9. Lim, C. H., Lee, P. J.: Security and performance of server-aided rsa computation protocols. In this Proceedings .
10. Matsumoto, T., Imai, H., Laih, C.-S., Yen, S.-M.: On verifiable implicit asking protocols for RSA computation. In Advances in Cryptology – Proceedings of Auscrypt '92 (1993) Lecture Notes in Computer Science vol. 718 Springer-Verlag pp. 296–307.
11. Matsumoto, T., Kato, K., Imai, H.: Speeding up secret computation with insecure auxiliary devices. In Advances in Cryptology – Proceedings of Crypto '88 (1989) Lecture Notes in Computer Science vol. 403 Springer-Verlag pp. 497–506.
12. NIST: FIPS 186 for Digital Signature Standard (DSS).
13. Pfitzmann, B., Waidner, M.: Attacks on protocols for server-aided RSA computation. In Advances in Cryptology – Proceedings of Eurocrypt '92 (1993) Lecture Notes in Computer Science vol. 658 Springer-Verlag pp. 153–162.
14. Quisquater, J.-J., De Soete, M.: Speeding up smart card RSA computation with insecure coprocessors. In Proceedings of Smart Cards 2000 (1989) pp. 191–197.
15. Rivest, R., Shamir, A., Adleman, L.: A method for obtaining digital signatures and public-key cryptosystems. Communications of the ACM 21 (1978) pp. 120–126.
16. Yen, S.-M., Laih, C.-S.: More about the active attack on the server-aided secret computation protocol. Electronic Letters (1992) p. 2250.

Security and Performance of
Server-Aided RSA Computation Protocols

Chae Hoon Lim and Pil Joong Lee

Department of Electrical Engineering
Pohang University of Science and Technology (POSTECH)
Pohang, 790-784, KOREA
E-mail : lch@baekdu.postech.ac.kr ; pjl@vision.postech.ac.kr

Abstract. This paper investigates various security issues and provides
possible improvements on server-aided RSA computation schemes, mainly
focused on the two-phase protocols, RSA-S1M and RSA-S2M, proposed
by Matsumoto et al. [4]. We first present new active attacks on these protocols when the final result is not checked. A server-aided protocol is then
proposed in which the client can check the computed signature in at most
six multiplications irrespective of the size of the public exponent. Next
we consider multi-round active attacks on the protocol with correctness
check and show that parameter restrictions cannot defeat such attacks.
We thus assume that the secret exponent is newly decomposed in each
run of the protocol and discuss some means of speeding up this preprocessing step. Finally, considering the implementation-dependent attack,
we propose a new method for decomposing the secret and performing
the required computation efficiently.

1 Introduction

Smart cards in popular use throughout the world do not have any dedicated
crypto-engine and thus are not powerful enough to perform complicated computations, such as modular exponentiation, required for most public key cryptosystems (e.g., signature generation with RSA and signature verification in
ElGamal-like signature schemes). To speed up such computations by a weak
power smart card, much research has been conducted under the subject called
the server-aided secret computation (SASC) [1-12]. In the SASC protocol, the
client (the smart card) wants to perform a secret computation (e.g., RSA signature generation) by borrowing the computing power of an untrusted powerful
server without revealing its secret information.

A lot of (passive and active) attacks have been developed on the server-
aided RSA computation protocols (called RSA-S1 and RSA-S2) proposed by
Matsumoto, Kato and Imai [1] (e.g., see, [7-9,11]). Matsumoto, Imai, Laih and
Yen [4] also proposed two-phase versions of the basic protocols (called RSA-S1M
and RSA-S2M) to gain resistance against passive attacks, mainly to counter the
passive attack proposed by Pfitzmann and Waidner [8]. On the other hand,
Quisquater and De Soete [2] and Kawamura and Shimbo [5] proposed rather
different protocols for server-aided RSA computation. In these protocols, the

D. Coppersmith (Ed.): Advances in Cryptology - CRYPTO '95, LNCS 963, pp. 70-83, 1995.
© Springer-Verlag Berlin Heidelberg 1995

secret exponent is decomposed using a fixed basis and the messages sent to the server are independent of the secret. Consequently, they are as secure against passive attacks as RSA but require too much computation and communication. Recently, Béguin and Quisquater [13] developed another protocol using the fast exponentiation algorithm due to Brickell, Gordon, McCurley and Wilson [14].

The purpose of this paper is to investigate various security issues on server-aided RSA computation protocols and provide possible improvements. Our attention is mainly focused on the two-phase protocols, RSA-S1M and RSA-S2M. We first show that these protocols in fact do not have any resistance against active attacks, contrary to the previous claim [10]. The proposed attacks seem applicable to any server-aided protocol for RSA computation, including the protocols in [2] and [5]. Thus it seems essential that the final result should be checked by the client before being sent to the server. We then present a server-aided protocol into which signature verification capability is integrated. As a result, the computation result can be efficiently verified in any server-aided RSA computation scheme, irrespective of the size of the public exponent.

Though the client only outputs the correct signature, there still exists another threat, the multi-round active attack such that the server deduces some partial information on the involved secret in each trial of attack by observing whether or not the client outputs its computation result (e.g., see [8,11,12]). We show that the parameter restrictions suggested in [11] are not effective against more sophisticated attacks. In fact, it seems impossible that every possible attack could be detected just by placing restrictions on the parameter selection. Thus we also assume that the secret exponent is newly decomposed in each run of the protocol and discuss some means of speeding up this preprocessing step.

The last security issue in these protocols is to prevent the implementation-dependent attack such as the server infers some information on the secret vectors by monitoring the duration of the client's computation. To avoid this type of attack, we have to require that the client should spend the same amount of time on each computation step or perform the required computation after receiving and storing all communications from the server. Under the latter condition, we propose a new method for decomposing the secret and performing each computation step fast. The resulting protocols are shown to be quite efficient and practical.

2 Proposed Active Attacks

In server-aided RSA computation protocols, the client wants to compute $y = x^d$ mod n, where n is the product of two large primes p and q, with the aid of the powerful server. This section presents new active attacks on the protocols RSA-S1M and RSA-S2M proposed by Matsumoto et al. [4]. Similar attacks can be devised on the protocols in [2,5] as far as the computed signature is not checked.

One point needs to be mentioned on the passive attack. Unlike to the assertion in [4], the birthday-type attack [8] can be applied to the two-phase protocols as well. Of course, its complexity is much increased in this case, compared to single-phase protocols. But it is still much lower than that for a simple exhaustive search. For details, see the complexity analysis of Section 6.

2.1 Attack on RSA-S1M

The following protocol, RSA-S1M, is a two-phase version of RSA-S1 proposed to counter the passive attack devised by Pfitzmann and Waidner [8].

0) (Preprocessing) The client chooses an integer vector $\mathbf{D} = \{d_i\}_{i=1}^{M}$ over $\mathbf{Z}_n$ and binary vectors $\mathbf{F} = \{f_i\}_{i=1}^{M}$ and $\mathbf{G} = \{g_i\}_{i=1}^{M}$ such that $d = f \cdot g \bmod \lambda(n), f = \sum_{i=1}^{M} f_i d_i \bmod \lambda(n)$ and $g = \sum_{i=1}^{M} g_i d_i \bmod \lambda(n)$, where $\mathbf{F}$ and $\mathbf{G}$ have Hamming weight $\leq W$. The client also randomly picks an integer $r \in \mathbf{Z}_n$ and computes $t = r^{-g} \bmod n$.
1) The client sends $n, \mathbf{D}$ and x to the server.
2) The server returns $\mathbf{Z} = \{z_i\}_{i=1}^{M}$ such that $z_i = x^{d_i} \bmod n$.
3) The client computes and sends to the server $z = r \cdot \prod_{i=1}^{M} z_i^{f_i} = r \cdot x^f \bmod n$.
4) The server returns $\mathbf{V} = \{v_i\}_{i=1}^{M}$ such that $v_i = z^{d_i} \bmod n$.
5) The client computes the final result y as $y = t \cdot \prod_{i=1}^{M} v_i^{g_i} = t \cdot z^g \bmod n$.

Anderson [9] devised a simple one-round active attack on RSA-S1 and later Yen and Laih [10] claimed that the above two-phase protocol is highly resistant against Anderson's attack. However, we show that Anderson's attack can be extended to RSA-S1M (in fact, to any server-aided RSA computation scheme).

The server's strategy is to manipulate its transmissions in such a way that the final result y only consists of the product of known numbers. For this, the server supplies $z_i' = x^{ed_i} \bmod n$ in step 2) and $v_i' = p_i \cdot (z')^{d_i} \bmod n$ in step 4) respectively, where p_i's are small primes whose product is less than n. Now, when receiving $y' = t \cdot \prod_{g_i=1} v_i' \bmod n$ from the client, the server computes $y' \cdot x^{-1} = \prod_{g_i=1} p_i \bmod n$. The server can then factor this number and obtain the secret vector $\mathbf{G}$ (hence, the secret number g). Once g is obtained, the server can find the other secret vector $\mathbf{F}$ using the identity $x = (x^{eg})^f \bmod n$ in about about $N \log_2 N$ operations with $N = \sum_{k \leq \lceil W/2 \rceil} \binom{M}{k}$ [8]. This attack can also be applied to the non-binary version of the protocol if $\prod_{g_i \neq 0} p_i^{g_i}$ is less than n.

The above attack can be avoided if the client sends a blinded x in step 1) and cancels out the blinding factor in step 5). However, we can also devise a variation of the attack for this case : The server returns in step 4) $v_i' = \gamma_i \cdot z^{d_i} \bmod n$ with random γ_i and, when receiving y', computes $(y')^e \cdot x^{-1} = \prod_{g_i=1} \gamma_i^e \bmod n$. Now this equation can be used to find the secret vector $\mathbf{G}$ using the birthday paradox. Consequently, the whole attacking complexity in this case can be reduced to about two times $N \log_2 N$ operations with $N = \sum_{k \leq \lceil W/2 \rceil} \binom{M}{k}$.

2.2 Attack on RSA-S2M

The following is (non-binary) RSA-S2M, a variant of RSA-S1M which employs the Chinese remainder theorem to speed up the client's computation.

0) (Preprocessing) The client chooses integer vectors $\mathbf{D} = \{d_i\}_{i=1}^{M}$ $(d_i \in \mathbf{Z}_n)$, $\mathbf{F_j} = \{f_{ji}\}_{i=1}^{M}$ and $\mathbf{G_j} = \{g_{ji}\}_{i=1}^{M}$ $(j = 1,2)$ such that $d = f \cdot g \bmod \lambda(n)$, $f = \sum_{i=1}^{M} f_{1i}d_i \bmod p - 1$, $f = \sum_{i=1}^{M} f_{2i}d_i \bmod q - 1$, $g = \sum_{i=1}^{M} g_{1i}d_i \bmod p - 1$ and $g = \sum_{i=1}^{M} g_{2i}d_i \bmod q - 1$, where the vectors $\mathbf{F_j}$'s and $\mathbf{G_j}$'s consist of small integers and their components satisfy $f_{1i} = f_{2i} \bmod 2$ and $g_{1i} = g_{2i} \bmod 2$ for all values of i. The client also randomly picks an integer $r \in \mathbf{Z}_n$ and computes $t = r^{-g} \bmod n$.

1) The client sends $n, \mathbf{D}$ and x to the sever.

2) The server returns $\mathbf{Z} = \{z_i\}_{i=1}^{M}$ such that $z_i = x^{d_i} \bmod n$.

3) The client computes $x^f \bmod n$ as $x^f = x_p^f w_p + x_q^f w_q \bmod n$, where $x_p^f = \prod_{i=1}^{M} z_i^{f_{1i}} \bmod p$, $x_q^f = \prod_{i=1}^{M} z_i^{f_{2i}} \bmod q$, $w_p = q(q^{-1} \bmod p)$ and $w_q = p(p^{-1} \bmod q)$. The client then sends $z = r \cdot x^f \bmod n$ back to the server.

4) The server returns $\mathbf{V} = \{v_i\}_{i=1}^{M}$ such that $v_i = z^{d_i} \bmod n$.

5) The client computes $z^g \bmod n$ as $z^g = z_p^g w_p + z_q^g w_q \bmod n$, where $z_p^g = \prod_{i=1}^{M} v_i^{g_{1i}} \bmod p$ and $z_q^g = \prod_{i=1}^{M} v_i^{g_{2i}} \bmod q$. The client then computes the final result y as $y = t \cdot z^g \bmod n$.

Shimbo and Kawamura [7] developed a factorization attack on RSA-S2, and claimed that their attack could be prevented by the parameter restriction such as adopted in step 0) in the above protocol (i.e., $f_{1i} = f_{2i} \bmod 2$ and $g_{1i} = g_{2i} \bmod 2$). However this is not true. We present a generalized version of the attack which can be applied to any server-aided RSA computation protocol using the Chinese remainder theorem (CRT).

The proposed attack is applied to the second phase of the protocol. Instead of returning $v_i = z^{d_i} \bmod n$ in step 4), the server sends back $v_i' = \gamma \cdot z^{d_i} \bmod n$ with γ chosen at random. Let $s_1 = \sum_{i=1}^{M} g_{1i}$ and $s_2 = \sum_{i=1}^{M} g_{2i}$) and, for the moment, assume that $s_1 \neq s_2$. Then the client's computation in step 5) will end up with the value y' given by

$$y' = t \cdot (\gamma^{s_1} z_p^g \cdot w_p + \gamma^{s_2} z_q^g \cdot w_q) \bmod n, \qquad (1)$$

where $\gamma^{s_1} z_p^g$ and $\gamma^{s_2} z_q^g$ are actually numbers reduced mod p and mod q respectively. (Note, however, that the equation still holds even if they are not reduced.) By raising y' to the e-th power, the server obtains

$$(y')^e = t^e \cdot (\gamma^{es_1} z^{eg} + (\gamma^{es_2} - \gamma^{es_1})z_q^{eg} w_q) \bmod n, \qquad (2)$$

where we used the fact that $w_p^e = w_p \bmod n$, $w_q^e = w_q \bmod n$ and $w_p w_q = 0 \bmod n$. Therefore, on receiving y', the server can find the prime factor p by computing $\gcd(n, (y')^e - x\gamma^{es_1} \bmod n)$ since it is unlikely that q divides $t^e \cdot (\gamma^{es_2} - \gamma^{es_1})z_q^{eg} \bmod n$. The server may try all small numbers within a reasonable bound as candidates for s_1 since it does not know the exact value of s_1. But s_1 will be small in practice and thus the prime factor can be quickly found.

The above attack cannot be prevented by the simple restriction on the secret vectors $\mathbf{G_1}$ and $\mathbf{G_2}$ such that $\sum_{i=1}^{M} g_{1i} = \sum_{i=1}^{M} g_{2i}$, since the server may use a set of random numbers as γ and compute v_i''s as above with a number randomly picked in the set. This will increase the number of gcd computations, but the attack will be still effective due to the small size of integers used. In fact, unless the two secret vectors G_1 and G_2 are the same, the proposed attack has a high probability of success only with several values of γ. However, setting $G_1 = G_2$ and $F_1 = F_2$ makes RSA-S2M essentially equivalent to RSA-S1M. Therefore, we conclude that server-aided RSA computation based on the CRT can hardly be secure against the active attack unless the client checks the final result.

3 Integrating Server-Aided Verification

The active attacks described seem applicable to any server-aided protocol for RSA computation. Thus it is essential that the client check the correctness of the final result before sending it to the server. If e is chosen to be small such as 3, the last checking will not much increase the client's computational load.

However, the RSA system with a small public exponent may be dangerous in some circumstances (e.g., when used for encryption in network environments, see [15]). Thus what is more desirable is that no restriction needs to be placed on the choice of e. It is then quite natural to consider that the verification of the computed signature could be carried out with the aid of the same server. Several such protocols were proposed (for example, see [2,16]), but they seem to require too much amount of communication or computation. Thus it remains still open to design an efficient verification protocol.

The problem of integrating signature verification capability is nontrivial. A main difficulty arises from the fact that the two processes are reverse each other (i.e., encrypting a signed message reveals the plaintext). This makes it difficult for the client to detect the server's attempt to obtain a signature on message of its own choosing. We here propose an efficient protocol for RSA signature generation with the correctness check. We only describe a variant of RSA-S1M and the same technique can be adapted for any SASC protocol. Suppose that $\gcd(3, \lambda(n)) = 1$ and $e > 3$.

0) (Preprocessing) The client computes d_0 such that $e_0 d_0 = 1 \bmod \lambda(n)$ with $e_0 = 3$ and randomly decomposes $d_0 d \bmod \lambda(n)$ as in RSA-S1M. It also computes $t_1 = r_1^{-g} \bmod n$ and $t_2 = r_2^{e_0 e} \bmod n$ with random $r_1, r_2 \in Z_n$.

1) The client sends n, D and x to the server.

2) The server returns $Z = \{z_i\}_{i=1}^M$ such that $z_i = x^{d_i} \bmod n$.

3) The client computes and sends to the server $z = r_1 \cdot \prod_{f_i=1} z_i \bmod n$.

4) The server returns $V = \{v_i\}_{i=1}^M$ such that $v_i = z^{d_i} \bmod n$.

5) The client computes y_0 as $y_0 = t_1 \cdot \prod_{g_i=1} v_i = x^{d_0 d} \bmod n$. Then the client sends $v = y_0 \cdot r_2 \bmod n$ back to the server.

6) The server returns $w = v^e \bmod n$ to the client.

7) The client checks that $w^{e_0} = x \cdot t_2 \bmod n$. Only if the check succeeds, does the client send $y = y_0^{e_0} = x^d \bmod n$ to the server.

The client's on-line computational load increased due to the integration of signature verification is just six multiplications, irrespective of the size of e. This is a considerable advantage over direct verification if e is not very small. In step 7) the client may send y_0 directly to the server, which can then compute the signature y from y_0. Though this can save two multiplications, we would not like to recommand this variant since it may cause a problem in certain careless use of the protocol (see the footnote below). (It also requires that x^{d_0} mod n be of no meaning.)

To pass the check of step 7), the server has to know the e_0-th root of xt_2 mod n. This is infeasible if the server deviates the protocol. Note that d_0 and d are fixed but r_1 and r_2 are refreshed in each run of the protocol. This means that multi-round attacks on the signature verification part is of no use. Note also that the server may obtain the e_0-th root of t_2 (i.e., r_2^e mod n) by providing $z'_i = x^{e_0 d_i}$ mod n in step 2) and then computing $v^e x^{-1}$ mod n in step 6). However, there exists no way to obtain x^{d_0} mod n.[1] Thus we believe that the above protocol can detect any active attack mounted by the server.

4 Multi-Round Attacks Under Parameter Restriction

Even if the client checks the final result and only outputs the correct signature, there exists another threat to the server-aided protocol, i.e. the multi-round active attack such that the server changes a few values of its transmissions each time and decides, say parity or equality etc., on the corresponding elements of the secret vector based on whether or not the client gives the signature. Burns and Mitchell [11] described various attacks on RSA-S1 and RSA-S2, including multi-round active attacks, and proposed some means of parameter selection with which they claimed any attempt to deceive the client would be detected. However, this kind of parameter restriction seems not sufficient to detect every possible attack. As an example, we describe a new attack on RSA-S1. (This attack was in fact turned out to be a special case of the general attack described in [12], see also the footnote in the next page.)

In (non-binary) RSA-S1, d is simply decomposed as $d = \sum_{i=1}^{M} f_i d_i$ mod $\lambda(n)$ where f_i's are small positive integers. On request of the client, the server replies with $z_i = x^{d_i}$ mod n and the client then computes the signature y as $y = \prod_{i=1}^{M} z_i^{f_i}$ mod n. To counter Gollmann's attack (see [11] for details), Burns et al. proposed to restrict all f_i's to odd integers (hence no f_i is zero). Even under this restriction, the following attack can be successful.

The server randomly picks two integers, say j and k, in $[1, M]$ and sends back correct values for z_i for all values of i such that $i \neq j$ and $i \neq k$. In the latter two cases, the server provides $z_j = 2x^{d_j}$ mod n and $z_k = 2^{-1}x^{d_k}$ mod n. Then the client will get $y = 2^{f_j - f_k} x^d$ mod n and thus output it only if $f_j = f_k$.

---

[1] There is one thing to note when the client sends y_0 instead of y in step 7). If the client signs the same message x twice, then the server with knowledge of x^{d_0} mod n can obtain the signature on message x' of its own choosing. For this, the server returns $z'_i = (x')^{e_0 d_i}$ mod n in step 2), obtains $t_2^{d_0} = v^e (x')^{-1}$ mod n in step 6) and then replies with $w = (xt_2)^{d_0}$ mod n. Since this w passes the check of step 7), the server will get $y_0 = (x')^d$ mod n, the desired signature.

Therefore, by observing whether or not the client outputs the result, the server can determine whether f_j equals f_k. Similarly, in case that $f_j \neq f_k$, it can also determine whether $af_j = bf_k$ for some small odd integers a and b.

Repeating the above attack, the server can find all f_i's if it can meet the same client as many times as required. Or, using some partial information obtained from the attack, it can accelerate an exhaustive search. The only way to reduce this threat is to choose f_i's all distinct, but this is unacceptable in both efficiency and security. The parameter restriction on RSA-S2 seems more effective, but it can be easily seen that a similar strategy of comparing (in this case) two pairs of elements at a time can substantially reduce the security level. (It is possible to completely determine secret vectors for the binary case.)

Multi-round active attacks based on binary tests (even or odd, zero or not, equal or not, etc.) can be applied to any server-aided protocol as far as the involved secret vectors are fixed (see also [12]). [2] Thus the secret vectors should be randomly changed in each run (more practically, in limited runs) of the protocol. This will practically nullify any kind of active attacks on protocols using the random decomposition of the secret.

In this respect, protocols using a fixed basis, such as KS (Kawamura-Shimbo) and QS (Quisquater-De Soete) protocols, seem rather disadvantageous, in spite of their advantage of obvious security against passive attacks. One possible alternative to defeating multi-round active attacks in this kind of protocols is to construct two-phase variants, like RSA-S1M, using the decomposition $d = fg$ mod $\lambda(n)$. In this case, one of f and g can be chosen arbitrarily small without loss of security, as far as it can provide large enough possibilities of decomposition. This will somewhat reduce the increase of complexity caused by two times serial execution of the original protocol. More preferably, d may be decomposed as $d = fg + h$ mod $\lambda(n)$, where f and g are chosen to be of size $|\lambda(n)|/2$, and h is computed as $h = d - fg$ mod $\lambda(n)$. Then h needs to be publicly transmitted.

5   Speeding up the Preprocessing Step

Assuming that the use of the same secret vectors is limited to a fixed times of protocol execution to counter multi-round active attacks, we need some means of speeding up the preprocessing step. Note that the client may use the same secret vectors repeatedly unless the failure count associated with the secret vectors exceeds a fixed threshold (say, five). However, it is clear that random blinding factors must be refreshed in each protocol execution.

---

[2] Kawamura [12] has already demonstrated the vulnerability of server-aided protocols to this kind of attack. He showed that in one of the KS protocols the secret exponent d could be deduced in 680 trials of attack for base $b = 16$ and, as a practical precaution, suggested that in order to limit the information leakage, the client refuse to interact with the same server if the final check fails more than a fixed threshold value. However, we have to note that although information leakage may be controlled by limiting the number of failures with the same server, a number of servers may conspire to find the secret of a specific client. Since partial information can add up to information sufficient to derive the secret, this precaution seems still unsatisfactory.

5.1 Precomputation of Blinding Factors

Let us first consider a method of accelerating the precomputation of r^{-g} mod n for random r and g. (We assume that $r^{e_0 e}$ mod n for random r is computed directly since e is not so large in most cases.) Note first that neither r nor $t = r^{-g}$ mod n is disclosed since they are used as $z = r \cdot x^f$ mod n and $y = t \cdot z^g$ mod n (see RSA-S1M). This means that r can be chosen at random in a restricted domain. One possible way, which we will adopt in this paper, is to use preprocessing algorithms for random exponentiation such as the one proposed by Schnorr [17,18]. For this, the prime factors p, q of n are chosen so that a prime β divides both $p - 1$ and $q - 1$ and then a base α of order β mod n is randomly generated. Now the client securely stores a small number of pairs $\{s, \alpha^s \text{ mod } n\}$ with random $s \in \mathbf{Z}_\beta$. Then, during an idle time the client can compute and store several values of α^k mod n using a preprocessing algorithm. One of these values, say α^{k_r} mod n, can be used as r and another, say α^{k_t} mod n, as t. Then g is determined by $g = -k_r^{-1} k_t$ mod β.

For security and performance of the above preprocessing method, several remarks need to be stated.

- Since r and t are never revealed and are refreshed in every protocol run[3], it seems unnecessary to choose β large. For example, β can be a prime of size 64 $\sim$ 80. Due to the same reason, de Rooij's attacks [19,20] on Schnorr's preprocessing algorithm cannot not be applied to our application.
- Due to the small order of α, one can easily find β if any power of α is known. If β is known, one can find f using $z^\beta = x^{\beta f}$ mod n. Thus it is essential to keep α and β secret. However, note that there exists no known algorithm for factoring n (for $|n| \geq 512$) faster with knowledge of β of size 64 $\sim$ 80.
- From the above remarks, we can see that Schnorr's algorithm can be safely used even with smaller sizes of security parameters than those given in [17]. Thus it is possible to compute one value of α^k mod n in around 10 multiplications mod n. For a practical implementation, all computations mod n can be performed mod p and mod q and then the results can be combined using the CRT. This will almost halve the computational amount.

5.2 Random Decomposition of the Secret

Since the secrecy of d does not rely on the randomness of d_i's, we may choose the integer vector $\mathbf{D}$ in some convenient ways. One attractive example (similar to Strategy C in [11]) is to decompose d as $d = cfg + h$ mod $\lambda(n)$. Note that g is determined during the precomputation of r^{-g} mod n. Suppose the case of non-binary RSA-S1M (see Section 6.2 for the case of RSA-S2M). The client

1. generates random numbers $d_i (1 \leq i \leq M - 1)$ of l-bit size (say, $l = 80$),
2. determines secret integer vectors $\mathbf{F} = \{f_i\}_{i=1}^M$ and $\mathbf{G} = \{g_i\}_{i=1}^M$,

---

[3] The server may obtain r by returning $z_i = 1$ for all values of i. Thus the client must check that $z \neq 1$ before sending z in step 3).

3. computes $d_M = (g - \sum_{i=1}^{M-1} g_i d_i) g_M^{-1} \bmod \beta$ (adds some multiples of β if $|\beta| < l$) and $f = \sum_{i=1}^{M} f_i d_i$ and

4. finally computes h as $h = d - cfg \bmod \lambda(n)$ with random $c \in [0, \lambda(n))$.

Here $d_i (1 \leq i \leq M)$, c and h are public numbers transmitted in step 1). To reduce the communication complexity, the client may generate $d_i (1 \leq i \leq M-1)$ and c from a small initial seed using a common pseudorandom number generator (e.g., a linear congruential generator). Then it will suffice for the client to send just the seed, d_M and h. Differences from the original RSA-S1M are that the server returns $z_i = x^{cd_i} \bmod n$ in step 2) and one more value $z_h = x^h \bmod n$ in step 4). (Note that z_h is needed at the end of step 5).)

This method of decomposition has several advantages : smaller number of pseudorandom bit generation, no inversion mod $\lambda(n)$ and dramatic reduction of the server's computation complexity. Note that evaluating many exponentials with the same base can be substantially speeded up by using precomputation [14,21]. However, this decomposition definitely gives up keeping about $|\lambda(n)| - 2l$ bits of information on d and thus may be subject to some attacks exploiting this fact. One possibility is to apply Wiener's attack [22] on short RSA secret keys using continued fractions. But this kind of attack seems not applicable to the above case. Note that if c is chosen in the interval $[0, \lceil \lambda(n)/fg \rceil)$, then Wiener's attack may be successful unless e and $L = \gcd(p-1, q-1)$ are large.

To see this, let $h = ch_q + h_r$ ($0 \leq h_r < c$) and note that $ed = ec(fg + h_q) + eh_r = 1 + \frac{K}{L}(n - p - q + 1)$ where a common factor of K and L can be cancelled out. Let $A = nK + (1 - eh_r)L$. Then we get $\frac{ec}{A} = \frac{1}{(fg+h_q)L}(1 - \delta)$ with $\delta = \frac{K(p+q-1)}{A}$. Thus the continued fraction attack will be successful if $\delta < \frac{2}{3(fg+h_q)L}$. This condition holds as far as $|fgL|$ is a little less than $|n|/2$. Here we need to find the correct value of A by guessing $K(< e)$ and L. But e is usually small and L is also small if not intentionally chosen to be large. Thus one can try all small values of K and L within certain bounds and apply the continued fraction algorithm to find the prime factors of n. This analysis shows that we had better choose c over $[0, \lambda(n))$ rather than taking other measures. Note, however, that in our decomposition c may be chosen over $[0, \lceil \lambda(n)/fg \rceil)$ since we have already chosen L large enough to defeat the above attack. On the other hand, we can avoid most weaknesses which may potentially exist by setting $c = 1$ (i.e., $2l \simeq |\lambda(n)|$) at the expense of more computations.

6 Improving Performance of Non-binary Schemes

We now assume that the client checks the computation result and that the preprocessing phase is carried out in each run of the protocol. Then the last threat to server-aided protocols will be the implementation-dependent attack (e.g., see [11]). Suppose that the computation of (say) $z = \sum_{f_i=1} z_i \bmod n$ should be done in a step-by-step manner, considering the storage limitation of typical smart cards. That is, the server supplies one value of z_i at a time and the client multiplies it into the partial result. This type of processing makes it possible for the server to monitor the client's computation time to deduce the weights of f_i's. Thus, to avoid this attack, the client has to spend the same amount of time on each computation step or it can compute z after receiving and storing all z_i's at a time. We adopt the latter strategy and propose a method for improving the performance of non-binary schemes.

6.1 The Case of RSA-S1M

Suppose that d is decomposed as described in Section 5.2. Let $f_i = \sum_{j=1}^{K} 2^{j-1} f_{ij}$ and $g_i = \sum_{j=1}^{K} 2^{j-1} g_{ij}$ with $f_{ij}, g_{ij} \in \{0, 1\}$. These K-bit integers are chosen so that the total binary weight of $\mathbf{F} = \{f_i\}_{i=1}^{M}$ and $\mathbf{G} = \{g_i\}_{i=1}^{M}$ respectively is at most W, i.e. Weight($\mathbf{F}$)=Weight($\mathbf{G}$) $\le W$. Then the value of z in step 3) (similarly y in step 5) can be computed as

$$z = r \cdot \prod_{j=1}^{K} \left(\prod_{i=1}^{M} z_i^{f_{ij}} \right)^{2^{j-1}} \mod n. \tag{3}$$

For this, the client receives all z_i's at a time (hence it needs M temporary registers for them) and then performs the following algorithm.

```
receive and store all z_i's;
z := 1;
for j := K to 1 step -1
    z := z² mod n;
    for each i, if(f_ij = 1)
        z := zz_i mod n;
z := zr mod n;
```

It can be easily seen that the above algorithm can compute z in at most $K+W-1$ multiplications ($K-1$ squarings $+ W$ multiplications). Note that if the total weight of secret integers are chosen to be less than W, then we need some simulation of multiplications to prevent the implementation-dependent attack.

We now consider the computational complexity for finding the secret d. Let $\mathbf{S_f} = \{\sum_{i=1}^{M} f_i d_i | \text{Weight}(\mathbf{F}) \le W\}$ and $\mathbf{S_g} = \{\sum_{i=1}^{M} g_i d_i | \text{Weight}(\mathbf{G}) \le \lceil W/2 \rceil\}$. The most promising way to find d would be an exhaustive search based on the equation $1 - eh - ecfg^\star = ecfg^\circ \mod \lambda(n)$. That is, the server computes $x^{1-eh-ecfg^\star} \mod n$ and $x^{ecfg^\circ} \mod n$ for all possible values of $f, g^\star$ and g° such that $f \in \mathbf{S_f}$ and $g^\star, g^\circ \in \mathbf{S_g}$ and then searches for equality by sorting all these values. This will reveal the secrets f and g. Therefore, the computational complexity for this attack will be about $N \log_2 N$ operations where N, the total number of values to be sorted (disregarding a factor 2), is given by

$$N = \sum_{w=1}^{W} \binom{MK}{w} \sum_{w=1}^{\lceil W/2 \rceil} \binom{MK}{w}. \tag{4}$$

For a practical estimation of the performance, we have chosen the attacking complexity of 2^{72} operations. Thus we have to choose parameters so that $N \log_2 N \geq 2^{72}$ (i.e., $|N| \sim 66$). Note that for the above attack the server also needs a storage of order N. Table 1 summarizes the resulting performance for some small M's. The number of multiplications required of the client, $COMP$, is given by $2(K + W) - 1$. The number of $|n|$-bit blocks to be communicated, $COMM$, is $2M + 5$, where for simplicity we counted the seed, d_M and h altogether as two blocks (see Section 5.2). For these figures we did not take into account the final check. Thus if $e = 3$, $COMP$ needs to be increased by 2. Or if the protocol in Section 3 is used, $COMP$ and $COMM$ should be increased by 6 and 2 respectively.

M	3	4	5	6	7	8	9
K	15	13	11	9	8	7	8
W	15	13	13	13	13	13	11
$COMM$	11	13	15	17	19	21	23
$COMP$	59	51	47	43	41	39	37

Table 1 : Selected parameters for RSA-S1M

The actual performance needs to be evaluated by considering the communication speed of the client (e.g., 9.6 Kbps for typical smart cards, but there exist smart cards with much faster interfaces, from 19.2 up to 115.2 Kbps) and the storage (RAM) available (typically $128 \sim 512$ bytes). Note that the client may store M z_i's in EEPROM if it does not have a sufficient RAM space, since they need not be updated during the computation.[4] In this case the client has to write $2M$ $|n|$-bit numbers into EEPROM. Finally note that the server has to perform M exponentiations with l-bit exponents in each phase and two full size exponentiations (for c and h, see Section 5.2). The PC with a DSP accelerator card (e.g., see [23]) seems powerful enough as a practical server.

6.2 The Case of RSA-S2M

Applying the same technique to RSA-S2M, we can further reduce the client's computation time. For the sake of simplicity, we here assume that reduction of $|n|$-bit number to $\frac{|n|}{2}$-bit number and multiplication of two $\frac{|n|}{2}$-bit numbers take the same time. Then, even for the case where the client computes z and y using the CRT, it can be seen that the signature can be generated in the equivalent of $\frac{COMP+M}{2} + 3$ multiplications mod n. Let us consider the performance of non-binary RSA-S2M in more details.

---

[4] Typical EEPROM has a byte-write time of 5 ms (2 ms with recent technology) and thus writing a 512 bit number consumes about 320 ms (128 ms, resp.). However, with somewhat complicated coding it is possible for the smart card to perform EEPROM writes in parallel with other operations.

The secret d can be decomposed as $d = fg + h \bmod \lambda(n)$ as follows. Without loss of generality, we assume that $p < q$. The client

1. generates $M - 1$ random numbers $d_i (1 \leq i \leq M - 1)$ of size $|n|/2$,
2. selects secret integer vectors $\mathbf{F_j} = \{f_{ji}\}_{i=1}^{M+1}$ and $\mathbf{G_j} = \{g_{ji}\}_{i=1}^{M}$ $(j = 1, 2)$ such that Weight$(\mathbf{F_1} \cup \mathbf{F_2}) \leq W$ and Weight$(\mathbf{G_1} \cup \mathbf{G_2}) \leq W$,
3. computes $d_M = (g_0 - g')g_{1M}^{-1} \bmod \beta + k\beta$ where g_0 is a value determined during precomputation of t, $g' = \sum_{i=1}^{M-1} g_{1i}d_i \bmod p - 1$ and $k \in [0, p/\beta)$,
4. computes $d_{M+1} = g - \sum_{i=1}^{M} g_{2i}d_i \bmod q - 1$ where $g = g' + d_M g_{1M} \bmod p - 1$,
5. computes $d_{M+2} = f - \sum_{i=1}^{M+1} f_{2i}d_i \bmod q - 1$ where $f = \sum_{i=1}^{M+1} f_{1i}d_i \bmod p - 1$ and finally $h = d - fg \bmod \lambda(n)$.

The client can send $x, n, d_M, d_{M+1}, d_{M+2}, h$ and the seed used to generate $d_i (1 \leq i \leq M - 1)$. The remaining part of the protocol should be clear from the above decomposition and the original protocol RSA-S2M.

To get the attacking complexity, let $\mathbf{S_g} = \{\sum_{i=1}^{M} g_{1i}d_i | \text{Weight}(\mathbf{G_1}) \leq \lceil W/2 \rceil\}$ and $\mathbf{S_f} = \{\sum_{i=1}^{M+1} f_{1i}d_i | \text{Weight}(\mathbf{F_1}) \leq \lceil W/2 \rceil\}$. The attacking server computes $\gcd(x^{egf^\star} - x^{1-eh-egf^\circ} \bmod n, n)$ for all possible values of g, $f^\star$ and f° : If Weight$(\mathbf{G_1}) \leq \lceil W/2 \rceil$, then the server can find the prime factor p with $g \in \mathbf{S_g}$ and $f^\star, f^\circ \in \mathbf{S_f}$. Otherwise, i.e. if Weight$(\mathbf{G_2}) \leq \lceil W/2 \rceil$, we have $g \in \{\mathbf{S_g} + d_{M+1}\}$, $f^\star \in \mathbf{S_f}$ and $f^\circ \in \{\mathbf{S_f} + d_{M+2}\}$ and thus the prime factor q can be found. Note that in either case $f^\star + f^\circ$ covers all possible values of f. The number of possible values that $x^{egf^\star} \bmod n$ can take is given by

$$N = \sum_{w=1}^{\lceil W/2 \rceil} \binom{MK}{w} \sum_{w=1}^{\lceil W/2 \rceil} \binom{(M+1)K}{w}. \tag{5}$$

This birthday-type attack needs about $N(\log_2 N)^2$ operations and a storage of order N (see Section 5.2 in [13]).[5]

M	3	4	5	6	7	8	9
K	11	9	8	7	7	6	5
W	21	19	19	17	17	17	17
$COMM$	16	18	20	22	24	26	28
$COMP$	26.5	24.0	23.5	22.0	22.5	22.0	21.5

Table 2 : Selected parameters for RSA-S2M

Table 2 shows the resulting performance of RSA-S2M for the complexity of 2^{72} operations (i.e., $|N| \sim 60$). Here $COMP$ denotes the equivalent number of

---

[5] A naive approach to performing the required gcd computations would require about N^2 operations. We are indebted to Béguin and Quisquater [13] for knowing that there exists an algorithm with complexity of $N(\log_2 N)^2$ operations.

multiplications mod n, evaluated under the previous assumption on multiplication and modular reduction, which can be shown to be $K + 0.5(M + W) + 3.5$. $COMM$ is computed as $2M + 10$ (counting the seed and d_M as one block). As for the three values, $z_{M+1} = x^{d_M+1} \bmod n$, $v_{M+2} = z^{d_M+2} \bmod n$ and $z_h = x^h \bmod n$, the client may request each of them at the time when it is needed, since they are used at the end of some computations. Thus it suffices for the client to store only M values during the computation of each phase.

7 Concluding Remarks

In this paper we have investigated various security issues on server-aided RSA computation protocols, mainly focused on the two-phase protocols, RSA-S1M and RSA-S2M, and provided possible improvements.

We described new one-round active attacks that can be applied to any protocol for server-aided RSA computation if the final result is not checked. As a practical countermeasure, we presented an efficient protocol for server-aided validation of the computed signature. This protocol allows the client to check the correctness of the computation result at most six modular multiplications, irrespective of the size of the public exponent. We also showed that the preprocessing step should be carried out in each protocol execution to counter multi-round active attacks and discussed possible means of speeding up such preprocessing. Finally, using a new method for selecting (secret and public) parameters, we proposed modifications of RSA-S1M and RSA-S2M and analyzed their performance. The resulting protocols seem to be quite efficient.

There may be a slight disadvantage in two-phase protocols : the requirement of precomputation. Though precomputation for each signature generation is not much expensive (e.g., about the equivalent of 10 multiplications mod n, see Section 5.1), the client has to store a certain (predetermined) number of precomputed values so that several signatures can be generated successively without time delay due to precomputation. If this is still undesirable in practical applications, we may use a modification of RSA-S2 by the same technique, in which such precomputation is unnecessary. However, in this case the computational load of the client will be somewhat increased, compared to RSA-S2M. The protocol proposed by Béguin and Quisquater [13] has some advantage over two-phase protocols in this connection, since it also uses no precomputation. But it can be seen that using our proposed technique will give better efficiency in both computation time and storage usage than using the algorithm in [14].

Acknowledgement : The authors would like to thank Philippe Béguin, Tsutomu Matsumoto and Shin-ichi Kawamura for providing some corrections and helpful comments on an earlier version of this paper.

References

1. T.Matsumoto, K.Kato and H.Imai, Speeding up secret computations with insecure auxiliary devices, In *Proc. of Crypto'88*, Springer-Verlag, LNCS 403, 497-506 (1990).

2. J.J.Quisquater and M.De Soete, Speeding up smart card RSA computation with insecure coprocessors, In *Proc. Smart Card 2000*, North-Holland, 191-197 (1991).

3. C.S.Laih, S.M.Yen and L.Harn, Two efficient server-aided secret computation protocols based on addition chain sequence, In *Proc. of Asiacrypt'91*, S.V., LNCS 739, 450-459 (1993).

4. T.Matsumoto, H.Imai, C.S.Laih and S.M.Yen, On verifiable implicit asking protocols for RSA computation, In *Proc. of Auscrypt'92*, S.V., LNCS 718, 296-307 (1993).

5. S.Kawamura and A.Shimbo, Fast server-aided secret computation protocols for modular exponentiation, *IEEE JSAC*, 11(5), 778-784 (1993).

6. S.Kawamura and A.Shimbo, Performance analysis of server-aided secret computation protocols, *Trans. IEICE*, 73(7), 1073-1080 (1990).

7. A.Shimbo and S.Kawamura, Factorization attack on certain server-aided secret computation protocols for the RSA secret transformation, *Elect. Lett.*, 26(17), 1387-1388 (1990).

8. B.Pfitzmann and M.Waidner, Attacks on protocols for server-aided RSA computation, In *Proc. of Eurocrypt'92*, S.V., LNCS 658 (1993).

9. R.J.Anderson, Attack on server-aided authentication protocols, *Elect. Lett.*, 28(15), 1473 (1992).

10. S.M.Yen and C.S.Laih, More about the active attack on the server-aided secret computation protocol, *Elect. Lett.*, 28(24), 2250 (1992).

11. J.Burns and C.J.Mitchell, Parameter selection for server-aided RSA computation schemes, *IEEE Trans. Computers*, 43(2), 163-174 (1994).

12. S.Kawamura, Information leakage measurement in a distributed computation protocol, *IEICE Trans. Fundamentals*, E78-A(1), 59-66 (1995).

13. P.Béguin and J.J.Quisquater, Fast server-aided RSA signatures secure against active attacks, In *this proceedings*.

14. E.F.Brickell, D.M.Gordon, K.S.McCurley and D.B.Wilson, Fast exponentiation with precomputation, In *Proc. of Eurocrypt'92*, S.V., LNCS 658, 200-207 (1993).

15. J.Hastard, On using RSA with low exponent in a public key network, In *Proc. of Crypto'85*, S.V., LNCS 218, 403-408 (1986).

16. T.Matsumoto, K.Kato and H.Imai, How to ask and verify oracles for speeding up secret computations (Part 2), *IEICE TR, IT89-24* (1989).

17. C.P.Schnorr, Efficient identification and signatures for smart cards, In *Proc. of Crypto'89*, S.V., LNCS 435, 239-252 (1990).

18. C.P.Schnorr, Efficient signature generation by smart cards, *J. Cryptology* 4 (3), 161-174 (1991).

19. P.de Rooij, On the security of the Schnorr scheme using preprocessing, In *Proc. of Eurocrypt'91*, S.V., LNCS 547, 71-78 (1991)

20. P.de Rooij, On Schnorr's preprocessing for digital signature schemes, In *Proc. of Eurocrypt'93*, S.V., LNCS 765, 435-439 (1994).

21. C.H.Lim and P.J.Lee, More flexible exponentiation with precomputation, In *Proc. of Crypto'94*, S.V., LNCS 839, 95-107 (1994).

22. M.J.Wiener, Cryptanalysis of short RSA secret exponents, *IEEE Trans. Inform. Theory*, IT-36, 553-558 (1990).

23. S.R.Dusse and B.S.Kaliski Jr., A cryptographic library for the Motorola DSP 5600, In *Proc. of Eurocrypt'90*, S.V., LNCS 473, 230-244 (1991).

Efficient Commitment Schemes with Bounded Sender and Unbounded Receiver

Shai Halevi

MIT – Laboratory for Computer Science,
545 Technology Square, Cambridge, MA 02139
shaih@theory.lcs.mit.edu

Abstract. In this paper we address the problem of commitment schemes where the sender is bounded to polynomial time and the receiver may be all powerful. We present a scheme for committing to a (possibly long) string. Our scheme is efficient in the following three ways:

ROUND EFFICIENCY: Each part of the scheme consists of a single round.

LOW COMMUNICATION: The number of bits required for the commitment equals the security parameter of the system, regardless of the length of the string which is being committed to.

FAST IMPLEMENTATION: The time taken to commit to a string is linear in the length of the string and almost linear in the security parameter of the system.

1 Introduction

In this paper we address the problem of commitment schemes for (possibly long) messages. The problem arises when Alice has a message which Bob does not know, and they want to simulate (by means of electronic communication) the effect of delivering the message to Bob in a sealed envelope (or better yet, in a locked box): Alice wants to prevent Bob from knowing anything about the message in the box until such time in the future when she decides to give him the key. Bob, on the other hand, wants to prevent Alice from changing the message in the box after he has already received it.

COMMITMENT SCHEMES. A protocol for implementing such a simulation is called a commitment scheme. It consists of two phases. The first phase simulates the delivery of the locked box. When this phase is completed, Bob does not know the message yet, but Alice can not change it any more. The second phase simulates the delivery of the key. Bob can now see the message and verify that it is indeed the message to which Alice is committed.

EXAMPLE. As a simple example of a commitment scheme, consider a public-key encryption function $E(\cdot)$. To commit to a message σ, Alice sends the encryption $c = E(\sigma)$ to Bob. After this is done, Bob still does not know what σ is (provided that he can not break the encryption) but Alice can not change it anymore since there is no other message which encrypts to c. When Alice wants to reveal

D. Coppersmith (Ed.): Advances in Cryptology - CRYPTO '95, LNCS 963, pp. 84-96, 1995.

her message, she just send it to Bob, who encrypts it and checks that it really encrypts to c.

COMMITMENT SCHEMES WITH COMPUTATIONALLY UNBOUNDED BOB. The above scheme can work only if Bob is computationally bounded, so he can not break the encryption. If Bob has unbounded computational power, then Alice needs to commit in a way that yields no information (in the information theoretic sense) about her message. Of course, in this case there are many different messages that correspond to the same commitment. Thus, Alice must be computationally bounded, so she can not find any of the other messages. This is the case that we address in this paper.

RUNNING-TIME. Although it is known that commitment schemes exists based on "weak" assumptions, the implementations of such schemes are still based on either the hardness of factoring or the hardness of discrete log. In the discrete-log based implementations, a typical operation requires a modular exponentiation which is a relatively expensive operation. Thus these implementations are usually less efficient than the factoring-based ones, where a typical operation requires only modular multiplication. The scheme that we present in this paper is of the latter kind.

1.1 Previous Work

Commitment schemes were first formulated by Blum in the context of flipping coins over the telephone. The problem which is considered in this context is committing to a single bit. The first implementation of a bit-commitment scheme for unbounded Alice and bounded Bob (which was based on the hardness of discrete log) was suggested in [BM81] (as cited in [Blu82]). A different technique based on the hardness of factoring was embedded in [GM84].

The first bit-commitment scheme for the case where Bob is computationally unbounded was described by Blum in [Blu82]. This scheme is based on the hardness of factoring. In the same paper Blum also introduced the use of "Blum integers" (i.e., product of two primes, both congruent to 3 mod 4), which were used in many other papers since, including this one.

Since then there has been a large body of research regarding the bit commitment problem. In particular, it was shown that such bit commitment schemes exist in various models, based on various assumptions. For example, see [Nao90, Dam90, BC91, DPP94, IOS94].

In [Nao90], Naor also considered the problem of committing to long messages in the case of unbounded Alice and bounded Bob. Based on the existence of pseudo random generators, he describes a very elegant commitment scheme for long strings which only uses $O(n)$ bits to commit to a string of length n (where the constant in the $O(\cdot)$ does not depend on the security parameter of the system).

In addition, there have been much work on using various bit-commitment schemes within cryptographic protocols and zero-knowledge proofs. For example, see [BM84, GMW91, BCC88, BMO90, NOVY92].

1.2 Contributions of This Paper

In this paper we address the problem of committing to (possibly long) messages where Alice is bounded and Bob is unbounded. Although it is possible to use bit-commitment schemes to commit to a longer message by committing to each bit separately, the performance of such schemes is quite bad: The protocols for bit commitment require k bits of commitment for every message bit (in a system with security parameter k). If the message is long, then sending and storing such a large commitment may be a problem.

COMMUNICATION, ROUNDS AND RUNNING-TIME EFFICIENCY. We present a scheme where the length of the commitment string *does not depend on the length of the message*. The number of bits it takes Alice to commit to any string equals the security parameter of the system, regardless of the length of that string. Our scheme is also efficient in terms of round complexity. Each part of the scheme consists of a single round.

The scheme we present in this paper uses the Goldwasser-Micali-Rivest claw-free permutation pairs ([GMR88]) which are based on the hardness of factoring. As was mentioned above, each operation in the scheme consists of a modular multiplication, which can be performed in time almost linear in the size of the numbers involved. The scheme requires one or two multiplications for every character in the message.

SIMPLE INITIALIZATION. Our scheme has an advantage over other factoring-based schemes even when committing to just one bit. In many of the known factoring-based schemes, the composite numbers which are used in the scheme must be "Blum integers" (i.e., they have to be products of two primes, both congruent to 3 mod 4). If the numbers are not of the right form, then the security of both parties may be compromised.

Therefore these schemes require additional tools to ensure that the numbers are of the right form (such as using zero-knowledge proofs). These tools are typically very expensive, so the schemes become less efficient.

We present a new technique which eliminates the need for such expensive initialization steps. Our scheme is unique in that the use of "Blum integers" effects only the security of Bob. Thus, we can simply let Bob choose the number and send it to Alice, knowing that the security of Alice does not depend on which number was chosen.

1.3 Organization of the Paper

The rest of this paper is organized as follows: In Sect. 2 we define the notion of a commitment scheme. In Sect. 3 we present our factoring-based scheme which uses the claw-free permutation pairs due to [GMR88]. We first present a very simple implementation and then show how it can be modified to allow simple initialization.

In Sect. 4 we show how we can generalize our scheme and implement it using any construction of claw-free families of permutations.

2 Commitment Schemes

THE SYNTACTIC STRUCTURE OF A COMMITMENT SCHEME. A commitment scheme is a two phase protocol between two parties, Alice and Bob. Both parties share a common input, 1^k (for some integer k) which indicates the security parameter of the system. Besides 1^k, Alice also has another input, σ, which is the message string to which she wants to commit herself. When used inside some other protocol, the parties may also have other inputs which represent their history at the point where the commitment scheme in being invoked.

The scheme itself consists of two phases: The *commit* phase and the *reveal* phase. The parties execute the commit phase first and the reveal phase at some later time. Typically, when used in another protocol, there will be some other parts of that protocol between the commit and the reveal phases.

During the commit phase Alice sends to Bob a commitment string c and during the reveal phase Alice sends to Bob a reveal string r. From c and r Bob computes the message σ and then checks that σ is consistent with c and r.

In the construction which we present below we need an "initialization phase" before we can use the scheme. This phase is independent of the message σ which Alice wants to commit to. In fact, we can execute the initialization phase only once and then use the system to commit to many different messages. Alternatively, we can add the initialization to the commit phase and execute it as part of the protocol. In the implementations that we discuss in this paper, the initialization can be executed quite efficiently.

THE SEMANTICS OF A COMMITMENT SCHEME. Intuitively, the commit phase has the effect of sending the message from Alice to Bob in a locked box. Bob does not yet know anything about the contents of the message, but Alice can not alter the message anymore. The reveal phase has the effect of giving Bob the key and revealing the message inside the box.

The definition of what it means for Bob "not to know anything about σ", and for Alice "not to be able to alter σ" depends on the computational power of the parties. In the context of this paper, Alice is bounded to probabilistic polynomial-time and Bob has unbounded computational power. Thus, we require the following properties

Meaningfulness: If both Alice and Bob follow their parts in the protocol, then the message σ which Bob computes from (c, r) after the reveal phase is equal to Alice's input message.

Security: The communication between Alice and Bob in the commit phase gives no information (in the information-theoretic sense) about σ.

Non-Ambiguity: It is computationally infeasible for Alice to generate a commitment string c and two reveal strings r, r' such that in the reveal phase, Bob would compute one message σ from (c, r), a different message $\sigma' \neq \sigma$ from (c, r') and would accept both (σ, c, r) and (σ', c, r').

This means that for any probabilistic polynomial-time algorithm, the probability of generating c, r, r' as above when given the input 1^k (the security parameter) is negligible.

3 A Factoring-Based Implementation

In this section we present a specific implementation which uses the claw-free permutation families due to [GMR88].

3.1 The Goldwasser-Micali-Rivest Claw-Free Permutation Pairs

Let p and q be two primes such that $p = 3 \pmod 8$, $q = 7 \pmod 8$, and denote $N \stackrel{\text{def}}{=} p \cdot q$. We start by defining two functions

$$f_{N,0}(x) \stackrel{\text{def}}{=} x^2 \pmod N \quad \text{and} \quad f_{N,1}(x) \stackrel{\text{def}}{=} 4x^2 \pmod N$$

Then, for any string $s = b_1 b_2 \cdots b_n$ we define $f_{N,s}(x) \stackrel{\text{def}}{=} f_{N,b_1}(\cdots f_{N,b_n}(x) \cdots)$. It is easy to see that both $f_{N,0}$ and $f_{N,1}$ are permutations over the squares mod N, which implies that for any s the function $f_{N,s}$ is also a permutations over the squares mod N.

3.2 Using the GMR Construction for Commitment

The following is a simple commitment scheme that uses the GMR construction. We assume that Alice and Bob uses some standard encoding function Enc, with the property that for no two messages $\sigma \neq \sigma'$ is $Enc(\sigma)$ a prefix of $Enc(\sigma')$.

Initialization: Alice and Bob "choose at random" a composite N with k bits of the above form. We discuss this phase in more details below.

Commit phase: Given a message σ, Alice computes $s = Enc(\sigma)$. Then she picks a random element $x \in Z_N^*$ and sends $y = f_{N,s}(x^2)$ to Bob.

Reveal Phase: Alice sends both σ and x to Bob. Bob computes $s = Enc(\sigma)$ and verifies that $y = f_{N,s}(x^2)$.

To show that this is a commitment scheme we need to show two things:

Claim 1. *The value of y does not give any information about σ.*

Proof. (sketch) Since both $f_{N,1}$ and $f_{N,0}$ are permutations, then so is $f_{N,s}$ for any s. Thus, for every y (which is a square mod N) and every s there exists exactly one square mod N x such that $y = f_{N,s}(x)$. This implies the claim. ☐

Claim 2. *If it is infeasible to factor composite numbers of the above form, then it is infeasible for Alice to generate on input N two strings s, s' (none of which is a prefix of the other) and $x, x' \in Z_N^*$ such that $f_{N,s}(x^2) = f_{N,s'}(x'^2)$.*

This claim was proven in [GMR88] (Theorem 1) and a generalization of it was proven in [Dam88] (Theorem 2.8). ☐

3.3 Efficiency of the Scheme

The amount of communication in the commit phase is independent of σ. Alice always send exactly k bits to Bob (where k is the number of bits in N). In the reveal phase, Alice sends the message σ and k more bits.

In terms of running time, to compute the commitment string Alice needs to perform one or two modular multiplications for every bit in s (which presumably has about the same length as σ). Using construction similar to [Dam88], we can use larger families of permutations to reduce the number of multiplication to one or two per byte (or even word) of s. However, we pay for this by having to keep many more bits to describe these larger families of permutations, and by having to choose one of these families in the initialization phase.

3.4 Implementing the Initialization Phase

The main problem with the above scheme is the implementation of the initialization phase. Clearly, it is important to choose the composite number N in such a way that Alice will not be able to factor it easily. Notice that it doesn't matter whether Bob knows the factorization of N or not.

One idea is to let Bob choose N in the appropriate way and send it to Alice. But if Alice doesn't know the factorization of N, how can she verify that N it is really a product of two primes which are 3 mod 4 (which is the property that makes the functions $f_{N,0}, f_{N,1}$ permutations) ?

At first glance this may not look like a real problem. After all, Alice can choose the starting point x at random, so she may be able to hide σ from Bob even if these functions are not permutations. Unfortunately, this is not the case.

Consider for example $N = 5$ and a message of one bit b. It is easy to see that for any element $x \in Z_5^*$ we have $f_{5,0}(x^2) = 1$ and $f_{5,1}(x^2) = 4$. Thus Bob can recover the message from the commitment string.

To solve this problem Bob can choose N and then prove (by means of a zero-knowledge proof) to Alice that it is of the right form. However this zero-knowledge proof can be expensive in terms of both running time and communication. Moreover, some zero-knowledge proofs use commitment schemes as basic primitives.

It will therefore be desirable to have a system where choosing a "bad N" does not help Bob getting any information about σ. We present such a system below.

3.5 A Modification of the GMR-Based Scheme

The only difference between the following scheme and previous one is that after computing $y = f_{N,s}(x^2)$, Alice squares y k more times (where k is the number of bits in N) and sends the result to Bob. The new scheme is:

Initialization: Bob picks at random an odd k-bit composite number N which is a product of two large primes, one congruent to 3 mod 8 and the other congruent to 7 mod 8. Bob sends N to Alice, who just verifies that N is odd.

Commit phase: Given a message σ, Alice computes $s = Enc(\sigma)$. Then she picks a random element $x \in Z_N^*$ and sends $y = f_{N,0^k s}(x^2)$ to Bob.

Reveal Phase: Alice sends both σ and x to Bob. Bob computes $s = Enc(\sigma)$ and verifies that $y = f_{N,0^k s}(x^2)$.

It is easy to see that if Bob picks N according to the protocol then it is still infeasible for Alice to find two different messages with the same commitment string (if factoring is hard).

The hard part is to show that even if Bob tries to "cheat" by picking a "bad" N, he still does not get any information about σ from the commitment string.

3.6 Proof of Security for the Modified Scheme

Let N be an odd integer and denote the number of bits in N by k. We model Bob's view of the protocol as an experiment in which Alice picks a string $\sigma \in \{0,1\}^*$ according to some distribution D (D represents the knowledge that Bob has about σ). Then Alice computes $s = Enc(\sigma)$, picks at random an element $x \in Z_N^*$ and sends $y = f_{N,0^k s}(x^2)$ to Bob.

Denote by S, X the random variables which take on the values of s, x respectively in the experiment above. The following lemma asserts that y does not give any information about s if we do not know x.

Lemma 3. *For any element* $y \in Z_n^*$ *such that* $\Pr[f_{N,0^k S}(X^2) = y] > 0$ *and for any string s we have*

$$\Pr_{S,X}\left[S = s \mid f_{N,0^k S}(X^2) = y\right] = \Pr_{S}[S = s]$$

See Appendix A for a detailed proof. The idea is that all the information which $y = f_{N,s}(x^2)$ gives about s depends only on a property that we call the "tag of y mod N". Moreover, by repeated squaring we force the tag of $y = f_{N,0^k s}(x^2)$ to be some constant which does not depend on s or x. Thus, y does not give any information about s.

Unfortunately, the formal proof is somewhat lengthy. On the up side, it contains some number-theoretic lemmas which may be interesting in their own right.

4 Using General Claw-Free Permutation Families

In this section we show how the above scheme can be generalized to use any construction for claw-free permutation families.

4.1 Claw-Free Families of Permutations

The notion of claw-free permutations families which we use here is a little more general than the one in [GMR88] but still not as general as in [Dam88]. A construction of claw-free permutation families consists of the following components:

1. A constant r which indicates the number of permutations in each family.
2. A set $INDEX_k$ for all $k \in \mathcal{N}$. Every string in $INDEX_k$ is an index of a family with security parameter k. $INDEX_k$ is polynomial-time samplable given 1^k.
3. For every k and every index $\tau \in INDEX_k$ there is a domain Dom_τ that is associated with τ. It is convenient to assume that if $\tau \in INDEX_k$ then $Dom_\tau \subseteq \{0,1\}^k$. Dom_τ is polynomial-time samplable given τ.
4. For every k and every index $\tau \in INDEX_k$ we have a family of functions $\mathcal{F}_\tau = \{f_{(\tau,0)}, f_{(\tau,1)}, \cdots, f_{(\tau,r-1)}\}$ such that all the $f_{(\tau,i)}$'s are permutations over Dom_τ, and there is an efficient algorithm $COMPUTE(\tau, i, x)$ which computes $f_{\tau,i}(x)$ given τ, i and an element $x \in Dom_\tau$.
5. What makes these families claw-free is that the following task is infeasible: Given 1^k and a random element $\tau \in INDEX_k$, find $i \neq j$ and two elements $x, y \in Dom_\tau$ such that $f_{\tau,i}(x) = f_{\tau,j}(y)$.

Notice that the set "legal indexes" in the above definition may or may not be polynomial-time recognizable (i.e. $INDEX = \bigcup_k INDEX_k$ may or may not be in BPP). For example, for the GMR construction it is not known whether $INDEX \in BPP$. On the other hand, it is easy to come up with a simple construction based on the hardness of discrete-log for which $INDEX \in BPP$. As it turns out, if we have a construction for which $INDEX \in BPP$, then the initialization phase for the scheme which we present below becomes much simpler.

4.2 Commitment-Schemes and Claw-Free Permutation Families

Assume that we have a construction of claw-free families of permutations. For any index τ and string $s = b_1 b_2 \cdots b_n$ we denote by $f_{\tau,s}$ the function $f_{\tau,s}(x) \stackrel{\text{def}}{=} f_{\tau,b_1}(\cdots f_{\tau,b_n}(x) \cdots)$. Here is how we use the claw-free families to implement a commitment scheme. On common input 1^k:

Initialization: Alice and Bob pick a family of permutations with security parameter k (by choosing a random index $\tau \in INDEX_k$). We discuss ways to implement this phase below.

Commit phase: Given a message σ, Alice computes $s = Enc(\sigma)$. Then she picks a random element $x \in Dom_\tau$ and sends $y = f_{\tau,s}(x)$ to Bob.

Reveal Phase: Alice sends both σ and x to Bob. Bob computes $s = Enc(\sigma)$ and verifies that $y = f_{\tau,s}(x)$.

Notice that if we have families with more than two permutations, we can use techniques similar to those in [Dam88] to save time by viewing s as a string over an alphabet with more than two symbols: For example, if we have 256 permutations in each family we can view s as a sequence of bytes. This way we only need to apply $f_{(\tau,\cdot)}$ once for every byte in s rather than once for every bit.

The proof that this is indeed a commitment scheme is similar to the proofs of Claims 1 and 2.

4.3 Implementing the Initialization Phase

We consider two different cases here:

Case 1: $INDEX \in BPP$ for this construction. In this case Bob can simply choose a random index $\tau \in INDEX_k$ and send it to Alice. Alice can verify that τ is indeed an index of some permutation family.

Notice that Alice doesn't care how τ was chosen. The fact that all the functions $f_{\tau,i}(\cdot)$ are permutations over Dom_τ is enough to ensure that Bob does not get any information about s from the commitment string. On the other hand, it is in the best interest of Bob to pick τ at random, since the infeasibility condition only applies when τ is chosen at random.

Case 2: $INDEX \notin BPP$ for this construction. In this case Alice can not verify that the functions $f_{\tau,i}(\cdot)$ are permutations, so Bob may choose τ so as to be able to extract information about s from $f_{\tau,s}(x)$.

In this case the parties either need to rely on a trusted party that will pick τ for them, or Bob can prove to Alice that τ is indeed in $INDEX_k$. Another possibility is to modify the scheme itself (as we did in the GMR-based implementation) to eliminate this problem.

5 Acknowledgments

I thank Silvio Micali for many helpful ideas and discussions and Shafi Goldwasser for several very helpful comments. I also thank the MIT dental-service for making me wait for an hour on the dentist chair, in which time I came up with the idea for this paper.

References

[BC91]   G. Brassard and C. Crépeau. Quantum bit commitment and coin tossing protocols. In A.J. Menezes and S. A. Vanstone, editors, *Proceedings CRYPTO 90*, pages 49–61. Springer-Verlag, 1991. Lecture Notes in Computer Science No. 537.

[BCC88] G. Brassard, D. Chaum, and C. Crépeau. Minimum disclosure proofs of knowledge. *JCSS*, 37(2):156–189, 1988.

[Blu82] M. Blum. Coin flipping by telephone. In *Proc. IEEE Spring COMPCOM*, pages 133–137. IEEE, 1982.

[BM81]   M. Blum and S. Micali. Coin flipping into a well. Unpublished, 1981.

[BM84]   M. Blum and S. Micali. How to generate cryptographically strong sequences of pseudo-random bits. *SIAM J. Computing*, 13(4):850–863, November 1984.

[BMO90] M. Bellare, S. Micali, and R. Ostrovsky. The (true) complexity of statistical zero-knowledge. In *Proc. 22nd ACM Symposium on Theory of Computing*, pages 494–502, Baltimore, Maryland, 1990. ACM.

[Dam88] I.B. Damgård. Collision free hash functions and public key signature schemes. In David Chaum and Wyn L. Price, editors, *Proceedings of EUROCRYPT 87*, pages 203–216. Springer-Verlag, 1988. Lecture Notes in Computer Science No. 304.

[Dam90]   I.B. Damgård. On the existence of a bit commitment schemes and zero-knowledge proofs. In G. Brassard, editor, *Proceedings CRYPTO 89*, pages 17–29. Springer-Verlag, 1990. Lecture Notes in Computer Science No. 435.

[DPP94]   I.B. Damgård, T.P. Pedersen, and B. Pfitzmann. On the existence of statistically hiding bit commitment schemes and fail-stop signatures. In Douglas R. Stinson, editor, *Proceedings CRYPTO 93*, pages 250–265. Springer, 1994. Lecture Notes in Computer Science No. 773.

[GM84]   S. Goldwasser and S. Micali. Probabilistic encryption. *JCSS*, 28(2):270–299, April 1984.

[GMR88]   S. Goldwasser, S. Micali, and R. Rivest. A digital signature scheme secure against adaptive chosen-message attacks. *SIAM J. Computing*, 17(2):281–308, April 1988.

[GMW91]   O. Goldreich, S. Micali, and A. Wigderson. Proofs that yield nothing but their validity or all languages in NP have zero-knowledge proof systems. *Journal of the ACM*, 38(1):691–729, 1991.

[IOS94]   T. Itoh, Y. Ohta, and H. Shizuya. Language dependent secure bit commitment. In Yvo G. Desmedt, editor, *Proceedings CRYPTO 94*, pages 188–201. Springer, 1994. Lecture Notes in Computer Science No. 839.

[Nao90]   M. Naor. Bit commitment using pseudo-randomness. In G. Brassard, editor, *Proceedings CRYPTO 89*, pages 128–137. Springer-Verlag, 1990. Lecture Notes in Computer Science No. 435.

[NOVY92] M. Naor, R. Ostrovsky, R. Venkatesan, and M. Yung. Perfect zero-knowledge arguments for np can be based on general complexity assumptions. In Ernest F. Brickell, editor, *Proceedings CRYPTO 92*, pages 196–214. Springer-Verlag, 1992. Lecture Notes in Computer Science No. 740.

A   A Proof of Lemma 3

Before we can prove Lemma 3 we need to develop some number-theoretic tools and notations.

A.1   Tags of Elements Modulo Prime-Powers

Let $p = q^e$ be an odd prime-power (that is, q is an odd prime and e is a positive integer). Denote by $\phi(p)$ the order of the multiplicative group Z_p^* and denote by m the largest integer such that 2^m divides $\phi(p)$. Also, let g be some "canonical" generator in Z_p^* (e.g. the smallest generator in Z_p^*).

Definition 4. Let z be an element in Z_p^* and let ℓ be the discrete log of z base g. (i.e., $g^\ell \equiv z \bmod p$). The *tag of $z \bmod p$* is the residue of $\ell \bmod 2^m$. We denote it by $TAG_p(z)$. That is,

$$\text{for all } 0 \le \ell < \phi(p), \ TAG_p(g^\ell) = [\ell]_{2^m}$$

where $[x]_y$ denotes the residue of $x \bmod y$. The following properties are immediate from the definition of the tag:

Claim 5. *Let p be an odd prime power and m be the largest integer such that 2^m divides $\phi(p)$.*

1. If $w, x, y, z \in Z_p^*$ such that $TAG_p(w) = TAG_p(x)$, $TAG_p(y) = TAG_p(z)$ then $TAG_p(wy) = TAG_p(xz)$.
2. An element $z \in Z_p^*$ is a square mod p iff it has an even tag.
3. If $TAG_p(z) = 0$ then so is $TAG_p(z^2)$.
4. For any $z \in Z_p^*$ and any $i \geq m$, $TAG_p(z^{2^i}) = 0$.
5. Let z be a square mod p and denote its tag by $TAG_p(z) = 2^i r$ where r is some odd integer and $1 \leq i \leq m$. Then, one square root of z has tag $2^{i-1} r$ and the other has tag $\left[2^{i-1} r + 2^{m-1} \right]_{2^m}$.

The following corollaries describe the behavior of the tags under the functions $f_{p,s}$ (and their inverses):

Corollary 6. If p is an odd prime-power, x, y are elements in Z_p^* such that $TAG_p(x) = TAG_p(y)$ and s is any string, then $TAG_p(f_{p,s}(x)) = TAG_p(f_{p,s}(y))$.

Corollary 7. If p is an odd prime-power and z is an element in Z_p^* then the tags of the pre-images of z under both $f_{p,0}(\cdot)$ and $f_{p,1}(\cdot)$ (if there are any) depend only on the tag of z.

Corollary 8. If p is a prime power, $p < 2^k$, then for any element $x \in Z_p^*$, any string s and any $i \geq k$, $TAG_p \left(f_{p,0^i s}(x) \right) = TAG_p \left(f_{p,s}(x)^{2^i} \right) = 0$

The following is the main technical lemma in the proof

Lemma 9. Let p be an odd prime-power and let $y, z \in Z_p^*$, so that $TAG_p(y) = TAG_p(z)$. Then for any string s we have

$$\#\{x \in Z_p^* \; : \; f_{p,s}(x) = y\} = \#\{x \in Z_p^* \; : \; f_{p,s}(x) = z\}$$

where $\#A$ denotes the number of elements in the set A.

Proof. Let $s = b_0 \cdots b_{n-1}$ be a string and consider the "pre-images-tree" w.r.t. s that is rooted at an element z (i.e., the children of z are its pre-images under f_{p,b_0}, their children are their pre-images under f_{p,b_1} etc.). An element in this tree is an internal node if it is a square mod p and its distance form the root is less than n. Otherwise, it is a leaf.

Notice that this is indeed a tree in the sense that all the elements at distance i from the root are distinct. We will now show that if y and z have the same tag then their trees w.r.t. s are isomorphic. This means, in particular, that the number of elements at distance n from the root is the same in both trees, which implies Lemma 9. The notion of pre-images-tree becomes formal in the following definition:

Definition 10. Let p be an odd prime power, let $z \in Z_p^*$ and let $s = b_0 \cdots b_{n-1} \in \{0,1\}^n$. The pre-images-tree w.r.t. s which is rooted at z is a directed graph $T_{z,s} = (V, E)$ where

$$V = \{\langle i, x \rangle \; : \; 0 \leq i < n, \; x \in Z_p^*, \text{ and } f_{p,b_0 \cdots b_{i-1}}(x) = z\}$$
$$E = \{\langle i, y \rangle \rightarrow \langle i+1, x \rangle \; : \; y = f_{p,b_i}(x)\}$$

To see that this is a directed rooted tree, notice that $\langle 0, z \rangle$ has in-degree 0, every other node has in-degree 1, and there is a path from $\langle 0, z \rangle$ to every other node in the graph. Now we can prove Lemma 9 by proving a stronger lemma

Lemma 11. *For every string s and every two elements $y, z \in Z_p^*$ such that $TAG_p(y) = TAG_p(z)$, there is an isomorphism between $T_{y,s}$ and $T_{z,s}$ which also preserves the tags. That is, there exists a function $I : T_{y,s} \to T_{z,s}$ which satisfies the following properties:*

1. *I is an isomorphism between the graphs $T_{y,s}$ and $T_{z,s}$ (notice that this implies that I always maps nodes in level i in $T_{y,s}$ to nodes in level i in $T_{z,s}$).*
2. *If $I(\langle i, x \rangle) = \langle i, x' \rangle$ then $TAG_p(x) = TAG_p(x')$.*

In particular, it follows that the number of nodes in level n in both trees is the same, which implies Lemma 9.

Proof. The proof is by induction over n (the number of bits in s). It consists of a straightforward implementation of Corollary 7 above. *details omitted.* □

A.2 Tags of Elements Modulo Composites

Definition 12. Let N be an odd integer, and let $z \in Z_N^*$. Denote N's prime factorization by $N = p_1 \cdots p_\ell$ where $p_1, \cdots, p_\ell$ are powers of distinct primes. The *tag of z mod N* is the vector $\langle t_1, \cdots, t_\ell \rangle$ where t_i is the tag of z mod p_i.

Notice that for an element $x \in Z_N^*$ and a string s we have $[f_{N,s}(x)]_{p_i} = f_{p_i,s}([x]_{p_i})$ for all i. Therefore, from Corollary 8 we get

Corollary 13. *If $N < 2^k$, then for any element $x \in Z_N^*$, any string s and any $i \geq k$ we have $TAG_N\left(f_{N,0^i s}(x)\right) = \langle 0, 0, \cdots, 0 \rangle$.*

and from Lemma 9 we get

Claim 14. *Let N be an odd integer and let $y, z \in Z_N^*$, so that $TAG_N(y) = TAG_N(z)$. Then for any string s we have*

$$\#\{x \in Z_N^* \; : \; f_{N,s}(x) = y\} \; = \; \#\{x \in Z_N^* \; : \; f_{N,s}(x) = z\}$$

Proof. Follows since for all $y \in Z_N^*$

$$\#\{x \in Z_N^* \; : \; f_{N,s}(x) = y\} \; = \; \prod_i \#\{x \in Z_{p_i}^* \; : \; f_{p_i,s}([x]_{p_i}) = ([y]_{p_i})\}$$

□

From Corollary 13 and Claim 14 we get

Lemma 15. *Let N be an odd k-bit integer, and let s be any string. Then, for every element $y \in Z_N^*$ with tag $\langle 0, \cdots, 0 \rangle$ we have*

$$\#\{x \in Z_N^* \; : \; f_{N,0^k s}(x^2) = y\} \; = \; \frac{\phi(N)}{T_0}$$

where $\phi(N)$ is the order Z_N^* and T_0 is the number of elements in Z_N^* with tags $\langle 0, \cdots, 0 \rangle$.

Proof. From Corollary 13 we know that $f_{N,0^k s}(x^2)$ has tag $\langle 0, \cdots, 0 \rangle$ for all $x \in Z_N^*$. Therefore the pre-images of all the y's with tag $\langle 0, \cdots, 0 \rangle$ cover all Z_N^*. From Claim 14 we know that all these pre-images have the same size, which mean that this size is exactly $\phi(N)/T_0$ (we can apply Claim 14 since $f_{N,0^k s}(x^2) = f_{N,0^k s0}(x)$). $\qquad\square$

A.3 Back to Lemma 3

Recall the experiment of Lemma 3. Alice has an k-bit odd integer N. She picks a string $\sigma \in \{0,1\}^*$ according to some distribution D, computes $s = Enc(\sigma)$, picks at random an element $x \in Z_N^*$, and computes $y = f_{N,0^k s}(x^2)$.

We denoted by S, X the random variables which take on the values of s, x respectively. The lemma asserts that the value of y does not give any information about s. Since Alice picks x randomly in Z_N^* regardless of s, then S and X are independent. Therefore, from Lemma 15 we get

Corollary 16. *For any element y such that $TAG_N(y) = \langle 0, \cdots, 0 \rangle$ and any string s*

$$\Pr_{X,S} \left[S = s \ \text{ and } \ f_{N,0^k S}(X^2) = y \right]$$

$$= \Pr_S [S = s] \cdot \Pr_X \left[X \in \{x \in Z_N^* \ : \ f_{N,0^k s}(x^2) = y \} \right] \ = \ \Pr_S [S = s] \cdot \frac{1}{T_0}$$

Notice also that if $TAG_N(y) \neq \langle 0, \cdots, 0 \rangle$ then $\Pr_{X,S} \left[f_{N,0^k S}(X^2) = y \right] = 0$. This implies that for every y for which $\Pr_{S,X} \left[f_{N,0^k S}(X^2) = y \right] > 0$ we have

$$\Pr_{X,S} \left[S = s \mid f_{N,0^k S}(X^2) = y \right] \ = \ \Pr_S [S = s]$$

which is what we need. $\qquad\blacksquare$

Precomputing Oblivious Transfer

Donald Beaver

317 Pond Lab, Penn State University, University Park, PA 16802, (814) 863-0147;
beaver@cse.psu.edu.

Abstract. Alice and Bob are too untrusting of computer scientists to let
their privacy depend on unproven assumptions such as the existence of
one-way functions. Firm believers in Schrödinger and Heisenberg, they
might accept a quantum OT device, but IBM's prototype is not yet
portable. Instead, as part of their prenuptial agreement, they decide to
visit IBM and perform some OT's in advance, so that any later divorces,
coin-flipping or other important interactions can be done more conve-
niently, without needing expensive third parties.

Unfortunately, OT can't be done in advance in a direct way, because
even though Bob might not know what bit Alice will later send (even if
she first sends a random bit and later corrects it, for example), he would
already know which bit or bits he will receive. We address the problem of
precomputing oblivious transfer and show that OT can be precomputed
at a cost of $\Theta(k)$ prior transfers (a tight bound). In contrast, we show
that variants of OT, such as one-out-of-two OT, can be precomputed
using only one prior transfer. Finally, we show that all variants can be
reduced to a single precomputed one-out-of-two oblivious transfer.

1 Introduction

Oblivious transfer [Rab81], a process by which Alice sends Bob a bit over a noisy
channel without knowing whether it arrives, is a fundamental and ubiquitous
tool for cryptographic protocols. It comes in several varieties, but one thing
is common to all implementations: they require intensive online computation
relying on unproven intractability assumptions [Rab81, BM89, HL90, Boe91,
Bea92], or they rely on not-yet-available hardware whose reliability is based
on well-demonstrated physical facts, such as Heisenberg's Uncertainty Principle
[BC90, BBCS91].

In this paper, we reduce online oblivious transfer to precomputed oblivious
transfer. These reductions provide two immediate benefits. First, online computations such as generating large Blum integers or computing modular square
roots can be performed in advance, using machine cycles and network bandwidth
at off-peak times. (Note that we are distinctly *not* referring to *preprocessing*, such
as computing a bunch of moduli and storing them for later use, but rather to
executing the transfers already, by whatever means available, and storing the
single-bit *results* to build later transfers when the desired bits are known.) Second, unproven assumptions can be avoided entirely by visiting Billes Grassard,[1]

---

[0] Supported in part by NSF grant CCR-9210954.
[1] Names have been changed to protect the innocent.

D. Coppersmith (Ed.): Advances in Cryptology - CRYPTO '95, LNCS 963, pp. 97-109, 1995.
© Springer-Verlag Berlin Heidelberg 1995

$\leq$	Precomputed		
	OT	$\frac{1}{2}$OT	$\binom{2}{1}$OT
OT	$\Theta(k)$	1	1
$\frac{1}{2}$OT	$\Theta(k)$	1	1
$\binom{2}{1}$OT	$\Theta(k)$	1	1

Fig. 1. Tight costs of implementing one OT variant using another precomputed variant, to achieve $1 - O(2^{-k})$ "security." Cost is measured as the number of invocations of the precomputed primitive.

asking for a couple of hours on the quantum OT machine [BBCS91], and transferring a set of bits for later use.

For subtle reasons, however, it is not at all immediate that OT can be precomputed. If Alice wishes to reserve her choice of a given bit b to send until tomorrow, then for whatever bits she sends today, Bob will already know which bits have arrived. For example, if she goes ahead and obliviously transfers a random bit r today, with the intent of sending $b \oplus r$ on a direct channel tomorrow, then even though Bob can't learn b until tomorrow, he does get some unfair information: he knows *today* whether he will get the result tomorrow or not!

We show that it is indeed possible to precompute the original OT introduced by Rabin [Rab81]. To achieve a $1 - O(2^{-k})$ level of information-theoretic security, our protocols use $\Theta(k)$ prior transfers. The parameter k arises regardless of whether the precomputed OT's are implemented through cryptographic means or via a quantum channel. Our bound is tight: we also show that $\Omega(k)$ prior transfers are necessary.

We also consider the following variants:

- OT, or "standard" OT, permits Alice to send a bit b to Bob. Bob receives b with 50-50 probability and knows whether he received the bit. Alice does not know whether Bob received the bit.
- $\frac{1}{2}$OT, or "one-out-of-two" OT, permits Alice to send one of two bits, b_0 or b_1, to Bob [EGL82]. Bob receives either b_0 or b_1, with equal probability, and he knows which one he received. Alice does not know which bit Bob received.
- $\binom{2}{1}$OT, or "chosen one-out-of-two" OT, is similar to $\frac{1}{2}$OT, except that Bob chooses an index c and receives b_c. Alice does not learn c.

In contrast to OT, $\frac{1}{2}$OT and $\binom{2}{1}$OT can be precomputed at no additional cost, *i.e.* requiring only one prior transfer per future transfer. Moreover, any of the three variants can be reduced to a single precomputed $\frac{1}{2}$OT (or $\binom{2}{1}$OT), with no chance of error.

Figure 1 illustrates our results, demonstrating the costs of implementing any of the three standard variants using a primitive that is only temporarily available. Clearly, the optimal number of invocations made by an *online* reduction can be no larger than that by a precomputed reduction (else the precomputation would be done online!). Except for the precomputed self-reduction for OT, the costs of precomputation are no more than those of *online* reductions (*cf.*

[BCR86a, BCR86b, Cré87]; as part of our analysis, we also introduce simple online reductions between ($\binom{?}{1}$)OT and $\frac{1}{2}$OT).

These results shed some light on the nature of the three variants. Intuitively, OT requires us to maintain a degree of flexibility so that the fate of a future bit – *ie.* whether it will arrive or not – remains undetermined. Unfortunately, the fact of whether a bit arrived or not cannot be reversed later on. In contrast, the as-yet undetermined choices made in $\frac{1}{2}$OT and ($\binom{?}{1}$)OT can be arbitrarily made and later reversed.

The price of inflexibility in the OT case is a need for additional prior transfers. Given that the arrival of a bit is an *irrevocable* event, it is remarkable that such flexibility can be accommodated at all. We show that these additional transfers are, in fact, necessary.

2 Preliminaries and Definitions

We use "$c \leftarrow \$$" to denote assigning c the value 0 or 1 with equal probability. Local variables are sometimes referred to in a (C++)-like notation to distinguish them from similarly-named local variables held by another agent: for example, $P : x$ denotes P's local variable x. The notation "$P \rightarrow Q : x$" denotes the event that P sends $P{:}x$ to Q. The notation "$Q \leftarrow P : y$" denotes the event that Q receives a message from P and stores it as $Q{:}y$. (We sometimes omit explicit mention of message reception.)

For clarity, we ignore blatantly detectable misbehavior (*e.g.* sending syntactically incorrect messages) by assuming that an honest party rewards it by refusing further interaction with the offender. Including such non-instructive cases makes for unreadable code.

The reader well-versed in OT may skip to §2.3 on first reading.

2.1 The Variants of Oblivious Transfer

We describe oblivious transfer formally in terms of a three-party protocol involving Alice, Bob, and an absolutely trusted third party, communicating over synchronized and absolutely secure channels. Essentially, the third party represents the desired primitive, and the protocol describes how the primitive is used. Let $\hat{A}$ be Alice and $\hat{B}$ be Bob; then our specification of Rabin's original concept would be the following protocol for for $\hat{A}$, $\hat{B}$, and third party OT:

Oblivious Transfer		
$\hat{A}$:	input b	
$\hat{A} \rightarrow$ OT:	b	
OT:	$?b \leftarrow \$$	
OT $\rightarrow \hat{B}$:	$(?b, ?b \cdot b)$	//either $(0,0)$ or $(1, b)$

The value of $?b$ records whether Bob receives a valid bit or not. The other standard variations on OT involve the special third parties $\frac{1}{2}$OT and ($\binom{?}{1}$)OT:

One-Out-Of-Two Oblivious Transfer	
$\hat{A}$:	input b_0, b_1
$\hat{A} \rightarrow \frac{1}{2}\text{OT}$:	(b_0, b_1)
$\frac{1}{2}\text{OT}$:	$c \leftarrow \$$
$\frac{1}{2}\text{OT} \rightarrow \hat{B}$:	b_c

Chosen One-Out-Of-Two Oblivious Transfer	
$\hat{A}$:	input b_0, b_1
$\hat{B}$:	input c
$\hat{A} \rightarrow \binom{2}{1}\text{OT}$:	(b_0, b_1)
$\hat{B} \rightarrow \binom{2}{1}\text{OT}$:	c
$\binom{2}{1}\text{OT} \rightarrow \hat{B}$:	(c, b_c)

2.2 Secure Implementations

A full formal framework with details of the tedious proofs included in our submitted version is omitted in this presentation – undoubtedly to the relief of many. Details will appear again in the final version. We sketch some of the main aspects, but the reader is invited to skip to §2.3.

In particular, we would like to map an *implementation*, namely a three-party protocol, $\langle A(x_a, k), B(x_b, k), P(k) \rangle$, to a *specification*, namely another three-party protocol, $\langle \hat{A}(x_a, k), \hat{B}(x_b, k), \hat{P}(k) \rangle$, and show that the implementation achieves results asymptotically indistinguishable from the specification (as $k \rightarrow \infty$). Here, x_a and x_b are the inputs of Alice and Bob, and P suggests a *primitive*, namely a trusted third party such as OT, $\frac{1}{2}$OT, or $\binom{2}{1}$OT.

The implementation should model the results of the specification even when attacked by an adversary, adv. In particular, attacks on the implementation should be mapped to "equivalent" attacks on the specification. We provide such a mapping by way of an *interface* that envelops adv, converting adv's attacks to equivalent actions in the specification protocol, and converting the information gained in the specification protocol to appear as though it came from the implementation, for adv's sake.

Let k be a security parameter. When adv attacks implementation $\langle A(x_a, k), B(x_b, k), P(k) \rangle$, let $[\text{adv}, A, B, P]$ denote the distribution on the inputs/outputs $((y_{\text{adv}}, k), (x_A, y_A, k), (x_B, y_B, k))$ of adv, A and B, respectively.

Definition 1. Protocol $\langle A, B, P \rangle$ is a $(1 - O(2^{-k}))$-**secure implementation** of protocol $\langle \hat{A}, \hat{B}, \hat{P} \rangle$ if there exists an interface $\mathcal{I}$ such that for all adversaries adv, for all x_{adv}, x_a and x_b,

$$[\text{adv}, A, B, P] \approx^{O(2^{-k})} [\mathcal{I}(\text{adv}), \hat{A}, \hat{B}, \hat{P}].$$

We consider *adaptive adversaries*, namely adversaries who can corrupt any number of players (except trusted primitives) at any point during the protocol.

The details of the interfaces will not appear here (the various sequences of events it must handle are numerous, particularly when handling adaptive adversaries), but we remark on two important aspects.

First, when the specification can be shown information-theoretically private (perfectly or statistically), the job of the interface reduces to supplying "faked" samples to adv, using a distribution that is independent of the inputs to which the interface has no access (namely, those held by uncorrupted players). For the protocols we have designed, this simulation is easy to do in polynomial time, typically consisting of mere coin flips or exclusive-or's with earlier messages.

Second, for the protocols presented here, the interface must convert messages from adv (who plays the part of a corrupt A or B) into "equivalent" messages sent by a corrupt $\hat{A}$ or $\hat{B}$ to the trusted third party, $\hat{P}$. The interface does so without looking "inside" the adversary, but rather by performing simple computations on the messages adv sends. (Subtly but importantly, this demonstrates that adv "knows" what effective messages it sends [Bea92].) We thus avoid any need to derive an "intent" or a "strategy" of the adversary in order to achieve the same effects when attacking the secure specification protocol.

2.3 Precomputed Reductions

Let $\langle \hat{A}, \hat{B}, \hat{P} \rangle$ be a specification in which $\hat{A}$ and $\hat{B}$ invoke a primitive, $\hat{P}$, precisely once, and then output their results. The three OT protocols listed in §2.1 are such protocols.

A precomputed reduction from one primitive, $\hat{P}$, to another primitive, P, is a two-stage protocol $\langle A, B, P \rangle$ that implements $\langle \hat{A}, \hat{B}, \hat{P} \rangle$, $O(2^{-k})$-securely. In the first, or *precomputation* stage, A and B can invoke P freely. At the start of the second, or *online* stage, A and B receive their respective inputs x_a and x_b, but no further communication with P is possible. If there is a precomputed reduction from $\hat{P}$ to P, we write $\hat{P} \overset{\text{pre}}{\leq} P$.

An **online** reduction has no precomputation stage. If there is an online reduction from $\hat{P}$ to P, we write $\hat{P} \overset{\text{onl}}{\leq} P$.

The **complexity** of an implementation is the worst-case number of times the protocol invokes P.

3 Precomputed $\frac{1}{2}$OT or $\binom{2}{1}$OT

We first describe implementations of all three variants using a single precomputed $\frac{1}{2}$OT (or a single precomputed $\binom{2}{1}$OT).

3.1 Precomputed Reduction: $\frac{1}{2}$OT to $\frac{1}{2}$OT

Figure 2 presents the simplest and cheapest precomputed reduction, $\frac{1}{2}$OT $\overset{\text{pre}}{\leq}$ $\frac{1}{2}$OT, which raises some subtle issues despite its simplicity. Roughly speaking, Alice initially transfers two random bits, r_0 and r_1. Bob receives one of them,

$$\tfrac{1}{2}OT \overset{pre}{\leq} \tfrac{1}{2}OT$$

Precomputation:		Online Stage:	
A:	$r_0 \leftarrow \$,\ r_1 \leftarrow \$$	A:	input $b_0,\ b_1$
$A \to \tfrac{1}{2}OT$:	(r_0, r_1)		$c \leftarrow \$$
$\tfrac{1}{2}OT$:	$d \leftarrow \$$	$A \to B$:	$(c,\ b_0 \oplus r_c,\ b_1 \oplus r_{\bar{c}})$
$\tfrac{1}{2}OT \to B$:	(d, r_d)	B:	receive (c, x_0, x_1)
			output $(d \oplus c,\ r_d \oplus x_{d \oplus c})$

Fig. 2. Precomputed self-reduction for $\tfrac{1}{2}OT$.

r_d, and knows which one it is. Later, when Alice has decided on b_0 and b_1, she sends the corrections $(b_0 \oplus r_0, b_1 \oplus r_1)$ to Bob. More accurately, Alice flips a coin, c, and may instead send the corrections $(b_0 \oplus r_1, b_1 \oplus r_0)$ to Bob. Bob thus gets $(b_0 \oplus r_c, b_1 \oplus r_{\bar{c}})$, and, knowing r_d, he can calculate $b_{d \oplus c}$. The other bit, $b_{\overline{d \oplus c}}$, remains masked by the unknown value $r_{\bar{d}}$.

The inclusion of c in this reduction is a very subtle point. If omitted, which one may be apt to do without careful consideration, then Bob learns in advance which bit he will receive – b_0 or b_1 – even though he does not learn the bit's value until the online stage.

This leak seems innocuous at first. But say that Bob the gambler is awaiting the answer b_0 to the question, "Will the roulette wheel in Las Vegas give black or red at 12:00 noon?" or the answer b_1 to the question, "Will the roulette wheel in Monte Carlo give black or red at 12:00 noon?" He has an inside contact in each casino who will let him know the local answer at 11:55, but one of those insiders will surely end up nabbed by the police (or other unsundry enforcers). At best, Bob can wait either in Las Vegas or in Monte Carlo and hope that the insider at his chosen location can give him the answer. If Bob doesn't know which insider will survive, he has at best a 50-50 chance of getting the prediction. But if Bob does find out which insider will survive, he simply waits in the right casino and is guaranteed an easy profit.

Indeed, without c, this reduction bears some similarity to the "non-interactive" OT implemented in [BM89, HL90]. (Such methods employ complexity-theoretic assumptions to transfer sequences of bits, and seek weaker[2] goals for the purpose of supporting particular tasks such as non-interactive proof systems.) If such a coin flip had been included, the non-interactive zero-knowledge proof system of [BM89] would have been secure against the very subtle attack proposed in [CHL94]. In contrast, the protocol described here is provably secure:

Theorem 2. $\tfrac{1}{2}OT \overset{pre}{\leq} \tfrac{1}{2}OT$ with complexity 1. That is, $\tfrac{1}{2}OT$ can be implemented using 1 prior invocation of $\tfrac{1}{2}OT$.

---

[2] Bob's reception of each successive bit is not independent of his reception of earlier bits. Alice may derive correlations in reception patterns, although she does not learn the precise patterns themselves.

Proof Sketch. Our goal is to convert attacks on the implementation (fig. 2) into equally-powered attacks on the specification (§2.1).

In the precomputation stage, Alice receives nothing. If adv corrupts her, we (the interface/simulator $\mathcal{I}$) simply keep track of what adv generates on her behalf. Bob receives the pair (d, r_d), which we need to generate for him only if adv has corrupted him. If so, we generate a uniformly random (d, r_d).

In the online stage, if adv chooses to corrupt Alice, we choose to corrupt $\hat{A}$ (the Alice in the specification protocol), thereby obtaining inputs b_0 and b_1. We generate a random r_0 and r_1, then hand r_0, r_1, b_0, b_1 to adv as the "faked" view of Alice. (If Alice was already corrupted in the precomputation stage, then we just hold onto the (r_0, r_1) she gave us earlier, hand her b_0, b_1, and continue as described below.)

Then, adv generates (c, x_0, x_1) on faulty Alice's behalf. We calculate Alice's *effective* bits as $\beta_0 = x_0 \oplus r_c$, $\beta_1 = x_1 \oplus r_{\bar{c}}$. (These bits may or may not be b_0, b_1. If adv is merely curious but honest, they indeed will be b_0, b_1. But whatever clever or mysterious methods adv uses to come up with (c, x_0, x_1), we can easily determine what the message really "means," *in terms of messages carrying equivalent information and influence in the specification.*) Having corrupted $\hat{A}$ in the specification protocol (see §2.1), we send (β_0, β_1) to $\frac{1}{2}$OT on behalf of $\hat{A}$. As a result, $\hat{B}$ will output either β_0 or β_1, which is exactly what Bob will do in the implementation. (Technically, let $\hat{\gamma}$ be $\frac{1}{2}$OT:c in the specification and let γ be adv:$c \oplus \frac{1}{2}$OT:d in the implementation. Then strictly speaking, $\hat{B}$ outputs $(\hat{\gamma}, \beta_{\hat{\gamma}})$ while Bob outputs (γ, β_γ), but in either scenario, $\hat{\gamma}$ or γ is uniformly random and fully independent of $\hat{A}$/Alice's view.)

The case in which adv corrupts Bob is simple, because Bob's role is essentially passive. All we need do is to corrupt $\hat{B}$ and discover the value (D, b_D) that the online $\frac{1}{2}$OT sends on to $\hat{B}$ in the specification. We don't need to discover the value $b_{\bar{D}}$, a fact which incidentally demonstrates that the implementation does not leak the value of the unreceived bit. Using the faked, uniformly random values (d, r_d), we set $c = d \oplus D$ (to make $D = d \oplus c$ be the index of the received bit), set $x_D = b_D$, and choose $x_{\bar{D}}$ at random. Finally, we hand Bob the triple (c, x_0, x_1). A curious but law-abiding adversary might then calculate Bob's output as $b_D = x_D$. In any case, a corrupt Bob receives precisely the same information in the implementation as in the specification.

This outline of an interface-based mapping can be filled in to show formally that any adversarial attack on the implementation can be equated with an attack on the specification, with perfectly identical distributions (on inputs/outputs of honest players and views of adversaries). □

3.2 Precomputed Reduction: $\binom{2}{1}$OT to $\frac{1}{2}$OT

Reducing $\binom{2}{1}$OT to precomputed $\frac{1}{2}$OT is slightly more complicated, particularly since is it somewhat surprising that a chosen reception can be obtained using a random reception.

$$\binom{2}{1}\text{OT} \overset{\text{pre}}{\le} \tfrac{1}{2}\text{OT}$$

Precomputation:		Online Stage:	
A:	$r_0 \leftarrow \$,\ r_1 \leftarrow \$$	A:	input b_0, b_1
$A \to \tfrac{1}{2}\text{OT}$:	(r_0, r_1)	B:	input c
$\tfrac{1}{2}\text{OT}$:	$d \leftarrow \$$	$B \to A$:	$e = c \oplus d$
$\tfrac{1}{2}\text{OT} \to B$:	(d, r_d)	$A \to B$:	$(b_0 \oplus r_e, b_1 \oplus r_{\bar{e}})$
		$B \leftarrow A$:	(x_0, x_1)
		B:	output $x_c \oplus r_d$

Fig. 3. Precomputed reduction from $\binom{2}{1}\text{OT}$ to $\tfrac{1}{2}\text{OT}$.

Fig. 3 describes the implementation. Again, Alice initially transfers two random bits r_0 and r_1 via the initial $\tfrac{1}{2}\text{OT}$. Bob receives one of them, namely r_d, and he knows which one he received.

Later, after Alice has chosen her desired bits b_0 and b_1, she sends corrections like those in the $\tfrac{1}{2}\text{OT} \overset{\text{pre}}{\le} \tfrac{1}{2}\text{OT}$ protocol. This time, Bob selects which sort of correction she should send, for otherwise he would have no control over which b_i he received. In particular, if Bob wants to discover b_c, he sends Alice $e = c \oplus d$. She provides him with $(b_0 \oplus r_e, b_1 \oplus r_{\bar{e}})$, and using r_d, Bob can decode $b_{e \oplus d}$, namely b_c. The other bit, $b_{\bar{c}}$, remains masked by the unknown bit $r_{\bar{d}}$.

The question remains whether Alice learns anything from the message e sent by Bob. From Alice's point of view, if Bob requests $e = 0$, there is an equal chance that $d = 0$ and Bob wants b_0 or that $d = 1$ and Bob wants b_1. Likewise, if Bob requests $e = 1$, there is an equal chance that $d = 0$ and Bob wants b_1 or that $d = 1$ and Bob wants b_0. The secret random bit d provided by the $\tfrac{1}{2}\text{OT}$ primitive is enough to protect Bob's privacy.

Theorem 3. $\binom{2}{1}OT \overset{\text{pre}}{\le} \tfrac{1}{2}OT$ *with complexity* 1. *That is,* $\binom{2}{1}OT$ *can be implemented using* 1 *prior invocation of* $\tfrac{1}{2}OT$.

Proof. Similar to the proof of Theorem 2. As before, a key ingredient is the extraction of Alice's *effective* bits by computing $(x_0 \oplus r_e, x_1 \oplus r_{\bar{e}})$ (which are certainly (b_0, b_1) if Alice is honest) and handing these to $\binom{2}{1}\text{OT}$ (the primitive available in the specification protocol), so that $\hat{B}$ sees the same distribution on chosen bits as adv induces Bob to calculate in the implementation. Another key ingredient consists of randomly faking (r_0, r_1) and e for B, and later compensating for this by faking the later message (x_0, x_1) with an appropriate distribution, even though we have access to only *one* of Alice's bits (by way of $\binom{2}{1}\text{OT}$). □

3.3 Other Precomputed Reductions

To facilitate the discussion of other precomputations, we first consider a few online reductions. While online reductions from OT to $\tfrac{1}{2}\text{OT}$ or to $\binom{2}{1}\text{OT}$ are trivial, an online reduction from $\tfrac{1}{2}\text{OT}$ to $\binom{2}{1}\text{OT}$ requires care, despite its simplicity.

Lemma 4. $\frac{1}{2}OT \overset{onl}{\leq} \binom{2}{1}OT$, *with complexity 1. That is, there is an online reduction from $\frac{1}{2}OT$ to $\binom{2}{1}OT$ that invokes $\binom{2}{1}OT$ precisely once.*

Proof Sketch. First, Alice simply switches b_0 and b_1 randomly (by transferring $(b_d, b_{\bar{d}})$ for $d \leftarrow \$$), before allowing Bob to choose. Second, Bob makes a *random* choice $c \leftarrow \$$, thereby obtaining $b_{d \oplus c}$.

The subtle aspect is in requiring Bob to make a random choice. Clearly, there is an unsubtle privacy issue: Alice must be prevented from learning which bit Bob gets. More subtly, however, there is a correctness issue: Alice must not be able to cause Bob to receive a deterministic selection of bits. The random coin toss of the $\frac{1}{2}OT$ primitive is effectively the exclusive-or of Alice's reversal choice with Bob's choice, and if Bob's choice is not random, the implementation will not correspond to the specification. In order for honest Bob to get a random selection – which could conceivably be crucial to some higher-level protocol[3] – he must randomize the choice himself, to protect against Alice's failing to do so.

(As a side remark, we note that this aspect becomes obvious when one tries to find an interface-based comparison between implementation and specification. In particular, the distribution obtained by honest $\hat{B}$ in the specification always contains the uniformly random c chosen by $\binom{2}{1}OT$. If Bob's random choice in the implementation is overlooked, then c is whatever Alice sent him, and the implementation's results are easily distinguished from the specification's results.) $\square$

Online reductions facilitate the development of precomputed reductions:

Lemma 5. *Let P, Q and R be (not necessarily distinct) primitives from $\{OT, \frac{1}{2}OT, \binom{2}{1}OT\}$. Then:*

(A) If $P \overset{onl}{\leq} Q$ and $Q \overset{pre}{\leq} R$ then $P \overset{pre}{\leq} R$.
(B) If $P \overset{pre}{\leq} Q$ and $Q \overset{onl}{\leq} R$ then $P \overset{pre}{\leq} R$.

Proof. (A) follows from incorporating the reduction $P \overset{onl}{\leq} Q$ into the online stage of the precomputed reduction demonstrating $Q \overset{pre}{\leq} R$. (B) follows from incorporating the reduction $Q \overset{onl}{\leq} R$ into the precomputation stage of the precomputed reduction demonstrating $P \overset{pre}{\leq} Q$. $\square$

We can now start to fill in the columns of Fig. 1.

Lemma 6. $OT \overset{pre}{\leq} \frac{1}{2}OT$, $\frac{1}{2}OT \overset{pre}{\leq} \frac{1}{2}OT$, *and* $\binom{2}{1}OT \overset{pre}{\leq} \frac{1}{2}OT$, *each with complexity 1.*

---

[3] Imagine a protocol in which Bob makes use of the c's he obtains from $\frac{1}{2}OT$ to run a primality test, for example. While this may be odd, stupid, and contrived, it is a mathematically acceptable use of the results of the $\frac{1}{2}OT$ primitive.

Proof. The first two cases follow from Theorems 2 and 3. Applying lemma 5 to Theorem 2 and the trivial online reduction $OT \overset{onl}{\leq} \frac{1}{2}OT$ gives the third. Each reduction (whether online or precomputed) requires only one invocation of its respective primitive. □

Lemma 7. $OT \overset{pre}{\leq} \binom{2}{1}OT$, $\binom{2}{1}OT \overset{pre}{\leq} \binom{2}{1}OT$, and $\frac{1}{2}OT \overset{pre}{\leq} \binom{2}{1}OT$, each with complexity 1.

Proof. Using lemmas 6, 4, and 5,

$$OT \overset{pre}{\leq} \tfrac{1}{2}OT \overset{onl}{\leq} \binom{2}{1}OT \Rightarrow \quad OT \overset{pre}{\leq} \binom{2}{1}OT$$
$$\tfrac{1}{2}OT \overset{pre}{\leq} \tfrac{1}{2}OT \overset{onl}{\leq} \binom{2}{1}OT \Rightarrow \quad \tfrac{1}{2}OT \overset{pre}{\leq} \binom{2}{1}OT$$
$$\binom{2}{1}OT \overset{pre}{\leq} \tfrac{1}{2}OT \overset{onl}{\leq} \binom{2}{1}OT \Rightarrow \binom{2}{1}OT \overset{pre}{\leq} \binom{2}{1}OT$$

□

4 Precomputed OT

We now turn to the precomputed reduction from OT to itself. The precomputed reductions from $\frac{1}{2}OT$ and $\binom{2}{1}OT$ to OT are similar in content and cost to this preprocessed self-reduction.

Figure 4 shows a preprocessed reduction from the standard OT to itself, using $15k = \Theta(k)$ invocations of OT in the preprocessing stage. The protocol is adapted from Crépeau's online reduction [Cré87] from $\frac{1}{2}OT$ to OT. (The necessity of using $\Omega(k)$ OT's is argued below; see Theorem 9.) In these $15k$ transfers, Bob will receive between $5k + 1$ and $10k - 1$ bits, with probability exceeding $1 - O(2^{-k})$. He can thus find two disjoint $5k$-sets U_0 and U_1 containing indices of received and unreceived bits, respectively. He can't find two *disjoint* sets containing only *received* bits, however. Alice uses these sets of indices to construct two masks, v_0 and v_1, by taking the exclusive-or over all r_i for a given set. Having received everything indexed by U_0, Bob can calculate either v_0 or v_1, but not both. In the online stage, Alice sends b, masked with either v_0 or v_1, randomly, giving Bob a 50-50 chance to calculate b. Since Alice does not know which of the two sets U_0 and U_1 is the set that Bob received, she does not know whether Bob receives b.

Theorem 8. $OT \overset{pre}{\leq} OT$ with complexity $\Theta(k)$. That is, OT can be implemented using $\Theta(k)$ prior invocations of OT.

Proof Sketch. The proof is similar to that of Theorem 2, but complicated by the (negligible) possibility of error. Conditioned on the event that the trusted OT party transmits between $5k + 1$ and $10k - 1$ bits to Bob, the interface-mapped attacks on the specification (§2.1) achieve results identical to those obtained by attacks on the implementation (Fig. 4). The chance that this event

$$OT \overset{pre}{\leq} OT$$

Precomputation:			$A \leftarrow B$:	(V_0, V_1)
A:	$r_i \leftarrow \$$	$(i = 1..15k)$	A:	reject if $V_0 \cap V_1 \neq \emptyset$
$A \rightarrow OT_i$:	r_i	$(i = 1..15k)$		$v_0 = \oplus_{i \in V_0} r_i$
$OT_i \rightarrow B$:	$(?r_i, r_i)$	$(i = 1..15k)$		$v_1 = \oplus_{i \in V_1} r_i$
B:	$R = \{i\| ?r_i = 1\}$		Online Stage:	
	Choose random $U_0 \subseteq R$, $\|U_0\| = 5k$		A:	input b
	Choose random $U_1 \subseteq \overline{R}$, $\|U_1\| = 5k$			$d \leftarrow \$$
	$u = \oplus_{i \in U_0} r_i$		$A \rightarrow B$:	$(d, b \oplus v_d)$
	$f \leftarrow \$$		B:	receive (d, x)
$B \rightarrow A$:	$(U_f, U_{\overline{f}})$			if $d = f$ then
				output $(1, u \oplus x)$
				else output $(0, 0)$

Fig. 4. Precomputed self-reduction for OT.

fails to occur is at most $10k[15k!/(10k!)(5k!)]2^{-15k}$. Using Stirling's formula, $15k!/(10k!)(5k!) \leq e^{9.55k} \leq 2^{13.8k}$. Thus the chance that the event fails is bounded by $10k \cdot 2^{13.8k} 2^{-15k} = O(2^{-k})$. Removing the condition that this event occurs, we see that the results are $O(2^{-k})$-indistinguishable. □

Theorem 9. *Any precomputed reduction from OT to OT that is $(1 - O(2^{-k}))$-secure against static 1-adversaries (even honest-but-curious ones) must use $\Omega(k)$ prior transfers.*

Proof. Assume there exists a $(1 - 2^{-k})$-secure implementation that does not use $\Omega(k)$ transfers. Then for infinitely many k, at most $k/10$ transfers are used. We create a static 1-adversary **adv** that corrupts Bob at the outset, but **adv** directs Bob to follow his program precisely. With probability at least $2^{-k/10}$, Bob receives *all* bits in the precomputation stage, in which case **adv** predicts with virtually 100% accuracy that it will indeed discover the b sent later on. In the specification protocol for OT, **adv** has no such ability to predict whether Alice's actual bit will arrive, and the $2^{-k/10}$-correlation of **adv**'s prediction with the reception of Alice's bit is enough to show that the protocol is not 2^{-k}-secure.

In slightly more detail, consider random variables describing Alice's internal random choices (α), Bob's (β), the OT reception pattern $(R \in \{0,1\}^{k/10}$, with a "1" indicating reception), and the conversation (C) in the precomputation stage. Call C *fair* if $\Pr[\text{Bob outputs } (1, b) \text{ later} | C] > .49$ and $\Pr[\text{Bob outputs } (0, 0) \text{ later} | C] > .49$. Then $\Pr[C \text{ unfair}] \leq 2^{-k}$, else the protocol is not $(1 - 2^{-k})$-secure (since even an honest Alice would learn the fate of her bit). Furthermore, $\Pr[C \text{ unfair} | R = 111 \cdots 1] \leq 2^{-k/2}$, else $\Pr[C \text{ unfair}] > 2^{-k/10} 2^{-k/2} > 2^{-k}$.

Now, **adv**'s only unusual behavior is to make a prediction when it sees $R = 111 \cdots 1$ (*ie.* Bob received all OT's). Even if Bob outputs $(0, 0)$ (Bob is law-abiding, after all), with high probability C is fair and **adv** can inspect Bob's records to determine a different β, consistent with C, that would have caused Bob to output $(1, b)$. (Naturally, the tiny error rate does admit the possibility of β's causing Bob to output $(1, \bar{b})$. If C is fair, the probability of this is at most .02.)

Thus, not only does **adv** obtain Alice's bit, but it is able to predict in advance that it will obtain the bit. In particular, in the assumed implementation, **adv**'s advantage is at least the following (fair and correct, minus fair and incorrect, minus unfair):

$$2^{-k/10}(1 - 2^{-k/2})(.98) - 2^{-k/10}(1 - 2^{-k/2})(.02) - 2^{-k/10}(2^{-k/2}).$$

This exceeds the best-possible 0 advantage in the specification protocol by far more than 2^{-k}, thus the implementation is not $(1 - 2^{-k})$-secure. It is a simple matter to extend $1 - 2^{-k}$ to $1 - O(2^{-k})$. □

The proof generalizes to show that $\omega(\log k)$ transfers are necessary to achieve $(1 - k^{-\omega(1)})$ security (*i.e.* "statistical" security), and in particular, that using $O(1)$ transfers admits a constant probability of error. It also implies lower bounds for *online* reductions from $\frac{1}{2}$OT or $\binom{2}{1}$OT to OT, showing that Crépeau's methods are asymptotically optimal [Cré87].

5 Conclusions

Putting together Lemmas 6 and 7 and Theorems 8 and 9, we conclude:

Theorem 10. *All variants of oblivious transfer can be reduced to precomputed invocation(s) of any variant, with tight complexity bounds as given in the table in Figure 1.*

In particular, one-out-of-two oblivious transfer can be performed in advance and used to implement any variant of OT later on, at a cost of one prior transfer per future transfer. Chosen one-out-of-two OT also suffices. Unfortunately, Rabin's original OT cannot be used in this manner, as it requires $\Omega(k)$ prior tranfers per future transfer.

These reductions are information-theoretically secure and permit prior execution of oblivious transfers, whether the execution involves quantum channels, Blum integers, or any other construct.

It is not hard to see that the reductions described in §3 and §4 apply to string transfers, letting variables b, b_0 and b_1 represent strings and interpreting the $\oplus$ symbol as bitwise exclusive-or, as appropriate.

Moreover, the discrete-log based "non-interactive" oblivious transfer channels of [BM89, HL90, Har91] can now be used correctly to implement *any* variant of OT, at a cost of 1 bit per transfer. The proof of this claim requires significantly greater formalism to address computational security and is beyond the scope of this work.

References

[Bea92]   D. Beaver. "How to Break a 'Secure' Oblivious Transfer Protocol." *Advances in Cryptology – Eurocrypt '92 Proceedings,* Springer–Verlag LNCS 658, 1993, 285–296.

[BM89]   M. Bellare, S. Micali. "Non-Interactive Oblivious Transfer and Applications." *Advances in Cryptology – Crypto '89 Proceedings,* Springer–Verlag LNCS 435, 1990, 547–557.

[BBCS91] C. Bennett, G. Brassard, C. Crépeau, M. Skubiszewska. "Practical Quantum Oblivious Transfer." *Advances in Cryptology – Crypto '91 Proceedings,* Springer–Verlag LNCS 576, 1992, 351–366.

[Boe91]   B. den Boer. "Oblivious Transfer Protecting Secrecy." *Advances in Cryptology – Eurocrypt '91 Proceedings,* Springer–Verlag LNCS 547, 1991, 31–45.

[BC90]   G. Brassard, C. Crépeau. "Quantum Bit Commitment and Coin Tossing Protocols." *Advances in Cryptology – Crypto '90 Proceedings,* Springer–Verlag LNCS 537, 1991, 49–61.

[BCR86a] G. Brassard, C. Crépeau, J. Robert. "All or Nothing Disclosure of Secrets." *Advances in Cryptology – Crypto '86 Proceedings,* Springer–Verlag LNCS 263, 1987.

[BCR86b] G. Brassard, C. Crépeau, J. Robert. "Information Theoretic Reductions among Disclosure Problems." *Proceedings of the 27th FOCS,* IEEE, 1986, 168–173.

[CHL94]   Y-H. Chen, T. Hwang, C-M. Li. "On the Zero Knowledge Proof Systems Based on One-out-of-two Non-Interactive Oblivious Transfers." Manuscript, 1994.

[Cré87]   C. Crépeau. "Equivalence Between Two Flavours of Oblivious Transfers." *Advances in Cryptology – Crypto '87 Proceedings,* Springer–Verlag LNCS 293, 1988, 350–354.

[CK88]   C. Crépeau, J. Kilian. "Achieving Oblivious Transfer Using Weakened Security Assumptions." *Proceedings of the 29th FOCS,* IEEE, 1988, 42–52.

[EGL82]   S. Even, O. Goldreich, A. Lempel. "A Randomized Protocol for Signing Contracts." *Proceedings of Crypto 1982,* Springer–Verlag, 1983, 205–210.

[HL90]   L. Harn, H. Lin. "Noninteractive Oblivious Transfer." *Electronics Letters* 26:10 (May 1990), 635–636.

[Har91]   L. Harn. "An Oblivious Transfer Protocol and Its Application for the Exchange of Secrets." *Advances in Cryptology – Asiacrypt '91 Proceedings,* Springer-Verlag LNCS 739, 1993, 312–320.

[Kil88]   J. Kilian. "Founding Cryptography on Oblivious Transfer." *Proceedings of the 20th STOC,* ACM, 1988, 20–29.

[Rab81]   M. Rabin. "How to Exchange Secrets by Oblivious Transfer." TR-81, Harvard, 1981.

Committed Oblivious Transfer
and
Private Multi-Party Computation

Claude Crépeau   Jeroen van de Graaf   Alain Tapp

Université de Montréal
Département d'informatique et de Recherche Opérationnelle
email:{crepeau, jeroen, tappa}@iro.umontreal.ca

Abstract. In this paper we present an *efficient* protocol for "Committed Oblivious Transfer" to perform oblivious transfer on committed bits: suppose *Alice* is committed to bits a_0 and a_1 and *Bob* is committed to b, they both want *Bob* to learn and commit to a_b without *Alice* learning b nor *Bob* learning $a_{\bar{b}}$. Our protocol, based on the properties of error correcting codes, uses Bit Commitment (BC) and one-out-of-two Oblivious Transfer (OT) as black boxes. Consequently the protocol may be implemented with or without a computational assumption, depending on the kind of BC and OT used by the participants. Assuming a Broadcast Channel is also available, we exploit this result to obtain a protocol for Private Multi-Party Computation, without making assumptions about a specific number or fraction of participants being honest. We analyze the protocol's efficiency in terms of BCs and OTs performed. Our approach connects Zero Knowledge proofs on BCs, Oblivious Circuit Evaluation and Private Multi-Party Computations in a conceptually *simple* and *efficient* way.

1 Introduction

Committed Oblivious Transfer (COT) is the natural fusion of one-out-of-two Oblivious Transfer (OT) [13] and Bit Commitment (BC). At the start of the protocol *Alice* is committed to bits $\boxed{a_0}$, $\boxed{a_1}$ and *Bob* to bit $\boxed{b}$. At the end *Bob* is committed to $\boxed{a_b}$ and knows nothing about $a_{\bar{b}}$. *Alice* learns nothing about b.

The current paper presents an *efficient* COT protocol. This protocol makes no assumption on the type of BCs and OTs that are used. For instance, with OT and BC based on the Quantum Channel [2, 3] we can perform COT without any computational assumption. Our protocol uses some elements of coding theory and simple Zero-Knowledge sub-protocols. It uses $O(n)$ OTs and $O(n^2)$ BCs, where n denotes the security parameter, but uses only $O(n)$ BCs if they have a special XOR-property. The global running time is $O(n^2)$ in the first case and $O(n)$ in the second (excluding the time necessary to build the code).

COT is a very powerful tool that can be used to perform general cryptographic tasks such as Oblivious Circuit Evaluation (OCE)[19] or Mental Games [16, 17], and Distributed Computation [18]. Such tasks have been achieved before [19]

D. Coppersmith (Ed.): Advances in Cryptology - CRYPTO '95, LNCS 963, pp. 110-123, 1995.

regardless of COT from generic BC and OT, but unfortunately the solution was not only complicated but very inefficient. At EUROCRYPT '89, Crépeau [8] introduced COT under the label "Verifiable Oblivious Transfer" and used it in a simpler protocol for OCE based on the work of Goldreich and Vanish [17]. Unfortunately this protocol for COT used $\Omega(n^3)$ OTs, which is still rather inefficient.

An apparently more efficient protocol for COT (using only $O(n)$ OTs) was presented in [18] under the label "Preprocess-Oblivious-Transfer". Unfortunately, it is very easy to misbehave in that protocol in a way that allows Bob to get both bits with non-negligible probability and for Alice to learn b with a non-negligible probability. These problems can be fixed easily by straightforward techniques, but then the resulting protocol becomes more or less equivalent to that of [8] using $\Omega(n^3)$ OTs. The best that we were able to get from protocols in the spirit of [8] and [18] is a protocol using $\Omega(n^2)$ OTs.

In the second part of this paper we use the COT protocol to obtain an efficient protocol for Private Multi-Party Computation (PMPC). The problem of computing a function when the participants want to keep their input private has been considered by many researchers under different models and various names (for an excellent overview see [15]). Distinguishing features of the several models are: the number of participants ($P = 2$ vs $P \geq 3$), the presence or absence of some unproven intractability assumption, the communication model (private channels between each pair of participants, broadcast channels, oblivious transfers or a combination of these) and the capacity of dealing with malicious participants.

We assume that a Broadcast Channel and an Oblivious Transfer Channel are available between each pair of participants. A protocol is a *correct* Multi-Party Computation if the output of the protocol to each participant is the same as that of the function it is emulating. Furthermore we say that such a protocol is *fair* if the the fact that one participant learns the output implies that every participant does. It is *honest* if an honest participant knows when he does not learn the output of the function and it is *private* if no coalition of less than P participants can learn information about the private inputs of other participants, other than what the output of the function logically implies.

Our result can now be stated as follows. Suppose that P participants wish to evaluate a boolean circuit F consisting of m boolean gates. Then there exists a *correct*, *private*, *honest* and *fair* protocol to evaluate F using $O(P^2 n^2 m)$ BC s and $O(P^2 nm)$ OT s. If instead of having BC and OT as basic primitives we only have the simpler Oblivious Transfer of Rabin [26], then the number of such primitives required is $O(P^2 n^3 m)$.

It is rather unfortunate that a single dishonest participant can make the PMPC protocol abort. Nevertheless, this is unavoidable, since to guarantee that no dishonest coalition can have any advantage over an honest participant, the protocol requires everybody's cooperation. In section 5 of [18], Goldwasser and Levin describe a protocol (henceforth called GL) under the same assumptions as ours. Though we adopt many of their ideas (some of which are common knowledge), the protocol presented here is far more efficient. For instance, to implement an AND gate for P participants GL uses computations on polynomials, and many conditions that have to be verified while the main protocol is in progress are verified through rather inefficient sub-protocols. In this paper we demonstrate a conceptually much simpler way to accomplish this same task using and extending techniques introduced in [21, 16].

This also allows us to give a proper analysis of the complexity (in terms of the underlying cryptographic primitives) of our protocol, which in GL is quite difficult. In other words, whereas GL shows that a protocol for PMPC exists under the current assumptions, we provide a much simpler and more efficient implementation.

The remainder of this paper starts with a section elaborating on the assumptions, and introducing some useful notations. Section 3 contains the protocol and proof for COT, whereas Section 4 extends the tolls of Section 2 and finally Section 5 outlines the protocol and proof for PMPC.

2 Preliminaries

2.1 One-out-of-two Oblivious Transfer and Bit Commitment

In this paper OT and BC are used as black boxes. Since these are well-known protocols, we only describe them briefly.

In a one-out-of-two Oblivious Transfer *Bob* has to choose between learning bit a_0 or a_1 prepared by *Alice* but she does not learn his choice b. *Bob* learns a_b and obtains no information about $a_{\bar{b}}$. Implementations of OT can only exist under some assumption. For instance, OT can be constructed if trapdoor functions exist [16], from a noisy channel [11, 12], or from a quantum channel [2, 10]. It is also a well-known fact that using $O(n)$ of Rabin's Oblivious Transfers [26] one can construct one-out-of-two Oblivious Transfer [7].

In a Bit Commitment *Alice* sends a committed bit $\lceil a \rceil$ to *Bob* in such a way that she is able to reveal it later in a *unique* way (a) but *Bob* is *not able to find* its value by himself. *Alice* cannot change her mind and open $\lceil a \rceil$ as $\bar{a}$.

BC is impossible without making an assumption. It is easy to convert any version of Oblivious Transfer into a BC. Using error-correcting codes this is done at a cost of $O(n)$ OTs per BC [3]. BC can also be implemented under the assumption of the existence of a one-way function (or equivalently a pseudo-random bit generator) by a result of Naor [25], from a noisy channel [12], or from a quantum channel [3].

2.2 Bit Commitments with XOR

In this subsection we show how to prove that some BCs satisfy an XOR-relation without giving away their values. For this purpose we use special Bit Commitments (BCX) and proof techniques described by Kilian [21][22] (partly attributed to Rudich and Bennett). To commit to b using a BCX $\boxed{b}$, *Alice* uses $2n$ plain BCs, $(\lceil b_{1L} \rceil, \lceil b_{1R} \rceil, ..., \lceil b_{nL} \rceil, \lceil b_{nR} \rceil)$ such that for each $i \in \{1...n\} : b_{iL} \oplus b_{iR} = b$. To open $\boxed{b}$, *Alice* opens its $2n$ plain BCs. *Bob* accepts only if indeed all pairs of BCs XOR to the same value b.

If *Alice* is committed to several BCXs $\boxed{b^1}$, $\boxed{b^2}$, ..., $\boxed{b^k}$ she can prove to *Bob* that $\bigoplus_{j=1}^{k} \boxed{b^j} = c$ for some public value c without disclosing the b^j's as follows. First, *Bob* specifies to *Alice* k random permutations to shuffle the n pairs b_{iL}^j, b_{iR}^j of each BCX. Then, for each $i \in \{1...n\}$ *Alice* announces $c_{iL} = \bigoplus_{j=1}^{k} b_{iL}^j$ and $c_{iR} = \bigoplus_{j=1}^{k} b_{iR}^j$. Finally, for each $i \in \{1...n\}$ *Bob* randomly asks *Alice* to open either all the b_{iL}^j's or the b_{iR}^j's, and verifies that the revealed values are consistent with c_{iL} or c_{iR}. If this is the case, *Bob* is convinced that the BCXs satisfy the linear relation announced by *Alice*, since her probability of convincing him of a bad relation without detection is exponentially small in n. Note that this technique can be used to prove that two BCXs $\boxed{a}$ and $\boxed{b}$ are equal (different) simply by proving $\boxed{a} \oplus \boxed{b} = 0$ ($\boxed{a} \oplus \boxed{b} = 1$).

Since each such proof destroys the BCX involved, we must copy them first. Suppose *Alice* is committed to $\boxed{b}$ and she wants two instances of this commitment, i.e. $\boxed{b^1}$ and $\boxed{b^2}$ such that $b = b^1 = b^2$. *Alice* creates $3n$ pairs of BCs such that each pair XORs to b. Then *Bob* randomly partitions these $3n$ pairs in three subsets of n pairs, thus obtaining $\boxed{b^0}$, $\boxed{b^1}$ and $\boxed{b^2}$ and asks *Alice* to prove that $\boxed{b^0} = \boxed{b}$ as suggested above. This destroys $\boxed{b^0}$ and $\boxed{b}$, but if *Alice* succeeds *Bob* is convinced that $\boxed{b^1}$ and $\boxed{b^2}$ are two BCXs with the same value as $\boxed{b}$.

Because of the complexity of these constructions, for the remaining of this paper we consider as a unitary BCX operation: creation of a BCX, opening of a BCX, or proof that a constant number of BCXs satisfy a given linear relation.

3 Committed Oblivious Transfer

Suppose that *Alice* is committed to bits $\boxed{a_0}$, $\boxed{a_1}$ and *Bob* is committed to bit $\boxed{b}$. After running $\text{COT}(\boxed{a_0}, \boxed{a_1})(\boxed{b})$ *Bob* will be committed to $\boxed{a} = a_b$. *Alice*, whatever she does, cannot use the protocol to learn information on b and *Bob*, whatever he does, cannot use the protocol to learn information on $a_{\bar{b}}$.

3.1 Coding Theory

In our protocol for COT, a code is required. The code is not used to correct transmission errors (although it could if necessary) but in a more elaborate way for efficiency and security reasons.

Let σ and ϵ be some positive constants such that *Bob* can choose an $[n, k, d]$ linear code for which $k > (1/2 + 2\sigma)n$ and $d > \epsilon n$. The code must be efficiently decodable. The number of errors corrected by the decoding algorithm affects the efficiency of the protocol. It suffices that the algorithm corrects $\Omega(n)$ errors to guarantee the asymptotic security of the protocol.

Although Concatenated codes [14] are the most common codes with these properties, we recommend using the *Superconcentrator Codes* of Spielman [27] for their remarkable efficiency: they can be coded and decoded in linear time. In our protocol, we construct codewords, decode codewords and prove that some words are codewords. All these operations can be achieved in time $O(n)$. In particular, to prove that a committed word is a codeword it is sufficient to prove a linear number of statements each involving only a constant number of committed bits XORed together. For more details on coding theory in general we refer the reader to [24].

3.2 Informal Protocol

Intuitively $\text{COT}(\boxed{a_0}, \boxed{a_1})(\boxed{b})$ is performed by doing an imperfect "One-out-of-two Committed Oblivious String Transfer" of two random codewords c_0 and c_1. This transfer is imperfect in the sense that even the honest *Bob* learns information about both words. He learns all the bits of c_b and some bits of $c_{\bar{b}}$. We use this trick because *Bob* is not only interested in getting a_b and committing to $\boxed{a} = a_b$, but he also wants to know that if he had ran the protocol with $\bar{b}$ it would have worked as well. This is to prevent *Alice* from misbehaving on some part of her protocol so to learn b as a result of *Bob* aborting or not.

In the beginning *Alice* is committed to $\boxed{a_0}$ and $\boxed{a_1}$ and *Bob* to $\boxed{b}$. *Bob* chooses a code with suitable properties. *Alice* commits to $\boxed{c_0}$ and $\boxed{c_1}$, two randomly chosen codewords, and proves that this is the case. She Obliviously Transfers c_0 and c_1 to *Bob* in such a way that for each pair of bits *Bob* can choose to learn one of c_0 or c_1, but not both. *Bob* reads the bits to learn mostly c_b and a small part of $c_{\bar{b}}$. As suggested above he must gain a little bit of information on $c_{\bar{b}}$ to prevent *Alice* from learning which word he is interested in (and thus b). A cheating *Alice* could use very different words c_0, c_1 in the OTs and in the BCs. To prevent this, *Alice* is forced to open a small fraction of the committed bits on both sides, at *Bob*'s choosing, to let him check that the bits he received through the OT are the same. If *Alice* cheats a lot for one of the two words (or both) she will be caught without *Bob* revealing which one he was interest in. If *Alice* passes the test, only a small number of inconsistencies can remain between the bits *Bob* received and the bits *Alice* is committed to. *Bob* uses the fact that his word w should be a codeword to correct these.

Bob commits to $\boxed{w} = c_b$ and proves to *Alice* that it is a codeword. It is now *Bob*'s turn to convince *Alice* that he is committed to what he actually received through the OT (he is not committed to an arbitrary codeword, but actually to c_b). *Alice* opens a random set of positions and *Bob* shows that the bits he is committed to are consistent with the bits *Alice* reveals and $\boxed{b}$. This is sufficient to convince *Alice* that *Bob* is committed to c_b without her learning b.

Although *Bob* is committed to $\boxed{w} = c_b$, he gets some information on $c_{\bar{b}}$ revealed by the fact that it is a codeword and by the bits opened by *Alice* in the checks. To get rid of that information, *Alice* chooses a random privacy amplification [1] function h such that $a_b = h(c_b)$ and $a_{\bar{b}} = h(c_{\bar{b}})$ and proves these two relations to *Bob*. *Bob* commits to $\boxed{a} = h(w)$ and proves this to *Alice*.

3.3 Formal protocol

Before starting the protocol, *Alice* is committed to $\boxed{a_0}$, $\boxed{a_1}$ and *Bob* to $\boxed{b}$. After the protocol *Bob* will be committed to $\boxed{a_b}$. Let σ, ϵ be two positive constants as defined in Section 3.1, and let n denote the security parameter. At any point in the protocol, a failure in a proof or check leads the participants to abort.

Protocol 3.1 (COT $\left(\boxed{a_0}, \boxed{a_1}\right)$ $\left(\boxed{b}\right)$)

1: *Bob* chooses and announces to *Alice* a decodable $[n, k, d]$ linear code C
   with $k > (1/2 + 2\sigma)n$ and $d > \epsilon n$, efficiently decoding $t \in \Omega(n)$ errors.

2: *Alice* randomly picks $c_0, c_1 \in C$, commits to $\boxed{c_0^i}$ and $\boxed{c_1^i}$, for $i \in \{1 \ldots n\}$,
   and proves that $\boxed{c_0^1}\boxed{c_0^2}\ldots\boxed{c_0^n} \in C$ and $\boxed{c_1^1}\boxed{c_1^2}\ldots\boxed{c_1^n} \in C$.

3: *Bob* randomly picks $I_0, I_1 \subset \{1 \ldots n\}$, with $|I_0| = |I_1| = \sigma n$, $I_1 \cap I_0 = \emptyset$
   and sets $b^i \leftarrow \bar{b}$ for $i \in I_0$ and $b^i \leftarrow b$ for $i \notin I_0$.

4: *Alice* runs $\text{OT}(c_0^i, c_1^i)(b^i)$ with *Bob* who gets w^i, for $i \in \{1, 2, \ldots, n\}$.

5: *Bob* tells $I = I_0 \cup I_1$ to *Alice* who opens $\boxed{c_0^i}$ and $\boxed{c_1^i}$, for each $i \in I$.

6: *Bob* checks that $w^i = c_b^i$, for $i \in I_0$ and that $w^i = c_b^i$, for $i \in I_1$,
   sets $w^i \leftarrow c_b^i$, for $i \in I_0$, corrects w using C's decoding algorithm,
   commits to $\boxed{w^i}$, for $i \in \{1 \ldots n\}$, and proves that $\boxed{w^1}\boxed{w^2}\ldots\boxed{w^n} \in C$.

7: *Alice* randomly picks and announces a subset $I_2 \subset \{1 \ldots n\}$
   with $|I_2| = \sigma n$, $I_2 \cap I = \phi$ and opens $\boxed{c_0^i}$ and $\boxed{c_1^i}$, for $i \in I_2$.

8: *Bob* proves that $\boxed{w^i} = c_b^i$, for $i \in I_2$.

9: *Alice* randomly picks and announces a privacy amplification function
   $h : \{0, 1\}^n \to \{0, 1\}$, such that $a_0 = h(c_0)$ and $a_1 = h(c_1)$
   and proves $\boxed{a_0} = h\left(\boxed{c_0^1}\boxed{c_0^2}\ldots\boxed{c_0^n}\right)$ and $\boxed{a_1} = h\left(\boxed{c_1^1}\boxed{c_1^2}\ldots\boxed{c_1^n}\right)$.

10: *Bob* sets $a \leftarrow h(w)$, commits to $\boxed{a}$ and proves $\boxed{a} = h\left(\boxed{w^1}\boxed{w^2}\ldots\boxed{w^n}\right)$.

3.4 Zero-Knowledge proofs

In the protocol, *Alice* and *Bob* make a number of zero-knowledge proofs. All these proofs are easily achieved if we are able to perform XOR proofs on committed bits. This is why the protocol uses BCXs and not plain BCs.

In Step 2 and 6 a party must show to the other that a set of committed bits form a codeword. In any linear code, this operation can be achieved by showing that the syndrome of the word is the zero-vector. This takes $O(n^2)$ BCX operations. In the case of Superconcentrator Codes, it is sufficient to prove that each bit is the XOR of a constant number of (publicly known) other bits of the word. Therefore it takes only $O(n)$ BCX operations.

In Step 8 *Bob* must prove to *Alice* that $\boxed{w^i} = c_b^i$, for $i \in I_2$. First, for each position in which $c_0^i = c_1^i$, *Bob* can simply open $\boxed{w^i}$ because in this case w^i contains no information on b. For the other positions, where $c_0^i \neq c_1^i$, one of c_0 or c_1 must be 0 and the other must be 1. When $c_0 = 0$ and $c_1 = 1$ we have $c_b^i = b$, therefore *Bob* proves $\boxed{w^i} \oplus \boxed{b} = 0$. In the opposite situation $c_b^i = \bar{b}$, therefore

Bob proves $\boxed{w^i} \oplus \boxed{b} = 1$. Both of these proofs give no information on b and take constant time. The total for all the positions in I_2 is $O(n)$ BCX operations.

In Step 9 and 10 a party must show to the other that $z = h\left(\boxed{x^1}\boxed{x^2}\cdots\boxed{x^n}\right)$ for some committed bit z and word x. If h is a random linear function specified by a public random subset H of the bits of x it defines a universal hash function [4]. In this case it suffices to show $z \oplus \bigoplus_{i \in H} x^i = 0$ which takes $O(n)$ BCX operations.

3.5 Validity of the Protocol

There are several ways in which *Alice* and *Bob* may misbehave when they execute our protocol. At any point *Alice* or *Bob* can decide to stop cooperating and they can commit to values different from those used in the oblivious transfer. The main point is that *Bob* cannot claim to be committed to a_b if he is not, and that even if the protocol aborts no unintended information leaks.

Bob cannot commit to $\overline{a_b}$. First observe that we cannot force *Bob* to commit to a_b after he has learned this value in Step 9. However, this is equivalent to the situation in which *Bob* refuses to open $\boxed{a}$ after the protocol since an unopened $\boxed{a}$ is useless to *Alice*. So the only important fact is that if the protocol has completed without complaints and *Bob* is committed to a bit, this must be a_b.

We show informally why this is true. Suppose *Bob* commits to $\boxed{a} = \overline{a_b}$. This means that he has proved at Step 10 that $\overline{a_b} = h(w)$. Since $a_b = h(c_b)$ this implies that $w \neq c_b$. *Bob* has proved at Step 6 that w is a codeword, so the Hamming distance between w and c_b is greater than ϵn or, equivalently, they differ in more than ϵn positions. But *Alice* has asked *Bob* to open σn positions chosen at random and for all these positions the bits were equal to their counterparts in c_b. This cannot happen except with a probability exponentially small in n.

Alice learns nothing about b. It is not hard to see that if *Alice* performs COT honestly she learns nothing about b. Furthermore, regardless of how *Alice* cheats, she gains no information about b. A straightforward analysis of each step of the protocol shows that only in Step 1, 5, 6, 8 and 10 *Bob* sends actual information to *Alice*. In Step 1 and 5 the information is independent of b. For Step 6,8 and 10 the definition of BC and the fact that the proofs are Zero-knowledge guarantee that *Alice* learns nothing about b. The only other way *Alice* could learn information would be to misbehave in such a way that *Bob* has to abort if $b = 0$ (or $b = 1$) in Steps 7 or 9. This way she could learn b and *Bob* would catch her only if $b = 1$ (or $b = 0$).

In Step 10 *Bob* will always succeed since he calculates a himself. If *Bob* fails the proofs at Step 6 or 8, this implies that $\boxed{w^1}\boxed{w^2}\cdots\boxed{w^n} \neq \boxed{c_b^1}\boxed{c_b^2}\cdots\boxed{c_b^n}$. But since c_b is a codeword (as proved in Step 2) this means that their Hamming distance is at least t because otherwise the decoding algorithm would have led to c_b. Since $t \in \Omega(n)$ the check of Step 6 has only an exponentially small probability of success. Thus, if the check of Step 6 succeeds, the proofs of Step 6 and 8 will fail with exponentially small probability whatever b is. And since $|I_0| = |I_1|$ the check of Step 6 has the same probability of success whatever b is.

Bob learns nothing about $a_{\bar{b}}$. *Alice* does not want a dishonest *Bob* to learn $a_{\bar{b}}$. In fact it is not sufficient that *Bob* learns no information about $a_{\bar{b}}$. We want that for all dishonest *Bob'* the amount of information he has about $a_{\bar{b}}$ after interacting with *Alice* will be the same as before, even if he knew a_b to start with. This is to ensure that *Bob* does not learn any correlation between a_0 and a_1 through our protocol, for instance $a_b \oplus a_{\bar{b}}$.

Note that the information on a_0 and a_1 is completely uncorrelated. Because *Alice* picks c_0 and c_1 at random, no information about $a_{\bar{b}}$ is sent before she reveals h. For *Bob* to learn h he must pass *Alice*'s test of Step 8 and convince her that he is committed to $\boxed{w} = c_b$. If *Bob* wishes to have no more than $2^{\sigma n/2}$ candidates for a_b by the time he commits at Step 6, he must have acquired on side b at least $n/2 + 3\sigma n/2$ of the $n + 2\sigma n$ bits available to him through Step 4 and 5. This is the case because there are $2^{n/2 + 2\sigma n}$ codewords, and thus after learning $n/2 + 3\sigma n/2$ extra bits, $2^{\sigma n/2}$ of them remain equally possible. If *Bob* gets $n/2 + 3\sigma n/2$ bits on side b, it leaves him with no more than $n/2 + \sigma n/2$ bits on side $\bar{b}$ of the $n + 2\sigma n$ bits available. In this case there are also $2^{\sigma n/2}$ candidates for $c_{\bar{b}}$ even after learning σn extra bits at Step 7.

In conclusion, if *Bob* gains as few as $n/2 + 3\sigma n/2$ bits on side b, he has probability no more than $2^{-\sigma n/2}$ of committing to the right codeword at Step 6, and any wrong codeword will have at least ϵn differences with the correct one. In this case the test of Step 8 would have an exponentially small probability in n of succeeding. On the other hand, if *Bob* gains as few as $n/2 + \sigma n/2$ bits on side $\bar{b}$ then, by a theorem of Bennett, Brassard and Robert [1], his probability of obtaining any information whatsoever about $h(c_{\bar{b}})$ is less than $2^{-\sigma n/2}$ provided h is chosen at random.

3.6 Complexity of the protocol

The exact complexity of the protocol depends on the kind of BCs and OTs used by the participants. We thus make our analysis based on the number of BCs and OTs that are required to perform the protocol. If the OT and BC being used require a constant number of communication rounds, then all the OTs and proofs can be done in parallel and the COT protocol also requires a constant number of rounds.

The protocol uses $O(n)$ OTs which are performed in Step 2, when c_0 and c_1 are transferred. It also requires $O(n)$ BCXs for both *Alice* and *Bob* to commit to c_0, c_1 resp. w. All the proofs on the BCXs can be done with only $O(n)$ BCX operations.

If the BCXs are not available one can use the trick described in Section 2.2. To perform COT, $O(n^2)$ BCs are used to perform the $O(n)$ necessary BCX operations. In this case, the overall complexity is $O(n)$ OTs and $O(n^2)$ BC operations.

4 Extension of the primitives to multiple participants

The second goal of this paper is to describe a protocol for Multi-Party Computation. To reach this goal, we enhance BCX and COT (prefixed with G for "global"), allowing other participants (who do not provide an input) to act as "verifier" to check that the active participant(s) behave honestly. Secondly we show how COT can be used to obtain a protocol for evaluating a partial AND gate between two participants, a mandatory step for PMPC.

4.1 Global Bit Commitment with XOR proof (GBCX)

A Global Bit Commitment with XOR (GBCX) is a bit commitment to a *group* of participants. A GBCX is conceptually equivalent to $P - 1$ BCXs, one to each participant, all with the same value. GBCXs are constructed such that a dishonest participant $\mathcal{A}$ cannot open it as a different value to two honest participants $\mathcal{B}_j, \mathcal{B}_k$. An obvious, but very inefficient way to achieve this is that $\mathcal{A}$ proves to each pair of participants that the two BCXs they hold are equal.

We show how to do this more efficiently. To commit to a bit a using a GBCX, $\mathcal{A}$ makes $2n(P - 1)$ pairs of (plain) BCs, $2n$ to each of the other participants $\mathcal{B}_1...\mathcal{B}_{P-1}$, such that for $i \in \{1...2n\}$ and $j \in \{1...(P-1)\}$, $a_{iL}^j \oplus a_{iR}^j = a$. Then each $\mathcal{B}_j$ chooses and broadcasts a random permutation of size $2n$ and renames his $2n$ pairs accordingly. For $i \in \{1...n\}$ $\mathcal{A}$ announces $V_i^0 \leftarrow \{j : (a_{iL}^j, a_{iR}^j) = (0, a)\}$ and $V_i^1 \leftarrow \{j : (a_{iL}^j, a_{iR}^j) = (1, \bar{a})\}$ in a random order. Then all participants flip n fair coins together (requiring $O(nP)$ BCs) to construct a random tuple $(S_1...S_n) \in \{L, R\}^n$. For $i \in \{1...n\}$, $j \in \{1...P - 1\}$ $\mathcal{A}$ opens $a_{iS_i}^j$. Finally, each $\mathcal{B}_j$ verifies that the values of the $a_{iS_i}^j$ opened to him correspond with those broadcasted, and that $W_i \leftarrow \{j : a_{iS_i}^j = 0\}$ equals either V_i^0 or V_i^1.

After this protocol $\mathcal{A}$ has n pairs of untouched BCs with each $\mathcal{B}_j$ (those with $i \in \{(n+1)...2n\}$), which constitute the new GBCX. It is easy to verify that if $\mathcal{A}$ tries to commit to two different bits with two honest participants, or if he tries to construct inconsistent commitments, he will be caught by each pair of honest participants except with exponentially small probability.

Although the role of $\mathcal{A}$ (the "*active*" participant) is different from the role of the $\mathcal{B}_j$s (the "*passive*" participants) in the protocol, the cost of creating a GBCX is the same for everybody, $O(Pn)$ BCs. Once created, operations on GBCXs, such as proving linear relations and copying, are reduced to P operations between $\mathcal{A}$ and $\mathcal{B}_j$ using BCXs. For $\mathcal{A}$ the cost of one such operation is $O(Pn)$, but for the $\mathcal{B}_j$ it is only $O(n)$ (since after the creation they do not need to interact with other users).

4.2 Global Committed Oblivious Transfer (GCOT)

Global Committed Oblivious Transfer (GCOT) is the extension of COT to a group. For *Alice* and *Bob* GCOT achieves exactly the same functionality as COT but it allows the passive participants to be convinced that *Alice* and *Bob* do not conspire, i.e. that indeed after $\text{GCOT}(\boxed{a_0}, \boxed{a_1})(\boxed{b})$ *Bob* is committed to $\boxed{a_b}$.

Some modifications are necessary to change COT to GCOT: (1) The error-correcting code C of Step 1 must be chosen by all participants; (2) BCXs are replaced by GBCXs and all commitment openings and proofs are done to each participant; (3) the subset I_2 in Step 7 must be chosen by all participants.

In order to choose I_2, only $O(n \log(n))$ random coins have to be flipped, at the expense of $O(Pn \log(n))$ plain BCs for each participant. When many GCOTs on different inputs are performed in parallel, as will be the case in the evaluation of AND gates in the final protocol, the same I_2 may be used. Therefore we do not take this cost into account in the next paragraph, but deal with it later.

For one GCOT *Alice* and *Bob* have to perform $O(n)$ *active* GBCX operations (resulting in a total of $O(Pn^2)$ BCs) and $O(n)$ OTs. The others have to perform $O(n)$ *passive* GBCX operations (resulting in $O(n^2)$ BCs).

4.3 PAND and GPAND

As a step towards PMPC we introduce a two-party protocol, called PAND, that takes the BCXs $\boxed{a}$ from *Alice* and $\boxed{b}$ from *Bob* as inputs. After the execution of PAND *Alice* is committed to $\boxed{a'}$ and *Bob* to $\boxed{b'}$ such that $a \wedge b = a' \oplus b'$, and neither participant learns the other's input value. Since the output is composed of the pair a' and b', we call this protocol PAND for Pair-AND.

We implement PAND using COT. *Alice* randomly chooses a bit a', commits to $\boxed{a''} = \boxed{a'} \oplus \boxed{a}$ and proves it. Then *Alice* performs $\text{COT}(\boxed{a'}, \boxed{a''})(\boxed{b})$ with *Bob* who gets $\boxed{b'}$. (Notice that $b = 0 \Rightarrow b' = a' \Rightarrow a' \oplus b' = 0$ and that $b = 1 \Rightarrow b' = a' \oplus a \Rightarrow a' \oplus b' = a$, so $a \wedge b = a' \oplus b'$.)

GPAND is a generalization of the PAND to a group: two active participants want to make a PAND, while the passive participants want to be sure that indeed $\boxed{a} \wedge \boxed{b} = \boxed{a'} \oplus \boxed{b'}$. The GPAND is done as a PAND, with COT replaced by GCOT. The cost of one GPAND is of the same order as the cost of one GCOT, both for the active and the verifying participants.

5 Private Multi-Party Computation (PMPC)

In this section we show how, given OT and BC between each pair of participants and a reliable Broadcast Channel, P participants can perform a *correct, honest, fair* and *private* Multi-Party Computation, where we make no assumptions about the number of honest participants. The protocol consists of three steps: *initialization*, *computation* and *revelation*. At each step we make extensive use of the primitives previously defined.

5.1 Initialization

In the *initialization* step all participants agree on the circuit F to be evaluated and on a security parameter n. They also agree on a parameter σ and a code C for the GCOT sub-protocol.

In the PMPC protocol each bit involved is represented by a Distributed Bit Commitment (DBC) consisting of P shares, each share being a GBCX created by a different participant. The value of a DBC is the XOR of the value of its shares, therefore even $P - 1$ participants are not able to reconstruct the value of the DBC if one participant refuses to cooperate.

As a second part of the initialization step each participant uses DBCs to commit to his input bits. To create a DBC of value a, $\mathcal{A}$ asks each participant to commit to a random bit using a GBCX, and to open it to $\mathcal{A}$ only. Then $\mathcal{A}$ creates a GBCX such that the XOR of all the GBCXs equals a.

5.2 Computation

In the *computation* step the participants evaluate the circuit consisting of AND and NOT gates, one gate after the other, where the input and output bits of each gate are DBCs.

Since the output bit of a gate can be an input bit to several other gates we must be able to copy DBCs. To make r copies of a DBC each participant makes r copies of his share (a GBCX). The NOT operation on a DBC is equivalent to a copy, except that one designated participant inverts the value of his GBCX.

To evaluate an AND gate on two DBCs of value a and b we observe that $(\bigoplus_{i=1}^{n} a_i) \wedge (\bigoplus_{j=1}^{n} b_j) = \bigoplus_{i,j=1}^{n} a_i \wedge b_j$. In other words, one Distributed AND can be reduced to P^2 GPANDs, one between each pair of participants. After all the GPANDs have been done each participant chooses a GBCX that equals the XOR of all his shares and he proves this. This GBCX is his share of the DBC whose value equals $a \wedge b$.

To evaluate one AND gate $O(P^2)$ GPANDs and so $O(P^2)$ GCOTs are executed. Each participant will be actively involved in P GCOTs, each requiring $O(n)$ OTs and $O(n)$ *active* GBCXs operations. Each participant also verifies $O(P^2)$ GCOTs between other participants. For one such GCOT verification he is involved in $O(n)$ *passive* GBCX operations. Therefore each participant is involved in $O(Pn)$ *active* GBCXs and $O(P^2n)$ *passive* GBCXs. This results in $O(P^2n^2)$ BCs and in $O(Pn)$ OTs.

We must also take into account the cost for creating the subset I_2 used in every GCOT. Only one random I_2 suffices for many parallel GCOTs, and since the creation of a single I_2 requires $O(n \log(n))$ coin flips, it requires a total of $O(Pn \log(n))$ BCs per participant. This does not change the previously stated complexity.

5.3 Revelation

At the end of the *computation* step each output bit of the circuit is hidden in a DBC. In order for each honest participant to learn an output bit of F, every participant could open his share of the DBC representing the answer, but obviously a dishonest participant could quit when he has more information than others. Better solutions, which achieve *fairness*, appear in the literature [6][18]. They can easily be incorporated into our protocol.

5.4 The validity of the PMPC protocol

We do not claim to provide a full proof that the protocol presented here is secure. The security of the final protocol relies on the security of all the sub-protocols that constitute it. Due to space restrictions, we only give a brief and sketchy proof that PMPC satisfies the four desired properties mentioned in the introduction.

Correctness: If all participants are honest the output of our protocol should equal the output of the function to be emulated. This is verified because each sub-protocol outputs the value it is designed for.

Privacy: Observe that from a participant's point of view the GBCX is the smallest "unit" to which he can commit and that no coalition can open it if the owner does not cooperate. Because of the way each DBC is constructed each participant holds a share of each bit of the computation, so for all sub-protocols no information about the values of the DBCs is revealed as long as at least one participant is honest. Therefore, before the revelation of the output bits no coalition of less than P participants can obtain any information whatsoever on the input bits, intermediate bits or output bits of the computation, except for what can be deduced from the information already known by the coalition. A more formal analysis along these lines could benefit from the definition of privacy given in [23].

Fairness: This is an issue in the revelation step only. We want to prevent one or several parties from learning more information on the output bits before the others. For the fairness of our protocol we totally rely on the model and definitions given by Goldwasser and Levin [18]. Though the implementation of their protocol differs from ours their model still applies, in particular their notion of fairness.

Honesty: Note that no coalition of participants can cheat on the resulting output DBCs of any sub-protocol. For the full protocol this means that, once the input bits have been committed to, no coalition of less than P players can change the values of an output DBC without being caught (except with exponentially small probability). In other words, even though any coalition can disrupt the protocol by preventing it from completing, no coalition can make an honest participant accept the output of F when any of the output DBCs has been tampered with.

5.5 Complexity of the PMPC protocol

It is easy to see that to evaluate a circuit F composed of m gates $O(m)$ DBC operations are needed. The most expensive operation on DBCs is the evaluation of an AND gate, so the overall complexity of the PMPC protocol is $O(P^2n^2m)$ BCs and $O(Pnm)$ OTs. Since BC and OT can be implemented using $O(n)$ of Rabin's Oblivious Transfers, PMPC can be performed using $O(P^2n^3m)$ of them.

Acknowledgments

We thank Joe Kilian and Matt Franklin for providing references. We give special thanks to Mélanie and Marie-France for their patience and understanding.

References

1. C.H. Bennett, G. Brassard, J.-M. Robert, *Privacy Amplification by Public Discussion*, SIAM Journal on Computing, Vol. 17, No.2, 1988, pp. 210–229.
2. C.H. Bennett, G. Brassard, C. Crépeau, M.-H. Skubiszewska, *Practical Quantum Oblivious Transfer*, Advances in Cryptology - CRYPTO'91, Springer-Verlag, 1992, pp. 351–366.
3. G. Brassard, C. Crépeau, R. Jozsa, D. Langlois, *A Quantum Bit Commitment Scheme Provably Unbreakable by both Parties*, 34th IEEE Symposium on Foundation of Computer Science, 1993, pp. 362–371.
4. J. L. Carter and M. N. Wegman, *Universal Classes of hash function*, Journal of Computer and System Sciences, Vol. 18, 1979, pp. 143-154.
5. D. Chaum, I. Damgård and J. van de Graaf, *Multiparty computations ensuring privacy of each party's input and correctness of the result*, Advances in Cryptology - CRYPTO'87, Springer-Verlag, 1988, pp. 87–119.
6. R. Cleve, *Controlled Gradual Disclosure Schemes for Random Bits and Their Applications*, Advances in Cryptology - CRYPTO'89, Springer-Verlag, 1991, pp. 573–590.
7. C. Crépeau, *Equivalence Between Two Flavours of Oblivious Transfer*, Advances in Cryptology - CRYPTO'87, Springer-Verlag, 1988, pp. 350-354.
8. Crépeau, C., *Verifiable Disclosure of Secrets and Applications*, Advances in Cryptology - Eurocrypt'89, Springer-Verlag, 1990, pp. 181–191.
9. C. Crépeau, *Correct and Private Reductions Among Oblivious Transfers*, Ph.D. thesis, MIT, 1990.
10. C. Crépeau, *Quantum Oblivious Transfer*, Journal of Modern Optics, vol. 41, No. 12, 1994.
11. C. Crépeau, J. Kilian, *Achieving Oblivious Transfer Using Weakened Security Assumptions*, 29th IEEE Symposium on Foundation of Computer Science, 1988, pp. 42–52.
12. C. Crépeau, *Cryptographic protocol based on noisy channel*, in preparation, 1995.
13. S. Even, O. Goldreich and A. Lempel, *A Randomized Protocol for Signing Contracts*, Communications of the ACM, Vol 28, 1985, pp. 637-647.
14. Forney, G. D., *Concatenated Codes*, The M.I.T. Press, 1966.
15. M. Franklin, *Complexity and security of distributed protocols*, Ph. D. thesis, Computer Science Department of Columbia University, New York, 1993.
16. O. Goldreich, S. Micali, and A. Wigderson, *How to play any mental game, or: A completeness theorem for protocols with honest majority*, 19th ACM Symposium on Theory of Computing, 1987, pp. 218–229.
17. O. Goldreich, R. Vainish, *How to solve any protocol problem — an efficiency improvement*, Advances in Cryptology - CRYPTO'87, Springer-Verlag, 1988, pp. 73–86.
18. S. Goldwasser, L. Levin, *Fair computation of general functions in presence of moral majority*, Advances in Cryptology - CRYPTO'90, Springer-Verlag, 1991, pp. 77–93.

19. J. Kilian, *Founding cryptography on Oblivious transfer*, 20th ACM Symposium on Theory of Computation, 1988, pp. 20–31.

20. J. Kilian, *Uses of Randomness in Algorithms and Protocols*, MIT Press, 1990.

21. J. Kilian, *A note on efficient zero-knowledge proofs and arguments*, 24th ACM Symposium on Theory of Computation, 1992, pp. 723–732.

22. J. Kilian, *On the complexity of bounded-interaction and noninteractive zero-knowledge proofs*, 35th IEEE Symposium on Foundations of Computer Science, 1994, pp. 466-477.

23. E. Kushilevitz, S. Micali, R. Ostrovski, *Reducibility and completeness in multi-party private computations*, 35th IEEE Symposium on Foundations of Computer Science, 1994, pp. 478-489.

24. F. J. MacWilliams, N.J.A. Sloane, *The Theory of Error-Correcting Codes*, North-Holland, 1977.

25. M. Naor, *Bit Commitment using Pseudo-Randomness*, Advances in Cryptology - CRYPTO'89, Springer-Verlag, 1989, pp. 128–136.

26. M. Rabin, *How to exchange secrets by oblivious transfer*, Tech. Memo TR–81, Aiken Computation Laboratory, Harvard University, 1981.

27. D. Spielman, *Linear-Time Encodable and Decodable Error-Correcting Codes*, 27th ACM Symposium on Theory of Computing, 1995, pp. 388–397.

On the Security of the Quantum Oblivious Transfer and Key Distribution Protocols

Dominic Mayers

Département IRO, Université de Montréal
C.P. 6128, succursale "A",Montréal (Québec), Canada H3C 3J7.
e-mail: mayersd@iro.umontreal.ca.

Abstract. No quantum key distribution (QKD) protocol has been proved fully secure. A remaining problem is the eavesdropper's ability to make *coherent* measurements on the joint properties of large composite systems. This problem has been recently solved by Yao in the case of the security of a quantum oblivious transfer (QOT) protocol. We consider an extended OT task which, in addition to $Alice$ and Bob, includes an eavesdropper $\mathcal{E}$ve among the participants. An honest $\mathcal{E}$ve is inactive and receives no information at all about Alice's input when Bob and $Alice$ are honest. We prove that the security of a QOT protocol against Bob implies its security against $\mathcal{E}$ve as well as the security of a QKD protocol.

1 Introduction

The goal of quantum cryptography is to design cryptographic protocols that are secure against unlimited quantum or classical computational power. At present, the quantum protocols that have been designed are commitment [BC, BCJL], oblivious transfer [Cr87, Cr94, BBCS, MS, Yao], key distribution [BB84, BBBSS, BBBW] and identification [CS]. Furthermore, prototypes for implementing some of these protocols have been built [BBBSS, MT, To94, TRT1, TRT2].

However, the full security of some of these protocols has not yet been proved. One of the difficulties in providing a full security proof is the cheaters' ability to execute *coherent* measurements on many photons at a time. At present, security against coherent measurements has been obtained in the case of commitment [BCJL] and bit oblivious transfer [Yao]. The security of QKD against coherent measurements has not yet been addressed in the literature and it is not clear whether the techniques used by Yao in [Yao] for a QOT protocol may be easily used for a QKD protocol. In any case, we do not use Yao's techniques. We show that the security against Bob of a QOT protocol implies its security against eavesdropping and, as a corollary, the security of a key distribution protocol. The level of security against eavesdropping that we obtain for QOT (and QKD) depends upon the level of security of QOT against Bob, and, in particular, full security against Bob implies full security against eavesdropping.

The security of a QOT protocol against an eavesdropper is interesting in itself because it allows the protocol to be executed securely over a long quantum channel by an honest $Alice$ and an honest Bob. The above implication works

D. Coppersmith (Ed.): Advances in Cryptology - CRYPTO '95, LNCS 963, pp. 124-135, 1995.
© Springer-Verlag Berlin Heidelberg 1995

with a string QOT protocol, that is, a QOT protocol that transfers a string rather than only a single bit. The implication requires that the QOT protocol tolerates errors in the quantum channel and that the classical announcements in the QKD protocol are made on a faithful *public* channel between $\mathcal{A}$lice and $\mathcal{B}$ob. It does not require any unrealistic physical assumption such as zero error in the quantum channel.

2 The QOT protocol and the security of OT

There are two types of OT: the ordinary OT and the $\binom{1}{2}$-OT. We consider the string version of both types. In the ordinary OT, $\mathcal{A}$lice inputs a string s, $\mathcal{B}$ob receives a random bit $c \in \{0,1\}$ and, if $c = 0$, the string s. In the $\binom{1}{2}$-OT, $\mathcal{A}$lice inputs two string s_1 and s_2, $\mathcal{B}$ob inputs a bit c_B and receives the string s_{c_B}.

In this paper, from the security against $\mathcal{B}$ob and tolerance against errors of an ordinary QOT protocol, we obtain its security against $\mathcal{E}$ve and, as a corollary, the security of a QKD protocol. This is significant in particular because Yao has proved the security against $\mathcal{B}$ob of an ordinary QOT protocol [Yao].

2.1 The protocol

In the discussion below, a dishonest $\mathcal{B}$ob and a dishonest $\mathcal{E}$ve, have been included. Both appear in the same description, but the security of the protocol against any one of them is based upon the assumption that the other is inactive.

For $b, \theta \in \{0,1\}$, let $|b\rangle_\theta$ be the state of a photon polarized at $b \times 90 + \theta \times 45$ degrees. In the BB84 coding scheme, b is the bit coded in the state $|b\rangle_\theta$ and θ determines the basis used to code this bit: $\theta = 0$ corresponds to the basis $\{0°, 90°\}$ whereas $\theta = 1$ corresponds to the basis $\{45°, 135°\}$.

Protocol $OT(s)$

1 Honest $\mathcal{B}$ob: He chooses and commits to a *random* string $\hat{\theta} = (\hat{\theta}_1, \ldots, \hat{\theta}_{4n}) \in_R \{0,1\}^{4n}$.
   Dishonest $\mathcal{B}$ob: He cannot gain any advantage from being dishonest at this step.
2 Until $4n$ pulses are detected by $\mathcal{B}$ob:
   2.1 $\mathcal{A}$lice: She sends a pulse to $\mathcal{B}$ob in which a random bit is coded using a random base in the BB84 coding scheme.
   2.2 Dishonest $\mathcal{E}$ve: She transfers some information from this pulse into her own quantum system and she uses that information to modify the residual state of the pulse which is sent to $\mathcal{B}$ob. The entire operation may be represented by a single unitary transformation on the product state of the photon and $\mathcal{E}$ve's system.
   2.3 Let us assume that, thus far, $i-1$ pulses have been detected by an honest $\mathcal{B}$ob or declared as such by a dishonest $\mathcal{B}$ob.
   Honest $\mathcal{B}$ob: He executes on this pulse a von Neumann measurement

in the basis $\hat{\theta}_i$ and, if the pulse is detected, he obtains a bit $\hat{b}_i$ that he commits to Alice.

Dishonest Bob: He executes a coherent measurement on this pulse and the previous pulses in order to determine:

- whether or not he declares this pulse as detected and, if he declares this pulse as detected,
- the bit $\hat{b}_i$ that he commits to Alice.

Typically, Bob executes an incomplete measurement.

The string of bits coded in these $4n$ detected pulses is $b \in_R \{0, 1\}^{4n}$ and the associated string of bases is $\theta \in_R \{0, 1\}^{4n}$.

3 Alice: She chooses a random string $open \in_R \{0, 1\}^{4n}$ and publicly announces it. For each i, if $open_i = 1$ she asks Bob to open the commitments $\hat{\theta}_i$ and $\hat{b}_i$. She publicly announces the string $error$ where

$$error_i = \begin{cases} 1 \text{ if } \theta_i = \hat{\theta}_i \wedge b_i \neq \hat{b}_i \wedge open_i = 1 \\ 0 \text{ otherwise} \end{cases}$$

If $\#error$ and the number of undetected pulses (another kind of error) are not too large, the remainder of protocol is executed, and $Pass$ is set to 1 otherwise Alice refuses to continue and $Pass$ is set to 0.

4 Alice: She publicly announces the string $\theta = (\theta_1, \ldots, \theta_{4n})$.

5 Honest Bob: He chooses a random bit c_B. He deterministically computes an ordered pair (e_0, e_1) such that $e_0 \cup e_1 = \{i | open_i = 0\}$, $|\#e_0 - \#e_1| \leq 1$ and

$$(\forall i \in e_{c_B}) \, \theta_i = \hat{\theta}_i \quad \vee \quad (\forall i \in e_{\bar{c}_B}) \, \theta_i \neq \hat{\theta}_i,$$

and publicly announces this ordered pair. For our proof, it is convenient to consider that Bob's deterministic algorithm to compute (e_0, e_1) returns the same output if $\hat{\theta}$ and c_B are both complemented (this is easy to accomplish). Dishonest Bob: Having learnt the string θ, he executes a first post-test measurement of his choice and uses the outcome to compute an ordered pair (e_0, e_1) such that $e_0 \cup e_1 = \{i | open_i = 0\}$ and $|\#e_0 - \#e_1| \leq 1$, and publicly announces the ordered pair.

For all $d \in \{0, 1\}$, the string coded by Alice in e_d is denoted w_d.

6 Alice: She chooses and publicly announces a random bit c_A and a hash function g from $\{0, 1\}^{\#e_{c_A}}$ to $\{0, 1\}^r$. The integer r is the length of the string to be sent via QOT. She also publicly announces $a = g(w_{c_A}) \oplus s$ and $Syn(w_{c_A})$, the syndrome of w_{c_A} which is needed by Bob for error correction.

7 Honest Bob: Let $c = c_A \oplus c_B$ (this is the c that appears in the description of the task). If $c = 0$, he uses $Syn(w_{c_A})$ to correct the error in $w_{c_B} = w_{c_A}$ and then he computes $s = g(w_{c_B}) \oplus a$.

Dishonest Bob: Using the information obtained at step 6, he executes a second and final post-test measurement and obtains the outcome j_{Bob}. This provides information about $s = a \oplus g(w_d)$, for $d = 0, 1$.

8 Dishonest Eve: She measures her system and obtains the outcome j_{Eve}. This provides information about $s = a \oplus g(w_d)$, for $d = 0, 1$.

We adopt the following notation: the random values $s, b, \theta, \hat{b}$, etc. associated with an execution of the protocol are values taken by random variables $S, B, \Theta, \hat{B}$, etc.

The remainder of the section contains the formal definitions of security that we use in our proof. As for the definition of security for $\binom{1}{2}$-OT found in [Cr94], our definitions are formulated in terms of the amount of information received by a given participant. Any initial information about s that may have this participant, Bob in sections 2.2 and 2.4 and Eve in section 2.3, corresponds to an apriori probability distribution on S which is implicit in our definitions.

Due to their relative complexity, we understand that the reader may have the impression that the following definitions are more complicated than necessary. However, these are the most simple and yet complete definitions that we could express in terms of mutual information. A more complete discussion on this subject, including a connection with definitions expressed in terms of statistical indistinguishability, will appear in another paper.

2.2 Security of OT against Bob

Let V_{Bob} represents all the information received or generated by Bob in the protocol. A QOT protocol is secure against Bob if $(\exists \alpha > 0)(\exists n_0)$ such that, $(\forall n > n_0)$, for every Bob, for every Channel, there exists a binary random variable $\tilde{C}$ (defined when $Pass = 1$) such that

$$I(S; V_{Bob} | \tilde{C} = 1 \wedge Pass = 1) \times \Pr(Pass = 1) \leq 2^{-\alpha n} \tag{1}$$

$$\Pr(\tilde{C} = 1 | Pass = 1) = 1/2 \tag{2}$$

$$I(S; V_{Bob} | Pass = 0) \times \Pr(Pass = 0) \leq 2^{-\alpha n} \tag{3}$$

$$I(S; \tilde{C}, Pass) \leq 2^{-\alpha n} \tag{4}$$

Let us remark that at step 5 a dishonest Bob does not even have to choose a bit C_B. If Bob does not choose a bit C_B, the bit $C = C_A \oplus C_B$ associated with an honest Bob is meaningless. Therefore, in the above definition, $\tilde{C}$ has, in general, nothing to do with the bit C associated with an honest Bob.

Statement 1 says that, if Bob passes the test with a significant probability, then, in the context where Bob passes the test and $\tilde{C} = 1$, Bob learns almost nothing about S. Statement 2 says that, in the context where Bob passes the test, $\tilde{C}$ is perfectly random. Statement 3 says that, if Bob fails the test with a significant probability, then, in the context where Bob fails the test, Bob learns almost nothing about S. Statement 4 says that the information $(\tilde{C}, Pass)$, where $\tilde{C}$ is not given to Bob in the protocol but could eventually be given to Bob outside the protocol, says almost nothing about S.

2.3 Security of the extended OT against Eve

Let V_{Eve} represents all the information that is available to an eavesdropper Eve. The protocol is secure against Eve, if $(\exists n_0)$ such that, $(\forall n > n_0)$, for every Channel, for every Eve,
$$I(S; V_{Eve}) \leq 2^{-\alpha n}.$$

2.4 Tolerance against errors in OT

The protocol is tolerant against errors (the tolerated error rate being indirectly specified by the test) if, $(\exists \alpha > 0)$ $(\exists n_0)$ such that, $(\forall n > n_0)$, for every Channel, if $\Pr(Pass = 1) > 2^{-\alpha n}$, then

$$I(S; V_{\mathcal{B}ob}|C = 0 \wedge Pass = 1) > r - 2^{-\alpha n} \tag{5}$$
$$\Pr(C = 0|Pass = 1) > 1/2 - 2^{-\alpha n} \tag{6}$$

where C is the bit that is received by an honest $\mathcal{B}$ob.

The condition $\Pr(Pass = 1) > 2^{-\alpha n}$ is needed because, if the expected rate of errors in the quantum channel is so high that the probability that $\mathcal{B}$ob passes the test is almost zero, then the protocol does not have to compensate for errors, even in the rare cases where $\mathcal{B}$ob does pass the test. Statement 5 says that, in the context where $C = 0$ and $\mathcal{B}$ob passes the test, $\mathcal{B}$ob must receive almost everything about the string S. This means that the protocol compensates for errors in the quantum channel. Statement 6 says that, in the context where $\mathcal{B}$ob passes the test, the bit C must almost be perfectly random.

3 From $\mathcal{B}$ob to $\mathcal{E}$ve

In this section we prove the following theorem.

Theorem 1. *The security against $\mathcal{B}$ob and tolerance against errors of the above protocol implies its security against $\mathcal{E}$ve.*

Looking ahead to an extension of this result to QKD, we shall be generous and assume that $\mathcal{E}$ve receives $\hat{\theta}$ and C_B at the same time as she receives the pair (E_0, E_1) (which is thus redundant). The following general purpose lemma is useful.

Lemma 2. *Let $\mathcal{A}$, $\mathcal{B}$,$\mathcal{C}$ be any random variables. We have*

$$I(\mathcal{A}; \mathcal{B}) = I(\mathcal{A}; \mathcal{C}) + I(\mathcal{B}; \mathcal{C}) - I(\mathcal{A}, \mathcal{B}; \mathcal{C}) + \sum_c I(\mathcal{A}; \mathcal{B}|\mathcal{C} = c) \Pr(\mathcal{C} = c).$$

The proof is left to the reader. When we refer to this lemma, we say that the mutual information $I(\mathcal{A}; \mathcal{B})$ is partitioned over $\mathcal{C}$. Note that, if $\mathcal{C}$ is a function of $\mathcal{B}$, we obtain $I(\mathcal{B}; \mathcal{C}) = I(\mathcal{A}, \mathcal{B}; \mathcal{C}) = H(\mathcal{C})$ and, therefore,

$$I(\mathcal{A}; \mathcal{B}) = I(\mathcal{A}; \mathcal{C}) + \sum_c I(\mathcal{A}; \mathcal{B}|\mathcal{C} = c) \Pr(\mathcal{C} = c).$$

Proof of Theorem 1. Let α and n_0 be the parameters for the security against $\mathcal{B}$ob. Let α' and n'_0 be the parameters for the tolerance. Let $\alpha_m = Max\{\alpha, \alpha'\}$ and n_m be such that

$$r \times 2^{-(\alpha_m/3)n_m} < \frac{1}{6}. \tag{7}$$

We shall see that $n_0'' = Max\{n_0, n_0', n_m\}$ and $\alpha'' = \alpha_m/3$ are adequate parameters for the security against $\mathcal{E}$ve. Partitioning $I(S; V_{\mathcal{E}\mathrm{ve}})$ over $Pass$ we obtain

$$
\begin{aligned}
&I(S; V_{\mathcal{E}\mathrm{ve}})\\
&= I(S; V_{\mathcal{E}\mathrm{ve}}|Pass = 0)\Pr(Pass = 0)\\
&\quad + I(S; V_{\mathcal{E}\mathrm{ve}}|Pass = 1)\Pr(Pass = 1)\\
&\quad + I(S; Pass)
\end{aligned}
$$

Using 3 and 4 and the fact that $V_{\mathcal{E}\mathrm{ve}}$ is a subset of $V_{\mathcal{B}\mathrm{ob}}$ we obtain

$$
I(S; V_{\mathcal{E}\mathrm{ve}}) = 2 \times 2^{-\alpha_m n} + I(S; V_{\mathcal{E}\mathrm{ve}}|Pass = 1)\Pr(Pass = 1).
$$

We only have to take care of the last term. Partitioning the last term over $C = C_A \oplus C_B$, we obtain

$$
\begin{aligned}
&I(S; V_{\mathcal{E}\mathrm{ve}}|Pass = 1)\Pr(Pass = 1)\\
&= \frac{1}{2}I(S; V_{\mathcal{E}\mathrm{ve}}|C = 0)\Pr(Pass = 1)\\
&\quad + \frac{1}{2}I(S; V_{\mathcal{E}\mathrm{ve}}|C = 1)\Pr(Pass = 1)
\end{aligned}
$$

We now use the two following propositions.

Proposition 3. *For every $\mathcal{E}$ve, for every Channel, $(\forall n > n_0'')$,*

$$
I(S; V_{\mathcal{E}\mathrm{ve}}|C = 1) \times \Pr(Pass = 1) \leq 2^{-\alpha'' n}.
$$

Proposition 4. *For every $\mathcal{E}$ve, for every Channel, $(\forall n > n_0'')$,*

$$
I(S; V_{\mathcal{E}\mathrm{ve}}|C = 0) \times \Pr(Pass = 1) \leq 2^{-\alpha'' n}.
$$

Using propositions 3 and 4 we obtain the security against $\mathcal{E}$ve. We shall prove these propositions in the remainder of this section.

Proof of Proposition 3. Let us consider any integer $n > n_0''$, any Channel and any $\mathcal{E}$ve. Let us consider a $\mathcal{B}$ob that executes $\mathcal{E}$ve's actions in addition to his honest actions. Because $V_{\mathcal{E}\mathrm{ve}}$ is a subset of $V_{\mathcal{B}\mathrm{ob}}$, it will be enough to show that

$$
I(S; V_{\mathcal{B}\mathrm{ob}}|C = 1) \times \Pr(Pass = 1) \leq 2^{-\alpha'' n}.
$$

The basic idea of the proof is simply that, because tolerance against error implies that $\mathcal{B}$ob must receive S each time that $C = 0$ and security against $\mathcal{B}$ob implies that he cannot receive S more than half of the time, then $\mathcal{B}$ob cannot receive S when $C = 1$. The remainder of the proof expresses this idea more formally in a way that takes care of additional points related to the test. By contradiction, let us assume that

$$
I(S; V_{\mathcal{B}\mathrm{ob}}|C = 1) \times \Pr(Pass = 1) > 2^{-\alpha'' n} = 2^{-(\alpha_m/3)n}.
$$

This implies

$$Pr(Pass = 1) > (1/r)2^{-(\alpha_m/3)n} \tag{8}$$

and

$$I(S; V_{Bob}|C = 1) > 2^{-(\alpha_m/3)n} \tag{9}$$

To obtain the contradiction, we show that

$$I(S; V_{Bob}|Pass = 1) \geq (r/2) + \frac{11}{24}2^{-(\alpha_m/3)n} \tag{10}$$

and

$$I(S; V_{Bob}|Pass = 1) < (r/2) + \frac{6}{24}2^{-(\alpha_m/3)n}. \tag{11}$$

First, we show 10. If we partition $I(S; V_{Bob}|Pass = 1)$ over C, we obtain

$$I(S; V_{Bob}|Pass = 1)$$
$$\geq \frac{I(S; V_{Bob}|C = 1)}{2}$$
$$+ \frac{I(S; V_{Bob}|C = 0)}{2}.$$

Formula 7 and 8 give us $Pr(Pass = 1) > 2^{-\alpha_m n}$ which is the required hypothesis in tolerance against errors. Formula 5 and 6 give us that

$$I(S; V_{Bob}|C = 0)Pr(C = 0|Pass) > (r/2) - (r + \frac{1}{2})2^{-\alpha_m n}. \tag{12}$$

Using equation 9, we obtain

$$I(S; V_{Bob}|C = 1)Pr(C = 0|Pass) \geq \frac{1}{2}2^{-(\alpha_m/3)n}. \tag{13}$$

Summing inequalities 12 and 13, one easily obtain 10. Now, we show 11. Let $\tilde{C}$ be the random bit whose existence is required by the security against Bob. Partitioning $I(S; V_{Bob}|Pass = 1)$ over $\tilde{C}$ and using 2, we obtain

$$I(S; V_{Bob}|Pass = 1)$$
$$= \frac{I(S; V_{Bob}|\tilde{C} = 0 \wedge Pass = 1)}{2}$$
$$+ \frac{I(S; V_{Bob}|\tilde{C} = 1 \wedge Pass = 1)}{2}$$
$$+ I(S; \tilde{C}|Pass = 1)$$

Clearly,

$$\frac{I(S; V_{Bob}|\tilde{C} = 0 \wedge Pass = 1)}{2} \leq \frac{r}{2}. \tag{14}$$

Also, using 1, we obtain

$$\frac{I(S; V_{Bob}|\tilde{C} = 1 \wedge Pass = 1)}{2}Pr(Pass = 1) < \frac{1}{2}2^{-\alpha_m n}.$$

from which, using 8, we get

$$\frac{I(S; V_{Bob}|\tilde{C} = 1 \wedge Pass = 1)}{2} < \frac{r}{2} 2^{-(2\alpha_m/3)n} \leq \frac{1}{12} 2^{-(\alpha_m/3)n}. \qquad (15)$$

Now, partitioning $I(S; \tilde{C}, Pass)$ over $Pass$, we obtain that

$$I(S; \tilde{C}, Pass) \geq I(S; \tilde{C}|Pass = 1) \Pr(Pass = 1).$$

Therefore, using 8 and 4, we obtain

$$I(S; \tilde{C}|Pass = 1) \frac{2^{-(\alpha_m/3)n}}{r} \leq 2^{-\alpha_m n}$$

which implies

$$I(S; \tilde{C}|Pass = 1) \leq r 2^{-(2\alpha_m/3)n} \leq \frac{1}{6} 2^{-(\alpha_m/3)n}. \qquad (16)$$

Summing inequalities 14, 15 and 16, one easily obtains 11. This concludes the proof of proposition 3.

To prove proposition 4, the following lemma is useful.

Lemma 5. *Let $\mathcal{A}$, $\mathcal{B}, \mathcal{C}$ and $\mathcal{D}$ be any random variables such that $\mathcal{C}$ is a function of $\mathcal{D}$. For every d, we have*

$$I(\mathcal{A}; \mathcal{B}, \mathcal{C}|\mathcal{D} = d) = I(\mathcal{A}; \mathcal{B}|\mathcal{D} = d).$$

The proof of this lemma is left to the reader.

Proof of Proposition 4. Let us consider any $n > n_0''$, any eavesdropper $\mathcal{E}ve_0$ and any channel $\mathcal{C}hannel$. Our proof consists of finding an eavesdropper $\mathcal{E}ve_1$ such that

$$I(S^{(1)}; V_{\mathcal{E}ve}^{(1)}|C^{(1)} = 1) = I(S^{(0)}; V_{\mathcal{E}ve}^{(0)}|C^{(0)} = 0),$$

where the upper index (i) on a random variable means that it is associated with the eavesdropper $\mathcal{E}ve_i$. Let

$$X = (Open, \hat{\Theta}, C_B),$$

$$U = (B, \Theta, E_0, E_1, C_A, G, S),$$

$$Y = (Error, J_{\mathcal{E}ve})$$

and

$$Z = (\Theta, Error, C_A, G, Syn(W_{C_A}), A, J_{\mathcal{E}ve}).$$

Note that $\mathcal{E}ve$'s view on the execution is $V_{\mathcal{E}ve} = (X, Z)$. Let F be the transformation that maps $x = (open, \hat{\theta}, c_B)$ into $x' = (open, \hat{\theta}', \bar{c}_B)$ where

$$\hat{\theta}_i' = \begin{cases} \hat{\theta}_i & \text{if } open_i = 1 \\ \hat{\theta}_i \oplus 1 & \text{if } open_i = 0 \end{cases}.$$

Let $p_n = \Pr(X = x) = \frac{1}{2 \times 4^{4n}}$. For every $\mathcal{E}ve_1$, using a partition over X and the relation $I(S, X | C = c) = 0$ to obtain the first equality, lemma 5 to obtain the second equality and the bijectivity of F on X to obtain the third equality, we have:

$$
\begin{aligned}
I(S^{(1)}; V_{\mathcal{E}ve}^{(1)} | C^{(1)} = 1) \\
= \sum_x I(S^{(1)}; V_{\mathcal{E}ve}^{(1)} | X^{(1)} = x \wedge C^{(1)} = 1) \times p_n \\
= \sum_x I(S^{(1)}; Z^{(1)} | X^{(1)} = x \wedge C^{(1)} = 1) \times p_n \\
= \sum_x I(S^{(1)}; Z^{(1)} | X^{(1)} = F(x) \wedge C^{(1)} = 1) \times p_n
\end{aligned}
$$

Similarly, we have

$$
\begin{aligned}
I(S^{(0)}; V_{\mathcal{E}ve}^{(0)} | C^{(0)} = 0) \\
= \sum_x I(S^{(0)}; Z^{(0)} | X^{(0)} = x \wedge C^{(0)} = 0) \times p_n
\end{aligned}
$$

Note that (S, Z) is a function of (U, Y). Therefore, we are done if we may define $\mathcal{E}ve_1$'s strategy at steps 2 and 8 such that the distribution of $(U^{(0)}, Y^{(0)})$ given $(X, C)^{(0)} = (x, 0)$ is identical to the distribution of $(U^{(1)}, Y^{(1)})$ given $(X, C)^{(1)} = (F(x), 1)$. Let us consider an execution under $\mathcal{E}ve_0$ where $(X, C)^{(0)} = (x, c)$ and an execution under $\mathcal{E}ve_1$ where $(X, C)^{(1)} = (F(x), \bar{c}) = F(x, c)$. For every $\mathcal{E}ve_1$'s strategy, because $\mathcal{A}lice$ acts exactly in the same way in both executions and U is invariant under F, we have that $U^{(0)}$ in $\mathcal{E}ve_0$ execution is identical to $U^{(1)}$ in $\mathcal{E}ve_1$ execution. Now, we fix the value of U and consider the random variable Y. We must construct $\mathcal{E}ve_1$'s strategy such that the distribution of $Y^{(0)}$ given $(U, X, C)^{(0)} = (u, x, c)$ is the same as the distribution of $Y^{(1)}$ given $F(U, X, C)^{(1)} = (u, x, c)$. At step 2, we define $\mathcal{E}ve_1$ such that she executes the same transfer of information as $\mathcal{E}ve_0$. This is a natural choice because, at this step, (Y, C) is unknown and $\mathcal{E}ve_1$ cannot make use of the difference between the above conditions. We obtain that the random variable $Error$ behaves in the same way in both executions because

- $\mathcal{B}ob$'s outcomes at positions that are used for the test are independent of $\mathcal{B}ob$'s choice of bases at positions that are not used for the test and
- $\mathcal{E}ve_1$ has tampered the photons in the same way as $\mathcal{E}ve_0$.

We now fix the value of $Error$. At step 8, $\mathcal{E}ve_1$ with the view $V_{\mathcal{E}ve}^{(1)}$ executes what $\mathcal{E}ve_0$ executes with the view $F(V_{\mathcal{E}ve}^{(1)}) = V_{\mathcal{E}ve}^{(0)}$. In other words, in these two distinct executions, $\mathcal{E}ve_0$ and $\mathcal{E}ve_1$ act in exactly the same way. The distribution of the random variable $J_{\mathcal{E}ve}$ must be the same in both case, because they have executed the same transfer of information and the same measurement, and $\mathcal{A}lice$ has sent the same state. This concludes the proof of proposition 4 and theorem 1.

4 Security of QKD

The QKD protocol is exactly the QOT protocol, where Bob announces $\hat{\Theta}$ and C_B, and Alice always chooses $C = 0$ ($C_A = C_B$). The security of this QKD protocol is a direct consequence of proposition 4.

5 Conclusion

In this paper, we have shown that the security of an *ordinary QOT* protocol and its tolerance against error implies its security against eavesdropping and, as a corollary, the security of a QKD protocol. In the $\binom{1}{2}$-OT case, security against an eavesdropper means that, if Alice and Bob are honest, Eve cannot find out anything new about (s_1, s_2). A $\binom{1}{2}$-QOT protocol is similar to an ordinary OT protocol, except that Alice transfers two random strings w_0 and w_1 using the sets e_0 and e_1 respectively. One may wonder, if we could obtain the security against eavesdropping of a $\binom{1}{2}$-QOT protocol via a similar approach. This would be interesting because, if efficiency is a concern, the $\binom{1}{2}$-OT task is more powerful than the ordinary OT task: one execution of a $\binom{1}{2}$-OT protocol is enough to construct an ordinary OT protocol, but Kn executions of an ordinary OT protocol, for some $K > 0$, is required to construct a $\binom{1}{2}$-OT protocol [Cr87].

Unfortunately, there is an additional problem in the $\binom{1}{2}$-OT case which is related to the fact that Eve may be aware of some correlation between s_1 and s_2 before the protocol begins. This correlation becomes a correlation between w_0 and w_1 at the time Eve measures her system and, in principle, this may help her to execute a better measurement. In a more elaborate version of this paper, we shall provide a proof for the $\binom{1}{2}$ case where s_1 and s_2 are independent in Eve's initial information.

It would have been reasonable to better explain our formal definitions of security that appear in section 2. Ideally, we should have explained the connection between these definitions and previous definitions found in the literature such as those found in [Cr90]. As mentioned before, an analysis of these definitions will be presented in a subsequent paper.

Finally, now that we know that security may be obtained, it will be useful to determine the maximal error rate that can be tolerated and, for a given error rate, how much resource is required to guarantee a desired level of security. We need this information to find out what kind of technology is required to realize quantum protocols that are efficient and secure. To our knowledge, some theoretical work remains yet to be done at this level, at the least for QOT and QKD.

Acknowledgement

The author gratefully acknowledge fruitful discussions with Charles Bennett, Gilles Brassard, Claude Crépeau and Louis Salvail, and the support of the Institute for Scientific Interchange at Torino under whose auspice much of this work was done. This research was partially supported by Québec's FCAR.

References

[BB84] C.H. Bennett, G. Brassard, Quantum Cryptography: Public key distribution and coin tossing, *Proc. of IEEE International Conference on Computers, Systems, and Signal Processing*, Banglore, India, December 1984, pp. 175–179.

[BBBSS] C.H. Bennett, F. Bessette, G. Brassard, L. Salvail, and J. Smolin, Experimental quantum cryptography, *Journal of Cryptology*, vol. 5, no. 1, 1992, pp. 3–28. Previous version in *Advances in Cryptology* — Eurocrypt '90 Proceeding, May 1990, Springer–Verlag, pp. 253–265.

[BBBW] C.H. Bennett, G. Brassard, S. Breidbart and S. Wiesner, Quantum cryptography, or unforgeable subway tokens, *Advances in Cryptology*: Crypto '82 Proceeding, August 1982, Plenum Press pp. 267–275.

[BBCS] C.H. Bennett, G. Brassard, C. Crépeau, M.-H. Skubiszewska, Practical Quantum Oblivious Transfer, In *proceedings of CRYPTO'91*, Lecture Notes in Computer Science, vol. 576, Springer–Verlag, Berlin, 1992, pp. 351–366.

[BC] G. Brassard and C. Crépeau, Quantum bit commitmemt and coin tossing protocols, in *Advances in Cryptology: Proceeding of CRYPTO'90*, Lecture Notes in Computer Science, vol. 537, Springer–Verlag, Berlin, 1991, pp. 49–61.

[BCJL] G. Brassard, C. Crépeau, R. Jozsa, D. Langlois, A quantum bit commitment scheme provably unbreakable by both parties, in *Proceeding of the 34th annual IEEE Symposium on Foundations of Computer Science*, November 1993, pp. 362–371.

[Cr87] C. Crépeau, Equivalence Between Two Flavors of Oblivious Transfers, *Advances in Cryptology* — Crypto '87 Proceeding, August 1987, Springer–Verlag, pp. 350–354.

[Cr90] C. Crépeau, Correct and Private Reductions among Oblivious Transfers, Ph.D. Thesis, Massachusetts Institute of Technology, 1990.

[Cr94] C. Crépeau, Quantum oblivious transfer, *Journal of Modern Optics*, vol. 41, no. 12, December 1994, pp. 2445–2454.

[CS] C. Crépeau, and L. Salvail, Quantum Oblivious Mutual Identification, *Advances in Cryptology: Proceedings of Eurocrypt'95*, May 1995, Springer–Verlag, to appear.

[MS] D. Mayers and L. Salvail, Quantum Oblivious Transfer is Secure Against All Individual Measurements, *Proceedings of the workshop on Physics and Computation*, PhysComp '94, Dallas, Nov 1994, pp. 69–77.

[MT] Marand, C. and P. Townsend, Quantum key distribution over distances up to 30km, *Optics Letters*, to appear.

[To94] P.D. Townsend, Secure key distribution system based on quantum cryptography, *Electronics Letters*, Vol. 30, no. 10, 12 May 1994.

[TRT1] P.D. Townsend, J.G. Rarity and P.R. Tapster, Single pulse interference in a 10 km long optical fibre interferometer, *Electronics Letters*, vol. 29, no. 7, 1 April 1993, pp. 634–635.

[TRT2] P.D. Townsend, J.G. Rarity and P.R. Tapster, Enhanced single photon fringe visibility in a 10 km-long prototype quantum cryptography channel, *Electronics Letters*, vol. 29, no. 14, 8 July 1993, pp. 1291–1293.

[Yao]   A. Yao, Security of Quantum Protocols Against Coherent Measurements, in *Proceedings of the 26th Symposium on the Theory of Computing*, June 1995, to appear.

How to Break Shamir's Asymmetric Basis

Thorsten Theobald*

Fachbereich IV - Informatik, Universität Trier
D-54286 Trier, Germany
theobald@ti.uni-trier.de

Abstract. At Crypto 93, Shamir proposed a family of signature schemes using algebraic bases. Coppersmith, Stern and Vaudenay presented an attack on one variant of the cryptosystem. Their attack does not recover the secret key. For one of the variants proposed by Shamir we show how to recover the secret key. Our attack is based on algebraic methods which are also applicable to many other instances of polynomial equations in the presence of some trapdoor condition.

1 Introduction

In 1984, Ong, Schnorr and Shamir [OSS84] proposed a very efficient signature protocol based on a quadratic equation in two variables modulo a number of unknown factorization. The scheme was broken by Pollard [PS87], who developed an efficient algorithm to solve these congruences. His algorithm was a great breakthrough in determining the computational complexity of modular quadratic congruences.

Recently, there have been two attempts [Na93], [Sh93b] to repair the weaknesses of the OSS-protocol by inserting additional non-linear structures into the framework of polynomial equations. The low computational requirements of the resulting schemes were especially attractive for devices with restricted computational power. In order to demonstrate the strength of his scheme, Shamir [Sh93a] developed a formal theory concerning the difficulty of polynomial factorization modulo a composite number. From all the members of a very general family, Shamir recommended two variants of his signature scheme: a symmetric one and an asymmetric one.

The symmetric variant could be broken by Coppersmith, Stern and Vaudenay [CSV93]. They showed how to forge a signature, but the secret key is not revealed. However, the main starting point of their attack very strongly relies on the specific symmetry of the underlying algebraic basis[2]. It has been an open problem whether other members of the family are also vulnerable or not.

In this paper we present an attack on the asymmetric basis. Our results are quite unexpected. The asymmetric basis is not only weak in the sense that

---

* Supported by DFG-Graduiertenkolleg "Mathematische Optimierung". This work was done while the author was at the University of Frankfurt.

[2] An algebraic basis is a set of polynomials with some additional properties (see [Sh93b]).

D. Coppersmith (Ed.): Advances in Cryptology - CRYPTO '95, LNCS 963, pp. 136-147, 1995.
© Springer-Verlag Berlin Heidelberg 1995

signatures can be forged. Moreover, it is even possible to recover the secret key within a very short time.

Our attack is based on a detailed analysis of the asymmetric basis polynomials. This characterization makes it possible to apply some ideas of [CSV93] even in the absence of symmetry: Algebraic conditions are transformed into polynomial equations. We then present a method that makes it even possible to solve the specific polynomial equations explicitly. From the solutions of the equations it is only a small step to recover the secret key.

Our attack does not only completely break Shamir's signature scheme. The presented methods also form general guidelines to attack polynomial equations in the presence of some trapdoor information. The connection of our methods with the above mentioned attacks on previous signature schemes provides powerful cryptanalytic tools. We hope that applications of these tools lead to further developments in the area of polynomial equations and in cryptographic research.

2 The signature scheme using algebraic bases

Let $k \geq 3$ and n be the product of two large secret primes p and q. All computations will be done in $\mathbb{Z}_n$, the ring of numbers modulo n. The set of polynomials $\{u_1^2, u_1 u_2, u_2 u_3, \ldots, u_{k-1} u_k\}$ is called the *asymmetric basis* ([Sh93b]).

Two secret invertible matrices $A, B \in \mathbb{Z}_n^{k,k}$ are used to mix up the polynomials. The matrix A transforms the original variables $u_1, \ldots, u_k$ into new variables $y_1, \ldots, y_k$:

$$u_i = \sum_{j=1}^{k} a_{ij} y_j, \quad 1 \leq i \leq k. \tag{1}$$

The matrix B defines k quadratic forms $v_1(u_1, \ldots, u_k), \ldots, v_k(u_1, \ldots, u_k)$ which are linear combinations of the basis elements:

$$v_i(u_1, \ldots, u_k) := b_{i1} u_1^2 + \sum_{j=2}^{k} b_{ij} u_{j-1} u_j, \quad 1 \leq i \leq k. \tag{2}$$

Applying the variable transformation (1) to the polynomials of (2) yields k homogenous quadratic forms $v_i'(y_1, \ldots, y_k)$, $1 \leq i \leq k$. The public key consists of the quadratic forms $v_i'(y_1, \ldots, y_k)$, $1 \leq i \leq k-1$. $v_k'(y_1, \ldots, y_k)$ is not published in order to prevent unique signatures.

A message m is represented by $k-1$ hash values $h_1(m), \ldots, h_{k-1}(m)$. Due to the basis property, each assignment of values (from $\mathbb{Z}_n$) to the basis elements y_1^2, $y_1 y_2, \ldots, y_{k-1} y_k$ implies unique assignments to all homogeneous polynomials of degree 2 in $y_1, \ldots, y_k$. Such an assignment to $y_1^2, y_1 y_2, \ldots, y_{k-1} y_k$ forms a valid signature for m if

$$v_i'(y_1, \ldots, y_k) = h_i(m), \quad 1 \leq i \leq k-1.$$

For a shorter notation, we write v_i instead of $v_i(u_1, \ldots, u_k)$ and v_i' instead of $v_i'(y_1, \ldots, y_k)$.

3 The attack

We address the algebraic problem of finding matrices $A', B' \in \mathbb{Z}_n^{k,k}$, which produce a given public key. The description of the attack refers to a prime modulus. We will justify at the end, why the methods also work in case of a composite modulus.

Throughout the attack, we will not consider degenerated cases which appear with probability of order $O(2^{-l(n)})$, where $l(n)$ is the bitlength of the modulus.

To make the signature scheme vulnerable by algebraic methods, the following definitions are helpful (see e.g. [Ja74]). With every homogeneous quadratic form $q(x_1, \ldots, x_k) \in \mathbb{Z}_n[x_1, \ldots, x_k]$, we associate a $k \times k$-matrix Q (Q is called the *matrix of q*):

$$Q_{j,l} := \begin{cases} \text{coefficient of } x_j x_l \text{ in } q, & j = l \\ \frac{1}{2} \cdot \text{coefficient of } x_j x_l \text{ in } q, & j \neq l \end{cases}.$$

Q is a symmetric matrix which satisfies

$$q(x_1, \ldots, x_k) = (x_1, \ldots, x_k) \, Q \, (x_1, \ldots, x_k)^T.$$

The *rank* of a homogeneous quadratic form is defined as the rank of the associated matrix. It can easily be checked that the rank of a quadratic form is invariant with respect to invertible linear variable transformations. For a matrix Q, the determinant of a $j \times j$-submatrix is called a *minor of order j*. The *row domain* of a matrix is the vector space spanned by the row vectors of the matrix.

3.1 Exploiting the trapdoor

A matrix, that is associated to a linear combination of the basis elements, has only very few nonzero entries. This fact will turn out to be the main weakness of the scheme. A precise classification of the linear combinations with low rank is given in the following lemma. The classification can be exploited in order to establish conditions about the rank of quadratic forms. A rigorous proof for lemma 1 can be found in the appendix.

Lemma 1. *Let $k \geq 4$. A linear combination of the basis elements u_1^2, $u_1 u_2$, ..., $u_{k-1} u_k$ is a quadratic form of rank not greater than 2 if and only if it is of the form*

$$
\begin{array}{lll}
& \alpha_1 u_1 u_2 + \beta_1 u_2 u_3 & \text{(type 1)}, \\
& \alpha_2 u_2 u_3 + \beta_2 u_3 u_4 & \text{(type 2)}, \\
& \quad\quad\vdots & \quad\quad\quad (3) \\
& \alpha_{k-2} u_{k-2} u_{k-1} + \beta_{k-2} u_{k-1} u_k & \text{(type $k-2$)} \\
or & \alpha_{k-1} u_1^2 + \beta_{k-1} u_1 u_2 & \text{(type $k-1$)}
\end{array}
$$

with $\alpha_i, \beta_i \in \mathbb{Z}_n$, $1 \leq i \leq k-1$.

Linear combinations of the polynomials $v_1, \ldots, v_{k-1}$ are linear combinations of the basis elements $u_1^2, u_1 u_2, \ldots, u_{k-1} u_k$. The characterization from lemma 1 can be used to fix $k-1$ important coefficient tuples in these linear combinations.

Lemma 2. *The following system of $k-1$ equations with unknowns α_i, β_i, δ_i, $\epsilon_{3,i}, \ldots, \epsilon_{k-1,i}$, $i \in \{1, \ldots, k-1\}$, has a unique solution:*

$$\alpha_1 u_1 u_2 + \beta_1 u_2 u_3 = v_1 + \delta_1 v_2 + \sum_{j=3}^{k-1} \epsilon_{j,1} v_j,$$

$$\vdots$$

$$\alpha_{k-1} u_1^2 + \beta_{k-1} u_1 u_2 = v_1 + \delta_{k-1} v_2 + \sum_{j=3}^{k-1} \epsilon_{j,k-1} v_j.$$

Proof. It is sufficient to show: For each $i \in \{1, \ldots, k-1\}$ there exists exactly one pair $(\alpha_i, \beta_i) \in \mathbb{Z}_n^2$ and exactly one $(k-2)$-tuple $(\delta, \epsilon_3, \ldots, \epsilon_{k-1}) \in \mathbb{Z}_n^{k-2}$, such that the quadratic form of type i from (3) is equal to $v_1 + \delta v_2 + \sum_{j=3}^{k-1} \epsilon_j v_j$.

Consider the linear space V of the linear combinations $a_1 u_1^2 + \sum_{j=2}^k a_j u_{j-1} u_j$ for arbitrary $a_j \in \mathbb{Z}_n$, $1 \leq j \leq k$. The vector space V is of dimension k. The linear combinations of type i for arbitrary $\alpha_i, \beta_i \in \mathbb{Z}_n$ form a two-dimensional subspace U_i of V.

The quadratic forms $v_1 + \delta v_2 + \sum_{j=3}^{k-1} \epsilon_j v_j$ with arbitrary coefficients $\delta, \epsilon_3, \ldots, \epsilon_{k-1} \in \mathbb{Z}_n$ form an affine subspace T of dimension not greater than $k-2$ of V. The quadratic forms $v_1, \ldots, v_{k-1}$ are linearly independent, because the matrix B from (2) is invertible. Therefore $\dim T = k-2$. In the non-degenerated case, $U_i \cap T$ is of dimension zero. $\square$

Consider now the algebraic condition

$$v_1 + \delta v_2 + \sum_{j=3}^{k-1} \epsilon_j v_j \text{ has a rank not greater than 2.} \tag{4}$$

A matrix is of rank not greater than 2 if and only if all minors of order 3 vanish. If the quadratic forms $v_i = v_i(u_1, \ldots, u_k)$, $1 \leq i \leq k-1$, were known, then condition (4) could be expressed as the vanishing of several determinants with unknowns $\delta, \epsilon_3, \ldots, \epsilon_{k-1}$.

Of course, the polynomials $v_i = v_i(u_1, \ldots, u_k)$, $1 \leq i \leq k-1$, are not part of the public key. However, due to the invariance of the rank with respect to invertible linear variable transformations the condition (4) is equivalent to

$$v_1' + \delta v_2' + \sum_{j=3}^{k-1} \epsilon_j v_j' \text{ has a rank not greater than 2.} \tag{5}$$

Condition (5) can be transformed into a system of polynomial equations: For the matrix of the quadratic form in (5), each minor of order 3 vanishes. The minors are polynomials in $\delta, \epsilon_3, \ldots, \epsilon_{k-1}$. It is possible to use algebraic standard techniques like resultants and the Gaussian elimination algorithm (see e.g. [Mi93]) in order to obtain

1. polynomial expressions for $\epsilon_3, \ldots, \epsilon_{k-1}$ in terms of δ.

2. a polynomial $P(\delta)$ of degree k with simple zeros $\delta_1, \ldots, \delta_{k-2}$ and a double zero δ_{k-1}, where $\delta_1, \ldots, \delta_{k-1}$ are defined as in lemma 2.

For further details of the technical elimination process, see [The95].

Remark. The essential reason why δ_{k-1} must be a double zero of the computed polynomial $P(\delta)$ is the following: In the matrix associated with the quadratic form of type $k-1$ from (3) each submatrix of order 3 consists of at least one row and one column in which only zeros appear. Therefore, if $\epsilon_3, \ldots, \epsilon_{k-1}$ are expressed as polynomials in δ, then δ_{k-1} is a double zero of all the minors of order 3. The variable transformation (1) preserves the double zero.

The explicit value for δ_{k-1} can be obtained by computing the greatest common divisor of P and P'.

3.2 Characterization of the variable transformation

The condition (5) does not distinguish between the $k-1$ different values for $\delta_1, \ldots, \delta_{k-1}$. We will now establish algebraic conditions between different values $\delta_1, \ldots, \delta_{k-1}$.[3] We will explicitly state these conditions for the cases $k = 5$ and (later on) for $k = 4$. With regard to the security and the computational requirements of the scheme, these seem to be the most interesting cases. The conditions can be established analogously for all $k \geq 6$.

Our aim is to find the matrix A' of the variable transformation. If the variable transformation A' is known, the matrix B' of the linear combinations can be easily computed. We reduce the problem of finding the variable transformation to the problem of finding the coefficients $\delta_1, \ldots, \delta_{k-1}$.

As $\epsilon_3, \ldots, \epsilon_{k-1}$ can be expressed in terms of δ, the matrix associated with

$$v_1' + \delta v_2' + \sum_{j=3}^{k-1} \epsilon_j v_j'$$

can be computed in terms of δ. Let Y_i be the row domain of this matrix at δ_i, $1 \leq i \leq k-1$, where $\delta_1, \ldots, \delta_{k-1}$ are defined as in lemma 2. Y_i is a subspace of the vector space $\mathbb{Z}_n^k$. In the following, $\overline{u}_i$ denotes the coefficient vector of the linear function that links in (1) the variable u_i to the variables $y_1, \ldots, y_k$, i.e. $\overline{u}_i := (a_{i1}, \ldots, a_{ik})$, $1 \leq i \leq k$. Each coefficient vector $\overline{u}_i$ is an element of the vector space $\mathbb{Z}_n^k$.

The linear spaces $Y_1, \ldots, Y_{k-1}$ will be characterized by the linear combinations of lemma 2. The vectors $\overline{u}_1, \ldots, \overline{u}_k$ or equivalently the matrix A will be expressed by the linear spaces $Y_1, \ldots, Y_{k-1}$.

---

[3] Some of the following ideas are due to D.Coppersmith [Co94].

The following fact can easily be checked. Let f, g be linear functions in $y_1, \ldots, y_k$, and let C be the symmetric $k \times k$-matrix associated with the quadratic form $f \cdot g$. The row domain of C is spanned by the coefficient vectors which are canonically associated with the linear functions f and g.

Lemma 3. *For $k = 5$ the linear subspaces $Y_1, \ldots, Y_4$ satisfy*

$$
\begin{aligned}
Y_1 &= span(\overline{u}_2, \alpha_1 \overline{u}_1 + \beta_1 \overline{u}_3), \\
Y_2 &= span(\overline{u}_3, \alpha_2 \overline{u}_2 + \beta_2 \overline{u}_4), \\
Y_3 &= span(\overline{u}_4, \alpha_3 \overline{u}_3 + \beta_3 \overline{u}_5), \\
Y_4 &= span(\overline{u}_1, \alpha_4 \overline{u}_1 + \beta_4 \overline{u}_2),
\end{aligned}
$$

where "span" denotes the set of linear combinations with respect to the vector space $\mathbb{Z}_n^k$.

Proof. The quadratic form of type 1 is $\alpha_1 u_1 u_2 + \beta_1 u_2 u_3 = u_2(\alpha_1 u_1 + \beta_1 u_3)$. Y_i is the matrix that is associated with this quadratic form after applying the variable transformation (1). The statement follows from the fact above. $Y_2, \ldots, Y_4$ can be treated analogously. □

Lemma 4. *The coefficient vectors $\overline{u}_1, \ldots, \overline{u}_5$ satisfy*

$$
\begin{aligned}
&\overline{u}_1 \in Y_4 \cap (Y_1 + Y_2) &&\text{(dimension 2)}, \\
&\overline{u}_2 \in Y_1 \cap (Y_2 + Y_3) \cap Y_4 &&\text{(dimension 1)}, \\
&\overline{u}_3 \in Y_2 \cap (Y_1 + Y_4) &&\text{(dimension 1)}, \\
&\overline{u}_4 \in Y_3 \cap (Y_2 + Y_1) \cap (Y_2 + Y_4) &&\text{(dimension 1)}, \\
&\overline{u}_5 \in Y_2 + Y_3 &&\text{(dimension 4)}.
\end{aligned}
\tag{6}
$$

Proof. The first statement follows from

$$
\begin{aligned}
Y_4 \cap (Y_1 + Y_2) &= span(\overline{u}_1, \alpha_4 \overline{u}_1 + \beta_4 \overline{u}_2) \\
&\quad \cap span(\overline{u}_2, \alpha_1 \overline{u}_1 + \beta_1 \overline{u}_3, \overline{u}_3, \alpha_2 \overline{u}_2 + \beta_2 \overline{u}_4) \\
&= span(\overline{u}_1, \overline{u}_2) \cap span(\overline{u}_1, \overline{u}_2, \overline{u}_3, \overline{u}_4) \\
&= span(\overline{u}_1, \overline{u}_2),
\end{aligned}
$$

the other statements can be verified analogously. □

3.3 Computing the variable transformation

So far, we do not know the explicit values for $\delta_1, \delta_2, \delta_3$. Therefore, it is not obvious, whether it is possible to distinguish algebraically between all the subspaces Y_1, Y_2, Y_3. We will show that the spaces Y_1, Y_2, Y_3 or equivalently the values $\delta_1, \delta_2, \delta_3$ are determined uniquely by linear relations. Consequently the values $\delta_1, \delta_2, \delta_3$ are accessible to an algebraic computation. The linear functions $\overline{u}_1, \ldots, \overline{u}_4$ are uniquely determined up to a multiplicative constant. The constants can be chosen arbitrarily, because they can be compensated by the second private transformation. The condition for $\overline{u}_5$ does not characterize $\overline{u}_5$ uniquely. Due to the explicit computation of δ_4, we already distinguished Y_4 from the set $\{Y_1, Y_2, Y_3\}$.

How to distinguish Y_1 from the set $\{Y_2, Y_3\}$:
It can be easily verified with lemma 3 and 4 that $Y_2 \cap Y_4 = \{0\}$, $Y_3 \cap Y_4 = \{0\}$. From (6) it follows that $\overline{u}_2 \in Y_1 \cap Y_4 \neq \{0\}$. Therefore, the row domain Y_1 can be distinguished from the set $\{Y_2, Y_3\}$.

How to distinguish Y_2 from Y_3:
From the observations $Y_3 \cap (Y_1 + Y_4) = \{0\}$, $\overline{u}_3 \in Y_2 \cap (Y_1 + Y_4) \neq \{0\}$, the linear spaces Y_2 and Y_3 can be distinguished.

As the intersections for $\overline{u}_2$, $\overline{u}_3$, $\overline{u}_4$ in (6) are of dimension 1, these coefficient vectors are uniquely determined up to a multiplicative constant.

How to determine $\overline{u}_1$ uniquely:
The intersection $Y_4 \cap (Y_1 + Y_2)$ is of dimension 2. From (6) it follows that $\overline{u_1}$ and $\overline{u_2}$ are elements of this intersection, i.e. $Y_4 \cap (Y_1 + Y_2) = \text{span}(\overline{u_1}, \overline{u_2})$. The division of the quadratic form $u_2(\alpha_1 u_1 + \beta_1 u_3)$ by the linear form u_2 yields the linear form $\alpha_1 u_1 + \beta_1 u_3$. We observe that $\overline{u}_1$ satisfies the condition

$$\overline{u}_1 \in \text{span}(\overline{u}_3, \alpha_1 \overline{u}_1 + \beta_1 \overline{u}_3),$$

whereas a linear combination $a \cdot \overline{u}_1 + b \cdot \overline{u}_2$ with $b \neq 0$ does not satisfy the condition. This condition serves to distinguish $\overline{u_1}$ among the space $\text{span}(\overline{u_1}, \overline{u_2})$. Consequently, $\overline{u}_1$ is uniquely determined.

How to determine $\overline{u}_5$:
For $\overline{u}_5' = a\overline{u}_3 + b\overline{u}_5 \in \text{span}(\overline{u}_3, \overline{u}_5)$ and a variable u_5' linked to the variables $y_1, \ldots, y_5$ via the coefficient vector $\overline{u}_5'$, it follows

$$a_1 \cdot u_1^2 + a_2 \cdot u_1 u_2 + a_3 \cdot u_2 u_3 + a_4 \cdot u_3 u_4 + a_5 \cdot u_4 u_5'$$
$$= a_1 \cdot u_1^2 + a_2 \cdot u_1 u_2 + a_3 \cdot u_2 u_3 + (a_4 + a\, a_5) \cdot u_3 u_4 + b\, a_5 \cdot u_4 u_5,$$

i.e. every linear combination of $u_1^2, \ldots, u_4 u_5$ is a linear combination of $u_1^2, \ldots, u_3 u_4, u_4 u_5'$ and vice versa. Therefore, $\overline{u}_5$ is not determined uniquely. However, it can be replaced by an element $\overline{u}_5'$ in $\text{span}(\overline{u}_3, \overline{u}_5)$ without changing the space of the linear combinations. Such an element can be obtained by dividing the quadratic form $u_4 \cdot (\alpha_3 u_3 + \beta_3 u_5)$ by u_4.

The process of distinguishing the subspaces Y_1, Y_2, Y_3 and the coefficient vectors $\overline{u}_1, \ldots, \overline{u}_5$ can be effectively done with algebraic standard methods, especially determinants and resultants. Of course, the realization of the algebraic condition will lead to high degree polynomials in the variables $\delta_1, \delta_2, \delta_3$. For technical and practical reasons the polynomials have to be reduced. Fortunately, due to the polynomial equation $P(\delta) = 0$ of degree 5, all polynomials can be reduced to degree 4 in each variable. When the double zero δ_4 has been computed, the polynomial $Q(\delta) := P(\delta)/(\delta - \delta_4)^2$ is of degree 3 and serves to reduce each polynomial to degree 4 in each variable. In fact, as $\delta_1, \ldots, \delta_4$ are pairwise different, the polynomials can even be reduced much better.

The ability to distinguish between the subspaces Y_1, Y_2, Y_3 means that we can also compute explicit values for $\delta_1, \delta_2, \delta_3$. These values can be used to evaluate the polynomial expressions for $\overline{u}_1, \ldots, \overline{u}_4, \overline{u}'_5$. The matrix A' that is formed by the rows $\overline{u}_1, \ldots, \overline{u}_4, \overline{u}'_5$ can replace the variable transformation from (1). The missing fifth polynomial can be replaced by the quadratic form

$$\overline{v}_5 := u_1^2 + \sum_{i=1}^{3} u_i u_{i+1} + u_4 u'_5.$$

By inverting the matrix A', the polynomials $v'_1, \ldots, v'_4, \overline{v}_5$ become polynomials in terms of $u_1, \ldots, u_5$. These polynomials are linear combinations of the basis elements and define a matrix B' according to (2). The pair of matrices (A', B') generates the given public key. Therefore we have found the secret key.

3.4 The case $k = 4$

We will now explain the modifications to the case $k = 5$ that are necessary to obtain an attack for $k = 4$. Most of the considerations are identical. It remains to show that all the spaces of Y_1, Y_2, Y_3 or equivalently $\delta_1, \delta_2, \delta_3$ can be distinguished.

When $k = 4$, the quadratic forms of a rank not greater than 2 are of the form

$$\begin{aligned}
\alpha_1 u_1 u_2 + \beta_1 u_2 u_3 \quad &(\text{type 1}), \\
\alpha_2 u_2 u_3 + \beta_2 u_3 u_4 \quad &(\text{type 2}), \\
\text{or} \quad \alpha_3 u_1^2 + \beta_3 u_1 u_2 \quad &(\text{type 3}).
\end{aligned}$$

With respect to the sum

$$v_1 + \delta v_2 + \epsilon_3 v_3,$$

the condition of type i defines δ_i, $1 \leq i \leq 3$. We obtain a polynomial $P(\delta)$ of degree 4. The double zero δ_3 can be extracted by computing the greatest common divisor of P and P'. The following two lemmas can be verified like in the case $k = 5$

Lemma 5. *For $k = 4$ the linear subspaces Y_1, Y_2, Y_3 satisfy*

$$\begin{aligned}
Y_1 &= span(\overline{u}_2, \alpha_1 \overline{u}_1 + \beta_1 \overline{u}_3), \\
Y_2 &= span(\overline{u}_3, \alpha_2 \overline{u}_2 + \beta_2 \overline{u}_4), \\
Y_3 &= span(\overline{u}_1, \alpha_3 \overline{u}_1 + \beta_3 \overline{u}_2).
\end{aligned}$$

Lemma 6. *For $k = 4$ the coefficient vectors $\overline{u}_1, \ldots, \overline{u}_4$ satisfy*

$$\begin{aligned}
\overline{u}_1 &\in Y_3 \cap (Y_1 + Y_2) \quad &(\text{dimension 2}), \\
\overline{u}_2 &\in Y_1 \cap Y_3 \quad &(\text{dimension 1}), \\
\overline{u}_3 &\in Y_2 \cap (Y_1 + Y_3) \quad &(\text{dimension 1}), \\
\overline{u}_4 &\in Y_2 + Y_3 \quad &(\text{dimension 4}).
\end{aligned}$$

δ_3 and therefore Y_3 is already known. Y_1 and Y_2 can be distinguished because of $\overline{u}_2 \in Y_1 \cap Y_3 \neq \{0\}$, $Y_2 \cap Y_3 = \{0\}$.

$\overline{u}_2$ and $\overline{u}_3$ are characterized by one-dimensional spaces. $\overline{u}_1$ can be distinguished from $\overline{u}_2$ in the same way as for $k = 5$. $\overline{u}_4$ can be replaced by the element $\overline{u}'_4 = \alpha_2 \overline{u}_2 + \beta_2 \overline{u}_4$. We further proceed like in the case $k = 5$.

3.5 Composite moduli

If n is a composite modulus of the form $p \cdot q$, there are $k^2 = 25$ zeros of the polynomial $P(\delta)$ modulo n. Both modulo p and modulo q, δ_{k-1} is a double zero. The sequence $\delta_1, \ldots, \delta_{k-1}$ is unique modulo p, and it is unique modulo q. Although there are $(k-1)^2$ different zeros of the polynomial modulo n, only one sequence $\delta_1, \ldots, \delta_{k-1}$ satisfies the uniqueness modulo p and modulo q. Therefore the Chinese remainder theorem guarantees that all computations work in the case of a composite modulus.

3.6 Experimental results

We implemented the attack for moderate key sizes of 50 bits using the package MATHEMATICA. The implementation breaks a given public key within 15 minutes on a HP workstation 735/50, although we did not aim at optimizing the program. For a larger modulus, the number of main steps in the attack does *not* increase. Of course, the cost of the elementary operations like polynomial addition and multiplication increase with the length of the modulus.

The recommended bit length for the modulus in [Sh93b] is 512. We estimate that an implementation for this key size runs in at most a few hours. Further details of an implementation as well as an example of the computation can be found in [The95].

4 Symmetric basis versus asymmetric basis

There are some remarkable differences between the attack on the symmetric basis $\{u_1 u_2, \ldots, u_{k-1} u_k, u_k u_1\}$ in [CSV93] and our attack on the asymmetric basis. These differences lead to a better insight into the attacks. In the symmetric case, there are several equivalent sequences for the coefficients $\delta_1, \ldots, \delta_k$. Therefore these coefficients cannot be computed. The sequence $\delta_1, \ldots, \delta_{k-1}$ is unique in the asymmetric case. Lemma 4 and 6 provide the necessary conditions to distinguish among the coefficients $\delta_1, \ldots, \delta_{k-1}$. All coefficients can be computed, and it is possible to discover the secret key.

Considering a composite modulus in the symmetric case, each (unknown) sequence of the coefficients modulo p can be combined with each (unknown) sequence of the coefficients modulo q. This is the essential reason why the computation of the secret key is at least as hard as factoring the modulus (see [Sh93b]). For the asymmetric basis, the sequence of the coefficients is unique even modulo

n. Our attack therefore shows that the asymmetric basis does not fit into the framework of difficult algebraic instances that was developed in [Sh93a].

From a practical point of view, we can mention the following results: Due to the ability to compute $\delta_1, \ldots, \delta_{k-1}$, the attack on the asymmetric basis can get rid of the time-consuming large polynomials. Therefore it takes much less time to attack the asymmetric basis than to attack the symmetric basis.

5 Open questions

The intention of [CSV93] and of our work was to break some specific proposed cryptosystems. For further research in the cryptographic applications of polynomials, it would be of interest to characterize the power of these cryptanalytic methods from a more general point of view. In a cryptosystem that is resistant against the presented cryptanalytic methods, the trapdoor condition should *not* influence the rank of quadratic forms. Otherwise, this influence is a promising starting point for an attack.

Acknowledgements

I wish to thank D. Coppersmith, C. P. Schnorr and S. Vaudenay for their encouragement and support.

References

[Co94]   D. Coppersmith: Private communication (1994).

[CSV93] D. Coppersmith, J. Stern and S. Vaudenay: Attacks on the Birational Permutation Signature Schemes. Proceedings of CRYPTO 93, Lecture Notes in Computer Science 773, 435-443 (1993).

[Ja74]   N. Jacobson: Basic Algebra. W. H. Freeman and Company, San Francisco (1974).

[Mi93]   B. Mishra: Algorithmic Algebra. Springer-Verlag, New York (1993).

[Na93]   D. Naccache: Can O.S.S. be Repaired ? - Proposal for a New Practical Signature Scheme. Proceedings of Eurocrypt 93, Lecture Notes in Computer Science 765, 233-239 (1993).

[OSS84] H. Ong, C. P. Schnorr, A. Shamir: A Fast Signature Scheme Based on Quadratic Equations. Proceedings 16th ACM Symposium on Theory of Computing, 208-216 (1984).

[PS87]   J. M. Pollard, C. P. Schnorr: An Efficient Solution to the Congruence $x^2 + y^2 = m \pmod{n}$. IEEE Transactions on Information Theory, Vol. 33, 702-709 (1987).

[Sh93a] A. Shamir: On the Generation of Multivariate Polynomials Which Are Hard To Factor. Proceedings 25th ACM Symposium on Theory of Computing, 796-804 (1993).

[Sh93b] A. Shamir: Efficient Signature Schemes Based on Birational Permutations. Proceedings of CRYPTO 93, Lecture Notes in Computer Science 773, 1-12 (1993).

[The95] T. Theobald: Digitale Unterschriften mittels birationaler Permutationen. Diplomarbeit, Universität Frankfurt (1995).

Appendix: Proof of Lemma 1

Proof. Let

$$a_1 u_1^2 + \sum_{i=2}^{k} a_i\, u_{i-1} u_i, \quad a_1, \ldots, a_k \in \mathbb{Z}_n \tag{7}$$

be a linear combination of the basis elements and Q the associated matrix. The matrix Q is (for better clearness zero-entries are omitted)

$$Q = \frac{1}{2}\begin{pmatrix} 2a_1 & a_2 & & & & \\ a_2 & & a_3 & & & \\ & a_3 & & a_4 & & \\ & & a_4 & & \ddots & \\ & & & \ddots & & a_k \\ & & & & a_k & \end{pmatrix}.$$

"$\Longleftarrow$" Obviously, a linear combination (7) of type $t \in \{1, \ldots, k-1\}$ is of rank not greater than 2.

"$\Longrightarrow$" We assume, that the linear combination (7) is not of the form (3).

Case 1: $a_1 = 0$. It follows

$$Q = \frac{1}{2}\begin{pmatrix} 0 & a_2 & & & & \\ a_2 & & a_3 & & & \\ & a_3 & & a_4 & & \\ & & a_4 & & \ddots & \\ & & & \ddots & & a_k \\ & & & & a_k & \end{pmatrix}.$$

For $j := \min\{i \in \{2, \ldots, k\} : a_i \neq 0\}$ and $l := \max\{i \in \{2, \ldots, k\} : a_i \neq 0\}$, we have $l - j \geq 2$. The 3×3-submatrix

$$\frac{1}{2}\begin{pmatrix} 0 & a_j & 0 \\ a_j & 0 & x \\ 0 & 0 & a_l \end{pmatrix} \quad \text{with } x := \begin{cases} 0, & \text{if } l - j > 2 \\ a_{j+1}, & \text{if } l - j = 2 \end{cases}$$

of Q has determinant $-(\frac{1}{2})^3 a_j^2 a_l \neq 0$. Therefore rank $Q \geq 3$.

Case 2: $a_1 \neq 0$. For $j := \min\{i \in \{3,\ldots,k\} : a_i \neq 0\}$ the upper left $j \times j$-submatrix of Q is

$$\frac{1}{2}\begin{pmatrix} 2a_1 & a_2 & & & & \\ a_2 & & 0 & & & \\ & 0 & & \ddots & & \\ & & \ddots & & 0 & \\ & & & 0 & & a_j \\ & & & & a_j & \end{pmatrix}.$$

The 3×3-submatrix

$$\frac{1}{2}\begin{pmatrix} 2a_1 & x & 0 \\ x & 0 & a_j \\ 0 & a_j & 0 \end{pmatrix} \quad \text{with } x := \begin{cases} 0, & \text{if } j > 3 \\ a_2, & \text{if } j = 3 \end{cases}$$

has determinant $-(\frac{1}{2})^2 a_1 a_j^2 \neq 0$. It follows rank $Q \geq 3$. $\qquad\qquad\square$

On the Security of the Gollmann Cascades

Sang-Joon Park   Sang-Jin Lee   Seung-Cheol Goh

Electronics and Telecommunications Research Institute
161 Kajong-Dong, Yusong-Gu, Taejon, 305-350, Korea
e-mail: goh@dingo.etri.re.kr

Abstract. The purpose of this paper is to evaluate the security of the Gollmann m-sequence cascades of k stages. We give some theoretical results, which can be utilized to construct the transition matrix T_n of the conditional probabilities between the input and output strings of a stage. And then, we describe an attack algorithm for guessing the initial state of the first LFSR with desired reliability, using the transition matrix $S_n = T_n^{k-1}$ of the conditional probabilities between the input string of the second stage and the output of the final stage of the given k-stage cascade. We finally evaluate the security of the cascades against this attack. Menicocci recently conjectured that there do not exist the complete analysis of the Gollmann cascades of more than 4 stages and it is infeasible to attack the 10-stage cascades with LFSRs of degree 100. Our experimental results show that the 9-stage cascades with LFSRs of degree 100 are completely breakable and the 10-stage cascades may be insecure.

1   Introduction

The purpose of this paper is to evaluate the security of the Gollmann m-sequence cascades of k stages[1].

A Gollmann m-sequence cascade of k stages consists of a series of k Linear Feedback Shift Registers(LFSRs), with primitive feedback polynomials of same degree n. It produces pseudo-random binary sequences of period $(2^n-1)^k$, linear equivalence exceeds $n(2^n-1)^{k-1}$[1]. The first LFSR is regularly clocked, whereas all registers except the first are clock controlled by their predecessors. A binary input bit a_t clocks a LFSR if $a_t = 1$, and then is added to the output from the LFSR to give output c_t, which becomes the input of the next.

$$c_t = a_t + b(h_t) \bmod 2, \ h_t = h_{t-1} + a_t \bmod 2^n-1, \ t = 0, 1, 2, \cdots \quad (1)$$

with the initial condition $h_{-1} = 0$, where $b(\cdot)$ is the sequence generated by the LFSR.

A weakness of the Gollmann cascades, called *lock-in effect*, was studied in [2]. But the attack by lock-in effect requires 10^{21} iterations to analysis the two-stage cascades with polynomials of degree 34. Menicocci proposed an efficient attack on the two-stage Gollmann m-sequence cascades, which utilized the correlations between the input and output strings of the final stage of cascades[4]. He also

D. Coppersmith (Ed.): Advances in Cryptology - CRYPTO '95, LNCS 963, pp. 148-156, 1995.

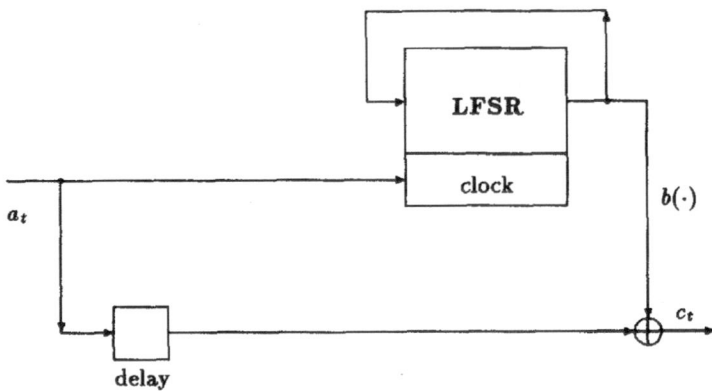

Fig 1: A stage of the m-sequence cascade; The input bit a_t clocks LFSR, and then is added to the output from the LFSR to give output c_t. The *delay* means the addition takes place after clock.

proposed an attack on the Gollmann cascade of k stages, and conjectured there do not exist algorithms for the complete analysis of the cascade of more than 4 stages, and it is infeasible to attack on the m-sequence cascade consists of more than 10 stages with primitive polynomials of degree greater than 100[5].

In this paper, we give some theoretical results, which can be utilized to construct the transition matrix T_n of the conditional probabilities between the input and output strings of a stage. And then, we describe an attack algorithm for guessing the initial state of the first LFSR with desired reliability, using the transition matrix $S_n = T_n^{k-1}$ of conditional probabilities between the input string of the second stage and the output of the final stage of the given k-stage cascade. We also evaluate the security of the cascades against this attack. We finally give experimental results on the cryptanalysis of the cascades with less than 10 stages. Our experimental results show that the 9-stage cascades with LFSRs of degree 100 are completely breakable and the 10-stage cascades may be insecure.

2 Constructions of Transition Matrices

Definition Let $w = x_0 x_1 \cdots x_{n-1}$ be a n-bit word. Then the value of w, denoted by $f(w)$, is

$$f(w) = \sum_{k=0}^{n-1} x_k 2^{n-1-k} \qquad (2)$$

Definition Let w and v be n-bit words such that $f(w) = i$, $f(v) = j$. And let $t^n[i, j]$ be the conditional probability that a given stage of the cascade produce v when w is applied to it. Then the $2^n \times 2^n$ matrix $T_n = (t^n[i, j])$, $0 \le i < 2^n$,

$0 \leq j < 2^n$, is called a transition matrix of the correlation probabilities between the input and output words of the stage.

Remark For a given cascade of k stages, $S^n = (T_n)^{k-1} = (s^n[i,j])$ is the transition matrix of the conditional probabilities, where $s^n[i,j]$ is the probability that the final stage generates $v = f^{-1}(j)$ when $w = f^{-1}(i)$ is applied to the second stage of the given cascade.

Example If $a_t = 0$, $c_t = b(S_t)$. The sequence $b(\cdot)$ is generated by LFSR, so we have $P(c_t = 0) = P(c_t = 1) = \frac{1}{2}$. If $a_t = a_{t+1} = 0$, then $c_{t+1} = b(S_{t+1}) = b(S_t) = c_t$, so we have $P(c_{t+1} = c_t = 0) = P(c_{t+1} = c_t = 1) = \frac{1}{2}$. Continuing this process, we will finally have

$$T_2 = \begin{pmatrix} \frac{1}{2} & 0 & 0 & \frac{1}{2} \\ \frac{1}{4} & \frac{1}{4} & \frac{1}{4} & \frac{1}{4} \\ 0 & \frac{1}{2} & \frac{1}{2} & 0 \\ \frac{1}{4} & \frac{1}{4} & \frac{1}{4} & \frac{1}{4} \end{pmatrix}, \quad (T_2)^2 = T_2 \circ T_2 = \begin{pmatrix} \frac{3}{8} & \frac{1}{8} & \frac{1}{8} & \frac{3}{8} \\ \frac{1}{4} & \frac{1}{4} & \frac{1}{4} & \frac{1}{4} \\ \frac{1}{8} & \frac{3}{8} & \frac{3}{8} & \frac{1}{8} \\ \frac{1}{4} & \frac{1}{4} & \frac{1}{4} & \frac{1}{4} \end{pmatrix}$$

Definition Let $S = \{t, t+1, \cdots, t+n-1\}$ be a set of n consecutive non-negative integers, and let $w = a_t a_{t+1} \cdots a_{t+n-1}$ be a n-bit word such that $f(w) = i$. Then, we may define a class P_i of subsets S_k by following procedure.

> **Procedure 1**
> Set $k = 1$ and $S_k = \{t\}$.
> For all $l = t+1, \cdots, t+n-1$
> If $a_l = 0$, then set $S_k = S_k \cup \{l\}$.
> Otherwise, set $k = k+1$ and set $S_k = \{l\}$.

Property 1 Let $w = a_t a_{t+1} \cdots a_{t+n-1}$ be a n-bit word such that $f(w) = i$. And let m be the number of ones in $\{a_t, a_{t+1}, \cdots, a_{t+n-1}\}$. Then the number of subsets $|P_i|$ of the class P_i is

$$|P_i| = \begin{cases} m & \text{if } a_t = 1 \\ m+1 & \text{otherwise} \end{cases} \tag{3}$$

Proof. By definition of the class P_i, it is trivial. □

Property 2 Assume that a stage of a cascade produces $v = c_t c_{t+1} \cdots c_{t+n-1}$ when $w = a_t a_{t+1} \cdots a_{t+n-1}$ is applied. If $P_i = \{S_1, \cdots, S_m\}$ is the class determined by Procedure 1, then, for all l in S_k,

$$a_l + c_l \bmod 2 = \text{const} \tag{4}$$

Proof. From (1), it is trivial. □

Property 3 For given integers i and j, where $0 \leq i < 2^n$, $0 \leq j < 2^n$, let $w = a_t a_{t+1} \cdots a_{t+n-1}$ and $v = c_t c_{t+1} \cdots c_{t+n-1}$ be the binary representations of i and j respectively. And let $P_i = \{S_1, \cdots, S_m\}$ be the class determined by Procedure 1. Then $t^n[i,j] = 0$ if and only if there exists an integer $0 \leq k \leq m$ such that $a_l + c_l \neq a_{l'} + c_{l'} \bmod 2$ for some l, l' in S_k,

Proof. By Property 2, it can be easily proved. □

Property 4 If $t^n[i,j] \neq 0$ for some $0 \leq i < 2^n$, $0 \leq j < 2^n$, then $t^n[i,j] = \frac{1}{2^{|P_i|}}$.

Proof. Let $w = a_t a_{t+1} \cdots a_{t+n-1}$ be the binary representations of i. And let P_i be the class determined by Procedure 1. If a stage produces $v = c_t c_{t+1} \cdots c_{t+n-1}$ when w is applied, then, by Property 2, there exist constants $b_1, b_2, \cdots, b_{|P_i|}$, such that $c_l = a_l + b_k \bmod 2$ for all l in S_k, where S_k is in P_i. This means that, when w is applied, all outputs are uniquely determined by $b_1, b_2, \cdots, b_{P_i}$, so that if $t^n[i,j] \neq 0$, $t^n[i,j] = \frac{1}{2^{P_i}}$. □

Property 5 For a given integer $0 \leq i < 2^n$, there are exactly 2^{P_i} number of nonzero elements in $T_{n,i}$, where $T_{n,i}$ is the i-th row of the transition matrix T_n.

Proof. By Property 4, trivial. □

Property 6 If $0 \leq j < 2^{n-1}$, $t^n[i,j] = t^n[i, 2^n - 1 - j]$ for all $0 \leq i < 2^n$.

Proof. Let $w = a_t a_{t+1} \cdots a_{t+n-1}$ and $v = c_t c_{t+1} \cdots c_{t+n-1}$ be n-bit words such that $f(w) = i$, $f(v) = j$. If the value of $\bar{v} = d_t d_{t+1} \cdots d_{t+n-1}$ is $2^n - 1 - j$, then we have, $d_l = c_l + 1 \bmod 2$ for all l, $l = t, \cdots, t+n-1$. Let $P_i = \{S_1, S_2, \cdots, S_{|P_i|}\}$ be the class determined by Procedure 1. If $t^n[i,j] = 0$, by Property 3, there is an integer k, $1 \leq k \leq |P_i|$, such that $a_l + c_l \neq a_{l'} + c_{l'} \bmod 2$ for some l, l' in S_k. Hence $a_l + d_l \neq a_{l'} + d_{l'} \bmod 2$, so $t^n[i, 2^n - 1 - j] = 0$. If $t^n[i,j] \neq 0$, by Property 3, we have, $a_l + d_l = a_l + c_l + 1 \bmod 2 = const$ for all l in S_k. Hence, by Property 4, $t^n[i,j] = t^n[i, 2^n - 1 - j] = \frac{1}{2^{|P_i|}}$. □

Property 7 For all $0 \leq i, j < 2^{n-1}$, $t^n[i,j] = t^n[2^{n-1} + i, 2^{n-1} + j]$.

Proof. We first note that $P_i = P_{2^{n-1}+i}$, where P_i and $P_{2^{n-1}+i}$ are the classes determined by Procedure 1. Let $w = a_t a_{t+1} \cdots a_{t+n-1}$, $v = c_t c_{t+1} \cdots c_{t+n-1}$. Then $w' = \bar{a}_t a_{t+1} \cdots a_{t+n-1}$ and $v' = \bar{c}_t c_{t+1} \cdots c_{t+n-1}$ are the binary representations of $2^{n-1} + i$ and $2^{n-1} + j$ respectively. Since $a_t + c_t = \bar{a}_t + \bar{c}_t$, we have $t^n[i,j] = t^n[2^{n-1} + i, 2^{n-1} + j]$. □

Similarly, we can prove:

Property 8 For all $0 \leq i, j < 2^{n-1}$, $t^n[2^{n-1} + i, j] = t^n[i, 2^{n-1} + j]$.

Property 9 In the case of $n \geq 3$, $t^n[i,j] = 0$ if $0 \leq i < 2^{n-2}$, $2^{n-2} \leq j < 2^{n-1}$.

Proof. Let $w = a_t a_{t+1} \cdots a_{t+n-1}$ and $v = c_t c_{t+1} \cdots c_{t+n-1}$ be n-bit words such that $f(w) = i$ and $(v) = j$. Then, we have, $a_t = a_{t+1} = c_t = 0$, $c_{t+1} = 1$. We note that $t, t+1$ belong to same set S_1 in the class P_i. But $a_t + c_t \neq a_{t+1} + c_{t+1} \bmod 2$ so $t^n[i,j] = 0$. □

Property 10 For all $0 \leq i, j < 2^{n-2}$, $t^n[i,j] = t^{n-1}[i,j]$.

Proof. Let P_i be the class of subsets of the set $S = \{t, t+1, \cdots, t+n-1\}$, and let P_i' be the class of subsets of the set $S = \{t+1, \cdots, t+n-1\}$ determined by Procedure 1, when i is applied. Note $|P_i| = |P_i'|$, $S_1' = S_1 - \{t\}$, $S_k' = S_k$, $k > 1$. Let $w = a_t a_{t+1} \cdots a_{t+n-1}$ and $v = c_t c_{t+1} \cdots c_{t+n-1}$ be the binary representations of i, j, respectively. If $t^n[i,j] \neq 0$, then $a_l + c_l \bmod 2 = const$ if l in S_k for all S_k in P_i. Hence $a_l + c_l \bmod 2 = const$ if l in S_k' for all S_k' in P_i', so that $t^{n-1}[i,j] \neq 0$.

Therefore, if $t^n[i,j] \neq 0$, then, by Property 4, $t^n[i,j] = t^{n-1}[i,j] = \frac{1}{2^{|P_i|}}$. In the case of $t^n[i,j] = 0$, we can prove similarly $t^n[i,j] = t^{n-1}[i,j] = 0$. $\square$

Property 11 For all $2^{n-2} \leq i < 2^{n-1}$, $t^n[i,j] = \frac{1}{2} t^{n-1}[i,j]$.

Proof. Let P_i and P_i' be the classes defined in the proof of Property 10. In a similar way, we can easily prove that $t^n[i,j] = 0$ if and only if $t^{n-1}[i,j] = 0$. We note that, by Property 1, $|P_i| = |P_i'|+1$. Hence, by Property 4, if $t^n[i,j] \neq 0$, $t^n[i,j] = \frac{1}{2} t^{n-1}[i,j]$. $\square$

Property 12 Let $w = a_t a_{t+1} \cdots a_{t+n-1}$ be the binary representation of i. Then $t^n[i,0] = 0$ if and only if there exist l, l' such that $t \leq l < l' \leq t+n-1$, $a_l = 1$ but $a_{l'} = 0$.

Proof. Assume that $a_l = 1$ but $a_{l'} = 0$ for some l, l' such that $t \leq l < l' \leq t+n-1$. Let $f(w) = i$. And let S_k be the subset of P_i which contains l'. Let us define $h = \max\{m \mid l \leq m < l', a_m = 1\}$. Then $h \in S_k$ and $a_h + 0 \neq a_{l'} + 0 \bmod 2$ so that $t^n[i,0] = 0$. Suppose that $t^n[i,0] = 0$. Then it is clear that $i \neq 0, 1$. Hence there exists at least one l such that $t \leq l < t+n-1$ and $a_l = 1$, say $p = \min\{l \mid t \leq l < t+n-1, a_l = 1\}$. In the case of $p = t$, if $a_l = 1$ for all $t+1 \leq l \leq t+n-1$, then each S_k, $k = 1, 2, \cdots, t+n-1$, has only one element so that $t^n[i,0] \neq 0$. So we may assume $p \geq t+1$ ie $a_t = 0$. If $a_l = 1$ for all l, $p \leq l \leq t+n-1$, then only S_0 may have more than one elements. Since all a_l in S_0 are 0 so $t^n[i,0] \neq 0$. $\square$

From Property 12, we have

Property 13

$$t^n[i,0] = t^n[i, 2^n-1] = \begin{cases} \frac{1}{2^{k+1}} & \text{if } i = 2^k-1 \text{ for some } k, \ 0 \leq k \leq n-1 \\ \frac{1}{2^n} & \text{if } i = 2^n-1 \\ 0 & \text{Otherwise} \end{cases}$$

Now we summarize Properties solved in this section to give Theorem 1.

Theorem 1. *For all $n \geq 3$, let $H_n = (h^n[i,j])$, $H_n' = (h'^n[i,j])$ be $2^{n-1} \times 2^{n-1}$ matrices of the probabilities such that $h'^n[i,j] = h^n[i, 2^{n-1}-1-j]$. Then*
1) the transition matrix $T_n = (t^n[i,j])$ can be represented as:

$$T_n = \begin{pmatrix} H_n & H_n' \\ H_n' & H_n \end{pmatrix} \tag{5}$$

2) the $2^{n-1} \times 2^{n-1}$ matrix $H_n = (h^n[i,j])$ can be represented as:

$$H_n = \begin{pmatrix} H_{n-1} & 0 \\ \frac{1}{2} H_{n-1}' & \frac{1}{2} H_{n-1} \end{pmatrix} \tag{6}$$

with the initial condition $H_1 = (\frac{1}{2})$.

3 The attack proposed

Our aim is to find all initial states of LFSRs of the given cascade, under the assumption that we know a sufficiently large number of consecutive bits of the output and all primitive polynomials.

For a given cascade of k stages, we define a matrix S^n by

$$S^n = (T_n)^{k-1} = (s^n[i,j]) \tag{7}$$

Then the component $s^n[i,j]$ is the probability that the final stage produces $v = f^{-1}(j)$ when $w = f^{-1}(i)$ is applied to the second stage of the cascade. This matrix S^n enables us to guess the initial state of the LFSR of the first stage with desired reliability.

Theorem 2. *For all i, $0 \leq i < 2^n$, let w_i be the n-bit word such that $f(w_i) = i$, and A_i the event that w_i is applied to the second stage of the cascade. And let E_0 be the event that the cascade generates a run of n consecutive zeros and E_1 a run of n consecutive ones. If $E = E_0 \cup E_1$, then*

$$P(A_i|E) = s^n[i,0] \tag{8}$$

where $P(A_i|E)$ denotes the conditional probability of A_i occurs given that E has occurred.

Proof. By Bayes's theorem, we have,

$$P(A_i|E_0) = \{P(A_i)P(E_0|A_i)\}/\{\sum_{m=0}^{2^n-1} P(A_m)P(E_0|A_m)\}$$

$$= P(E_0|A_i)/\sum_{m=0}^{2^n-1} P(E_0|A_m) = s^n[i,0]/\sum_{m=0}^{2^n-1} s^n[m,0] = s^n[i,0]$$

Hence, $P(A_i|E) = \frac{1}{2}P(A_i|E_0) + \frac{1}{2}P(A_i|E_1) = \frac{1}{2}(s^n[i,0]+s^n[i,2^n-1]) = s^n[i,0]$.

Corollary 3. *Let $w = a_t a_{t+1} \cdots a_{t+n-1}$. Then, for all k, $0 \leq k \leq n-1$,*

$$P(a_{t+k} = 0|E) = \sum_i s^n[i,0] \tag{9}$$

where i ranges over integers such that $i \wedge 2^{n-k-1} = 0$. The notation $\wedge$ denotes the bitwise AND of two integers. Hereafter, for all $0 \leq k \leq n-1$, we define q_k by $q_k = 1-P(a_{t+k} = 0|E)$.

For a precise description of our attack, let m be the same degree of primitive polynomials and let p be the desired error rate. In order to find the initial state of LFSR in the first stage, we first scan the given output bits to find all runs of at least n consecutive zeros or ones. Then we scan each run for k such that $q_k \leq p$, to set linear equations, which have the m unknowns for the initial state.

In this way, we can set a sufficiently large number, say l, of linear equations. Choose randomly m out of the l equations and solve the linear system. If it has a solution, say $\bar{x} = x_0, x_1, \cdots, x_{m-1}$, then examine how many equations hold when $\bar{x}$ is substituted. If it is the real initial state, the number of equations held is on average $l(1 - p)$. This is repeated for each except final stage. And then, apply the algebraic technique[6] to the final stage to guess the initial state of its LFSR. Test, finally, whether the cascade produces the given output sequence. If not, repeat again.

How many trials of selections m from l equations is needed to find the initial state of the first stage. By a similar way in [3], we can estimate the probability q that a trial successes, so q^{-1} is the expected number of trials.

$$q = (1-\frac{pl}{l})(1-\frac{pl}{l-1}) \cdots (1-\frac{pl}{l-m+1}) \geq (1-\frac{pl}{l-m+1})^m \qquad (10)$$

Example Consider a cascade of 2 stages with LFSRs of the same length 34. We are assumed to know 800 consecutive output bits, so that we can find, on average, 25 number of runs whose lengths are greater than or equal to 5. If we put $p = 2^{-4}$, we can set 50 linear equations so $q \approx 10^{-3}$. Hence about 1000 trials are expected to find the correct solution.

Algorithm: An attack on the Gollmann k-stage cascades

Step 1 Scan the given output bits, $a_h, a_{h+1}, a_{h+2}, \cdots\cdots$, to find all runs of at least n consecutive zeros or ones. This number n may be determined by the desired error rate p, the number of given output bits, and the degrees of LFSRs.

Step 2 Repeat the process of Steps 3 - 6 in order to guess the initial states of LFSRs of all but final stages with desired reliability $1-p$.

Step 3 Select a_t such that $P(a_t = 0|E) \geq 1-p$ to set a linear equation, which has the m unknowns for the initial state. In this way, we can set a sufficiently large number, say l, of linear equations.

Step 4 Select randomly m out of the l equations and solve the linear system.

Step 5 If it has no solution, goto Step 4.

Step 6 Examine how many equations hold when the solution is substituted. If the number is less than k, which is heuristically determined by $l(1 - p)$, goto Step 4.

Step 7 Apply the algebraic technique to the kth stage using the sequence generated by the $k-1$th stage.

Step 8 Test whether the cascade produces the correct final output sequences. If success terminate, else goto Step 2.

4 Security evaluation

The security of the Gollmann cascades may be evaluated by the number of trials for obtaining the correct initial state of its first LFSR and the number of the final output bits for obtaining runs whose lengths are sufficiently large enough to guarantee small error rate. From the view point of cryptanalysis, there may be trade-off between the number of trials and bits. When the length of runs is greater than 9, we could not calculate correctly the error rate and the number of trials due to the memory problem of computer. However, by some statistical observation, we expect that the runs of length 20 enable us to break completely the cascades of 10 stages with LFSRs of degree 100.

We have implemented the proposed attack in C on an Axil Hyundai workstation(80 MIPS), compatible with a SUN Sparc 10, in a UNIX environment. For the case of 2-stage cascades with primitive polynomials of degree 34, we have found all initial states of LFSRs within 2.8 CPU seconds on average.

Following table gives our experimental results. In those tables, Degree denotes the degree of primitive polynomials, Stage the numbers of stages, Bit the numbers of output bits, Run the lengths of runs, p desired error rates, Eqns the numbers of equations, Trial the numbers of trials, Theo the numbers theoretically calculated, Expe the numbers of experimental results, CPU the CPU seconds needed execution the algorithm.

Table: Experimental Results

Degree	Stage	Bit	Run	p	Eqns	Trial		CPU
						Theo	Expe	
34	2	800	5	0.0625	75	18	15	2.8
	3	1800	6	0.1269	82	423	119	4.5
	4	7000	8	0.1208	73	399	96	7.2
	5	35000	10	0.1145	110	146	241	49.4
44	2	1000	5	0.0625	95	46	23	8.7
	3	2300	6	0.1269	116	2033	626	17.4
	4	9000	8	0.1208	109	1576	436	28.4
	5	45000	10	0.1145	117	891	255	68.8
54	2	1200	5	0.0625	115	114	437	18.2
	3	3000	6	0.1269	127	16734	1843	56.7
	4	15000	8	0.0801	136	327	873	99.2
	5	60000	10	0.0801	121	412	271	155.3
100	3	20000	7	0.04699	304	350	896	123.0
	4	60000	9	0.05235	256	949	1892	265.1
	5	240000	11	0.05318	263	1006	3950	687.3
	6	500000	12	0.08440	242	99874	4282	721.0
	7	2000000	14	0.08960	230	267088	1741	828.0
	8	8000000	16	0.08562	262	88956	39147	2334.0
	9	30000000	18	0.07188	248	15340	51006	3322.0

5 Concluding Remarks

In this paper, we have evaluated the security of the Gollmann m-sequence cascades. We have described, in detail, an algorithm for guessing the initial state of the first LFSR with desired reliability. We have also given experimental results on the complete analysis of 9-stage cascades with polynomials of degree 100. We have finally defined the security of cascades by the number of trials for obtaining the correct initial state of its first LFSR and the number of the final output bits obtaining runs whose lengths are sufficiently large enough to guarantee small error rate. Now we are trying the complete analysis of 10-stage cascades with LFSRs of degree 100. It is expected to be successful with 2^{26} output bits of the cascades.

References

1. Gollmann, D.: Pseudorandom Properties of Cascaded Connections of Clock Controlled Shift Registers, Advances in Cryptology-Eurocrypt'84, Lecture notes in Computer Science **209**, (1985) 93–98
2. Chambers, W. G., Gollmann, D.: Lock in Effect in Cascades of Clock Controlled Shift Register, Advances in Cryptology-Eurocrypt'88, Lecture notes in Computer Science **330**, (1989) 331–343
3. Meier, W., Staffelbach, O.: Correlation Properties of Combiners with Memory in Stream Ciphers, Journal of Cryptology, 5, No. 1, (1992) 67–86
4. Menicocci, R.: Cryptanalysis of a Two-Stage Gollmann Cascade Generator, Proc. the 3rd Symposium of State and Progress of Research in Cryptography, Rome, (1993) 62–69
5. Menicocci, R.: Short Gollmann Cascade Generator May Be Insecure, presented at Fourth IMA Conf. on Cryptography and Coding, (1993)
6. Beker, H., Piper, F,: Cipher Systems: The Protection of Communications, Wiley-interscience, (1982)

Improving the Search Algorithm
for the Best Linear Expression

Kazuo Ohta, Shiho Moriai, and Kazumaro Aoki*

NTT Laboratories, 1–2356 Take, Yokosuka, Kanagawa, 238-03 Japan

Abstract. It is important to find the best linear expression to estimate the vulnerability of cryptosystems to Linear Cryptanalysis. This
paper presents a method to improve Matsui's search algorithm which
determines the best linear expression. This method is based on analyzing the dominant factor of search complexity. We introduce the *search
pattern* in order to reduce unnecessary search candidates, and apply the
proposed search algorithm to DES and FEAL. The n-round best linear
expressions of DES are found as fast as Matsui's algorithm for $n \leq 32$.
Those of FEAL are found much faster than his algorithm; the required
time is decreased from over three months to about two and a half days.
New results for FEAL are also described; we find the n-round best linear expressions ($n \leq 32$) with higher deviations than those derived from
Biham's 4-round iterative linear approximations.

1 Introduction

Linear Cryptanalysis was proposed by Matsui [M93] and is known to be one
of the most effective known plaintext attacks. In Linear Cryptanalysis, we find
the following linear approximation (equation (1)) which holds with probability
$p \neq 1/2$ for randomly given plaintext P and the corresponding ciphertext C, and
then determine one key bit $K[k_1, k_2, \ldots, k_c]$ by the maximum likelihood method.

$$P[i_1, i_2, \ldots, i_a] \oplus C[j_1, j_2, \ldots, j_b] = K[k_1, k_2, \ldots, k_c], \qquad (1)$$

where $i_1, i_2, \ldots, i_a$, $j_1, j_2, \ldots, j_b$, and $k_1, k_2, \ldots, k_c$ denote fixed bit locations and
$A[i_1, i_2, \ldots, i_a] = A[i_1] \oplus A[i_2] \oplus \cdots \oplus A[i_a]$ ($A[i]$ denotes the i-th bit of A). If
the success rate is fixed, equation (2) holds [M93, *Lemma 2*],

$$N |p - 1/2|^2 = c \text{ (fixed)}, \qquad (2)$$

where N denotes the number of the known plaintexts needed to attack the
cryptosystem. We call $|p - 1/2|$ the *deviation*. Equation (2) shows that it is
important to find the linear approximation with the highest deviation, the *best
linear expression*, to estimate the smallest value of N[†].

At EUROCRYPT'94, two important topics were discussed in [B94] and [M94];
1) the duality between Linear Cryptanalysis and Differential Cryptanalysis, and

---

* Affiliation during this work: Department of Mathematics, School of Science and
Engineering, Waseda University.
† Here we don't consider the reduction in N possible with multiple approxima-
tions [KR94].

D. Coppersmith (Ed.): Advances in Cryptology - CRYPTO '95, LNCS 963, pp. 157-170, 1995.
© Springer-Verlag Berlin Heidelberg 1995

2) a search algorithm for determining the best linear expression (differential characteristic) of DES-like cryptosystems. On 1), it was clarified how to construct the global linear expressions (differential characteristics) from the local ones in the field of Linear Cryptanalysis (Differential Cryptanalysis). On 2), it should be noted that confirming that the "best linear expression" is the best is difficult, for example, Biham described that "We have exhaustively verified that this iterative characteristic is the best among all the characteristics with at most one active S box at each round, ... (Matsui claims that his linear expression is the best without any restriction.)" in his paper [B94], which implies that there might be more effective linear expressions than that used by Matsui [M93]. To ensure his characteristic is the best without any restriction, Matsui developed a search algorithm where he introduced the temporary value of the best probability, $\overline{BEST_n}$, obtained under the restriction of limiting the number of "active" S-boxes at each round, and then got the best probability, $BEST_n$, without the restriction.

Matsui's search algorithm was applied to DES, s^2DES, and LOKI successfully, whose f functions have similar structures to that of DES [TSM94]. In applying this search algorithm to the above cryptosystems, we can find the best linear expressions with ease, because the speed-up technique using $\overline{BEST_n}$ is effective in pruning off the unnecessary branches.

Unfortunately, for some cryptosystems, his search algorithm is not fast enough and it takes over 3 months to find the best linear expression of FEAL–8 on a workstation (about 30 MIPS), for example. The main reason is that the search bound of his algorithm depends on the cryptosystem, and is so loose for some cryptosystems that an enormous number of candidates are searched for.

If we impose restrictions on the type of linear expression, for example, those constructed using the iterative linear approximations with small number of rounds, the search problem becomes easy. An efficient method was proposed in [K92], where the idea is to find effective differential characteristics of iterative type in Differential Cryptanalysis. His idea is expected to be also applicable to Linear Cryptanalysis because of the duality of these cryptanalyses. His approach might satisfy some attackers, since it helps them find an effective linear approximation with a large probability, but it doesn't satisfy the designers of cryptosystems. Thus, it is important to develop a search algorithm for the best linear expression with less complexity in order to estimate the vulnerability of cryptosystems to Linear Cryptanalysis.

We extend Matsui's search algorithm by analyzing the dominant factor of its complexity carefully, and propose a method to reduce the number of candidates, in other words, not to expand unnecessary nodes in the search tree. We can eliminate the unnecessary candidates using the set of possible values of the deviations of the linear approximations of f functions of all rounds, what we call the *search pattern*, before determining the linear approximations themselves.

We apply the improved search algorithm to DES and FEAL, and compare its search complexity against that of Matsui's algorithm. The n-round best linear expressions of DES are found as fast as Matsui's algorithm for $n \leq 32$. Those of

FEAL are found much faster than his algorithm; the time required is decreased from over three months to about two and a half days.

In truth, the best linear expressions obtained by Matsui's algorithm might not be the best (it doesn't mean that we don't consider the effect of multiple approximations), and those obtained by ours might not, either. This is because both algorithms calculate the deviation of the linear approximation of the f function by the *Piling-up Lemma* [M93] which assumes the linear approximations of the "active" S-boxes in the f function hold independently. The subject of this paper is to reduce the search time of Matsui's algorithm.

New results for FEAL are also obtained; we find the best linear expressions of 7-round with the deviation of $1.15 \times 2^{-8\ddagger}$, those of 15-round with 1.48×2^{-20}, and those of 31-round with 1.99×2^{-41}, while Biham's equivalent values were 2^{-11}, 2^{-23}, and 2^{-47}, respectively [B94].

This paper is organized as follows. In chapter 2 we introduce Matsui's search algorithm [M94]. Next, in chapter 3, after considering the search bounds and complexity of his search algorithm, we give two problems. In chapter 4 we propose an improved search algorithm to solve these problems. In chapter 5 we apply the improved search algorithm to DES and FEAL, and show how the search time is reduced. We also show the best deviations of DES and FEAL obtained from the search.

2 Preliminary

2.1 Linear Approximation of f Function

This paper discusses the security of iterated cryptosystems, where f is the round function. Let I_i, O_i, and K_i be the input data, the output data, and the subkey data of the i-th round f function. We define ΓX as the masking value of data X, and $X[\Gamma X]$ as the even parity value of $X \cdot \Gamma X$, where $\cdot$ represents a bitwise AND operation. We call equation (3) the *linear approximation of the i-th round f function*. This linear approximation may be abbreviated to the pair of masking values, $(\Gamma O_i, \Gamma I_i)$.

$$I_i[\Gamma I_i] \oplus O_i[\Gamma O_i] = K_i[\Gamma K_i] \tag{3}$$

This paper uses the term, *deviation*, which means the absolute value of the difference of the probability from $1/2$. The deviation of the linear approximation of the i-th round f function $(\Gamma O_i, \Gamma I_i)$, denoted as $p'_i(\Gamma O_i, \Gamma I_i)$, is defined as follows[§]. We may simply use p'_i for $p'_i(\Gamma O_i, \Gamma I_i)$.

$$p'_i(\Gamma O_i, \Gamma I_i) = |\, Prob\{I_i[\Gamma I_i] \oplus O_i[\Gamma O_i] = 0\} - 1/2 \,| \tag{4}$$

[‡] Ohta and Aoki showed 7-round linear approximations with the deviation of 1.15×2^{-8} in [OA94]. We have confirmed that they found some of the 7-round "best" linear expressions.

[§] Since in our search algorithm $p'_i(\Gamma O_i, \Gamma I_i)$ is calculated by the *Piling-up Lemma* using the deviations of the linear approximations of "active" S-boxes considering complexity, the value of $p'_i(\Gamma O_i, \Gamma I_i)$ might differ from that obtained by equation (4). This problem was also not solved in [M94]. See section 4.3.

2.2 Best Linear Expression

In this paper we use $BEST_n$ defined by equation (5), following the definition of $BEST_n^{LC}$ used in [TSM94]. We don't consider multiple approximations described in [N94]. We call $BEST_n$ the *n-round best deviation*. The n-round linear approximation with the n-round best deviation is called the *n-round best linear expression*.

$$BEST_n = \max_{\Gamma O_i = \Gamma O_{i-2} \oplus \Gamma I_{i-1}(3 \leq i \leq n), \; \Gamma P \neq 0} \{2^{n-1} \prod_{i=1}^{n} (p_i'(\Gamma O_i, \Gamma I_i))\} \qquad (5)$$

2.3 Matsui's Search Algorithm for Best Linear Expression

This section introduces the search algorithm for the best linear expression proposed in [M94]. It is based on the mathematical inductive method and derives $BEST_n$, which is the n-round best linear deviation, from the knowledge of all i-round best linear deviations, $BEST_i$ $(1 \leq i \leq n-1)$.

$\overline{BEST_n}$ is the temporary value of $BEST_n$ during the search. We have to pay attention when setting the initial value of $\overline{BEST_n}$. The search program can determine $BEST_n$ as long as $\overline{BEST_n} \leq BEST_n$ holds, but the farther from $BEST_n$ the initial value of $\overline{BEST_n}$ is, the more complex the search becomes. How close the initial value of $\overline{BEST_n}$ is to $BEST_n$ decides the efficiency of the search. The framework of Matsui's algorithm consists of the following recursive procedures. Here the value of $[p_1', p_2', \ldots, p_t']$ is defined as the following equation,

$$[p_1', p_2', \ldots, p_t'] = 2^{t-1} \prod_{i=1}^{t} p_i'. \qquad (6)$$

[Matsui's Search Algorithm] [M94]

Procedure Round-1:
  For each candidate for ΓO_1, do the following:
  ▷ Let $p_1' = \max_{\Gamma I} p_1'(\Gamma O_1, \Gamma I)$.
  ▷ If $[p_1', BEST_{n-1}] < \overline{BEST_n}$, then try another candidate for ΓO_1.
  ▷ Call *Procedure Round-2*.

  Let $BEST_n = \overline{BEST_n}$.
  Exit the program.

Procedure Round-2:
  For each candidate for ΓO_2 and ΓI_2, do the following:
  ▷ Let $p_2' = p_2'(\Gamma O_2, \Gamma I_2)$.
  ▷ If $[p_1', p_2', BEST_{n-2}] < \overline{BEST_n}$, then try another candidate for ΓO_2 and ΓI_2.
  ▷ Call *Procedure Round-3*.

  Return to the upper procedure.

Procedure Round-i: $(3 \leq i \leq n-1)$
  For each candidate for ΓI_i, do the following:

▷ Let $\Gamma O_i = \Gamma O_{i-2} \oplus \Gamma I_{i-1}$.

▷ Let $p'_i = p'_i(\Gamma O_i, \Gamma I_i)$.

▷ If $[p'_1, p'_2, \ldots, p'_i, BEST_{n-i}] < \overline{BEST_n}$, then try another candidate for ΓI_i.

▷ Call *Procedure Round-(i+1)*.

Return to the upper procedure.

Procedure Round-n:

Let $\Gamma O_n = \Gamma O_{n-2} \oplus \Gamma I_{n-1}$.

Let $p'_n = \max_{\Gamma I} p'_n(\Gamma O_n, \Gamma I)$.

If $[p'_1, p'_2, \ldots, p'_n] > \overline{BEST_n}$, then $\overline{BEST_n} = [p'_1, p'_2, \ldots, p'_n]$.

Return to the upper procedure.

3 Consideration of Matsui's Search Algorithm

3.1 Search Bound

Consider the search bound of Matsui's search algorithm that determines the range of p'_i, which is the deviation of the linear approximation of each round. In his search algorithm, the linear approximations whose deviation satisfies equation (7) become search candidates at the i-th round ($1 \leq i \leq n$).

$$[p'_1, p'_2, \ldots, p'_i, BEST_{n-i}] \geq \overline{BEST_n} \tag{7}$$

Equation (7) is transformed into equation (8) using equation (6).

$$p'_1 \times p'_2 \times \cdots \times p'_i \geq \frac{\overline{BEST_n}}{2^i \times BEST_{n-i}} \tag{8}$$

The above equation shows that the product of $p'_1, p'_2, \ldots, p'_i$ depends on the ratio of $\overline{BEST_n}$ to $BEST_{n-i}$. There is a possibility that the value of $\overline{BEST_n}/BEST_{n-i}$ is too small for some cryptosystems, even if the initial value of $\overline{BEST_n}$ can be set close to $BEST_n$. If the value of the right side of inequality (8) is small, an excessive number of candidates of the linear approximations with deviation p'_i are searched for. This is because the number of the linear approximations of the f function with deviation $\geq p'$ increases, as the value of p' decreases, as Table 1 shows.

p'	2^{-1}	2^{-2}	2^{-3}	2^{-4}	2^{-5}	2^{-6}	$\cdots$
DES	1	13	195	3803	40035	371507	$\cdots$
FEAL	16	1808	98576	3453200	774484304	1215648016	$\cdots$

Table 1. Number of linear approximations of f function with deviation $\geq p'$

3.2 Complexity

This section discusses two subjects on the search complexity of Matsui's search algorithm. One is the dominant factor of search complexity, and the other is the relation between search complexity and round number.

First, we show that the complexity of the search for the n-round best linear expression is dominated by the number of candidates in *Procedures Round-1* and *Round-2*. In *Procedure Round-1*, all the ΓO_1s that satisfy equation (9) are search candidates, and in the *Procedure Round-2*, all the $(\Gamma O_2, \Gamma I_2)$s that satisfy equation (10) are search candidates.

$$p_1' = \max_{\Gamma I} p_1'(\Gamma O_1, \Gamma I) \geq \frac{\overline{BEST_n}}{2 \times BEST_{n-1}} \tag{9}$$

$$p_2' = p_2'(\Gamma O_2, \Gamma I_2) \geq \frac{\overline{BEST_n}}{2^2 \times p_1' \times BEST_{n-2}} \tag{10}$$

In *Procedure Round-i* ($3 \leq i \leq n$), ΓO_i is fixed by the equation, $\Gamma O_i = \Gamma O_{i-2} \oplus \Gamma I_{i-1}$, and only the ΓI_is that satisfy equation (7) are search candidates. The number of candidates heuristically increases from *Procedure Round-1* to *Procedure Round-2* but then decreases in subsequent procedures. Thus the complexity of the search can be estimated from the number of the candidates in the first two procedures.

The search becomes more complex as the number of the candidates in the first two procedures increases; in other words, as the ratio of $\overline{BEST_n}$ to $BEST_{n-1}$ and that of $\overline{BEST_n}$ to $BEST_{n-2}$ decrease, according to inequalities (9) and (10).

Next, we show the relation between search complexity and the number of rounds. We define C_n as the estimated value of the *incremental* search complexity for the n-round best linear expression under the desirable condition that we know all $BEST_r$ ($r \leq n$) values, *i.e.* the least search complexity. As mentioned above, the search complexity can be estimated from the number of candidates in the first two procedures. Thus C_n is obtained using the data in Table 1 and calculating all the ΓO_1s and $(\Gamma O_2, \Gamma I_2)$s that satisfy both inequalities (9) and (10).

The relations between C_n and n for DES and FEAL are shown in Figure 1. Figure 1 shows that C_n of FEAL is much more than that of DES. (Note the scale of the vertical axis.) It is also clear that C_n isn't related to n.

3.3 Problems

This section describes two problems of Matsui's search algorithm. We begin by introducing the term, *search pattern*. We define the search pattern as the set of n deviations $(q_1', q_2', \ldots, q_n')$ that satisfies $p_1' = q_1'$, $p_2' = q_2'$, $\ldots$, and $p_n' = q_n'$ when we search for the n-round linear approximations.

Problem 1. Duplicate Candidates

The n-round linear approximation whose linear approximation of the i-th round f function $(\Gamma O_i, \Gamma I_i)$ is exchanged for that of the $(n - i + 1)$-th round

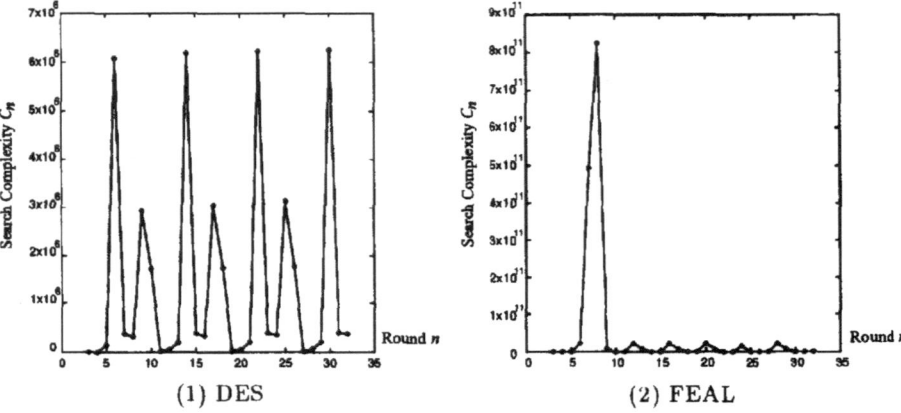

Fig. 1. Relation between C_n and n

f function (ΓO_{n-i+1}, ΓI_{n-i+1}) for all i ($1 \leq i \leq n$) has the same meaning as the original one, since we can exchange the roles of P and C, and also K_i and K_{n+1-i} in equation (1). In Matsui's search algorithm, when search pattern $(q'_1, q'_2, \ldots, q'_n)$ satisfies equation (7), so does its inverse pattern $(q'_n, q'_{n-1}, \ldots, q'_1)$. Two linear approximations with the same meaning from the viewpoint of Linear Cryptanalysis are searched for in his search algorithm, and one of them is unnecessary.

Problem 2. Nonexistent Candidates

In Matsui's search algorithm each search pattern $(q'_1, q'_2, \ldots, q'_n)$ satisfies all the following inequalities, which are derived from equation (7).

$$\begin{aligned}
[\, q'_1, q'_2, \ldots\ldots, q'_n] &\leq \overline{BEST_n} \\
[\, q'_2, q'_3, \ldots\ldots, q'_n] &\leq BEST_{n-1} \\
[\, q'_3, q'_4, \ldots, q'_n] &\leq BEST_{n-2} \\
&\vdots \\
[\, q'_n] &\leq BEST_1
\end{aligned} \tag{11}$$

These inequalities are not guaranteed to yield the conditions that *all* r-round ($1 \leq r < n$) linear approximations contained in the n-round linear approximation have the deviation less than or equal to $BEST_r$. In other words, there is a possibility that the linear approximation whose search pattern satisfies inequality (12) is searched for unnecessarily.

$$[q'_i, q'_{i+1}, \ldots, q'_{i+r-1}] > BEST_r \quad (1 \leq i \leq n,\ i+r-1 < n) \tag{12}$$

From the definition of $BEST_r$ (equation (5)), the linear approximations that satisfy inequality (12) don't exist, and need not be searched for.

4 Discussion

4.1 Solving the Problems

The problems explained in section 3.3 are solved as follows. Note that the problems concern only the deviations of the linear approximations of the f function, and are independent of the choice of $(\Gamma O_i, \Gamma I_i)$.

First we choose the search patterns $(q'_1, q'_2, \ldots, q'_n)$ that satisfy the following two conditions, and secondly decide $(\Gamma O_i, \Gamma I_i)$ for each round using the chosen search patterns. In order to cover all possible search patterns, we make use of the idea of Matsui's search algorithm and list all the $(q'_1, q'_2, \ldots, q'_n)$s that satisfy inequality (7). We can guarantee that the set of the chosen search patterns contains all the candidates needed to be searched for, since Matsui's search algorithm doesn't miss possible candidates.

Condition 1 (Deletion of Duplicate Candidates)
The search pattern $(q'_1, q'_2, \ldots, q'_n)$ must satisfy the following condition,

$$C(q'_1, q'_2, \ldots, q'_n) \leq C(q'_n, q'_{n-1}, \ldots, q'_1)$$

where $C(q'_1, q'_2, \ldots, q'_n)$ that satisfies $p'_1 = q'_1$, $p'_2 = q'_2$, $\ldots$, and $p'_n = q'_n$ is the complexity of the search for the n-round linear approximations.

As shown in section 3.2, $C(q'_1, q'_2, \ldots, q'_n)$ can be estimated well by the number of candidates in the first two procedures. Thus we might compare $C(q'_1, q'_2)$ with $C(q'_n, q'_{n-1})$ instead of the inequality above, where $C(a, b)$ denotes the number of the linear approximations with $p'_1 = a$ and $p'_2 = b$. $C(a, b)$ is also calculated using the data in Table 1.

Consider the following example; when we search for the 7-round best linear expression of FEAL, there are such search patterns as $(2^{-3}, 2^{-3}, 2^{-2}, 2^{-2}, 2^{-1}, 2^{-1}, 2^{-1})$ and its inverse pattern $(2^{-1}, 2^{-1}, 2^{-1}, 2^{-2}, 2^{-2}, 2^{-3}, 2^{-3})$. We choose the latter search pattern, because $C(2^{-3}, 2^{-3})$ is 9,364,045,824, and $C(2^{-1}, 2^{-1})$ is only 256.

Condition 2 (Deletion of Nonexistent Candidates)
For all i and r $(1 \leq i \leq n$, $i + r - 1 < n)$, the search pattern $(q'_1, q'_2, \ldots, q'_n)$ must also satisfy the following condition.

$$[q'_i, q'_{i+1}, \ldots, q'_{i+r-1}] \leq BEST_r$$

4.2 Improved Search Algorithm

The improved search algorithm is shown below. It determines $BEST_n$ and the best linear expression *i.e.* the set of n linear approximations of the f function $(\Gamma O_i, \Gamma I_i)$ $(1 \leq i \leq n)$. It consists of four routines, *Procedure Main*, *MakeList-round-i*, *PickandChoose* and *Search*.

In *Procedure MakeList-round-i*, the list of all the possible search patterns is made. In *Procedure PickandChoose*, the search pattern $Pattern_j(p''^j_1, p''^j_2, \ldots,$

$p_n^{\prime j}$) which satisfies *Conditions 1* and *2* is chosen from the list made in *Procedure MakeList-round-i*, and then *Procedure Search* is called with *Pattern$_j$* as a parameter. In *Procedure Search*, the linear approximations of the f function whose deviation of the i-th round p_i' equals $p_i^{\prime j}$ are searched for. The difference of *Procedure Search* from Matsui's search algorithm is that the deviation of the linear approximation of the f function is decided using *Pattern$_j$*($p_1^{\prime j}$, $p_2^{\prime j}$, ..., $p_n^{\prime j}$) chosen in *Procedure PickandChoose*, not equation (7).

[Our Search Algorithm] (for FEAL[¶])

Procedure Main:

Let $\overline{BEST_n} = 2 \times BEST_{n-1}$, and $BEST_n = 1$.
Do
  ▷ Let $\overline{BEST_n} = 2^{-1} \times \overline{BEST_n}$.
  ▷ Call *Procedure MakeList-round-1*.
  ▷ Call *Procedure PickandChoose*.
while $BEST_n \neq \overline{BEST_n}$.
Exit the program.

Procedure MakeList-round-i: ($1 \leq i \leq n$)

For each candidate for p_i', do the following:
  ▷ If $[p_1', p_2', \ldots, p_i', BEST_{n-i}] < \overline{BEST_n}$, then try another candidate for p_i'.
  ▷ If $i < n$, then call *Procedure MakeList-round-(i+1)*,
    else if $i = n$, then add $(p_1', p_2', \ldots, p_n')$ satisfying $[p_1', p_2', \ldots, p_n'] = \overline{BEST_n}$
    to the list of the search patterns.
Return to the upper procedure.

Procedure PickandChoose:

For each candidate for $(p_1', p_2', \ldots, p_n')$, do the following:
  ▷ If $^\exists i, ^\exists r$ ($1 \leq i \leq n, i + r - 1 < n$) satisfying $[p_i', p_{i+1}', \ldots, p_{i+r-1}'] > BEST_r$,
    then try another candidate for $(p_1', p_2', \ldots, p_n')$.
  ▷ If $\mathcal{C}(p_1', p_2') > \mathcal{C}(p_n', p_{n-1}')$, then try another candidate for $(p_1', p_2', \ldots, p_n')$.
  ▷ Let *Pattern$_j$*($p_1^{\prime j}, p_2^{\prime j}, \ldots, p_n^{\prime j}$) = $(p_1', p_2', \ldots, p_n')$.
  ▷ Call *Procedure Search* (*Pattern$_j$*).
Return to the upper procedure.

Procedure Search (*Pattern$_k$*):

Procedure Round-h: ($h = 1, 2$)

For each ΓO and ΓI *s.t.* $p_h'(\Gamma O, \Gamma I) = p_h^{\prime k}$,
  ▷ Let $\Gamma O_h = \Gamma O$, and $\Gamma I_h = \Gamma I$.
  ▷ Call *Procedure Round-(h+1)*.
Return to the upper procedure.

Procedure Round-i: ($3 \leq i < n$)

Let $\Gamma O_i = \Gamma O_{i-2} \oplus \Gamma I_{i-1}$.

[¶] This search algorithm cannot always find the best linear expression of DES faster than Matsui's algorithm. In the case of DES, the technique of improving the value of the right side of inequality (8) using the search patterns chosen in *Procedure PickandChoose* speeds up his algorithm.

For each ΓI s.t. $p_i'(\Gamma O_i, \Gamma I) = p_i'^k$,

▷ Let $\Gamma I_i = \Gamma I$.
▷ Call *Procedure Round-(i+1)*.

Return to the upper procedure.

Procedure Round-n:

Let $\Gamma O_n = \Gamma O_{n-2} \oplus \Gamma I_{n-1}$.
Let $\Gamma I_n = \Gamma I$ and $BEST_n = \overline{BEST_n}$ if $p_n'(\Gamma O_n, \Gamma I) = p_n'^k$.

Return to the upper procedure.

4.3 Revision of Best Deviation

Because both of Matsui's search algorithm and ours calculate the deviation of the linear approximation of the f function by the *Piling-up Lemma* for the sake of complexity, it might differ from the true one obtained from equation (4). We revise $BEST_n$ determined by the search program as follows; for all the best linear expressions obtained from the search, we calculate the true deviations of all rounds of linear approximations and determine the best deviations.

However, we might miss the *truly* best linear expression. Because the deviation of linear approximation of each round can be increased or decreased by revision, the truly best linear expression might come from the linear approximation with smaller deviation than $BEST_n$[||]. It is difficult to search all possible linear expressions for the truly best deviation because of too much complexity. The method to compute the true deviation of the linear approximation of the f function with less complexity will solve this problem. (See Appendix.)

5 Experimental Results

5.1 Search Time

We applied our search algorithm to DES and FEAL using a SPARCstation 10 (SuperSPARC/36MHz, 31MIPS). It takes the times shown in *Tables 2* and *3*, respectively, to find the n-round ($4 \leq n \leq 8^{**}$) best linear expression when we set the initial value of $\overline{BEST_n}$ equal to $BEST_n$. Note that the data with "∗" are theoretical estimations from the number of candidates.

The best linear expressions of DES are found with ease by Matsui's algorithm. One of the reasons is that DES doesn't have so many linear approximations of the f functions with large deviations, as *Table 1* shows. Another reason is that the technique that reduces candidates by limiting the number of "active" S-boxes in the f function is effective for DES [M93]. On the other hand, Matsui's algorithm takes too much time to search for those of FEAL. This is because FEAL has many more candidates than DES, and the technique of limiting the number of

[||] The 7- and 15-round linear expressions of FEAL with the best deviation we have found are in the linear approximations with the deviation of $2^{-1} \times BEST_n$ [MAO96].

[**] The search times of the other n-round ($9 \leq n \leq 32$) best expressions can be estimated from *Figure 1*.

Round	4	5	6	7	8
Our Algorithm	0.2 sec	1.2 sec	0.2 sec	0.2 sec	0.3 sec
Matsui's Algorithm	0.2 sec	1.2 sec	19.2 sec	1.3 sec	0.9 sec

Table 2. Search Time (DES)

Round	4	5	6	7	8
Our Algorithm	0.0 sec	0.4 sec	3.8 min	4.2 hr	2.3 day
Matsui's Algorithm	0.3 sec	23.1 sec	39.0 hr	*34 day	*58 day

Table 3. Search Time (FEAL)

"active" S-boxes is not effective for FEAL, since the linear approximation of each round in the n-round ($n \leq 32$) best linear expressions of DES has no "active" S-box or only one, but almost all S-boxes are "active" in the case of FEAL.

5.2 Best Deviations of DES and FEAL

Figure 2 shows the n-round best deviations of DES and FEAL ($1 \leq n \leq 32$). This is based on the data in *Table 4*, which are the revised data as explained in section 4.3. Note that each best deviation of DES is the same as $BEST_n$ determined by the search program, because the linear approximation of each round of the best linear expressions of DES has no "active" S-box or only one, and the deviation calculated by the *Piling-up Lemma* is the same as the true deviation. *Figure 2* also shows that all the n-round ($n \leq 32$) best deviations of FEAL we found are higher than those derived from Biham's 4-round iterative linear approximations [B94].

All the n-round ($10 \leq n \leq 32$) best linear expressions of FEAL that we obtained are based on the following type of 8-round iterative linear approximations. The following iterative linear approximation is one of them, and there are other variations of this type. The best linear expressions of DES are also based on a similar 8-round iterative linear approximation "−ACD−DCA" [M93]. Note that numbers in brackets are the true deviations obtained from the definition of p_i' (equation (4)). The difference occurs because p_i' is calculated by the *Piling-up Lemma* in the search algorithm (See section 4.3).

i	ΓO_i	ΓI_i	p_i' (true deviation)	
1	00000000	00000000	2^{-1} (1.00×2^{-1})	
2	02a40104	81010100	2^{-3} (1.00×2^{-3})	
3	81010100	00600000	2^{-4} (1.81×2^{-4})	
4	02c40104	81010100	2^{-2} (1.00×2^{-2})	2^{-13} (1.64×2^{-12})
5	00000000	00000000	2^{-1} (1.00×2^{-1})	
6	02c40104	81010100	2^{-2} (1.00×2^{-2})	
7	81010100	00600000	2^{-4} (1.81×2^{-4})	
8	02a40104	81010100	2^{-3} (1.00×2^{-3})	

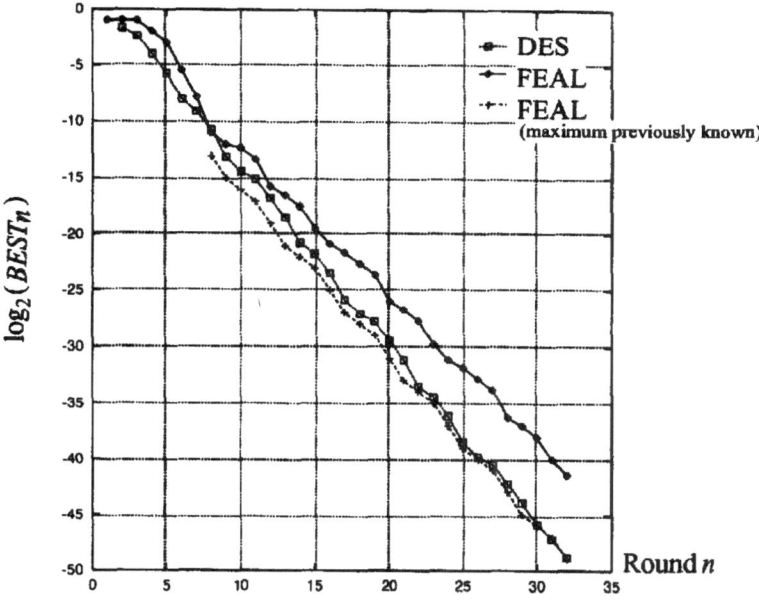

Fig. 2. The Best Deviation

6 Conclusion

We have improved Matsui's search algorithm by considering the dominant factor of search complexity. To discard many unnecessary search candidates before searching, we introduced the *search patterns*, which are preselected set of deviations of linear approximations of each round. Now we have two alternatives for finding the best linear expression. Matsui's search algorithm is easy to implement and works well for some cryptosystems. Our algorithm is equally effective and is suitable for those cryptosystems which cannot be readily searched by his algorithm.

Our search algorithm was applied to DES and FEAL. The n-round best linear expressions of DES we found as fast as Matsui's algorithm for $n \leq 32$. Those of FEAL we found much faster than his algorithm; the time required is decreased from over three months to about two and a half days. We have found the n-round best linear expressions of FEAL ($n \leq 32$) with higher deviations than those derived from Biham's iterative linear expressions.

Our idea is expected to be also effective in searching the differential characteristics in Differential Cryptanalysis.

Number of Round n	DES Best Deviation ($= BEST_n$)	FEAL Best Deviation	$BEST_n$	Biham's Results [B94]
2	1.25×2^{-2}	1.00×2^{-1}	2^{-1}	–
3	1.56×2^{-3}	1.00×2^{-1}	2^{-1}	–
4	1.95×2^{-5}	1.00×2^{-2}	2^{-2}	–
5	1.22×2^{-6}	1.00×2^{-3}	2^{-3}	–
6	1.95×2^{-9}	1.52×2^{-6}	2^{-6}	–
7	1.95×2^{-10}	1.15×2^{-8}	2^{-8}	–
8	1.22×2^{-11}	1.00×2^{-11}	2^{-11}	2^{-13}
9	1.91×2^{-14}	1.00×2^{-12}	2^{-12}	2^{-15}
10	1.53×2^{-15}	1.64×2^{-14}	2^{-14}	2^{-16}
11	1.91×2^{-16}	1.64×2^{-14}	2^{-15}	2^{-17}
12	1.19×2^{-17}	1.24×2^{-16}	2^{-18}	2^{-19}
13	1.49×2^{-19}	1.48×2^{-17}	2^{-19}	2^{-21}
14	1.19×2^{-21}	1.48×2^{-18}	2^{-20}	2^{-22}
15	1.19×2^{-22}	1.48×2^{-20}	2^{-21}	2^{-23}
16	1.49×2^{-24}	1.13×2^{-21}	2^{-24}	2^{-25}
17	1.16×2^{-26}	1.34×2^{-22}	2^{-25}	2^{-27}
18	1.86×2^{-28}	1.34×2^{-23}	2^{-26}	2^{-28}
19	1.16×2^{-28}	1.34×2^{-24}	2^{-27}	2^{-29}
20	1.46×2^{-30}	1.02×2^{-26}	2^{-30}	2^{-31}
21	1.82×2^{-32}	1.21×2^{-27}	2^{-31}	2^{-33}
22	1.46×2^{-34}	1.21×2^{-28}	2^{-32}	2^{-34}
23	1.46×2^{-35}	1.21×2^{-30}	2^{-34}	2^{-35}
24	1.82×2^{-37}	1.85×2^{-32}	2^{-36}	2^{-37}
25	1.42×2^{-39}	1.10×2^{-32}	2^{-37}	2^{-39}
26	1.14×2^{-39}	1.10×2^{-33}	2^{-38}	2^{-40}
27	1.42×2^{-41}	1.10×2^{-34}	2^{-39}	2^{-41}
28	1.78×2^{-43}	1.67×2^{-37}	2^{-42}	2^{-43}
29	1.11×2^{-44}	1.99×2^{-38}	2^{-43}	2^{-45}
30	1.12×2^{-46}	1.99×2^{-39}	2^{-44}	2^{-46}
31	1.78×2^{-48}	1.99×2^{-41}	2^{-46}	2^{-47}
32	1.11×2^{-49}	1.51×2^{-42}	2^{-48}	2^{-49}

Table 4. The Best Deviation

Acknowledgement

We are grateful to Dr. Mitsuru Matsui for providing his software implementation of his DES search algorithm and helpful discussions, and to Toshio Tokita for showing details of experiments made to search for the best linear expressions of DES, s^2DES and LOKI.

References

[B94]   E. Biham: "On Matsui's Linear Cryptanalysis (extended abstract)," Pre-
         proceedings of EUROCRYPT'94, 1994
[K92]   L. R. Knudsen: "Iterative Characteristics of DES and s^2-DES," Advances
         in Cryptology – EUROCRYPT'92, Springer-Verlag 658, 1993
[KR94]   B. S. Kaliski Jr. and M. J. B. Robshaw: "Linear Cryptanalysis Using Mul-
         tiple Approximations," Advances in Cryptology – CRYPTO'94, Springer-
         Verlag 839, 1994
[N94]   K. Nyberg: "Linear Approximation of Block Ciphers," Preproceedings of
         EUROCRYPT'94, 1994
[M93]   M. Matsui: "Linear Cryptanalysis Method for DES Cipher," Advances in
         Cryptology – EUROCRYPT'93, Springer-Verlag 765, 1994
[M94]   M. Matsui: "On Correlation between the order of S-Boxes and the Strength
         of DES (extended abstract)," Preproceedings of EUROCRYPT'94, 1994
[MAO96] S. Moriai, K. Aoki and K. Ohta: "The Best Linear Expression Search of
         FEAL," IEICE Trans. Fundamentals, Vol. E79-A, No. 1, 1996 (to appear)
[MSS88] S. Miyaguchi, A. Shiraishi and A. Shimizu: "Fast Data Encipherment algo-
         rithm FEAL-8," Review of Electrical Communication Laboratories, Vol. 36,
         No. 4, 1988
[OA94]   K. Ohta and K. Aoki: "Linear Cryptanalysis of the Fast Data Encipherment
         Algorithm," Advances in Cryptology – CRYPTO'94, Springer-Verlag 839,
         1994
[TSM94] T. Tokita, T. Sorimachi and M. Matsui: "Linear cryptanalysis of LOKI and
         s^2DES (extended abstract)," Preproceedings of ASIACRYPT'94, 1994

Appendix Open Problem

When more than two S-boxes are approximated in the f function, the *Piling-up Lemma* can't always calculate the true deviation of the f function. *Table 5* gives examples of the linear approximations of the f functions of DES and FEAL whose deviations calculated by the *Piling-up Lemma* differ from the true values.

We can't currently compute the true deviation of the f function except using equation (4) with exhaustive search. How to compute the true deviation of the linear approximation of the f function with less complexity is an open problem.

	Linear approximation of f function $(\Gamma O_i, \Gamma I_i)$	$p_i'(\Gamma O_i, \Gamma I_i)$ calculated by	
		Piling-up Lemma	Definition (Equation(4))
DES	(00140000, a10400c0)	0	1.25×2^{-3}
FEAL	(81010100, 00600000)	2^{-4}	1.81×2^{-4}

Table 5. Examples of Linear approximations of f functions whose deviations calculated by the *Piling-up Lemma* differ from the true values

On Differential and Linear Cryptanalysis of the RC5 Encryption Algorithm

Burton S. Kaliski Jr. and Yiqun Lisa Yin

RSA Laboratories, 100 Marine Parkway, Redwood City, CA 94065
burt@rsa.com, yiqun@rsa.com

Abstract. This paper analyzes the security of the RC5 encryption algorithm against differential and linear cryptanalysis. RC5 is a new block cipher recently designed by Ron Rivest. It has a variable word size, a variable number of rounds, and a variable-length secret key. In RC5, the secret key is used to fill an expanded key table which is then used in encryption. Both our differential and linear attacks on RC5 recover every bit of the expanded key table without any exhaustive search. However, the plaintext requirement is strongly dependent on the number of rounds. For 64-bit block size, our differential attack on nine-round RC5 uses 2^{45} chosen plaintext pairs (about the same as DES), while 2^{62} pairs are needed for 12-round RC5. Similarly, our linear attack on five-round RC5 uses 2^{47} known plaintexts (about the same as DES), and the plaintext requirement is impractical for more than six rounds. We conjecture that the linear approximations used in our linear cryptanalysis are optimal. Thus, we conclude that Rivest's suggested use of 12 rounds is sufficient to make differential and linear cryptanalysis of RC5 impractical.

1 Introduction

The RC5 encryption algorithm [9] is a new block cipher designed by Ron Rivest in 1994. RC5 has a variable word size, a variable number of rounds, and a variable-length secret key. A particular RC5 algorithm is designated as RC5-$w/r/b$, where w is the *word size* in bits ($w = 16, 32, 64$), r is the number of rounds ($0 \leq r \leq 255$), and b is the number of bytes in the secret key ($0 \leq b \leq 255$). RC5 has a two-word ($2w$-bit) input and output block size.

RC5 encryption consists of three primitive operations: (1) addition modulo 2^w denoted by "$+$," (2) bit-wise exclusive-OR denoted by "$\oplus$," and (3) rotation: the rotation of x left by y is denoted by $x \lll y$ (the $\log_2(w)$ low-order bits of y are used). Before encryption, the secret key is used to fill an expanded key table S with $2r + 2$ words. Let (A, B) denote the two words in both the input and output block. The encryption algorithm is described below.

$$A = A + S[0]$$
$$B = B + S[1]$$
for $i = 1$ **to** r **do**
$$A = ((A \oplus B) \lll B) + S[2i]$$
$$B = ((B \oplus A) \lll A) + S[2i + 1]$$

D. Coppersmith (Ed.): Advances in Cryptology - CRYPTO '95, LNCS 963, pp. 171-184, 1995.
© Springer-Verlag Berlin Heidelberg 1995

A distinguishing feature of RC5 is its heavy use of *data-dependent rotations*—the rotation amounts are random variables dependent on the input data, and they are not predetermined values. The security of RC5 relies on the rotation operation as well as the mixed use of different operations.

In this paper, we analyze the security of RC5 against differential cryptanalysis [1] and linear cryptanalysis [7]. We first briefly review both types of cryptanalysis. For differential cryptanalysis, the basic idea is that two chosen plaintexts P and P^* with a certain difference $P' = P \oplus P^*$ can encipher to two ciphertexts C and C^* such that $C' = C \oplus C^*$ has a specific value with non-negligible probability, and such a "characteristic" (P', C') is useful in deriving certain bits of the key. For linear cryptanalysis, the basic idea is to find linear approximations (parity relations among certain bits of plaintext, ciphertext, and key) which hold with probability $\neq 1/2$ (i.e., bias $= |p - 1/2| \neq 0$); such approximations can be used to obtain information about the key.

To attack RC5, one can try to find either the original secret key or the expanded key table S; both our differential attack and linear attack use the latter approach, and hence they are independent of the length of the secret key.

Our differential attack is quite effective on RC5 when the number of rounds r is relatively small, and it recovers every bit of the expanded key table S. The number of chosen plaintext pairs needed for RC5-32 is 2^{45} for $r = 9$ (about the same as 16-round DES), and the number of plaintext pairs is 2^{62} for $r = 12$. Hence, the number of plaintext requirement of our attack becomes impractical when the number of rounds is large, which is also the case for other block ciphers. We implemented our attack on for certain choices of w and r, and the actual number of plaintexts used matched our theoretical calculation.

A notable feature of our differential attack is that the type of the characteristics used is quite different from the characteristics used in attacks on other block ciphers, e.g. DES. In particular, for a given plaintext difference P' and ciphertext difference C', there are many possible paths (intermediate differences) from P' to C', each with the same probability. (In contrast, while there may be many paths for other block ciphers [4], generally one path dominates the rest.) These paths are of a similar form, allowing us to treat them uniformly. The feature has a boosting effect on the probability of getting a plaintext pair with specified P' and C'.

In our linear cryptanalysis of RC5, we construct linear approximations for RC5 and use them to compute every bit of the expanded key table S. Our linear attack on five-round RC5-32 uses 2^{47} known plaintexts (about the same as 16-round DES), and the plaintext requirement becomes impractical when $r > 6$. We also try to establish an upper bound on the bias of the best linear approximation. We prove that for each half-round of RC5, the best linear approximation that can be alternated with a trivial linear approximation holds with bias $1/2w$, and we conjecture that by alternating the two approximations, we indeed obtain the best approximation for RC5 for the proposed word sizes.

Both our differential and linear attacks on RC5 are very effective in determining key bits. After enough plaintext/ciphertext pairs are generated, the entire S

table is computed bit by bit without *any* exhaustive search. For examples, in the implementation of our differential attack, all the running time was used for generating plaintext/ciphertext pairs, and the time for key search (i.e., computing S) was negligible—less than a second on a Sun4.

We conclude that the nominal choice $r = 12$ for RC5-32 proposed by Rivest [9] provides good security against differential and linear cryptanalysis. Of course, the possibility remains that there are other ways to attack RC5, and further study is needed to fully determine the security of RC5 for any particular parameter values.

The remainder of the paper is organized into sections as follows. In §2, we introduce some useful notation. In §3, we present our differential attack on RC5, and in §4, we describe our linear attack on RC5. In §5, we summarize our recent progress in analyzing the security of RC5 with focus on the use of data-dependent rotations. In §6, we conclude with some future research directions.

2 Notation

In Rivest's description of RC5 [9], a round consists of two equations, and in each equation, one of A or B is modified while the other remains unchanged. We will refer to each equation as a *half-round*. So one half-round of RC5 is similar to a full-round in a Feistel cipher. For ease of discussion, we adopt the common notation for Feistel ciphers and rewrite RC5 as follows.

$$L_1 = L_0 + S_0$$
$$R_1 = R_0 + S_1$$
for $i = 2$ to $2r + 1$ do
$$L_i = R_{i-1}$$
$$R_i = ((L_{i-1} \oplus R_{i-1}) \lll R_{i-1}) + S_i$$

We will use the above description of RC5 throughout the paper. We will refer to the two equations which involve (L_{i-1}, R_{i-1}) and (L_i, R_i) as the ith half-round of RC5. Hence, the two initial equations ($L_1 = L_0 + S_0$ and $R_1 = R_0 + S_1$) together are considered as the *first* half-round, and RC5 contains $2r+1$ half-rounds in total. The input block is (L_0, R_0) and the output block is (L_{2r+1}, R_{2r+1}). For ease of use, we have changed $S[i]$ to S_i.

For a binary vector x of length w, we label the bit positions from the most significant bit to the least significant bit as $w-1, \ldots, 1, 0$. We use $x[s]$ to denote the sth bit of x and $x[s..t]$ ($s \geq t$) to denote the sth through t^{th} bits of x. We use $\lg(w)$ to denote $\log_2(w)$. Note that $x \bmod w = x[\lg(w) - 1..0]$ are the bits of x which are used to determine a rotation count. We use $x[s, t, \ldots, u]$ to denote $x[s] \oplus x[t] \oplus \cdots \oplus x[u]$.

3 Differential cryptanalysis of RC5

3.1 Structure of the differential attack

In this subsection, we first describe a general idea for attacking RC5 by analyzing the structure of the cipher. We then introduce the form of the chosen plaintext/ciphertext pairs that are used in our differential attack and outline the high-level structure of the attack.

We first observe that RC5 has an iterative structure. Such a structure allows us to reduce the problem of computing the entire S table to the problem of computing the last entry S_{2r+1}. We now consider the last half-round of RC5 which involves L_{2r}, R_{2r}, R_{2r+1}, and S_{2r+1}. Suppose an algorithm $\mathcal{B}$ can compute $L_{2r}[b]$ for some $b \in \{0, \ldots, w-1\}$ given plaintext/ciphertext pairs. Since R_{2r+1} and $R_{2r}(= L_{2r+1})$ are known, bit $L_{2r}[b]$ gives information about some bit of S_{2r+1} depending on the rotation amount $R_{2r} \bmod w$. (For instance, given $L_{2r}[0]$, we can compute $S_{2r+1}[0]$ if $R_{2r} \bmod w = 0$.) For $s = 0, \ldots, w-1$, the following pseudocode computes $S_{2r+1}[s]$ using $\mathcal{B}$ when $S_{2r+1}[s-1..0]$ has been obtained.

> **Select a plaintext/ciphertext pair** $(L_0, R_0)/(L_{2r+1}, R_{2r+1})$
> **such that** $(b + L_{2r+1}) \bmod w = s$
> **Use algorithm** $\mathcal{B}$ **to calculate** $L_{2r}[b]$
> **If** $s \geq 1$
> $\quad x[s-1..0] = R_{2r+1}[s-1..0] - S_{2r+1}[s-1..0]$
> $x[s] = L_{2r}[b] \oplus L_{2r+1}[b]$
> $S_{2r+1}[s..0] = R_{2r+1}[s..0] - x[s..0]$

Therefore, we reduce the problem of computing S table to computing some bit of L_{2r}. We remark that the basic idea described so far will be used in both our differential and linear cryptanalysis.

We will use the following type of the plaintext/ciphertext pairs for computing S_i in our differential attack. Let e_s denote the w-bit binary vector which is 1 in bit s and 0 everywhere else. A *good pair for S_i* consists of two plaintexts $P = (L_0, R_0), P^* = (L_0^*, R_0^*)$ and their ciphertexts $C = (L_i, R_i), C^* = (L_i^*, R_i^*)$ satisfying the following conditions:

$$P' = P \oplus P^* = \begin{cases} (0, e_{w-1}) & \text{if } i \bmod 3 = 0, \\ (e_{w-1}, 0) & \text{if } i \bmod 3 = 1, \\ (e_{w-1}, e_{w-1}) & \text{if } i \bmod 3 = 2, \end{cases} \tag{1}$$

$$C' = C \oplus C^* = (e_t, e_u \oplus e_v), \quad t \geq \lg(w), \ u > v. \tag{2}$$

Note that when computing S_i, we can verify if the conditions hold since $S_{i+1}, \ldots, S_{2r+1}$ are already known. We will show later that with high probability, a good pair for S_i allows us to recover $L_{i-1} \bmod w$, which yields $L_{i-1}[b]$ for $0 \leq b \leq \lg(w) - 1$. Based on the discussions above, a good pair for S_i is *useful for predicting $S_i[s]$* if

$$\exists b, \ 0 \leq b \leq \lg(w) - 1, \ \text{s.t. } (b + L_i) \bmod w = s.$$

The structure of our differential attack is given in the following pseudocode. At a high level, we compute the expanded key table S entry by entry[1] in reverse order and compute each entry S_i from the least significant bit to the most significant bit.

---

[1] The first three entries S_0, S_1, and S_2 cannot be computed in the uniform way as described in the pseudocode. Nevertheless, they can be easily computed by a simple algorithm when $S_3, \ldots, S_{2r+1}$ are known.

```
For i = 2r + 1 down to 3
    Obtain a set Gᵢ of good pairs for Sᵢ
    For s = 0 to w - 1
        Select a pair in Gᵢ that is useful for predicting Sᵢ[s]
        Compute Sᵢ[s]
```

3.2 Characteristics of RC5

In this subsection, we first describe some characteristics for a half-round of RC5. Then we show how to join them together to form characteristics that are useful to compute the expanded key table S.

A characteristic for a half-round is denoted by $\Omega = (\Omega_P, \Omega_T)$, where $\Omega_P = (L'_{i-1}, R'_{i-1})$ and $\Omega_T = (L'_i, R'_i)$. Generally speaking, if a pair of inputs to a half-round have different rotation amounts, then the pair of outputs from the half-round will differ in many bits. Consequently, we will focus on characteristics for which the pair of inputs have the same rotation amounts. In most of our characteristics, each half of Ω_P and Ω_T is either zero or e_s where $s \geq \lg(w)$, implying that the rotation amounts are the same.

We will calculate the probability associated with a half-round characteristic by averaging over both the pair of inputs and round key S_i for simplicity; there may be keys for which the probability is higher and others for which it is lower. However, assuming the key expansion of RC5 is good, each round key will be essentially independent, and hence the overall probability of a characteristic for $(2r + 1)$ half-rounds will be closed to what we would expect for nearly all keys. Our implementation results (see §3.4) also confirm this.

The following table lists five half-round characteristics that will be used in our attack. When analyzing the probabilities, we use the fact that for random inputs x and y such that $x \oplus y = e_s$ and random key S_i, the probability that $(x + S_i) \oplus (y + S_i) = e_s$ is at least $1/2$.

Ω	Ω_P	Ω_T	conditions	probability
Ω^1	$(0, e_s)$	(e_s, e_s)	$s \geq \lg(w)$	$p \geq (1/w) \cdot (1/2)$
Ω^2	(e_s, e_s)	$(e_s, 0)$	$s \geq \lg(w)$	$p = 1$
Ω^3	$(e_s, 0)$	$(0, e_t)$	$s, t \geq \lg(w)$	$p \geq (1/w) \cdot (1/2)$
Ω^4	$(0, e_s)$	(e_s, e_t)	$s, t \geq \lg(w), t \neq s$	$p \geq (1/w) \cdot (1/2)$
Ω^5	(e_s, e_t)	$(e_t, e_u \oplus e_v)$	$s, t \geq \lg(w), t \neq s, u > v$ $t - s = \pm(u - v) \bmod w$	$p \geq (1/w) \cdot (1/2) \cdot (1/2)$

We note that for characteristics Ω^3, Ω^4, and Ω^5, there are many possible output differences Ω_T for each input difference Ω_P. In particular, there are $(w - \lg(w))$ choices of parameter t for Ω^3, $(w - \lg(w) - 1)$ choices of parameter t for Ω^4, and w choices of parameters (u, v) for Ω^5 for each choice of Ω_P.

Two characteristics can be joined together if Ω_T from the first one and Ω_P from the second have the same value. For example, Ω^3 with parameters (s_1, t_1) can be joined with Ω^1 with parameter s_2 if $t_1 = s_2$. Therefore, the possible ways to join the above five characteristics are Ω^1-Ω^2, Ω^2-Ω^3, Ω^3-Ω^1, Ω^3-Ω^4, and Ω^4-Ω^5. In particular, $\bar{\Omega} = \Omega^1$-Ω^2-Ω^3 is a useful characteristic for three half-rounds since it can be repeatedly joined with itself.

For the first half-round, there are three characteristics that hold with probability 1:

$\Omega^{1'}$: $\Omega_P = \Omega_T = (0, e_{w-1})$, which may be joined with Ω^1,

$\Omega^{2'}$: $\Omega_P = \Omega_T = (e_{w-1}, e_{w-1})$, which may be joined with Ω^2, and

$\Omega^{3'}$: $\Omega_P = \Omega_T = (e_{w-1}, 0)$ which may be joined with Ω^3.

In what follows, we construct characteristics for $2r + 1$ half-rounds of RC5, which we denote by Ω_{2r+1}. Characteristic Ω_{2r+1} consists of a sequence of half-round characteristics. Since there are many possible values for the parameters of some of the half-round characteristics, there are many possible paths (i.e., intermediate differences (L'_i, R'_i) for $1 \le i \le 2r$) from P' to C' for Ω_{2r+1}, all of which have the same probability p. Let N denote the total number of possible paths for $\Omega_{2r+\cdot}$. We define the probability associated with Ω_{2r+1} as $p^{\Omega_{2r+1}} = Np$.

For different values of r, the following table lists the plaintext difference P', the sequence of half-round characteristics in Ω_{2r+1}, and the probability[2] associated with $p^{\Omega_{2r+1}}$.

$2r + 1$	P'	Ω_{2r+1}	$p^{\Omega_{2r+1}}$
$3m$	$(0, e_{w-1})$	$\Omega^{1'}$-$\bar{\Omega}$-$\cdots$-$\bar{\Omega}$-Ω^4-Ω^5	$\frac{w - \lg(w) - 1}{w} \left(\frac{w - \lg(w)}{(2w)^2} \right)^{m-1} \frac{1}{w}$
$3m + 1$	$(e_{w-1}, 0)$	$\Omega^{3'}$-Ω^3-$\bar{\Omega}$-$\cdots$-$\bar{\Omega}$-Ω^4-Ω^5	$(w - \lg(w) - 1) \left(\frac{w - \lg(w)}{(2w)^2} \right)^{m}$
$3m + 2$	(e_{w-1}, e_{w-1})	$\Omega^{2'}$-Ω^2-Ω^3-$\bar{\Omega}$-$\cdots$-$\bar{\Omega}$-Ω^4-Ω^5	$(w - \lg(w) - 1) \left(\frac{w - \lg(w)}{(2w)^2} \right)^{m}$

A *right pair* with respect to Ω_{2r+1} consists of two plaintexts P, P^* and their ciphertexts C, C^* such that for all $0 \le i \le 2r + 1$, the corresponding difference (L'_i, R'_i) has the proper form specified by sequence of the half-round characteristics for Ω_{2r+1}. We remark that for $i \le 2r$, characteristic Ω_i, its associated probability p^{Ω_i}, and a right pair with respect to Ω_i can be defined in a similar way.

Recall the definition of a good pair for S_i in §3.1. We see that a right pair with respect to Ω_i is a good pair for S_i. Hence, the expected number of chosen plaintext pairs to get a good pair for S_i is at most the expected number of chosen plaintext pairs to get a right pair with respect to Ω_i which is at most $\frac{1}{p^{\Omega_i}}$.

---

[2] (1) The factor $\frac{1}{4}$ in Ω^5 can be mostly eliminated by taking the carry effect into account when analyzing output differences. Hence the factor does not appear in $p^{\Omega_{2r+1}}$ in the table. (2) When $2r + 1 = 3m$, the probability associated with the first occurrence of the half-round characteristic Ω^1 is $\frac{1}{w}$ instead of $\frac{1}{2w}$ since the parameter $s = w - 1$.

3.3 Using a right pair to compute L_{2r} mod w

In this subsection, we show how to compute L_{2r} mod w using a right pair with respect to Ω_{2r+1}. Similarly, we can compute L_{i-1} mod w using a right pair with respect to Ω_i. Let Ω^4 and Ω^5 be the characteristics for the $(2r)^{th}$ and $(2r+1)^{th}$ half-rounds, respectively. Recall that the parameters in Ω^4 are s and t and the parameters in Ω^5 are $s, t, u,$ and v.

We consider the $(2r)^{th}$ half-round and obtain the following formula:

$$L_{2r} \bmod w = R_{2r-1} \bmod w = (t - s) \bmod w.$$

Given the ciphertext difference (L'_{2r+1}, R'_{2r+1}), the values of t, u, and v are easily obtained. So we need only compute s in order to get L_{2r} mod w. In the $(2r+1)^{th}$ half-round, the rotation amount L_{2r+1} mod w ($= R_{2r}$ mod w) is equal to either $(u-t)$ mod w or $(v-t)$ mod w. Since u, v, t, and L_{2r+1} are known, it is obvious which case holds. In the first case $s = (v - L_{2r+1})$ mod w and in the second case $s = (u - L_{2r+1})$ mod w, and the value of L_{2r} mod w follows.

3.4 Implementation of the differential attack

In this subsection, we estimate the expected number of chosen plaintext pairs required to mount our differential attack, and then we present some experimental results.

We first recall the structure of our attack described in §3.1. The expected number of plaintext pairs required for computing S_i is the product of (1) the number of good pairs required for S_i (i.e., $|G_i|$) and (2) the expected number of plaintext pairs to get a single good pair ($\leq \frac{1}{p^{\Omega_i}}$). From the form of the good pairs, we can see that the chosen plaintext pairs for computing S_i can be reused to compute S_j if $j = i$ mod 3. Hence, the expected number of plaintext pairs required for our attack is determined by

$$\sum_{i=2r-1}^{2r+1} |G_i| \times \frac{1}{p^{\Omega_i}}. \tag{3}$$

We now consider how many good pairs are needed, and we focus our discussions on RC5-32. We can show that $|G_i| = 2w$ good pairs are enough to guarantee a high success rate for our attack when $r \leq 11$. For $r = 12$, $8w$ good pairs are needed due to random noise. (A detailed discussion is included in the appendix.) For RC5-32, the number of chosen plaintext pairs determined by Equation 3 are listed for increasing r in the following table.

r	plaintext pairs	r	plaintext pairs	r	plaintext pairs	r	plaintext pairs
1	2^7	4	2^{21}	7	2^{36}	10	2^{50}
2	2^{10}	5	2^{25}	8	2^{39}	11	2^{54}
3	2^{16}	6	2^{31}	9	2^{45}	12	2^{62}

We have implemented our full attack for $w = 32, r \leq 6$ on a Sun4 workstation. The actual number of plaintexts used matched our theoretical calculation, and the success rate was very high. Each run took about 10 minutes for five rounds and about 12 hours for six rounds. Note that for each S_i, only 64 plaintext/ciphertext pairs were actually used for computing the key, and all other pairs were discarded right after they were generated. In addition, no exhaustive search is needed in our attack. Therefore, in our implementation, the time used for computing the S table was negligible (less than a second on the Sun4) after enough good pairs were generated.

4 Linear cryptanalysis of RC5

4.1 Linear approximations for a half-round

In this subsection, we study linear approximations for a half-round of RC5. We decompose the equation $R_i = ((L_{i-1} \oplus R_{i-1}) \lll R_{i-1}) + S_i$ into the following three equations, each of which involves only a single primitive operation, and we consider possible linear approximations for each of the equations. We will say that a linear approximation is *perfect* if it holds with bias $1/2$ (probability 1 or 0).

$$X = L_{i-1} \oplus R_{i-1}, \quad Y = X \lll R_{i-1}, \quad R_i = Y + S_i.$$

I. $X = L_{i-1} \oplus R_{i-1}$

The equation has numerous perfect linear approximations. In particular, all approximations involving the same bits of X, L_{i-1}, and R_{i-1} are perfect. All other approximations have bias zero.

II. $Y = X \lll R_{i-1}$

The linear approximations for this equation can be divided into two types depending on whether bits of R_{i-1} are involved. We first consider approximations in which no bits of R_{i-1} are involved. Any such approximation involving just one bit of X and Y holds with probability $1/2 + 1/2w$, since for one rotation amount, the bits are guaranteed to be equal and for the other $w - 1$ amounts, the bits are equal with probability $1/2$. In general, for $t = 0, \ldots, \lg(w)$, an approximation involving 2^t bits of X and 2^t bits of Y (in proper positions) holds with probability $1/2 + 2^t/w$.

We next consider approximations in which some bits of R_{i-1} are involved. Some of these approximations have a non-zero bias. For example,

$$Y[0] = X[0] \oplus R_{i-1}[0] \tag{4}$$

holds with probability $1/2 + 1/2w$, since when the rotation amount is 0, $R_{i-1}[0] = 0$ and $Y[0] = X[0]$, and when the amount is otherwise, the equation holds with probability $1/2$. We remark that an approximation will have bias zero if any bit $R_{i-1}[s]$ where $s \geq \lg(w)$ is involved.

III. $R_i = Y + S_i$

The best linear approximation for this equation is

$$R_i[0] = Y[0] + S_i[0], \tag{5}$$

which holds with probability 1. All other approximations are not perfect, and their biases are dependent on the key. For example, the bias of the approximation $R_i[s] = Y[s] + S_i[s]$ ranges from 0 to 1/2 and is averaged at 1/4 for $s \geq 1$.

For the first half-round which uses only the + operation, both approximations

$$L_1[0] = L_0[0] \oplus S_0[0] \quad \text{and} \quad R_1[0] = R_0[0] \oplus S_1[0]$$

hold with probability 1. We will denote them as C and D, respectively.

IV. Joining the linear approximations

Based on the above discussion, we can construct many possible linear approximations for a half-round of RC5 by joining the approximations for the three operations. For example, by joining $X[0] = L_{i-1}[0] \oplus R_{i-1}[0]$, approximation (4), and approximation (5), we obtain the following approximation for a half-round:

$$R_i[0] = L_{i-1}[0] \oplus S_i[0].$$

This holds with probability $1/2 + 1/2w$. We will denote it as E.

Since $L_i = R_{i-1}$ in a half-round of RC5, there are many trivial approximations which involve the same bits of L_i and R_{i-1} and hold with probability 1. The following trivial approximation

$$L_i[0] = R_{i-1}[0]$$

can be alternated with approximation E, and we will denote it as -.

4.2 The linear cryptanalytic attack

In this subsection, we show how to use half-round linear approximations C, D, E, and - to compute the expanded key table S. Based on our discussions in §3.1, we need only show how to compute the last entry S_{2r+1}. Similarly as in our differential attack, obtaining information about bits of L_{2r} is also very useful for computing S_{2r+1} in our linear attack.

We first note that D-E-E-...E- is a linear approximation for $2r$ half-rounds with the following form:

$$R_0[0] \oplus L_{2r}[0] = S_1[0] \oplus S_3[0] \oplus \cdots \oplus S_{2r-1}[0].$$

Since E appears exactly $r - 1$ times, by Matsui's "piling-up" lemma [7], the above approximation holds with probability $\frac{1}{2} + \frac{1}{2w^{r-1}}$. Let T denote the value $S_1[0] \oplus S_3[0] \oplus \cdots \oplus S_{2r-1}[0]$. Then T is fixed for a given expanded key table S, and the value of $R_0[0] \oplus L_{2r}[0]$ is always biased toward T.

In our linear attack on RC5, we will compute S_{2r+1} in three steps: (1) compute $S_{2r+1}[0]$, (2) compute T given $S_{2r+1}[0]$, and (3) for $s = 1, \ldots, w-1$, compute $S_{2r+1}[s]$ given T and $S_{2r+1}[s-1..0]$. (For $i \leq 2r$, we can compute S_i using similar techniques with the linear approximation D-E-E-...E- if i is even and the linear approximation CE-E-...E- if i is odd.)

Step 1: Compute $S_{2r+1}[0]$. We observe that for plaintext/ciphertext pairs such that $L_{2r+1} \bmod w = 1$, one of the following two approximations is perfect:

$$L_{2r}[0] = L_{2r+1}[0] \oplus R_{2r+1}[1] \qquad \text{if } S_{2r+1}[0] = 0$$
$$L_{2r}[0] = L_{2r+1}[0] \oplus R_{2r+1}[1] \oplus R_{2r+1}[0] \text{ if } S_{2r+1}[0] = 1$$

(These can be seen by observing the effect of the carry out from the least significant bit of the addition.) Moreover, the first approximation has zero bias if $S_{2r+1}[0] = 1$ and the second approximation has zero bias if $S_{2r+1}[0] = 0$. To compute $S_{2r+1}[0]$, we obtain N known plaintext/ciphertext pairs such that $L_{2r+1} \bmod w = 1$ and consider the two quantities

$$R_0[0] \oplus (L_{2r+1}[0] \oplus R_{2r+1}[1]) \quad \text{and} \quad R_0[0] \oplus (L_{2r+1}[0] \oplus R_{2r+1}[1] \oplus R_{2r+1}[0]).$$

Let U_0 be the number of plaintexts such that the first quantity is zero and U_1 be the number of plaintexts such that the second quantity is zero. If $|U_0 - N/2| \geq |U_1 - N/2|$, we predict $S_{2r+1}[0] = 0$; otherwise, we predict $S_{2r+1}[0] = 1$.

Step 2: Compute T given $S_{2r+1}[0]$. We observe that for plaintext/ciphertext pairs such that $L_{2r+1} \bmod w = 0$, the approximation

$$L_{2r}[0] = L_{2r+1}[0] \oplus R_{2r+1}[0] \oplus S_{2r+1}[0]$$

holds with probability 1 and the right-hand side is known. To compute T, we obtain N known plaintext/ciphertext pairs such that $L_{2r+1} \bmod w = 0$. Let U be the number of plaintexts such that

$$R_0[0] \oplus (L_{2r+1}[0] \oplus R_{2r+1}[0] \oplus S_{2r+1}[0])$$

is zero. If $U \geq N/2$, we predict $T = 0$; otherwise, we predict $T = 1$.

Step 3: For $s = 1, \ldots, w-1$, compute $S_{2r+1}[s]$ given T and $S_{2r+1}[s-1..0]$. For a given plaintext/ciphertext pair, let $y = R_{2r+1} - S_{2r+1}$ and let $carry(s)$ denote the carry out from $y[s-1..0] + S_{2r+1}[s-1..0]$. We observe that for plaintext/ciphertext pairs such that $L_{2r+1} \bmod w = s$, the approximation

$$L_{2r}[0] = L_{2r+1}[0] \oplus R_{2r+1}[s] \oplus S_{2r+1}[s] \oplus carry(s)$$

holds with probability 1. To compute $S_{2r+1}[s]$, we obtain N known plaintext/ciphertext pairs such that $L_{2r+1} \bmod w = s$. Let U be the number of plaintexts such that

$$(R_0[0] \oplus T) \oplus (L_{2r+1}[0] \oplus R_{2r+1}[s] \oplus carry(s))$$

is zero. If $U \geq N/2$, we predict $S_{2r+1}[s] = 0$; otherwise, we predict $S_{2r+1}[s] = 1$.

The total number of known plaintexts expected is determined by the number of plaintexts expected for computing S_{2r+1} since the plaintexts can be reused for each S_i. Note that the bias of the $2r$ half-round approximation D-E-E-...E- is $\frac{1}{2w^{r-1}}$. Standard techniques for the type of linear cryptanalysis in our attack require an amount of plaintexts approximately equal to the inverse square of the bias. Therefore, $N = 4w^{2(r-1)}$ known plaintexts are required for computing each bit $S_{2r+1}[s]$ for $s = 0, 1, \ldots, w-1$, which leads to $w \times 4w^{2(r-1)}$ known plaintexts for r-round RC5. For RC5-32, the number of plaintexts is 2^{37} for $r = 4$, 2^{47} for $r = 5$, and 2^{57} for $r = 6$.

4.3 A conjecture on the bias of the best linear approximations

We conjecture that for the proposed word sizes $w = 16, 32, 64$, the linear approximation for $2r$ half-rounds used in our linear attack in §4.2 is the best linear approximation for RC5. If the conjecture is correct, we would then be able to conclude that standard linear cryptanalysis is only effective for RC5 with a very small number of rounds.

We have strong evidence for the correctness of our conjecture. In particular, we show that E is a best half-round approximation that can be alternated with a trivial approximation.

Lemma 1. *Let set M contain all half-round approximations in which neither bits of R_{i-1} nor bits of L_i are involved. Then E is a best approximation among all approximations in M.*

Proof. Let F be an arbitrary approximation in M. Then F can be decomposed into three approximations, one for each operation. There may be many possible decompositions, and we consider the constraints on the three approximations for a given decomposition. The approximation for $Y = X \lll R_{i-1}$ cannot involve $R_{i-1}[s]$ with $s \geq \lg(w)$ since F has bias zero otherwise. Hence, the approximation for $X = L_{i-1} \oplus R_{i-1}$ cannot involve $X[s]$ with $s \geq \lg(w)$; otherwise, either F involves bits of R_{i-1} or it has bias zero. Any approximation for $Y = X \lll R_{i-1}$ involving only $X[s]$ with $s \leq \lg(w) - 1$ holds with bias at most $1/2w$ since there is only one rotation amount that can match the bit positions of X and Y. Therefore, F has bias at most $1/2w$. Since E holds with bias $1/2w$, it is a best approximation among all approximations in M. $\square$

We remark that the $\lll$ and $+$ operations are incompatible when constructing linear approximations for a half-round of RC5. It is clear that the bias gets larger for $\lll$ if more bits are involved in an approximation, and the bias gets smaller for $+$ if more bits are involved. Hence, the mixed use of the two operations provides good security against linear cryptanalysis.

5 Work in progress

In this section, we summarize some of our research progress [3] in analyzing the security of RC5. In particular, we focus our analysis on the use of data-dependent rotations.

We have studied how the use of data-dependent rotations helps prevent differential cryptanalysis. More specifically, we have analyzed the number of possible output differences of a half-round when the pair of inputs have different rotation amounts. We have proved that, for instance, if $\Omega_P = (e_s, e_s)$ for $s < \lg(w)$, then Ω_T is uniformly distributed in a large set containing at least $2^{w/2}$ distinct values. (Recall that for $s \geq \lg(w)$, Ω_T only has one possible value $(e_s, 0)$.) The results show that data-dependent rotations spread out bit differences in a pair of inputs in a drastic way when the differences affect the rotation amounts. Clearly, the more input bits that differ, the higher chance that one of them will affect the rotation amounts in the pair of outputs. So a good characteristic for RC5 should always keep the number of input bit differences in each half-round as small as possible. From this viewpoint, the characteristics used in our differential attack are the best possible since there is at most one bit difference in L_i' and at most one in R_i' (except in the last two half-rounds).

The notion of a "Markov cipher" was introduced by Lai, Massey, and Murphy [4] as a criterion for an iterative cipher to be resistant to differential cryptanalysis. Loosely speaking, an iterative cipher is *Markov* if there is a way of defining differences such that the probability of an output difference of the round function depends on only the input difference and is independent of the values of the inputs. We have shown that RC5 is not a Markov cipher with respect to the difference measures $\oplus$ and $-$. This fact, however, does not imply that RC5 is vulnerable to differential attacks, since the essential property that makes a Markov cipher secure against differential attacks is that every output difference will be roughly equally likely after sufficiently many rounds. For RC5, the output difference of a half-round ranges over a large set of possible values if the input differences affect the rotation amounts, and the probability that this will happen goes to one as the number of rounds increases. Hence, even though it may not be the case that *every* output difference will occur after many rounds of RC5, the large number of possible output differences would make a differential attack impossible.

We have also considered the impact of certain simple modifications to RC5 in an effort to appreciate which operations are essential for security. For instance, if all additions were change to exclusive-or, our differential attack would be more successful since the change increases the probability of some half-round characteristics by a factor of two.

6 Conclusions

In this paper, we have studied the security of RC5 using standard techniques from differential and linear cryptanalysis. We conclude that the choice $r = 12$ for RC5-32 proposed by Rivest provides good security against both types of attacks. As a next step, we will analyze RC5 based on more special techniques such as differential cryptanalysis with partial differentials and high-order differentials, linear cryptanalysis with multiple approximations [2], and differential-linear cryptanalysis [5].

The heavy use of data-dependent rotations is a distinguishing feature of RC5, which provides certain security against differential and linear cryptanalysis. We have also seen that, however, the rotation operation "helps" an attacker in the sense that information about a bit of L_{2r} can be spread by the rotation in the last half-round to give information about every bit of the key S_{2r+1}. This feature of RC5 may also be useful in other types of attacks on RC5.

One of the design goals of RC5 was an exceptional simplicity, with the objective of making analysis easier. In contrast with other block ciphers, all of our characteristics and linear approximations for RC5 were obtained analytically without any aid of computer experiments. The simple design of RC5 will help fully determine its security in a rather rapid way.

Acknowledgement
We would like to thank Paul Kocher, Ron Rivest, and Matt Robshaw for helpful discussions.

References

1. E. Biham and A. Shamir. *Differential Cryptanalysis of the Data Encryption Standard.* Springer-Verlag, New York, 1993.
2. B. S. Kaliski Jr. and M. J. B. Robshaw. Linear cryptanalysis using multiple approximations. In Y. G. Desmedt, editor, *Advances in Cryptology — Crypto '94*, pages 26–39, Springer Verlag, New York, 1994.
3. B. S. Kaliski and Y. L. Yin. *On the Security of the RC5 Encryption Algorithm.* Technical Report, RSA Laboratories. In preparation.
4. X. Lai, J. L. Massey, and S. Murphy. Markov ciphers and differential cryptanalysis. In D. W. Davies, editor, *Advances in Cryptology — Eurocrypt '91*, pages 17–38, Springer Verlag, Berlin, 1991.
5. S. K. Lanford and M. E. Hellman. Differential-linear cryptanalysis In Y. G. Desmedt, editor, *Advances in Cryptology — Crypto '94*, pages 17–25, Springer-Verlag, New York, 1994.
6. M. Matsui. The first experimental cryptanalysis of the Data Encryption Standard. In Y. G. Desmedt, editor, *Advances in Cryptology — Crypto '94*, pages 1–11, Springer-Verlag, New York, 1994.
7. M. Matsui. Linear cryptanalysis method for DES cipher. In T. Helleseth, editor, *Advances in Cryptology — Eurocrypt '93*, pages 386–397, Springer-Verlag, Berlin, 1994.
8. National Institute of Standards and Technology (NIST). *FIPS Publication 46-2: Data Encryption Standard.* December 30, 1993.
9. R. L. Rivest. The RC5 encryption algorithm. In *Proceedings of the Workshop on Cryptographic Algorithms*, K. U. Leuven, December 1994. To appear.

Appendix

In this appendix, we consider the number of good pairs needed to guarantee a high success rate in our differential attack on RC5-32. We will focus on $|G_{2r+1}|$, and the discussions also apply to any other $|G_i|$.

There are two issues concerning $|G_{2r+1}|$. The first issue is the signal/noise ratio. For a pair of randomly chosen plaintexts, the probability that the pair of ciphertexts have the difference C' defined in Equation 2 is

$$p^{noise} = \frac{(w - \lg w) \cdot w(w-1)/2}{2^{2w}}.$$

Such a random noise is much smaller than $p^{\Omega_{2r+1}}$ when $r \leq 11$ and can be ignored in the analysis.

However, since a good pair has a fixed plaintext difference P' satisfying Equation 1, there is a non-negligible probability that it is not a right pair due to the special difference P'. To see how this can happen, we recall the characteristics for the last five half-rounds in a right pair. The number of non-zero bits in (L_i', R_i') for $i = 2r - 3, \ldots, 2r + 1$ are the following:

$$(1, 1), (1, 0), (0, 1), (1, 1), (1, 2).$$

A pair of plaintexts with difference P' may follow the correct intermediate differences until the $(2r - 4)^{th}$ half-round and then have the following number of non-zero bits in the last five half-rounds:

$$(1, 1), (1, 2), (2, 1), (1, 1), (1, 2).$$

This happens for a fraction of the good pairs, and yields good pairs which are not right pairs. Implementation results and preliminary analytical results show that the fraction is no more than 10% for $w = 32$ and $r \leq 11$. Therefore, if we generate enough good pairs, we can expect to get many right pairs.

The second issue is how many right pairs are needed. In order to compute each $S_{2r+1}[s]$, G_{2r+1} must contain a right pair that is useful for predicting $S_{2r+1}[s]$. In other words, G_{2r+1} must have the following property:

For all s in $\{0, \ldots, w - 1\}$, there exists a right pair in G_{2r+1} and b in $\{0, \ldots, \lg(w) - 1\}$, such that $b + L_{2r+1} \bmod w = s$.

We know that a right pair recovers $\lg(w)$ bits of S_{2r+1} and assuming that the S table contains random values, the rotation amount $L_{2r+1} \bmod w$ is uniformly distributed in $\{0, \ldots, w - 1\}$ for a random plaintext pair with the proper P'. Hence, if we generate $2w$ good pairs, then on average, each bit $S_{2r+1}[s]$ gets $2w \lg(w)/w = 2 \lg(w)$ good pairs that are useful for predicting its value. With high probability, more than half of the good pairs are right pairs, so a majority vote will yield the correct value of $S_{2r+1}[s]$. Therefore, $2w$ good pairs are enough to guarantee a high success rate when p^{noise} is relatively small.

We remark that as r gets larger, $p^{\Omega_{2r+1}}$ will eventually become smaller than p^{noise} and more good pairs will be needed in our attack. For RC5-32, $r = 12$ is the starting point at which $p^{\Omega_{2r+1}}$ becomes smaller than p^{noise}. Simple calculations show that $8w$ pairs are enough to guarantee a high success rate for $r = 12$.

A Simple Method for Generating and Sharing Pseudo-Random Functions, with Applications to Clipper-like Key Escrow Systems

Silvio Micali[1] and Ray Sidney[2] *

[1] MIT Laboratory for Computer Science, 545 Technology Square,
Cambridge, MA 02139.
[2] Trusted Information Systems, Inc., 3060 Washington Rd. (Rt. 97),
Glenwood, MD 21738. e-mail: sidney@tis.com.

Abstract. We present a very simple method for generating a shared pseudo-random function from a poly-random collection of functions. We discuss the applications of our construction to key escrow.

1 Introduction

Much work has been done towards the general goal of enabling parties to share information and cryptographic capabilities. In particular, see Blakley's [4] and Shamir's [30] notion of *secret sharing*; Chor, Goldwasser, Micali, and Awerbuch's [8] notion of *verifiable secret sharing*; Desmedt and Frankel's [12] notion of *shared authenticators and signatures*; and De Santis, Desmedt, Frankel, and Yung's [9] notion of a *shareable function*. (A more extensive treatment of sharing cryptographic capabilities can be found in Desmedt's survey [13].)

In this paper, we describe a very simple method for generating and sharing a pseudo-random function $f(\cdot)$ among n players. That is, we generate a function $f(\cdot)$ such that:

- For all input x, any sufficiently large collection of players (at least u players) can compute and reveal $f(x)$.
- For all inputs x such that $f(x)$ has never been revealed, no sufficiently small collection of players (t or fewer players) is feasibly able to compute $f(x)$.

If the number of players is not large, our method is particularly attractive. Indeed, after a preprocessing step has been performed, each evaluation of f at an input x is obtained by essentially the following procedure:

1. Each player performs *locally* (i.e., noninteractively) some pseudo-random function evaluations.
2. These local results are revealed (in a single round of communication).
3. The party evaluating the function takes majority and exclusive-ORs of the revealed results.

Our method is readily applicable to key-escrow systems.

---

* This research was performed while this author was at MIT, supported by a Fannie and John Hertz Foundation Fellowship for Graduate Study.

D. Coppersmith (Ed.): Advances in Cryptology - CRYPTO '95, LNCS 963, pp. 185-196, 1995.
© Springer-Verlag Berlin Heidelberg 1995

2 Background: Poly-Random Functions

We discuss briefly Goldreich, Goldwasser, and Micali's notion of a *poly-random collection of functions* [18]. For any $j \in \mathbb{N}$, set $I_j = \{0,1\}^j$. Informally, a *poly-random collection of functions* is a collection $F = \{F_j | j \in \mathbb{N}\}$, where $F_j = \{f_x | x \in I_j\}$ is a collection of 2^j functions, each mapping I_j to itself, satisfying the following properties:

- There is a polynomial-time algorithm A which, given inputs x and y satisfying $|x| = |y|$, computes $f_x(y)$.
- If we choose $x \in I_j$ at random, then the function $f_x : I_j \rightarrow I_j$ behaves (as far as a tester limited to computational time which is polynomial in j can tell) exactly like a "random" function mapping I_j to itself.

The precise definition is in [18], as is an efficient construction of a poly-random collection of functions from any cryptographically strong bit generator (CSB generator). Since Levin [27] and Impagliazzo, Levin, and Luby [22] showed how to construct CSB generators from one-way functions, we can assume that we have a poly-random collection of functions at our disposal (to have secure encryption, we require a one-way function).

In practice, something simpler and even easier to compute is often used as a "random function" in implementing protocols. If we want a collection of random functions mapping $I_a \rightarrow I_b$, we take a secure hash function H with range I_b, and set $f_x(y) = H(x \circ y)$, where $\circ$ denotes concatenation. If H is indeed a good hash function, and x is chosen at random, then we would expect $f_x(y)$ to behave like a random function.

3 Generating a Shared Pseudo-Random Function

Parts of our algorithms bear a resemblance to protocols developed in [32], [6], [1], and [16], which actually aimed at a somewhat different goal: namely, achieving "common coin flipping" protocols in Byzantine agreement scenarios.

3.1 Resilient Collections

Our method is based on a combinatorial object which we call a *resilient collection*. A resilient collection is what one gets by taking a classic combinatorial *covering design* and complementing each set in it.

DEFINITION. Let $0 \leq t < u \leq n$. An (n, t, u)-*resilient collection* of sets is a collection $S = \{S_1, \ldots, S_d\}$ of subsets of $\{1, \ldots, n\}$ such that:

- $|S_i| = n - u + 1$ for $i = 1, \ldots, n$.
- If $S \subseteq \{1, \ldots, n\}$ and $|S| = t$, then there is an S_i such that $S_i \cap S = \emptyset$.

In section 3.3, we discuss how to produce resilient collections.

3.2 Constructing Shared Pseudo-Random Functions from Resilient Collections

We now outline our method. We shall fill in the details of our protocols in section 3.4.

Let F be a poly-random collection of functions, as discussed in section 2. Let $S = \{S_1, \ldots, S_d\}$ be an (n, t, u)-resilient collection, and let $j \in \mathbb{N}$. We wish to produce a pseudo-random function mapping I_j to I_j, shared among n players. We make the natural identification between players and elements of $\{1, \ldots, n\}$.

Seed Generation and Distribution. Each subset $S_i \in S$ of players jointly runs a protocol GEN-SEED(i) which chooses at random a j-bit secret, s_i, and makes it known to every player in S_i (but to no other players). We call s_i a "random function seed," or just "seed" for short.

Function Evaluation. The shared random function is just

$$f(y) = \bigoplus_i f_{s_i}(y).$$

That is, $f(\cdot)$ is the exclusive-OR of a group of pseudo-random functions (one function for each random function seed). The values $f_{s_1}(y), f_{s_2}(y), \ldots, f_{s_d}(y)$ are called the *pieces* of $f(y)$.

Note that, as desired, the definition of a resilient collection ensures that any u players have enough information to compute all values of $f(\cdot)$, but no t players have any information about the values of $f(\cdot)$. Even if the information that t players have is combined with knowledge of previously evaluated values of $f(\cdot)$, the properties of poly-random functions guarantee that the players have no *useful* information about new values of $f(\cdot)$.

What's required to complete our description are a method to select the collection S, and more precisely specified protocols.

3.3 Existence and Construction of Resilient Collections

We now discuss choosing the collection S used in our construction. Before we give any constructions for S, we first recall a lower bound on $d = |S|$ due to Schonheim. Define $(a)_b$ to be the falling factorial:

$$(a)_b = (a)(a-1)\cdots(a-b+1) = \frac{a!}{(a-b)!}.$$

Theorem 1. *If d is the number of subsets in an (n, t, u)-resilient collection, then*

$$d \geq \frac{(n)_t}{(u-1)_t}.$$

Proof. Let $\mathcal{S} = \{S_1, \ldots, S_d\}$ be an (n, t, u)-resilient collection. Consider a particular subset S_i. There are

$$\binom{n - |S_i|}{t} = \binom{n - (n - u + 1)}{t} = \binom{u - 1}{t}$$

subsets of $\{1, \ldots, n\}$ of size t which are disjoint from S_i. Since there are $\binom{n}{t}$ subsets of $\{1, \ldots, n\}$ of size t, each of which must be disjoint from some S_i, the number d of subsets in $\mathcal{S}$ must be at least

$$\frac{\binom{n}{t}}{\binom{u-1}{t}} = \frac{(n)_t}{(u - 1)_t}.$$

In the other direction, we now wish to derive an upper bound on how large d must be. Bounds of this type exist in the combinatorics literature (see, for instance, [31], for some asymptotic results), but for our purposes, we want the following specific result:

Theorem 2. *Set*

$$d = \left\lceil \frac{(n)_t}{(u - 1)_t} \ln \binom{n}{t} \right\rceil.$$

Then there exists an (n, t, u)-resilient collection S of d sets, $\mathcal{S} = \{S_1, \ldots, S_d\}$.

Proof. We perform a probabilistic construction. Choose the d subsets $S_1, \ldots, S_d$ independently and uniformly at random to be subsets of size $(n - u + 1)$ of $\{1, \ldots, n\}$. Now, let S be any fixed t-element subset of $\{1, \ldots, n\}$. Consider a particular subset, S_i.

$$\Pr(S \text{ and } S_i \text{ are disjoint}) = \frac{\binom{n-t}{n-u+1}}{\binom{n}{n-u+1}}.$$

Since there are d subsets S_i, all distributed independently,

$$\Pr(S \text{ is not disjoint from any } S_i) = \left[1 - \frac{\binom{n-t}{n-u+1}}{\binom{n}{n-u+1}} \right]^d.$$

Because there are $\binom{n}{t}$ possibilities for S,

$$\Pr(\text{There is an } S \text{ which intersects every } S_i) \leq \binom{n}{t} \left[1 - \frac{\binom{n-t}{n-u+1}}{\binom{n}{n-u+1}} \right]^d.$$

Taking natural logs of both sides of this inequality,

$$\ln \Pr(\text{There is an } S \text{ which intersects every } S_i)$$
$$\leq \ln \binom{n}{t} + d \cdot \ln \left[1 - \frac{\binom{n-t}{n-u+1}}{\binom{n}{n-u+1}} \right]$$

$$< \ln \binom{n}{t} + d \cdot \left[-\frac{\binom{n-t}{n-u+1}}{\binom{n}{n-u+1}} \right]$$

$$\leq \ln \binom{n}{t} - \left[\frac{(n)_t}{(u-1)_t} \ln \binom{n}{t} \right] \cdot \frac{\binom{n-t}{n-u+1}}{\binom{n}{n-u+1}}$$

$$= 0,$$

where the strict inequality arises from the fact that $e^x \geq 1 + x$, with equality if and only if $x = 0$.

Hence, with nonzero probability, no subset of $\{1, \ldots, n\}$ of size t intersects every S_i. So there must exist some particular S such that this is true; this S is a (n, t, u)-resilient collection.

It is worth mentioning that since $\binom{n}{t} < 2^n$, the lower bound and upper bound which we have just shown differ from each other by at most a factor of $(n \ln 2)$, and so our randomized proof of an upper bound is actually not too far from optimal. Now, it would be nice if, given legal values for n, t, and u, we could easily come up with an (n, t, u)-resilient collection of as few sets as possible. Unfortunately, we don't know how to do this, and it seems to be a difficult problem to solve in general. We give here two specific construction methods for producing families of resilient collections; in each construction, values for n and u can be specified, and an $(n, f(n, u), u)$-resilient collection is produced.

Method 1. Fix n and $1 \leq u \leq n$. Set

$$S_1 = \{1, \ldots, (n - u + 1)\}, \ S_2 = \{(n - u + 1) + 1, \ldots, 2(n - u + 1)\}, \ \ldots .$$

Here, each subset contains the next $(n - u + 1)$ numbers, and we take enough subsets that every number between 1 and n is in at least one subset. Note that unless $(n - u + 1)$ divides n, the last subset in this list will "wrap around" and include numbers which are already in S_1. It's not difficult to see that this method yields a collection of $d = \lceil \frac{n}{n-u+1} \rceil$ subsets which is either $(n, \lceil \frac{n}{n-u+1} \rceil - 2, u)$-resilient (if $(n - u + 1)$ doesn't divide n) or $(n, \frac{n}{n-u+1} - 1, u)$-resilient (if $(n - u + 1)$ does divide n).

Method 2. Again, fix n and $1 \leq u \leq n$. Set

$$S_1 = \{1, \ldots, (u - 1)\}^c, \ S_2 = \{(u - 1) + 1, \ldots, 2(u - 1)\}^c, \ \ldots ,$$

where A^c denotes the complement (with respect to the set $\{1, \ldots, n\}$) of A. In analogy with method 1, we take enough subsets that each number between 1 and n is not contained in at least one subset, and we "wrap around" if $(u - 1)$ doesn't divide n. This construction produces an $(n, 1, u)$-resilient collection consisting of $\lceil \frac{n}{u-1} \rceil$ subsets.

3.4 Protocols

We now give more specific protocols for generating and evaluating shared pseudo-random functions. We assume each player is capable of digitally signing its messages, and of communicating privately with every other player. We give protocols for dealing with a single specific random function seed, s_i; these protocols should be executed concurrently for each seed. We assume that all players behave properly during the seed generation and distribution phase[3].

Seed Generation and Distribution. We implement protocol GEN-SEED(i) (to generate a random seed s_i and distribute it to the set S_i of players) as follows. Fix some particular player $T \in S_i$ (for example, T could be the "first" player in S_i, in some ordering). T chooses s_i at random, and sends it to every player in S_i. Each player in S_i then takes the value s_i that it receives, signs the string $s_i \circ i$, and sends this signed message to every player in S_i.

Function Evaluation. To enable an entity E (E may or may not be one of the players who shared the pseudo-random function) to evaluate $f(\cdot)$ at a point y, each player in S_i simply sends the value $f_{s_i}(y)$ to E.

The definition of a resilient collection guarantees that E will receive each piece of $f(y)$. Nonetheless, this may not suffice to enable E to compute $f(y)$. If players can send spurious values for pieces of $f(y)$, then E might receive *multiple* candidates for a piece of $f(y)$. In general, in fact, E could receive up to $(n-u+1)$ distinct candidates for each piece, which would lead to up to $(n-u+1)^d$ potential values for $f(y)$. We give two solutions to this problem:

1. If $t < 2u - n$, then we can modify our construction by basing our protocols on an $(n, t, 2u - n)$-resilient collection, instead of on an (n, t, u)-resilient collection. If u players cooperate to reveal $f_{s_i}(y)$ to E, then this ensures that more than half of the players in S_i give E the *correct* value of $f_{s_i}(y)$. So for each piece of $f(y)$, E can just take the most popular value sent to it. This enables E to compute $f(y)$.
2. If we do not wish to use the above solution (either because $t \geq 2u - n$, or because we are unwilling to allow the increase in the number of subsets required to make an $(n, t, 2u - n)$-resilient collection), then we can have each player give E a *zero-knowledge proof* of the correctness of each alleged piece of $f(y)$ that it sends. This proof enables E to ascertain what the correct value for each key-piece is, and therefore, to reconstruct chip z's secret key.

The theoretical approach of the second solution requires some further elaboration. If a player $T \in S_i$ sends E the value c, how does he or she then give a zero-knowledge proof that $f_{s_i}(z) \circ i = c$? The answer lies in the fact that T has copies of $s_i \circ i$ signed by each player in S_i. Essentially, T needs to claim

---

[3] The second method in Appendix B can be used if this assumption is unwarranted.

possession of values s_i, V_1, V_2, ..., V_{n-u+1} such that $f_{s_i}(z) \circ i = c$ and V_a is a signature of $s_i \circ i$ relative to the public signing key of the player a in S_i, for all $1 \leq a \leq n - u + 1$. This claim is an NP statement, and hence, it is possible to perform a zero-knowledge proof of it, using the methods of Goldreich, Micali, and Wigderson [21], Brassard, Chaum, and Crépeau [7], or Feige, Fiat, and Shamir [14]. Any honest player can convincingly execute the zero-knowledge proof protocol, whereas a player who is trying to convince E of the correctness of some false value will only be able to do so with negligible probability (since it will not possess the digital signatures needed to perform the proof protocol correctly, and will be unable to forge them).

We observe that the complexity of the zero-knowledge based protocol can be reduced to one round by the adoption of *non-interactive zero-knowledge proofs* (See [10], [15], and [5] for more information), or by the idea of *hashing* to obtain challenges to the prover (see, for example, [17] or [2]).

Notice also that the zero-knowledge proofs need be used only if absolutely necessary. If all of the alleged values of $f_{s_i}(y)$ that E receives agree with one another, then E knows that it has the correct value for that piece of $f(y)$. It is only if E receives several distinct values for $f_{s_i}(y)$ that there is any need to engage in zero-knowledge proofs.

If the only way a player can fail to cooperate is to refuse to yield information (as opposed to being able to maliciously present false data), then we can, of course, forgo the zero-knowledge proofs. In addition, in this case, we do not need to implement the second round of communication in the seed generation and distribution phase, since there is no need for players to possess digitally signed seeds if any information they send is known a priori to be correct.

4 Shared Pseudo-Random Functions and Key Escrow

4.1 Background: Key Escrow

The past two decades of progress in hardware and software have made (presumably) secure cryptography feasible for a large segment of the population. Recently, this has generated much concern that governments and corporations will soon be unable to make use of certain tools that have traditionally been used to apprehend wrongdoers. In particular, it is feared that court-authorized wiretapping will become essentially useless as a means of law enforcement [11].

The obvious way to ensure that this doesn't happen is for governments or corporations to arrange to possess every user's secret key; however, even legitimate users then enjoy no privacy. A better alternative that has recently been proposed is *key escrow* (see [28], [29], [26], and [25] for examples). In a key escrow scheme, a user's secret key is somehow split into *shares* held by *trustees*. The intent is that these trustees may, under appropriate circumstances, enable the reconstruction of a given secret key; however, sufficiently few trustees, behaving maliciously, do not possess enough information to reconstruct any key.

The Clipper Chip. This scheme works in a "top-down" fashion. For every user, x, two trustees each generate, randomly and independently, a secret string. These two strings are then sent to x's chip, which exclusive-ORs them together to compute x's secret key, c. c is stored in tamper-proof memory, so that no one (not even x) has any information about what c is, except for the trustees. Whenever x wishes to send a message m to user y, using a common key K_{xy} (which x and y have previously agreed upon), x's chip sends not only the encryption of m with key K_{xy}, but also the encryption of K_{xy} with key c (the algorithm for this latter encryption is classified, and is referred to as *Skipjack*). When presented with authorization for a wiretap of x's communications, the trustees each reveal their share of c to the FBI, who is then able to decrypt first K_{xy}, and then m.

4.2 Bad Behavior of Trustees

We distinguish three types of bad trustees in a key escrow scheme:

1. *Gossipy* trustees, who try to procure information that they should not have about users' keys, or who spread users' key information about.
2. *Withholding* trustees, who do not cooperate with appropriate authorities to recover a user's secret key.
3. *Obstructive* trustees, who are so malicious that they may not even behave properly during the set-up phase of a key escrow scheme (when users generate or otherwise obtain their keys). We shall consider obstructive trustees to be both gossipy and withholding.

For $0 \leq t < u \leq n$, we define an (n, t, u)-*escrow scheme* to be a method of splitting a secret key among n trustees such that:

– If at most t trustees are gossipy, then reconstruction of a secret key is not feasible without a court order.
– If at most $(n - u)$ trustees are withholding, then reconstruction of a secret key can easily be accomplished with a court order.

With this terminology, the Escrowed Encryption Standard is a (2,1,2)-escrow scheme.

4.3 Our Escrow Scheme

Using our shared pseudo-random functions, we can easily create a Clipper-like (n, t, u)-escrow scheme, if no trustees behave obstructively[4]:

1. Each chip has a unique j-bit ID number.
2. The secret key of chip #z is the j-bit string $f(z)$, where $f(\cdot)$ is a pseudo-random function shared among the n trustees.

---

[4] We defer consideration of obstructive trustees until Appendix B.

3. To initialize a chip with its secret key, each trustee gives the chip all the pieces of $f(z)$ that it has (since no trustees are obstructive, no signatures or zero-knowledge proofs are necessary here).

4. To permit wiretapping of a chip's communications, the trustees perform a pseudo-random function evaluation protocol (with the FBI playing the part of the evaluator E). Two such protocols were presented earlier. However, if $(n - u + 1)^d$ is small (recall that this is an upper bound on the number of candidate values of the pseudo-random function), then an alternative protocol exists: have each trustee send all its key pieces to the FBI, and then let the FBI simply "cycle through" all possible secret keys, and see which one decrypts the user's messages.

5 Conclusion

We have presented a simple method of generating and sharing a pseudo-random function among a group of players. Our procedure can be used with any type of collection of pseudo-random functions: poly-random collections, hash functions, DES, etc. After an initial set-up phase, in which a shared function is generated, evaluating the function requires only one round of communication. Of course, communications grow in length as we increase the number of random function seeds; in some situations, exponentially many (in the number of trustees) seeds are needed.

Our shared pseudo-random functions can be readily applied to produce key escrow schemes. Since escrow schemes generally have a rather small number of trustees, the exponential number of seeds mentioned above is not a real problem for this application.

Like the Clipper Chip proposal, our escrow scheme also combines well with the key distribution scenarios suggested by Leighton and Micali in [26] to achieve a conventional cryptosystem which achieves many of the gains of public-key cryptography without requiring the complexity-theoretic assumptions needed for public-key cryptography.

References

1. B. Awerbuch, M. Blum, B. Chor, S. Goldwasser, and S. Micali. How to implement Bracha's $O(\lg n)$ Byzantine agreement algorithm. Submitted to 1985 Principles of Distributed Computing Conference.

2. M. Bellare and P. Rogaway. Random oracles are practical: A paradigm for designing efficient protocols. In *Proceedings of the 1st ACM Conference on Computer and Communications Security*. ACM, 1993.

3. J. Benaloh and J. Leichter. Generalized secret sharing and monotone functions. In *Advances in Cryptology— CRYPTO '88*, pages 27-35. Springer-Verlag, 1988.

4. G. R. Blakley. Safeguarding cryptographic keys. In *Proceedings of the AFIPS 1979 National Computer Conference*, pages 313-317, June 1979. New York, NY.

5. M. Blum, A. De Santis, S. Micali, and G. Persiano. Noninteractive zero-knowledge. *Siam Journal of Computing*, 20(6):1084-1118, December 1991.

6. G. Bracha. An $O(\lg n)$ expected rounds randomized Byzantine generals protocol. In *Proceedings of the 17th Annual ACM Symposium on Theory of Computing*, pages 316–326. ACM, 1985.

7. G. Brassard, D. Chaum, and C. Crépeau. Minimum disclosure proofs of knowledge. *Journal of Computer and System Sciences*, 37(2):156–189, 1988.

8. B. Chor, S. Goldwasser, S. Micali, and B. Awerbuch. Verifiable secret sharing and achieving simultaneity in the presence of faults. In *Proceedings of the 26th Annual IEEE Symposium on Foundations of Computer Science*, pages 383–395. IEEE, 1985.

9. A. De Santis, Y. Desmedt, Y. Frankel, and M. Yung. How to share a function securely. In *Proceedings of the 26th Annual ACM Symposium on Theory of Computing*, pages 522–533. ACM, 1994.

10. A. De Santis, S. Micali, and G. Persiano. Non-interactive zero-knowledge proof-systems. In *Advances in Cryptology— CRYPTO '87*, pages 52–72. Springer-Verlag, 1987.

11. D. Denning. To tap or not to tap. *Communications of the ACM*, 36(3):25–44, March 1993.

12. Y. Desmedt and Y. Frankel. Shared generation of authenticators and signatures. In *Advances in Cryptology— CRYPTO '91*, pages 457–469. Springer-Verlag, 1991.

13. Y. G. Desmedt. Threshold cryptography. *European Transactions on Telecommunications and Related Technologies*, 5(4):449–457, 1994.

14. U. Feige, A. Fiat, and A. Shamir. Zero-knowledge proofs of identity. *Journal of Cryptology*, 1(2):77–94, 1988.

15. U. Feige, D. Lapidot, and A. Shamir. Multiple non-interactive zero-knowledge proofs based on a single random string. In *Proceedings of the 31st Annual IEEE Symposium on Foundations of Computer Science*, pages 308–317. IEEE, 1990.

16. P. Feldman and S. Micali. Byzantine agreement in constant expected time (and trustine no one). In *Proceedings of the 26th Annual IEEE Symposium on Foundations of Computer Science*, pages 267–276. IEEE, 1985.

17. A. Fiat and A. Shamir. How to prove yourself: Practical solutions to identification and signature problems. In *Advances in Cryptology— CRYPTO '86*, pages 186–194. Springer-Verlag, 1986.

18. O. Goldreich, S. Goldwasser, and S. Micali. How to construct random functions. *Journal of the Association for Computing Machinery*, 33(4):792–807, October 1986.

19. O. Goldreich, S. Micali, and A. Wigderson. Proofs that yield nothing but their validity and a methodology of cryptographic protocol design. In *Proceedings of the 27th Annual IEEE Symposium on Foundations of Computer Science*, pages 174–187. IEEE, 1986.

20. O. Goldreich, S. Micali, and A. Wigderson. How to play any mental game or a completeness theorem for protocols with honest majority. In *Proceedings of the 19th Annual ACM Symposium on Theory of Computing*, pages 218–229. ACM, 1987.

21. O. Goldreich, S. Micali, and A. Wigderson. Proofs that yield nothing but their validity or all languages in NP have zero-knowledge proof systems. *Journal of the Association for Computing Machinery*, 38(1):691–729, July 1991.

22. R. Impagliazzo, L. Levin, and M. Luby. Pseudo-random generation from one-way functions. In *Proceedings of the 21st Annual ACM Symposium on Theory of Computing*, pages 12–24. ACM, 1989.

23. J. Kilian. Founding cryptography on oblivious transfer. In *Proceedings of the 20th Annual ACM Symposium on Theory of Computing*, pages 20–31. ACM, 1988.

24. J. Kilian. *Uses of Randomness in Algorithms and Protocols*. ACM Distinguished Dissertations. MIT Press, 1990.

25. J. Kilian and T. Leighton. Failsafe key escrow. Technical Report TR-636, MIT, August 1994.

26. T. Leighton and S. Micali. Secret-key agreement without public-key cryptography. In *Advances in Cryptology— CRYPTO '93*, pages 456–479. Springer-Verlag, 1993.

27. L. Levin. One-way functions and pseudorandom generators. In *Proceedings of the 17th Annual ACM Symposium on Theory of Computing*, pages 363–365. ACM, 1985.

28. S. Micali. Fair cryptosystems. Technical Report TR-579.b, MIT, November 1993.

29. National Institute for Standards and Technology. *Escrowed Encryption Standard (EES)*, 1994. Federal Information Processing Standards Publication (FIPS PUB) 185.

30. A. Shamir. How to share a secret. *Communications of the ACM*, 22(11):612–613, November 1979.

31. J. Spencer. Asymptotically good coverings. *Pacific Journal of Mathematics*, 118(2):575–586, 1985.

32. A. Yao. On the succession problem for Byzantine generals. Manuscript, Stanford University.

33. A. Yao. How to generate and exchange secrets. In *Proceedings of the 27th Annual IEEE Symposium on Foundations of Computer Science*, pages 162–167. IEEE, 1986.

Appendix A: Oblivious Trustees

In [28], Micali brings up an idea which he calls *oblivious trustees*— namely, that trustees who are cooperating with the FBI to reveal a user's secret key should not know *whose* secret key they are revealing. This prevents them from giving advance warning to the party whose line will be tapped. By using an *oblivious circuit evaluation* protocol[5], we can make our scheme have oblivious trustees. Note that oblivious circuit evaluation is a somewhat complicated process, and so this feature may be more theoretical than practical.

We assume that there is some court which is capable of signing authorizations for wiretaps. That is, if chip z is to be tapped, then the court signs z and gives this signature to the FBI. It is enough for us to consider how the FBI gets a hold of a single value $f_{s_i}(z)$ from a trustee $T \in S_i$.

Define

$$g(s_i, \text{SIGS}, z, \text{AUTH}) = \begin{cases} f_{s_i}(z) & \text{If SIGS consists of legal signatures of } s_i \\ & \text{by all trustees in } S_i \text{ and AUTH is a legal} \\ & \text{signature of } z \text{ by the court.} \\ \text{"ERROR"} & \text{Otherwise.} \end{cases}$$

---

[5] *Oblivious circuit evaluation* is a way for two parties, A and B, who hold private inputs x and y, respectively, to interact to compute a value $f(x, y)$. After they have finished interacting, B has learned $f(x, y)$, but nothing else about x, and A has learned nothing about y. See [33], [20], [23], and [24] for more information about oblivious circuit evaluation, including protocols for accomplishing it.

Then we perform an oblivious circuit evaluation in which the FBI computes g, where T supplies inputs s_i and SIGS, and the FBI supplies inputs z and AUTH. This gives the FBI precisely the information it deserves and needs to know.

Observe that to make our scheme have oblivious trustees, we needed to use the fact that users' secret keys are chosen using poly-random functions. If secret keys were not chosen in this fashion, there would be no underlying algebraic connection between them— that is, if users were given keys completely independently of each other, then there would be no small circuit which could compute chip z's secret key from the value z— and so trustees could not be made oblivious like this.

Appendix B: Obstructive Trustees

We now sketch methods for dealing with obstructive trustees. There are two possibilities to consider:

1. We merely want to be aware of the presence of obstructive trustees, so that we can abort setting up our cryptosystem (and possibly start over again, with different trustees). In this case, the protocols we presented work, as long as we insist that at each stage, each trustee and each chip checks the data it has and makes certain that it is consistent. For example, in the chip initialization phase, each chip should ensure that for each $1 \leq i \leq d$, it received a common value s_i from every trustee in S_i. If it didn't, then the chip should complain that there is a malicious trustee present.

2. We need to ensure that even if some trustees are obstructive, we can nevertheless go about setting up our cryptosystem. This is more complicated. In this case, we do something similar to what was done in the second solution from section 3.4. That is, we require that $t < 2u - n$, and base our key escrow scheme on an $(n, t, 2u - n)$-resilient collection. Since obstructive trustees are withholding (by definition), this means that in each S_i, the obstructive trustees are in the minority. Thus, the trustees in each S_i can use a general secure multi-party computation protocol (see [19], for example) to produce a common seed in the generate and distribute seeds phase. In addition, we modify the chip initialization phase so that for each seed i, the chip uses whichever alleged value has the most trustees claiming that it's correct; similarly, in the wiretapping phase, the FBI also takes the majority value for each individual key-piece. No digital signatures or zero-knowledge proofs are required.

A Key Escrow System with Warrant Bounds

Arjen K. Lenstra[1], Peter Winkler[2], Yacov Yacobi[3]

[1] MRE-2Q330, Bellcore, 445 South Street, Morristown, NJ 07960, U.S.A.
E-mail: lenstra@bellcore.com
[2] 2D-147, AT&T Bell Laboratories, 600 Mountain Ave., Murray Hill, NJ 07975, U.S.A.
E-mail: pw@research.att.com; Research done at Bellcore
[3] MRE-2Q338, Bellcore, 445 South Street, Morristown, NJ 07960, U.S.A.
E-mail: yacov@bellcore.com

Abstract. We propose a key escrow system that permits warrants for the interception and decryption of communications for arbitrary time periods, and with either one or two communicating parties specified as the target. The system is simple and practical, and affords reasonable protection against misuse. We argue that use of such a system can produce *both* greater privacy protection and more effective law enforcement than we now enjoy.

1 Background

The tug-of-war between law enforcement agencies and rights advocates regarding communications privacy is not necessarily a zero-sum game. We believe that with a well-designed key escrow system, it is possible to increase the effectiveness of electronic surveillance as an anti-crime measure, while affording better protection to the privacy of citizens.

Although U.S. law enforcement agencies such as the Federal Bureau of Investigation have complained that digital telephony and commercially available cryptography threaten the effectiveness of electronic surveillance, it is a fact that in many respects electronic surveillance is becoming easier.

Electronic surveillance is currently expensive; the average cost of installing and monitoring an intercept in 1993 was $57,256 [1]. There have been only about 900 intercepts ordered per year by state and federal authorities put together, with between 200,000 and 400,000 incriminating conversations recorded annually; the number of *non-incriminating* conversations recorded each year has increased to over 1.7 million. The non-incriminating conversations are weeded out "by hand" at the cost not only of time and money but of the privacy of innocent parties.

Increasingly, however, cordless and cellular telephony permit electronic surveillance without physical access; programmable switches can obviate the necessity for hardware altogether; and digital messaging permits automatic sifting of conversations (by destination, content etc.). Thus the potential exists for cheaper and more effective use of electronice surveillance, and the consequences for the privacy of citizens must be examined carefully.

D. Coppersmith (Ed.): Advances in Cryptology - CRYPTO '95, LNCS 963, pp. 197-207, 1995.

The availability of public-key cryptography, and the explosion of public awareness in cryptography in general, have put a powerful privacy-enhancing tool into the hands of citizens; conceivably the widespread use of encryption could cripple electronic surveillance as a law enforcement tool. In an effort to provide an alternative, the White House announced on April 16, 1993 the "Escrowed Encryption Initiative." Subsequently the National Institute of Standards and Technology (NIST) approved the *Escrowed Encryption Standard* (EES) for telephone systems [10].

The EES (known often by the name of its chip, "Clipper") caused a substantial outcry [12], partly from cryptologists opposed to the use of a secret algorithm, partly from rights advocates opposed to the whole idea of escrowed keys. The secret algorithm (SKIPJACK) is certainly unnecessary for an escrow system and excellent alternatives have been proposed, e.g., by Micali [9] and Kilian and Leighton [7].

The escrow issue itself is more troublesome. As presently constituted, EES calls for individual keys to be split in two pieces which are given to two "trustees" (namely, NIST and a branch of the Treasury Department) who, when served with a lawful warrant, will turn the key over to law enforcement authorities. The warrant itself will contain the usual limitations on target, content and time interval (usually a specified 30-day period), but these limitations do not apply to the key. Instead, the key is supposed to be "returned" (!) at the expiration of the warrant, but non-compliance with this or other Dept. of Justice procedures explicitly "shall not provide the basis for any motion to suppress or other objection to the introduction of electronic surveillance evidence lawfully acquired" [4]. Of course, such a disclaimer is understandable in view of the difficulty of proving (for example) that the FBI no longer has some target's key.

In effect, if citizens a and b give law enforcement authorities reason to believe that they have or will use the telephone to commit a crime, *each* of them gives up his or her "cryptographic rights" for all telephone conversations and for all time—past, present and future. Surely such a concession is unnecessary and excessive, even if one believes that law enforcement authorities have no intention of misusing a key. When automatic sifting of telephone conversations becomes possible, it will be increasingly tempting for the authorities to gather large quantities of data for possible later use when a key is held. But decrypting conversations prior to the start of a warrant would not be very useful since they could not be entered into evidence, and would in fact be illegal to collect.

We believe that a system in which the courts can enforce the terms of a warrant will not only help preserve privacy rights, but will also enable the courts to be more liberal in granting warrants so that federal and state authorities can make better use of electronic surveillance. Micali's system [9] already permits time-bound warrants, for which the law enforcement authorities receive only keys good for the warrant period. We go a step further, allowing the trustees to supply authorities with a key good for only a *pair* of conversants, when appropriate. Thus, for example, if some set of persons is suspected of conspiracy, the courts will be able to issue a warrant for surveillance only of conversations among the targeted parties. Our system will satisfy other requirements as well, detailed below.

2 Requirements

Following are requirements for a key escrow system. We begin with the warrant bounds.

time-boundedness:
> It must be possible for the courts to enforce the time-limits of a warrant, by supplying a key that will be effective only for a given period of time (most likely some set of days). As noted above, Micali's system [9] offers a time-bounded option; Clipper of course does not, nor does Kilian-Leighton [7].

target flexibility:
> It must be possible for the courts to permit either node surveillance (in which all communications involving a particular target a can be decrypted) or edge surveillance (in which only communications between a and b can be decrypted). None of Clipper, [9] or [7] offers target flexibility.

It is in fact not difficult to design an escrow system that is time-bounded and target-flexible, but we insist that other important characteristics not be sacrificed[4]. In particular, the following properties are desirable and (as we shall see) attainable as well.

non-circumventability:
> It should be impossible for a user to *unilaterally* alter his communication protocol in such a way as to obtain encryption from the system without exposing himself to decryption by the proper authorities. Obviously, one cannot stop persons from *colluding* to avoid key escrow, because they can always use their own system (or pieces of the given system); but the system should not provide an obvious way to collude in this manner. (As an example, a system with daily keys must avoid making it easy for persons to cheat by resetting their calendars.)

Roughly speaking, the system of Kilian and Leighton [7] enjoys this non-circumventability, as does Clipper except for the (presumably correctable) flaw found by Blaze [2]. Micali's scheme, however, appears to be unilaterally circumventable when used in a time-bounded manner.

security:
> The system should rely on familiar and tested cryptologic techniques. Ideally it should rely on *proven* techniques, but given that there are too

---

[4] In [5] a "balanced cryptosystem," based on shareable functions, is mentioned where each individual message can be revealed without affecting other messages. It remains to be seen if that cryptosystem applies to our situation, and which of our other properties is satisfied.

few of these, the system should at least avoid techniques that have not built up some empirical credibility.

The Micali and the Kilian-Leighton schemes offer reasonable security in this sense and Clipper probably would if the SKIPJACK algorithm were made public; as it is, the public has to rely on a panel report [3].

simplicity:

The system should be practical and understandable; in particular it should not rely on repeated contact between users and trustees, nor should it require many-round preliminaries between communicating parties. It should not offer any impediment to telephone, FAX, or e-mail communication. It should be explicable, in outline if not mathematically, to intelligent lay persons, e.g. the courts.

The Clipper and Kilian-Leighton systems are reasonably simple by this definition; some incarnations of Micali's system are less so, and in particular his time-bounded version requires considerable interaction.

Some additional properties will be discussed later. We now proceed to describe the proposed key escrow system.

3 Preliminaries

Let p and q be two large primes with $q|p-1$, and let $g \in \mathbf{Z}/p\mathbf{Z}$ be an element of order q. We make the obvious identification between $\mathbf{Z}/m\mathbf{Z}$ and $\{0, 1, \ldots, m-1\}$, for any integer m, and between $(\mathbf{Z}/p\mathbf{Z})^*$ and $\{1, 2, \ldots, p-1\}$.

All users of the key escrow system described here share the same p and g. Each user u has a public key $P(u) \in (\mathbf{Z}/p\mathbf{Z})^*$ and a secret key $S(u) \in \mathbf{Z}/q\mathbf{Z}$ such that $g^{S(u)} \equiv P(u) \bmod p$. It is assumed that for all u it is computationally infeasible to derive $S(u)$ from p, g, and $P(u)$—this assumption is based on the supposed difficulty of the discrete logarithm problem. The keys $P(u)$ and $S(u)$ are referred to as the *permanent keys* of user u. The trustees can get their verifiable shares of the permanent keys, with or without thresholds, as described in [11].

Let f be any good conventional block cipher, like (triple) DES. We designate by $c = f(k, m)$ the cryptogram that results from encrypting message m using f with secret key k, and $m = f^{-1}(k, c)$, where c, k, and m are bit strings. It is assumed that m can be derived efficiently from c if and only if k is known, but that k cannot be derived efficiently from c and m.

Furthermore, let $h : \mathbf{Z}/p\mathbf{Z} \times \mathbf{Z}/p\mathbf{Z} \to \mathbf{Z}/p\mathbf{Z}$ be a one way hash function, such that given d and $d_i \neq d$, $y_i = h(x, d_i)$ for any number of $i \in \mathbf{Z}$ and some unknown x, it is computationally infeasible to find $y = h(x, d)$. The hash function h has to satisfy several other requirements that are specific to our protocol; they will be discussed in the next section.

Both h and f are fixed throughout this paper.

4 The system

We assume that the users of the system agree upon a certain time (presumably UTC), which may for instance be provided by satellite. In what follows we assume that the desired granularity of time for warrants is daily, and we regard any day d as an element of $\mathbf{Z}/p\mathbf{Z}$.

Furthermore we assume that the provider of the communication links (the 'service-provider', e.g. a phone company) is the party that actually provides access to communications for a law enforcement agency who provides the service-provider with a valid warrant. All data obtained by the law enforcement agency from the service-provider should be immediately and unforgeably time-stamped and signed by the latter, so that law enforcement should never be able to pass communication data from one day for data from any other day.

Let $P(a)$, $S(a)$ and $P(b)$, $S(b)$ be the permanent public and secret keys of users a and b, respectively, as described above.

Protocol (for parties a and b on day d):

1. First a and b establish, non-interactively, their session key $k(a,b,d) = k(b,a,d)$, which is computed as $k(a,b,d) = h(P(b)^{S(a)}, d)$ by a and as $k(b,a,d) = h(P(a)^{S(b)}, d)$ by b.

2. Next, before the actual communication using the common key $k(a,b,d)$ takes place, a and b are required to exchange a message to enable law enforcement to compute $k(a,b,d)$ and to decypher the communication between a and b. This is done as follows.

   2a. Party a computes $S(a,d) = h(S(a), d)$, and
        party b computes $S(b,d) = h(S(b), d)$.

   2b. Party a computes $S(a,b,d) = h(S(a,d), P(b))$, and
        party b computes $S(b,a,d) = h(S(b,d), P(a))$.

   2c. Party a sends the message $c(a,b,d) = f(S(a,b,d), k(a,b,d))$ to b, and
        party b sends the message $c(b,a,d) = f(S(b,a,d), k(b,a,d))$ to a.

   (Note that a uses $S(a,b,d)$ as key to encrypt $k(a,b,d)$ using f. Therefore, party b cannot determine the common key $k(a,b,d)$ by decrypting $c(a,b,d)$, but has to compute the common key in Step 1. Similarly a cannot decrypt the message $c(b,a,d)$.)

3. Parties a and b communicate using the conventional block cipher f, with their common key $k(a,b,d)$ as key. All messages encrypted using f and $k(a,b,d)$ are required to have a certain fixed structure, for instance by pre-fixing them with a certain system dependent header.

Remarks. Note that computing $g^{S(a)S(b)}$ in Step 1 is the greatest part of the computation, and that this value only depends on the communicating parties but not on the day d. This allows users to precompute and store those 'expensive' values for frequent partners, and to do the 'cheap' computations involving h and f on a day to day basis.

Warrants. We now describe the various types of warrants law enforcement might obtain. Unless there is demonstrable cheating by a user, all warrants are time-bounded. In the absence of deliberate collusion, both parties to a conversation must obey the rules of Step 1 in order to create a common session key. This implies that non-collaborative cheating can only take place in Step 2, during which, conceivably, a corrupted c value might be sent. (Note: our description of the warrants assumes that there is a single trustee who knows the secret keys of all users of the escrow system. A threshold secret sharing scheme will be discussed in a later section.)

Possible warrants against a on day d:

1. Edge surveillance: the trustee provides law enforcement with $S(a, b, d)$, for all partners b of a to which the warrant applies.
2. Node surveillance: the trustee provides law enforcement with $S(a, d)$.
3. If a cheats in any part of Step 2: the trustee provides law enforcement with $S(a)$.

In each of these three cases it is clear from the protocol how law enforcement proceeds to obtain the session key(s) for day d. Note that if a cheats in any part of Step 2 and sends a corrupted $\bar{c}(a, b, d)$ to b, law enforcement will not be able to retrieve the correct session key. In this case, however, law enforcement should be able to convince the trustee that cheating took place, because it is highly unlikely that the ensuing communication between a and b decrypted with the wrong key will have the right fixed structure. Note, however, that the trustee will only provide law enforcement with the user's secret key, if law enforcement can prove, using the time-stamp, that the communication took place on the right day. This leaves the possibility that the user claims a non-malicious error due to noise; we assume that the lower level communication protocols are designed such that this happens with sufficiently low probability.

In principle, the messages in Step 2c do not have to be sent to b and a as long as some party (e.g. the provider of the communication links) records the messages, and forwards them to law enforcement, if appropriate. Sending them to b and a is the most natural solution, however, because the channel between a and b is the channel that will be monitored.

Remark. Because the key agreement part of this protocol is non-interactive, the protocol can be used in applications such as FAX and e-mail: party a carries out its portion of the protocol when sending a message to b, and b carries out its part only when, and if, it responds. For these possibly one-sided communications, however, the warrants have to be formulated differently. In the case of edge surveillance the trustee provides law enforcement not only with $S(a, b, d)$ but with $S(b, a, d)$ as well, for all partners b of a to which the warrant applies; if a only receives communications from any of these b's, they can be decrypted, which is not possible if law enforcement has only $S(a, b, d)$ at its disposal. Node surveillance still works if it is only intended for the decryption of outgoing messages (from a)—as soon as a receives a message from some party b to which a has not sent and will not send a message, law enforcement needs $S(b, a, d)$ as well.

Protection against a frame-up. A warrant should not enable anyone to "frame" or "impersonate" any of the parties affected by the warrant. Users are therefore supposed to sign their messages using other systems, with non-escrowed keys, to assure that they cannot be framed by law enforcement or a collusion of trustees. This is necessary in all escrow systems, because anyone with a legal warrant can use the resulting session key to encrypt any message "on behalf" of any user.

Additional requirements on the hash function h. All key escrow systems should be secure against all types of illegal intercepts—also by law enforcement agencies who attempt to exceed their warrant. In this context, known-key attacks against key agreement systems are relevant [13].

From the property of the hash function h mentioned in Section 3 it follows that law enforcement cannot predict any past or future session key given any feasible number of other session keys. Similarly, any collection of $S(a,d) = h(S(a),d)$ for any feasible number of d's in itself is not enough to derive $S(a,\tilde{d})$ for any other previous or future $\tilde{d}$, and neither can a collection of $S(a,b,d) = h(S(a,d),P(b))$ for some feasible number of b's be used to derive $S(a,\tilde{b},d)$ for any other $\tilde{b}$.

However, the context as well as the values themselves might be known. For example, to protect against finding $S(a)$ given $S(a,d)$, or $S(a,d)$ given $S(a,b,d)$, stronger conditions than are usually considered for hash functions are crucial. Namely, given $S(a,d)$ the secret $S(a)$ is not only a value for which $S(a,d) = h(S(a),d)$, but it is also a value for which the 'correctness' can be verified in another way—by checking whether $h(P(b)^{S(a)},d)$ unlocks the communication in Step 3 in the proper way. And, slightly more involved, given $S(a,b,d)$, the 'node surveillance' key $S(a,d)$ is not only a value for which $S(a,b,d) = h(S(a,d),P(b))$, but also a value for which an $S(a)$ exists for which the correctness can be checked by other means—again by attempting to unlock the communication in Step 3. This is related to so-called 'promise problems,' for which we refer to [6].

Even more complicated conditions have to be imposed on h to make it infeasible to derive useful values from any collection of session keys, $S(a,d)$'s, or $S(a,b,d)$'s. The details are hardly enlightening, however, and we do not elaborate. We note, however, that this is a common problem in applications of hashing functions that is by no means typical for our protocol. For example, the security of the Challenge-Handshake Authentication Protocol of the Point-to-Point Protocol as described in [8] implicitly depends on assumptions about hash functions that are very similar to the assumptions that we have to make. Most likely any popular hash function that provides a decent number of bits satisfies all conditions required. Note also that even if different hash functions are used in the different steps of the protocol, we still require that these functions have the more involved properties discussed here.

Security of the protocol. We have not been able to show that our protocol is provably secure or at least as hard to break as some well-established hard-to-solve problem. The following observations might, however, be helpful to shed some light on this matter.

Obviously, for any a and b it should be hard to derive $k(a, b, d) = h(P(b)^{S(a)}, d) = h(P(a)^{S(b)}, d)$ from h, d, $P(a)$, and $P(b)$. Otherwise, any eavesdropper would be able to decypher the traffic between a and b, even without having access to $c(a, b, d)$ or $c(b, a, d)$. If h were an invertible one-to-one function, then this cracking problem would be at least as hard to solve as the 'Diffie-Hellman problem' (i.e., finding g^{xy} given g^x and g^y). For a one-way hash function h the cracking problem certainly looks even more difficult, in general, but we are not aware of any rigorous results in this direction.

The picture gets considerably more complicated if we consider an eavesdropper who has access to $c(a, b, d)$ or $c(b, a, d)$, or a law enforcement agent who tries to exceed the bounds of a legal warrant. The worst situation, leading to the easiest cracking problem, is where law enforcement tries to exceed the bounds of some number of node surveillance warrants: given $P(a)$, $P(b)$, $k(a, b, d)$, $c(a, b, d)$, $c(b, a, d)$, and $S(a, d)$, for $b \in B$ and $d \in D$, (for some party a, some set B of parties communicating with a, and some set of days D), compute $k(a, b, d')$ for any $b \in B$ and $d' \notin D$.

We require that the hash function h be such that this cracking problem is hard on the average. Note that simple one-wayness or even ordinary 'claw-freeness' may be insufficient because of the dependencies between the various inputs to h and f.

In the next section we discuss the situation where there is more than a single trustee.

5 Threshold secret sharing

To allow more than one trustee in our protocol, we assume that there are $m > 1$ trustees, and that trustee i has a verifiable share $S_i(a)$ of a's permanent secret key $S(a)$, for $1 \leq i \leq m$. Furthermore, we assume that there is some n, $1 < n \leq m$, such that any subset of n trustees can recover $S(a)$ using their $S_i(a)$'s, i.e., we assume the existence of a public function T such that $T(S_{v_1}(a), S_{v_2}(a), \ldots, S_{v_n}(a)) = S(a)$ for any subset $\{v_1, v_2, \ldots, v_n\}$ of $\{1, 2, \ldots, m\}$. The function T is such that any subset of at most $n - 1$ trustees has no advantage over anyone else to compute $S(a)$.

If we allow that an n-subset of the trustees, upon receipt of a legal surveillance authorization, computes $S(a)$, and provides law enforcement either with $S(a, d)$, or $S(a, b, d)$, or $S(a)$, depending on the type of authorization and other considerations, then our protocol is unaffected. But for a this situation would be highly undesirable, because, even in the event of the most limited warrant against a, someone would get to see his permanent secret $S(a)$. Therefore, we assume that the trustees are not allowed to collude in this way, and that each trustee is only supposed to provide law enforcement with the trustee's relevant share of information. This requires the following changes in our protocol.

Secret sharing without thresholds. If we only have to allow $n = m$, i.e., if the shares of *all* trustees are required, the change is rather straightforward. Instead of computing $S(a, d)$, party a computes $S_i(a, d) = h(S_i(a), d)$ for $1 \leq i \leq m$ and $S(a, b, d)$ is computed by first computing $S_i(a, b, d) = h(S_i(a, d), P(b))$ and next

$$S(a, b, d) = T(S_1(a, b, d), S_2(a, b, d), \ldots, S_m(a, b, d)),$$

after which $c(a, b, d)$ is computed as usual. The changes for party b are similar.

Depending on the type of warrant, law enforcement either gets the $S_i(a, d)$'s (node surveillance) or the $S_i(a, b, d)$'s (edge surveillance) from all m trustees. In both cases law enforcement can derive the relevant $S(a, b, d)$ or $S(a, b, d)$'s by applying (h and) T. If user a cheats, the $S_i(a)$ are revealed by the trustees, which allows law enforcement to compute $S(a)$ using T.

Secret sharing with thresholds. In the more general case that only the shares of any n-subset of trustees are required, we have to make more extensive changes to our protocol. The solution we present is based on the verifiable threshold secret sharing scheme from [11]. Using this scheme, we may assume that for any ordered subset $V = \{v_1, v_2, \ldots, v_n\}$ of $\{1, 2, \ldots, m\}$ there are non-secret and easily computable constants $b_i(V)$ such that

$$S(a) = \sum_{i=1}^{n} b_i(V) S_{v_i}(a).$$

Without loss of generality we take $\{v_1, v_2, \ldots, v_n\} = \{1, 2, \ldots, n\}$, and we write b_i for $b_i(V)$.

Instead of computing $S(a, d)$ and $S(a, b, d)$ as in Step 2 of the protocol of Section 4, party a computes $S(a, d) = h(d, d)^{S(a)}$ and $S(a, b, d) = h(d, P(b))^{S(a)}$. Furthermore, party a is not only required to send $c(a, b, d) = f(S(a, b, d), k(a, b, d))$ to b, but also an additional message

$$\bar{c}(a, b, d) = f(S(a, d), k(a, b, d)).$$

The changes for party b are similar.

For a node surveillance warrant, law enforcement gets $S_i(a, d) = h(d, d)^{S_i(a)}$ for $1 \leq i \leq n$ (cf. our subset assumption). From these law enforcement can compute $S(a, d) = \prod_{i=1}^{n} S_i(a, d)^{b_i}$, so that $k(a, b, d)$ can be derived from $\bar{c}(a, b, d)$ for any b communicating with a.

For edge surveillance for the communication between parties a and b, law enforcement gets $S_i(a, b, d) = h(d, P(b))^{S_i(a)}$ for $1 \leq i \leq n$. From these law enforcement can compute $S(a, b, d) = \prod_{i=1}^{n} S_i(a, b, d)^{b_i}$, so that $k(a, b, d)$ can be derived from $c(a, b, d)$.

If user a cheats, the $S_i(a)$ are revealed by n trustees, which allows law enforcement to compute $S(a) = \sum_{i=1}^{n} b_i S_i(a)$.

6 Covert Channels

In [7] it is observed that in many systems, the escrow service can be abused by encoding "shadow" public keys in the data that normally represents ordinary public keys. For example, a long RSA public exponent may represent another public key for a Diffie-Hellman key agreement system. The secret key that corresponds to the shadow public key is never given to the trustees, and thus a covert channel may be established, to which the law enforcement authorities have no access. Yet, the user gets certification services from the system. This form of collusion can be overcome in our system by letting each secret key be a modular sum of two integers, one contributed by the user, the other by the trustees, as noted in [7].

7 Conclusion

A key escrow system is offered that permits cryptographic limits on warrants; in particular it allows the courts to enforce time-limits on surveillance and to target either individuals or pairs of communicating parties. The system is reasonably simple and preserves most of the desirable properties enjoyed by other proposed systems.

Acknowledgments. We gratefully acknowledge valuable suggestions made by Milt Anderson, Ernie Brickell, Stuart Haber, and James Kulp.

References

1. Administrative Office of the United States Courts, 1993, *Report on Applications for Orders Authorizing or Approving the Interception of Wire, Oral, or Electronic Communications (Wiretap Report)*, 1993.
2. M. Blaze, "Protocol Failure in the Escrowed Encryption Standard," Proceedings of the 2nd ACM Conference on Computer and Communications Security, November 1994, pp. 59–67.
3. E. Brickell, D. Denning, S. Kent, D. Maher, and W. Tuchman, "SKIPJACK Review: Interim Report, The SKIPJACK Algorithm," July 28, 1993, available electronically from cpsr.org.
4. Department of Justice Briefing re Escrowed Encryption Standard, Department of Commerce, February 4 1994, Washington, DC.
5. A. De Santis, Y. Desmedt, Y. Frankel, and M. Yung, "How to Share a Function Securely," Proc. STOC'94, 1994, pp. 522–533.
6. S. Even, A. Selman, and Y. Yacobi, "The Complexity of Promise Problems with Applications to Public Key Cryptography," *Information and Control*, Vol. 61, No. 2, May 1984.
7. J. Kilian and T. Leighton, "Failsafe Key Escrow," presented at Rump Crypto'94.
8. B. Lloyd, W. Simpson, "PPP authentications protocols", Internet Request for Comments **1334** (1992).
9. S. Micali, "Fair public key cryptosystems," Proc. Crypto'92.

10. National Institute of Standards and Technology, *Federal Information Processing Standards Publication 185, Escrowed Encryption Standard*, February 9 1994, Washington, DC.

11. T. P. Pedersen, "Distributed provers with application to undeniable signatures," Proc. Eurocrypt'91, Springer-Verlag LNCS 547, pp 221–238.

12. R. Rivest, "Responses to NIST's Proposal," *Communications of the ACM*, Vol. 35 (7), July 1992, pp. 41–47.

13. Y. Yacobi, "A key distribution "paradox"," Proc. Crypto'90, Springer-Verlag LNCS 537, 1991, pp. 268–273.

Fair Cryptosystems, Revisited

A Rigorous Approach to Key-Escrow

(Extended Abstract)

Joe Kilian[1] and Tom Leighton[2]

[1] NEC Research Institute, 4 Independence Way, Princeton, NJ 08540, USA.
joe@research.nj.nec.com

[2] Mathematics Department and Laboratory for Computer Science, MIT, Cambridge, MA 02139. ftl@math.mit.edu

Abstract. Recently, there has been a surge of interest in key-escrow systems, from the popular press to the highest levels of governmental policy-making. Unfortunately, the field of key-escrow has very little rigorous foundation, leaving open the possibility of a catastrophic security failure. As an example, we demonstrate a critical weakness in Micali's Fair Public Key Cryptosystem (FPKC) protocols. Micali's FKPC protocols have been licensed to the United States Government for use with the Clipper project, and were considered to be a leading contender for software-based key escrow. In the paper, we formally model both the attack and what it means to defend against the attack, and we present an alternative protocol with more desirable security properties.

1 Introduction

1.1 Background

In a Public Key Cryptosystem, each user is assigned or chooses a matching pair of keys (P, S), where P is the public key corresponding to the pair and S is the secret key. For ease of access, as well as authentication purposes, the public key for each user is catalogued and/or certified by a central authority (or authorities) so that other users in the system can retrieve the authentic public key for any individual. Public Key Cryptosystems can be used for many purposes, including encryption and/or digital signatures. For a survey of the extensive literature in this area, we refer the reader to [7, 25, 19, 4].

One problem with a PKC (and Cryptosystems in general) is that they may be abused by non-law-abiding users. For example, two criminals could communicate using a PKC established by the Government and law enforcement authorities would have no way to decrypt their message traffic, even if the authorities had received a court authorization to wiretap the communication. Such activity might take place even if the PKC were established solely for the purposes of digital signatures since the criminals might use the PKC for other purposes such as encryption.

D. Coppersmith (Ed.): Advances in Cryptology - CRYPTO '95, LNCS 963, pp. 208-221, 1995.

The general issue of the need for government wiretaps versus the need for individual privacy has been debated in society for decades. With the advent of inexpensive and fast cryptographic technology, this debate has intensified. This is because wiretaps can be effective against encrypted traffic only if the government can gain access to the secret key that is used to decrypt the traffic. In France, for example, all cryptographic material must be revealed to the government before it can be used. Even in Germany, where there is great sensitivity to government monitoring of individuals, the issue of government escrow of secret keys for the purposes of government wiretapping has been under discussion for many years [1].

The simplest method of key control is to have a trusted government agency (or agencies) simply escrow the secret key for each individual. Then, in the event of the proper authorization, the government can retrieve the secret key from storage and decipher the intercepted communications of a suspected criminal. In such a system, the government would have the same power that it had before the advent of public key cryptography, and the citizens would have no less privacy than before. (This is essentially the proposal made by Beth [1] to the German Parliament in 1990.)

As observed by Blakley [2], Shamir [20], Karnin-Greene-Hellman [13] and many others, however, it may be cheap to simply store copies of the secret keys, but such a solution can be corrupted. In an effort to prevent such corruption, Blakley and Shamir propose methods for splitting a secret key into n shares so that the secret can be reconstructed from any k of the shares. In addition, no information about the secret key is revealed given only $k - 1$ shares. By providing each government trustee with one share of each secret key, the chances for corruption of the escrow system are substantially reduced, since the secret key of an individual can be recovered if and only if k of the trustees reveal their shares.

Since the Blakley and Shamir schemes were first proposed in 1979, a wide variety of "secret sharing" schemes have been discovered (e.g., see the survey paper by Simmons [24]).

One difficulty with the secret sharing schemes discovered by Blakley and Shamir is that there is no provision for insuring that the trustees have received valid shares of each user's secret key. Indeed, when the trustees reveal their shares under a court order (say), the shares may be found to be useless because the criminal user did not provide proper shares of his or her secret key. This problem is partially resolved in [6], where it is shown how shares of a secret can be provided in a way so that each trustee can be assured that he or she has received a valid share of the secret.

A secret sharing scheme in which each trustee can be assured that he or she has a valid share of a secret is known as a *Verifiable Secret Sharing (VSS)* scheme. Many VSS schemes are known in the literature. Typical VSS schemes proceed by having the user choose a secret m, and then publish an encryption $E(m)$ of m. The user then splits m into shares for the trustees, and the trustees verify that they have valid shares by checking against the published value of $E(m)$.

In order to be useful in the context of key escrow, it is necessary and sufficient that the pair $(m, E(m))$ form a (secret key, public key) pair of the public key cryptosystem that is being used. (This is because the secret being shared is the secret *key* of the user.) Feldman [11] and Pedersen [18] describe such VSS methods where $E(m) = g^m$ is based on the discrete-log problem. The Feldman and Pedersen VSS schemes can thus be used to share secret keys in the Diffie-Hellman, DSS, El Gamal, and elliptic curve cryptosystems. Micali provides some alternative VSS schemes based on discrete logarithms in [16], but these methods are less efficient than the Pedersen scheme. Micali also provides a VSS scheme that can be used to share secret keys in the RSA system.

In [16], Micali proposes the Fair Public-Key Cryptosystem approach to key escrow. In the Micali FPKC approach, each user shares his or her secret key with the trustees using a VSS scheme that allows each trustee to verify that they have a share of the secret key for the user that corresponds to the public key for that user. A key claim about FPKCs is that they "cannot be misused by criminal organizations."

1.2 The results of this paper

Cryptanalysis of FPKCs Naively, it might seem that the Micali FPKC cannot be misused by criminals. The government-escrowed keys cannot be used for encryption without the government being able to listen in since the escrow agents can collaborate to reveal the secret key for any user. While the criminals can use alternative means for secure communication, they are not using the *government* key escrow system except, perhaps, for the purposes of authentication, and thus the government escrow system has not been "abused" per se. Moreover, even if criminals use the government key escrow system for authentication purposes during a protocol to exchange other secret keys, they would still have to go through some for of interactive secret key exchange protocol prior to the initiation of secure communications, thus losing the convenience of a noninteractive public-key system.

Unfortunately, this reasoning assumes that the criminals use the same secret keys that were provided to the trustee. However, there is no reason to believe that the criminal will be so cooperative. We in fact describe a very simple way that criminals can exploit the Micali method for Fair PKCs without fear of eavesdropping by the government.

We exploit one of the FPKCs advertised features – the ability of the user to choose his or her public and private keys. Indeed, the defining features of the Micali FPKC are that each user can have the security of selecting his or her own secret key and that the government can be assured that criminal users cannot use the escrowed keys in a manner that is secure against government wiretaps. We demonstrate that it is impossible to achieve both goals at the same time: any escrow system which allows users to select their own keys can be easily abused by criminal users.

Defining security We suggest a number of desirable properties for key-escrow systems. Most of these properties are well understood from the extensive work on verifiable secret sharing. However, we know of no "standard" property that implies immunity from the weakness we found in FPKCs. We give a simple example demonstrating that immunity from subliminal attacks is insufficient for our purposes. We give the first formal definition of being *shadow-public-key resistant*, i.e. being secure against untappable messages.

An alternative key-escrow protocol In the paper, we also describe an alternative approach to Fair Cryptosystems that is provably immune to such subliminal attacks. The approach, which we refer to as *Failsafe Key Escrow*, is characterized by the use of government-user interaction to select keys. Indeed, an important conclusion of our work is that only by having interaction between users and the government is it possible to attain the security features that are desired by both the users and the government. In particular, our protocol has the following five properties:

Property 1: Each user in the system should have sufficient control over his or her secret key to be sure that the key is chosen securely, even if all the trustees and central authorities are malicious.

Property 2: The central authority will also be guaranteed that the secret key for each user is chosen securely even if the user doesn't have access to a good random number generator or if the user fails to use the random number generator properly (e.g., by using a birthday or phone number instead of a random number).

Property 3: Each user will be guaranteed that his or her secret key will remain secret unless a sufficient number of trustees release their shares of the key to the central authority.

Property 4: The central authority needs to be assured that it can obtain the secret key for a user who is suspected of using his or her escrowed public key for encryption in the context of illegal activities by retrieving shares of the key from a certain number of trustees.

Property 5: The central authority needs to be assured that the escrow system will not be abused by criminals in a way that helps them to communicate without fear of court-authorized wiretapping. More precisely, if two criminals abuse the FKE by using the information contained in their public keys to communicate using any published public-key encryption algorithm, and the central authority is provided knowledge of the criminals' escrowed secret keys by the trustees, then one of the following two cases should hold:

1. it should as easy (at least on a probabilistic basis) for the central authority to decrypt the message traffic between the criminals as it is for the criminals themselves to decrypt that traffic, or
2. the criminals already had a way to communicate that could not be decrypted by the government.

One can never disallow the possibility that criminals will use a completely different means for covert communication, but one does not wish to assist them in this process.

Whereas the first four properties are well understood, the last property requires a more detailed discussion. Indeed, one section of our paper is devoted to making this property, which we call *shadow-public-key resistance*, well defined. Our main theorem is as follows:

Theorem: *Failsafe key-escrow is shadow-public-key resistant.*

We note that achieving this property requires complications to our protocol that would seem to be extraneous without a rigorous standard. This motivates further foundational work in this area.

In comparison, the Fair Public-Key Cryptosystem (FPKC) approach advocated by Micali [16] does not satisfy Properties 2 and 5, and at least one proposed variant of his FPKC does not satisfy Property 3.

Techniques used Our attack is based on the subliminal channel attacks developed by Simmons and Desmedt in the 1980s [21, 22, 23, 9, 8, 26]. Using such an attack, a government-sanctioned FPKC can be subverted by criminals or other users to form a "shadow" public key cryptosystem that is untappable by the government. In some cases, the shadow cryptosystem is even more secure against the government than the original cryptosystem is against nongovernmental adversaries. The shadow cryptosystem can be set up using only publicly available information, yet we know of no way for the government to prevent its use or to determine who is using it.

In our protocol, we also make important use of information theoretically secure bit commitments, first proposed by Brassard, Chaum and Crépeau [5]. In addition, we require protocols with the *chameleon* property. Informally, the chameleon property says that the recipient of a committed bit is able to open the bit as either a 0 or a 1. This property is necessary for our proof of security to go through. Such schemes are well known, in particular we can use the protocols of [3] based on the hardness of computing discrete logarithms.

Outline of the paper The remainder of the paper is partitioned into sections as follows. The flaw in the Micali FPKC is explained in Section 2. We describe the Failsafe approach to key escrow in Section 3. We formalize our attack and show the resistance of Failsafe escrow system to this attack in Section 4. Some applications of the new approach are discussed in Section 5 and its limitations are discussed in Section 6. We conclude with some acknowledgments in Section 7. For brevity, several proofs and details have been omitted or deferred to the longer version of the paper. This longer version is available at `ftp://theory.lcs.mit.edu/pub/ftl/failsafe.ps`.

2 The Flaw in the Micali FPKC

In what follows, we first give a high-level description of the attack, and then show how to apply it with varying degrees of efficiency to the most popular public-key cryptosystems.

2.1 Shadow public-key systems

Our attack is essentially a subliminal attack on a given public-key cryptosystem. A normal user generates a pair (P, S), publishes P and gives the government the ability to reconstruct S. In the simplest form of our attack, the attacker instead generates two key pairs (P, S) and (P', S'), where (P, S) is a proper (public-key, private-key) pair, (P', S') is a shadow key pair, and $P' = f(P)$ where f is an easily computed and publicly known function. The attacker uses (P, S) in the same way as would an ordinary user, but keeps S' reserved as his shadow secret key. In order for someone to send a truly secret message (i.e., one that cannot be deciphered by the government) to an attacker, the sender computes $P' = f(P)$ and then encrypts the message using P'. (The truly-encrypted message could then be superencrypted using P, if desired, so that it would appear as if the government FPKC were being used in the normal fashion.) The receiver of the message then decrypts it using S' (as well as S if superencryption by P was used).

The key to this approach is to find efficient ways of generating P, S, P' and S' along with an easy to compute f that generated P'. We call such a system a *shadow public-key cryptosystem*. (Note that since the attacker generates a valid (P, S) pair, and uses it in exactly the same way as does a legitimate user, the trustee verification protocols will not detect any cheating.)

2.2 A shadow public-key system based on RSA

Our attack is most straightforwardly implemented against the RSA cryptosystem. Recall that an RSA public key is of the form $P = (n, e)$ where $n = pq$ is a product of two primes and e is some exponent which is typically represented as a number mod n.[3] We first note that e is essentially unrestricted. Thus, given a security parameter k (e.g., where the k-bit product of two $k/2$-bit primes is considered hard to factor), one can encode k bits in e. This is already enough to encode the public key to Rabin's public-key cryptosystem or to public-key cryptosystems based on discrete logarithms (such as the Diffie-Hellman scheme), using the same security parameter k.

As observed by Desmedt [8], an attacker can publish roughly $k/2$ additional bits in the escrow system by suitably choosing n. Given a string m of approximately $k/2$ bits (we ignore small factors that will not affect the theoretical analysis or practical utility of the attack), an attacker can choose a random

---

[3] Mathematically, it is an element of $Z_{\phi(n)}$, but this is irrelevant to how it is represented, especially since $\phi(n)$ is secret.

$k/2$-bit prime p, and then divide p into $2^{k/2}m$ to obtain a q and r such that $pq + r = 2^{k/2}m$ and $r < 2^{k/2}$. If q is also prime, then choose $n = pq$, in which case m is contained in the higher order bits of n. Otherwise, start over with a new p. Making reasonable assumptions on the distribution of primes, $O(k)$ iterations will suffice to find a suitable n.

Thus, by choosing n and e correctly, the attacker can encode an arbitrary shadow public key of size $3k/2$ in the RSA key escrowed in the FPKC. While this isn't as many bits as was used to set up the RSA public key, it allows one to use a discrete-log based scheme or Rabin's scheme with a *higher* security parameter than the one supported by the government. He can simply choose an arbitrary (P', S') such that $|P'| = 3k/2$, and then generate $(n = pq, e)$ to encode P', publish (n, d) (where $d = e^{-1} \bmod \phi(n)$) and share e with the escrow agency.

We give more shadow-public key attacks in the longer version of this paper.

3 The Failsafe Key Escrow Approach

The flaw in the Micali FPKC is derived from the fact that it is possible for a user to choose a pair of keys (S, P) with the special properties that:

1) the trustees can be provided with valid shares of the secret key S, and

2) the FPKC public key P can be easily converted into a shadow public key P' (using a published algorithm) for a shadow cryptosystem for which the user has also precomputed a shadow secret key S'.

The criminal user can then communicate using the shadow cryptosystem and the shadow pair of keys. The central authority (with the aide of the trustees) can retrieve S but this will not be useful in deciphering traffic encrypted with S'. Moreover, the central authority may have no hope of discovering S'. Unfortunately, it appears that such an attack can be mounted against any escrow system in which the users are given the freedom to select their own keys.

The subliminal key attacks can be avoided by having the central authority or the trustees themselves select the pair of keys for each user. But schemes in which the central authorities select the secret key for each user may leave the user with no assurance that his key has been properly generated (so as to be secure). Such a scheme would not satisfy Property 1.

Several methods have been proposed in the literature for overcoming subliminal attacks. Desmedt [8], in particular, has proposed a general method for defending against subliminal attacks in public-key cryptosystems, and our methods have a number of similarities to his approach. In both cases, the user and the government collaborate to generate a fair key by a "coin-flipping" technique (first proposed by Blum) in which one side precommits its half of the final key. However, there are also a number of differences: The Desmedt scenario assumes a trusted center (warden) who can be relied on to make his bits random. Whereas we consider a key-escrow setting, in Desmedt's protocol, the secret key is completely reserved by the user. Also Desmedt's solution works in polynomial time, but is not practical. We more efficiently exploit the algebraic properties of our public-key cryptosystem to yield a practical system which is easily implementable in software. Finally, and most importantly, protection against subliminal channels from the user to the outside world is necessary *but not sufficient* for our security properties to hold. Indeed, some further technical subtleties seem to be required to guarantee that no attack on our system will succeed.

3.1 Why abuse-freeness is insufficient

The attack on the previous key-escrow scheme worked by *abusing* the protocol to set up a subliminal channel. The creation of abuse-free protocols [8, 10] is well understood, so it is tempting to simply require that the key-escrow protocols be abuse-free. Unfortunately, there exist abuse-free protocols which are nevertheless vulnerable to this attack. The example we give below is artificial, and would never be reasonably proposed, but illustrates that abuse-freeness is a technically insufficient for our purposes.

Given a canonical public-key cryptosystem for generating secret-key/public-key pairs (S, P), we first construct a new public-key cryptosystem as follows. To generate a pair in the new cryptosystem, one independently generates (S_1, P_1) and (S_2, P_2) in the canonical system. The secret-key/public-key will be $(S_1, P_1 P_2)$. To encrypt a message m according to public key $P_1 P_2$, the sender simply encrypts m with P_1 in the canonical manner, ignoring P_2.

Now consider the following key-escrow protocol: A genuinely trusted entity

1. Constructs (S_1, P_1) and (S_2, P_2) in the canonical manner,
2. Publishes $P_1 P_2$ as U's public key,
3. Shares S_1 among the escrow agents and
4. Sends (S_1, S_2) to U.

Note that for the purpose of this discussion, we assume that the agents have combined their shares, so Step 3 is equivalent to sending S_1 to G. Also, one can replace the genuinely trusted entity with a secure protocol without affecting our analysis.

The above protocol is clearly abuse-free, since U cannot influence the output in any way. However, there is an obvious shadow public-key system: O can simply encrypt his message according to P_2, which the government has no way of knowing. The point is that our ultimate goal is not just to keep information from leaving U but to keep information from being sent (in an untappable manner) to U.

3.2 An example of failsafe key-escrow

In what follows, we describe one example of the general Failsafe Key Escrow approach. This example is based on a Discrete-Log PKC such as Diffie-Hellman or DSS. Here we assume that a prime modulus Q and a generator g for Z_Q^* are publicly known. In this case the public key P that is escrowed for a user is $g^S \bmod Q$, where S is the secret key for the user. The escrow system that will be used in conjunction with the US Digital Signature Standard has this form.

The keys for a user are selected as follows:

Step 1: The trustees and/or the central authority select a random value B from the interval $[0, Q - 2]$ and commit to B with the user using an information-theoretically secure commitment protocol with the chameleon property.[4] One very simple family of protocols, based on the discrete-log problem is given in [3]. (In fact, depending on the security desired, each trustee i might select and commit to a B_i, with the value of B being formed by taking the XOR of the B_i's. Then only one trustee needs to be trustworthy for the user to be assured of security.

Step 2: The user picks a random secret value A from $[0, Q - 2]$ and announces the value of g^A mod Q to the trustees and/or the central authority.

Step 3: The user "shares" A with the trustees using a VSS scheme such as that described by Pedersen [18]. (The precise VSS scheme that is used depends on the degree to which the trustees can be trusted to behave properly and the degree to which the users distrust the trustees.) This requires X to send the shares of A to the trustees and it requires the trustees to verify that they received valid shares of A.

Step 4: The trustees and/or the central authority reveal B to the user (who verifies that it is the value previously committed to) and set the public key to be $P = (g^A)g^B$ mod Q. The value of B is escrowed with the public key for the user. The value of B is not released to the public.

Step 5: The user then sets his secret key to be $S = A + B$ mod $(Q - 1)$.

In what follows, we show that Properties 1–5 hold for this system. For simplicity, we will argue Properties 1–4 informally, since they are well understood. In the next section, we consider Property 5 in detail.

Verification of Property 1: Every user who follows the protocol can be sure that he or she has a randomly chosen secret key. This is because the user chooses A at random in $[0, Q - 2]$. The authority chooses B, but does so with no knowledge of A. In order to renege on the commitment, the authorities must break the discrete-log problem, in which case they could easily break the whole system anyway. This means that if A was selected at random by the user, then the user can be assured that the distribution on $S = A + B$ mod $(Q - 1)$ is indistinguishable from the uniform distribution on $[0, Q - 2]$. Dishonest authorities can skew the distribution slightly by, for example, trying to guess a discrete logarithm that allows them to break the commitment scheme, which will happen with positive but negligible probability. However, this will not measurably affect the security of the key.

---

[4] For slightly greater efficiency, a bit-commitment scheme without this simulatability property may be used, and we know of no major problems with such a protocol, but a formal security analysis becomes problematic.

Verification of Property 2: Even a user who fails to select the value of A correctly (e.g., by using a birthday instead of a random number generator) will get a random secret key. This is because the value of B is selected randomly by the authorities and it is revealed to the user after the user commits to the value of A. Hence, the authorities can be assured that $S = A + B \bmod (Q - 1)$ is a random integer in $[0, Q - 2]$.

Verification of Property 3: Each user can be assured that his or her secret key stays secret unless a sufficient number of trustees release their shares. This is because knowledge of A can be revealed only with the assent of a sufficient number of trustees by the properties of the VSS scheme. Even if B were to be public, this means that $A + B \bmod (Q - 1)$ will remain secret unless a sufficient number of trustees cooperate to reveal A.

Verification of Property 4: The central authority is guaranteed to be able to retrieve the secret key of any user provided that a sufficient number of trustees reveal their shares. This is because the properties of the VSS scheme assure that a sufficient number of trustees can collaborate to reveal A. Since B is escrowed, it is then a simple matter to compute $S = A + B \bmod (Q - 1)$.

Similar protocols can be developed for use with other PKCs such as RSA, but the details become more complicated since the authorities need to interact with the user to choose a "random" number with some special structure. For example, the public keys used with RSA need to be the product of a small number of primes. (If we relax the constraint of having to formally prove that the scheme is secure, then it may suffice for the trustees to multiply the RSA modulus supplied by the user by a random prime, and to add a random number to the RSA exponent.)

4 A formal foundation for security

The flaws in previous attempts at key-escrow highlight the need to put this area on a firmer theoretical foundation. We give a first step in formalizing the subtle security issues that surround key escrow. While the issue of protecting the user's privacy against improper subsets of escrow agents is well understood, protecting the government against shadow public-key attacks is somewhat more complicated. In this section we give present a more formal discussion of this problem.

First, let us state a reasonable objective. We cannot possibly prevent interactive participants from agreeing on a secret-key that is unavailable to the government. Nor can we guard against a user U and and outsider O from having a previously agreed upon convention or secret key that allows O to noninteractively send a message to U. What we *can* do is to ensure that O cannot exploit our key-escrow system to noninteractively send a single untappable message to U, *unless he could already do so.*

4.1 Modeling the participants' knowledge

It is often useful to model the participants' knowledge before and after the protocol. For example, in zero-knowledge proofs the verifier's auxiliary input [17] represents whatever information he may have had before the protocol began. In even the simplest of key-escrow scenarios we must represent many types of knowledge held by different subsets of the participants. For simplicity, we consider the case of a single outside entity user O who wishes to send a bit b to U in a manner untappable by the combined set of escrow agents, denoted G.

Definition 1. We define a *knowledge ensemble* $\mathcal{K}(k)$ to be a parameterized ensemble on 4-tuples (K_O, K_U, K_G, K_P), where K_O, K_U, K_G, K_P are of expected size polynomial in k.

The components of the knowledge vector correspond to information held by the outsider, the user, the government (once the escrow agents have combined their information) and the general public. We consider a parameterized ensemble to, among other things, model the fact that the key escrow protocol is being run with a security parameter k.

We don't make any assumptions about the joint distribution of the components of a knowledge ensemble. For example, one legal knowledge ensemble is where (K_O, K_U, K_G, K_P) uniformly distributed over

$$\{(0, 0, \emptyset, \emptyset), (1, 1, \emptyset, \emptyset)\},$$

corresponding to the user and the outsider possessing a single private random bit in common.

Running any sort of protocol causes the knowledge ensemble to evolve. Indeed, if U flips a fair coin this will cause K_U to change. Given a key-escrow protocol $\mathcal{P}$, we define the knowledge ensemble

$$\mathcal{K}^{\mathcal{P}} = (K_O^{\mathcal{P}}, K_U^{\mathcal{P}}, K_G^{\mathcal{P}}, K_P^{\mathcal{P}})$$

by the following procedure for sampling from $\mathcal{K}^{\mathcal{P}}$. Given parameter k:

1. Sample $(K_O, K_U, K_G, K_P) \leftarrow \mathcal{K}(k)$
2. U, with inputs k, K_U, K_P, and G, with input k, execute protocol $\mathcal{P}$, generating public-key P. Denote by View_U and View_G the views obtained by U and G.
3. Output $(K_O, K_U \mathsf{View}_U, K_G \mathsf{View}_G, K_P P)$.

4.2 Modeling the attack

What does it mean for O to send a (single-bit) message to U? We model this by a pair of probabilistic polynomial-time procedures $\mathsf{send}(b, K_O, K_P, k) \to M$ and $\mathsf{receive}(M, K_U, K_P, k) \to \{0, 1\}$. We model the government's attempt to tap this message by a probabilistic polynomial-time procedure $\mathsf{tap}(M, K_G, K_P, k) \to \{0, 1\}$. Given a security parameter k and b chosen uniformly from $\{0, 1\}$, we define (b_U, b_G) as the result of the following operations:

1. Choose $(K_O, K_U, K_G, K_P) \leftarrow \mathcal{K}(k)$,
2. Choose $M \leftarrow \mathsf{send}(b, K_O, K_P, k)$,
3. Choose $b_U \leftarrow \mathsf{receive}(M, K_U, K_P, k)$ and
4. Choose $b_G \leftarrow \mathsf{tap}(M, K_G, K_P, k)$.

Using this notation, we define our basic concepts.

Definition 2. We say that (send, receive) is *valid* with respect to $\mathcal{K}$ if for $b \leftarrow \{0, 1\}$,

$$Pr(b_U = b) > \frac{1}{2} + \frac{1}{k^c}$$

for some constant c and k sufficiently large.

Definition 3. We say that (send, receive) is *immune* to tap with respect to $\mathcal{K}$ if for $b \leftarrow \{0, 1\}$,

$$|Pr(b_U = b) - Pr(b_G = b)| > \frac{1}{k^c}$$

for some constant c and all sufficiently large k. Otherwise, we say that tap is successful against (send, receive) with respect to $\mathcal{K}$.

Definition 4. We say that (send, receive) is *untappable* with respect to $\mathcal{K}$ if it is *immune* to all probabilistic polynomial-time procedures tap. We say that $\mathcal{K}$ is *vulnerable* if there exist PPT procedures (send, receive) that are untappable with respect to $\mathcal{K}$.

Less formally, we want the receiver to receive some nonnegligible information about the message-bit b and to learn more than the person tapping the line. Note that untappability implies validity, since the send-receive pair must by immune against the "tapping" procedure that flips a fair coin. We purposefully define a very weak notion of untappability, since we will show that even this is impossible using our scheme.

Finally, we now give the key definition of resistance against shadow public-key attacks.

Definition 5. We say that protocol $\mathcal{P}$ is *shadow-public-key resistant* if for all $\mathcal{K}$, $\mathcal{K}^{\mathcal{P}}$ is vulnerable iff $\mathcal{K}$ is vulnerable.

We note that $\mathcal{K}^{\mathcal{P}}$ is certainly vulnerable if $\mathcal{K}$ is. Informally, $\mathcal{P}$ is shadow-key resistant if running it does not open up an opportunity for a covert message where none before existed.

Our main security theorem is as follows:

Theorem 6. *Failsafe key-escrow is shadow-public-key resistant.*

We sketch its proof in the longer version of this paper.

5 Applications

Failsafe Key Escrow systems can be used in conjunction with any PKC to protect the interests of both law enforcement and the users. FKE may prove to be particularly valuable in the context of the new US Digital Signature Standard (DSS). In particular, it will be important to insure that criminals are not able to use DSS keys for the purposes of encrypting communications in a way that is indecipherable to the Government. This issue is of particular concern in the context of DSS since DSS keys can be easily adapted for encryption. The FKE approach described in Section 3 prevents precisely this sort of abuse.

6 Limitations

It is also worth pointing out the limitations of the Failsafe Key Escrow Approach. Most importantly, the FKE approach does not prevent a pair of criminals from communicating securely using secret information or an alternative escrow system, or from using other protocols for secret key agreement. The main point of the FKE is to prevent criminals from abusing the public keys in the key escrow system. In other words, by designing the key escrow system in a failsafe fashion, the Government can be more assured that the escrow system will not make it any *easier* for criminals to communicate securely.

Our formal proof of Property 5 also requires that the precise name (and other header information) listed in the public file is easily computable given already publicly available information. Otherwise, one can subliminally hide information by declaring one's name to be John "2134fewlr4323423423423...." Doe. Similar restrictions must also be placed on other information available in the public file such as the number of keys for an individual, etc.

7 Acknowledgments

We would like to thank Bonnie Berger, Dorothy Denning, Yvo Desmedt, Ron Graham, John Rompel and Moti Yung for their help with this work.

References

1. T. Beth. Zur diskussion gestellt. *Informatik-Spektrum*, 13(4):204–215, 1990.
2. G. Blakley. Safeguarding cryptographic keys. *In AFIPS – Conference Proceedings*, 48:313–317, June 1979.
3. J. F. Boyar, S. A. Kurtz, and M. W. Krentel. A discrete logarithm implementation of perfect zero-knowledge blobs. *Journal of Cryptology*, 2(2):63–76, 1990.
4. Brassard, G. (1988). *Modern Cryptology; A Tutorial*. Lecture Notes in Computer Science, No. 325 Springer Verlag.
5. G. Brassard, D. Chaum and C. Crépeau. *Minimum Disclosure Proofs of Knowledge*. In *JCSS*, pages 156–189. 1988.

6. B. Chor, S. Goldwasser, S. Micali, and B. Awerbuch. Verifiable secret sharing and achieving simultaneity in the presence of faults. *Proceedings of the 26th IEEE Symposium on Foundations of Computer Science*, pages 383–395, 1985.

7. Denning, D. E. (1982). *Cryptography and Data Security*. Massachusetts: Addison-Wesley.

8. Y. Desmedt. Abuses in cryptography and how to fight them. *Crypto '88*, pages 375–389, August 1988.

9. Y. Desmedt, C. Goutier, and S. Bengio. Special uses and abuses of the Fiat-Shamir passport protocol. *Crypto '87*, pages 21–39, August 1987.

10. Y. Desmedt and M. Yung. *Minimal Cryptosystems and Defining Subliminal Freeness* In *Symposium on Information Theory*, 1994.

11. P. Feldman. A practical scheme for non-interactive verifiable secret sharing. *Proceedings of the 28th IEEE Symposium on Foundations of Computer Science*, pages 427–437, 1987.

12. G. Harper, A. Menezes and S. Vanstone. Public-key Cryptosystems with Very Small Key Lengths. *Eurocrypt '92*, pages 163–173, May 1992.

13. Karnin, Greene and Hellman. On secret Sharing Systems, *IEEE Transactions on Information Theory*, vol. 29, 1983.

14. T. Leighton. Failsafe key escrow systems. Technical Memo 483, MIT Lab. for Computer Science, August 1994.

15. T. Leighton and S. Micali. Secret key distribution without public-key cryptography. *Crypto '93*, August 1993.

16. S. Micali. Fair public-key cryptosystems. Technical Report 579, MIT Lab. for Computer Science, September 1993.

17. Oren. On the cheating power of cunning verifiers. *Proceedings of the 28th FOCS*, IEEE, 1987.

18. T. P. Pedersen. Distributed provers with applications to undeniable signatures. *Eurocrypt '91*, April 1991.

19. Schneier, B. (1993). *Applied Cryptography*. John Wiley.

20. A. Shamir. How to share a secret. *Communications of the ACM*, 22(11):612–613, 1979.

21. G. J. Simmons. The prisoners' problem and the subliminal channel. *Crypto '83*, pages 51–67, August 1983.

22. G. J. Simmons. The subliminal channel and digital signatures. *Eurocrypt '84*, pages 364–378, April 1984.

23. G. Simmons. A secure subliminal channel (?). Crypto '85, pages 33–41, August 1985.

24. G. J. Simmons. How to really share a secret. *Crypto '90*, pages 390–448, August 1990.

25. Simmons, G. (1991). *Contemporary Cryptology*. IEEE Press.

26. G. J. Simmons. Subliminal communication is easy using the DSA. Eurocrypt '93, pages 218–232, May 1993.

27. Y. Yacobi. Discrete-log with compressible exponents. *Crypto 90*, pages 639–643, August 1990.

Escrow Encryption Systems Visited: Attacks, Analysis and Designs

Yair Frankel[1] * and Moti Yung[2]

[1] Sandia National Laboratories, Albuquerque, NM
[2] IBM T. J. Watson Research Center, Yorktown Heights, NY

Abstract. The Escrow Encryption Standard and its realization – the Clipper chips – suggest a new type of encryption scheme. We present a few basic and somewhat subtle issues concerning escrow encryption systems. We identify and perform attacks on the actual Clipper and other recent designs (fair cryptosystems, TIS software escrow, etc.). We review requirements and concerns and suggest design approaches to systems with desired properties of key escrow.

1 Introduction

The Escrowed Encryption Standard (EES) and more specifically the Clipper chip family [18, 8] is the U.S. government's approach to technically bridging the gap between a users' need for cryptographically strong privacy and society's need for protection against criminal behavior. When key escrow was introduced, it resulted in heated discussions about the Clipper program (which includes recoverable keys from two trusted escrow agents when a court order is issued). The perspective we take here (from a technical stand point — leaving our personal views aside) is that there is a need for both criminal investigation mechanisms (fully handled by the judicial system) and full individual privacy (as protected by the Constitution). Further based on the above starting point, we advocate a sound approach and attempt a prudent analysis that covers both aspects.

The obvious threats to escrow systems considered already are those of dishonest escrow agents (which is why Clipper uses two agents) and a dishonest user (which is the purpose of the law enforcement access field (LEAF) in Clipper). We will assume here that Clipper's classified encryption algorithm, SKIPJACK, is strong enough to protect the user's privacy due to [6]. However, this may not be enough. It appears to us, in light of the findings of this paper, that the concept of escrow encryption for law enforcement purposes is not a well understood one. We analyze requirements and ask: are there any inherent subtle problems with existing systems? and, are there necessary security mechanisms lacking in existing designs? At the same time we attack existing designs (like the Clipper and fair cryptosystems). We hope that our findings and analysis will allow the escrow encryption systems designers and users to build better systems which satisfy both user's and law enforcement's needs.

---

* Research was performed while the author was at GTE Laboratories Incorporated.

D. Coppersmith (Ed.): Advances in Cryptology - CRYPTO '95, LNCS 963, pp. 222-235, 1995.
© Springer-Verlag Berlin Heidelberg 1995

Attacks: As part of our analysis we raise several concerns and attacks. We show that the actual Clipper chip based system allows a scenario where the potential criminal can attack and somewhat choose whose device key must be opened to read the criminal's own messages by *squeezing* LEAFs. Squeezing also enables an attack by two interoperable rogue entities within two way communication systems; its strength is that law enforcement cannot catch the fact that the entities do not follow the Clipper specification. We also demonstrate that a public key system on its own (e.g. as in Crypto'92 fair cryptosystems) is not good enough for key escrow. Moreover, in environments like the Internet with both fair cryptosystems and Clipper, two collaborating parties can spoof two legitimate users and can "frame" them, by faking their own session to "look like" it is between the legitimate parties.

Methodology: We take a three step approach in our investigation. 1) We identify basic design issues in escrow encryption cryptosystems and provide questions we believe designers should ask in determining requirements. Rather than speculating on what is needed, we use official documents as systems specifications. We are guided by the FBI requirements [17] (derived from the telephony law) and by the administration goals as stated in press releases [32]. We also follow the tradition of cryptographic design. 2) We then relate our questions and concerns to proposed technologies (e.g., Clipper [8], fair cryptosystems [25], TIS software key escrow [2], etc.), and 3) we discuss threats and demonstrate the attacks mentioned above on actual systems and designs when they fail to solve the basic issues. We also extract basic design issues from the analysis; further designs we deduced will be discussed in [19].

Overview– escrow encryption systems: Let us first review the basic key escrow systems which provide a third party – the escrow agency – with the capability to decrypt sessions of a registered user (device). To provide for added trust and security, the escrow agency is often treated as a distributed function (multiple entity collaboration). Most of the proposed schemes are based on one or combination of the following two approaches.

- The LEAF approach (Clipper is the prototypical system for this approach) transmits an encrypted session and key identifier with the ciphertext. Law enforcement, given the key identifier, obtains from escrow agents the key associated with the identifier and decrypts the encrypted session key [8].
- A threshold decryption [13] approach provides agents with a share of a user's key. The user encrypts a message and the agents are able to decrypt the message using their share of the user's key. Micali's fair cryptosystem [25] is the prototypical system for this approach.

Organization: Section 2 describes issues and attacks concerning identification of entities and proof of compliance that assure that messages can be retrieved by law enforcement. Section 3 describes specifically the squeezing attacks realization. Section 4 analyzes key management issues while Section 5 presents attacks on public-key based systems. Section 6 analyzes the system operation stage and general issues.

2 Entity Identification and Compliance Certification

Wiretapping over phone lines does not provide for strong authentication of the user and messages, so it is often used only to aid in collection of evidence rather as an evidence itself. Thus, strong authentication is not a wiretapping requirement. Yet, identification in escrow encryption systems is a more subtle issue than in a typical cryptographic protocol. Identification determines a binding between the device/ user and a key to be taken off escrow. The need for identification, therefore, surfaces naturally. Regarding this function we ask:

Issue A: Entity identification/ message authentication issues

1. Must there exist a secure binding between a sender (and/or receiver) identity with the ciphertext (in a general network environment)?
2. Should users insist that their key gets opened only when they are part of the action (session)?
3. Is it possible and practical to provide message authentication?

The Clipper chip transmits a chip ID field (as part of the LEAF) in order to provide law enforcement with the information to determine which key needs to be removed from escrow. Also fair cryptosystems and other suggested escrow encryption systems must provide a key identifier for a similar reason. Such a field has a potential of establishing the source or destination of the message. Is this required or does this provide law enforcement with a stronger tool than current wiretapping provides? In fact, this is required and assuring that wiretapping is performed w.r.t. the specified subject is a must (by the FBI's own specification)! We observe that the FBI specifies in Requirement 5 (Verification Information) [17]: *"Law enforcement agencies require (1) information from the service provider to verify the association of the intercepted communications with the intercept subject, and (2) information on the services features subscribed to by the intercept prior to intercept implementation and during the intercept"*

The FBI requirements further state: *"Specifically, court authorities require law enforcement agencies to verify that the communications facilities or service being intercepted correspond to the* **subject or subjects** *identified in the lawful authorization."* In fact currently for voice wiretapping the notion of "minimization" rules is used, namely the listening has to concentrate on the subject only (when identified by its voice characteristics) and not on its line that may be used by others. Moreover the U.S. Department of Justice (DOJ) in its letter discussing authorization procedures for release of encryption keys states that federal agents may obtain a certificate which shall [9] (requirement 3f): *"specify the identifier (ID) number of the key-escrow encryption chip providing such encryption ... "*. This, however, does not seem to be an absolute requirement[3].

---

[3] It states "...non-compliance with these procedures shall not provide the basis or motion to suppress or other objections to the introduction of electronic surveillance evidence lawfully acquired.

For wiretapping of data we need other means for identifying a subject. Further, for general network environment we cannot always rely on source identification (sources may be multiplexed). This could be one of the roles of the LEAF. However, it is crucial to point out that the law enforcement requirements [9, 17] are not in agreement with Clipper.

Claim 1 *There is an attack showing that the LEAF field in Clipper does not determine the identity of the chip which encrypted a message. Moreover, criminals can squeeze an honest person's LEAF in order to force escrow agents to open the honest person's escrow key.*

The above should be a concern to, both, users and law enforcement. If a sender is not identified in a secure fashion then he/she may be *framed* (e.g., suspected of an action that one did not do). Even if the government does not intend to make an implication about binding of user to device (which we expect is the government's current policy) this does not prevent the defendants from making such implications to generate an alibi by "proving" that they were in the locality of the supposed encrypting chip during the crime when they actually were not. Thus, the implication of identity and binding of message to chip is a dual-edge sword abusable by both law enforcement and criminals (this borrows a terminology from [32] where encryption is described as "dual-edge sword"). We believe that the fact that binding of chip identity and a ciphertext cannot serve as an alibi must get known to the courts!

Claim 1 also tells us that **users who have no part in some illegal action (whether knowing it or not) may have their keys opened and messages read**. This is specifically easy over packet switching networks (the Internet) and wireless networks. Such possibilities may defeat the purpose of the escrow agents as protectors of individual privacy. We believe that showing that in some cases wiretapping is possible only if a potential violation of privacy is allowed must be made clear to the courts!

One may claim that Clipper is intended only for voice (which provides for user authentication)– but we note that data seems to be an important goal of the key escrow program. In fact, a data processing device – the Tessera crypto card — is a PCMCIA version of Clipper [26]. Moreover, [9] specifically pertains to other electronic surveillance.

Message authentication (not included in Clipper) is more costly than identification. Such operation inside the tamper-proof device will enable user protection against law enforcement "planting" messages in wiretapping (and will thus make information gathered more admissible in courts, since in this case no one can tamper with the ciphertext). The tamperproof devices will insist on proper authentication prior to decryption. This, however, requires tremendous overhead of processing each message (in voice and real time applications we may be able to authenticate only parts of the message).

The issue of the tamper proof insisting on verification brings up the question of compliance certification: the fact that the system's operation is assured only when the user comply with its requirement (which we next discuss briefly):

Issue B: On compliance certification

1. What level of certification of the usage and compliance is necessary?
2. What level is technologically possible and practically acceptable?

It is always possible to bypass an escrow system's algorithm or use another encryption within it. There is nothing that can be done about that. Defining compliance (what is correct usage) and providing a certification mechanism (and by whom?) is a difficult problem; at least we can say that compliance should assure that: *a user employing solely the "encryption portion of the mechanism" for assuring secrecy, should also enable law enforcement readability of the message.* How hard is it to bypass compliance in Clipper? In fact, Blaze has shown that it is possible due to short integrity check (16 bits), and that collaborating rogue entities can avoid sending the LEAFs (where they can be caught by law enforcement). In fact, in Section 3 a self-squeezing attack is described. It can be viewed as a strengthening of Blaze's second attack [5]. The strengthening enables misbehaving entities to avoid being caught as such.

Claim 2 *There is an attack showing that compliance is not possible in Clipper given two rogue parties (even if the LEAF contains the identity of the parties' chips and proper integrity check).*

In theory, zero-knowledge techniques can certify compliance. In practice however it is hard; it is much easier (but costly) with tamper-proof devices checking for certification as part of their function. One suggested design in [25] assures that a key is properly distributed. This may not be enough for operation time. The solution in [25] for operation time is to make the use of other keys illegal (which is a non-technical issue), but a criminal may just as well risk violating the law about the key in order to hide the perhaps much bigger violation which is the actual crime discussed. Some solutions assure that some sampling of the use can be done by authorized tapping to verify that proper text is being sent [10].

3 Squeezing Attack: How to Use Other People's LEAFs

We now realize the attack of Claim 1. The compliance certification in Clipper attempts to assure Law Enforcement that a user receives the LEAF from the other user before the ciphertext can be opened as required by [18].

Our experiment: We performed the squeezing attack by experimenting with an actual Mykotronx MYK-78T which performs SKIPJACK in Output Feedback (OFB) mode [27], it was part of a board constructed in GTE labs. The chip works in the following manner. For encryption mode: an external session key is input and then the device outputs the IV and LEAF. Afterwards, encryption using SKIPJACK in OFB mode with the session key and IV can be performed. For decryption mode: the session key, IV and LEAF must be input and pass the identification test (to test for compliance) before decryption is allowed. (Observe the relationship between identification and compliance used here).

We had two MYK-78T chips which were given a session key KS and in the spirit of [18] the IV/LEAF tuple was transmitted with any ciphertext. Let the LEAF generating function be H. Chip A first generated the IV=IV_A and a LEAF$_A$=H(IV$_A$,KS,..) and then encrypted a message. The attacker's chip B having received the session key KS earlier, and the IV$_A$/LEAF$_A$ tuple and the ciphertext from A was then employed.

We attempted that B would encrypt its own message (ciphertext and LEAF$_B$) and then will decrypt the modified message (same ciphertext and LEAF$_A$). So B's output must be consistent with IV$_A$/LEAF$_A$ tuple to be concatenatable to B's ciphertext. However, B can not use encryption mode to accomplish this since in encryption mode the MYK-78T chip chooses its own IV=IV_B internally. This problem was, however, resolved by using chip B in decryption mode. So that we could enter in the IV$_A$ manually as IV$_B$ (assuming the IV$_A$ is chosen by the supposed "other party" to the "legal session"). We were able to verify our test showing that B was able to conversed with itself, using A's IV$_A$, replacing LEAF$_B$ to LEAF$_A$ before transmission. B then decrypted this transmission! To encrypt in OFB mode, an input of all 0 bits is input to the decrypting chip to retrieve the encrypting string (pad). This string can now be exclusive-ored with the plaintext to generate the ciphertext. Note that in ECB mode (as is expected to be provided in the MYK-78E which is in development), the IV problem is even simpler (as IV is ignored).

Discussion: What environment will enable our attack? In a typical attack, the LEAF will be squeezed from one device and be served in a session between the attacking device and a third party; similarly two devices can be attacked by two coordinated attackers. We used manual key distribution, RSA public key [30] can be used just as well – the common session key is used by both sides (as EES requires). If Diffie-Hellman key exchange were the approach then the attack becomes more complex, unless the Diffie-Hellman key exchange is used as a master key to encrypt session keys which again simplifies the problem. The difficulty in the attack is that the session key can not be uniquely controlled by the bad guy(s) since the good guy has an effect on the choice of keys when they start their session. For an attack to work in this case, the third guy has to be an accomplice and then again, both of the bad guys can use the same LEAF (i.e., the good guy will appear as both the sender and receiver in this case!); in addition, the key and LEAF must be shared outside the box as in MYK-78T and not under the tamper resistant device. A squeezing example is of two rogue government officials that talk with the president and the vice president, and employ their partners' LEAFs in an illegal private session between the two of them. Another example is a criminal organization that opens a 900 phone service and gets sessions (and LEAFs) of numerous privacy-conscious citizens which can be squeezed and used for simultaneous crime-related sessions.

Self-Squeezing- strengthening Blaze's attack: In one of Blaze's attacks [5] two rogue entities manage to avoid sending LEAFs. It is important to note that in telephony (two way communication) both parties must be rogue since [18] requires that both parties send a LEAF and use the same session key. (We, thus, feel that this attack may be more important than what is actually believed). In his attack, two entities can bypass the system in a real-time fashion and drop the LEAF from communication. When getting wire-tapped, the two entities are caught. Using squeezing they can do so and their communication will look like they follow the compliance certification. All law enforcement will notice is that the messages sent are randomly looking while they are not (the entities may claim they were merely testing their devices and they can reconstruct the randomly looking messages!). To do this, the rogue entities generate two keys in the key exchange phase (one is the key and the other is, say, a derived key which is a result of a one-way function application over the key). They each generate a LEAF using the derived key (call it a derived LEAF), then use the actual key to generate encryption and true LEAF, and before transmission just replace the generated true LEAF with the derived LEAF as in squeezing. Before decrypting, a user replaces the sent derived LEAF with its own (true) LEAF. This operation is done in real time. Note that this attack is what is relevant to two way communication, is independent of the size of the LEAF and its fields, and cannot be avoided by disabling devices that fail too many times (a measure suggested against other attacks of Blaze).

On prevention: The squeezing attack on the LEAF could have been prevented. One possible design would be to have the session key be a function (hashed using a globally chosen pseudorandom function known only to devices) of the sender and receiver id (ID-based key distribution). Compliance could be verified under the protection of the tamperproof device. This assures that the key is used between two specific users and not others. To protect against self-squeezing a stronger measure is needed (message and message originator and receiver authentication).

4   Analysis of Key Management Issues

Key management is a major issue in key escrow. We ask:

Issue C: General key management issues

1. Who generates the user's keys?
2. How active should the user, escrow agent and other government agents be in the key distribution/management process?
3. What is needed for international escrow systems? How to make an international escrow when countries may not trust each other?
4. How is the key disseminated to escrow agents and how do we prevent cloning of keys?
5. Is it necessary to share keys of users.

The purpose of an escrow system is to allow strong cryptography, therefore weak keys are unacceptable. Thus, the user may wish to play a strong role in the process of generating escrow and session keys (as in Clipper) whereas device generation of keys (as in Capstone) may assume reliance on the device designers (e.g. to use a proper pseudorandom generator).

Too much involvement of or shepherding by the escrow agents in various stages is highly undesirable (in particular when hardware is used). In [4] law enforcement gets involved in session key generation in order to provide a means to restrict law enforcement in time; such context limitation is needed. Also, international key escrow may require agents trusted by mutually distrusting countries and a key hierarchy for escrow.

We next note that **key cloning attack** may cause illegal wiretapping and may be possible by fooling the court. The scenario allows an entity who is a "an intentional future suspect" to copycat a key of another user (say in [25] system). Then when this entity misbehaves, a court may issue a wiretapping order, which will enable law enforcement to listen to the honest user as well (due to the cloning). Consider the espionage potential of such an attack in the context of an international key escrow system and how the international court can be fooled.

We ask: should user registration involve (the typically complicated) sharing of its key? In fact, by employing tamper-proof devices at the escrow agents we designed a technique where the user's key is not required to be shared to agents. The technique of *protected function sharing* has numerous other applications and is discussed in [19, 20] and briefly in Section 6.

4.1 Tamperproof devices verification

Clipper uses tamperproof devices created by the government to hide the secret algorithms encoded in the chips and to some degree to enforce compliance. The main user concerns with a system based on the use of tamperproof devices are:

Issue D: Users' concerns with tamperproof encryption device

1. How to verify that the "correct" algorithms are encoded? and how to minimize potential problems in a device?
2. What prevents cloned devices?

Correctness and uniqueness are valid concerns of users when they may not trust the builder of the equipment. This is especially true when some of the algorithms encoded are classified and correctness is therefore ill-defined; the problem of Capstone internal algorithms not being trusted is mentioned above (which justifies independent key distribution). Auditing by independent trusted entities may provide some assurances of the integrity of the devices and procedures.

5 Attacking Public-Key based Systems

5.1 Public-key issues

Public key reduces the number of user keys and appears to be the method of choice for software escrow encryption systems for this reason. We note that due to e.g. [24, 14], private key systems protected under a tamperproof devices exhibit properties of public key. We ask:

Issue E: Public keys

1. Which keys should be opened by escrow agents?
2. How can the public key be abused by the user?
3. Is ID-based public key useful?

It appears to us that the use of public keys for escrow encryption is not well understood. Indeed with any "pure" public key approach to provide for escrow encryption [25, 13, 10] implies that opening ciphertexts requires a situation that is as unacceptable as the U.S. postal service having the authority to open one's mail under the suspicion that criminals are sending, say, mail to people in the neighborhood. Thus, naturally, we have to augment public-key techniques when used in the context of key escrow systems. We specifically claim about public key escrow systems that:

Claim 3 *Fair cryptosystems are not fair.*

Discussion: Fair cryptosystem discusses the monitoring "of user's suspected of unlawful activities" yet it achieves only the monitoring of messages to and not **from** suspected criminals. In an El-Gamal based Fair cryptosystem, a user A has a public key Y_A where $Y_A = g^{X_A} \bmod p$, X_A is the shared private key. To send a message in a session the criminal B chooses a random element R and sends to A $PUB = \langle K = g^R, C = Y_A^R * M \bmod p \rangle$ and A can recover the message by computing $M = C * (K^{X_A})^{-1} \bmod p$. For the law enforcement to get the message from the public information available PUB, they have to remove A's key X_A from escrow. Thus, fair cryptosystems *require* the opening of the keys of individuals receiving messages from criminals rather than opening the keys of the suspected criminals sending the messages. A criminal may get in sessions with as many as possible users (preferably, high government officials and secret service officers) which will enforce opening of all their keys. ♣

Potential abuses in the stage of key generation and distribution are known in traditional systems and apply also to key-escrow systems. The potential use of "law enforcement trapdoor" in the key exchange phase is suggested in [4].

We note that Desmedt [11] presented a method to enforce compliance by the agents generating a public key based on the owner ID. (This is a novel variant on ID based cryptosystem, the ID association is visible to the escrow agents). With ID-based systems, the receiver is traceable by the public-key used, as he noticed. As mentioned above, receiver traceability is not enough to assure fairness in wiretapping in public-key systems.

5.2 Spoofing attacks

We now assume that a public-key system is not used "as is" in escrow systems since the key is asymmetric and the lack-of-fairness attack. Rather we assume the public-key is employed for session key exchange. As advocated in [1] a designer should check what may go wrong with its system in an operational environment and not merely treat it as a separate component. So, here we consider the Internet, as an example of a vulnerable setting representing potential problematic operational environments. The attackers are a pair of accomplices $\tilde{S}$ and $\tilde{R}$ who will spoof two honest users $\bar{A}$ and $\bar{B}$. $\tilde{S}$ sends from $\bar{A}$'s output queue (spoofing operation) a message routed to $\bar{B}$ which can be routed from $\bar{A}$ to $\bar{B}$. $\tilde{R}$ can read and remove the message from $\bar{B}$'s input queue (spoofing operation, again).

Claim 4 *Spoofing is easy with public-key based key escrow and the Clipper.*

Discussion:

Let $\tilde{S}$ and $\tilde{R}$ spoof $\bar{A}$ and $\bar{B}$. Consider a key-exchange based escrow encryption system. To dictate a key K, $\tilde{S}$ puts in $\bar{A}$'s output queue $< K^{e_B} \bmod n_B >$ where $< n_B, e_B >$ is $\bar{B}$'s RSA-based public key. This enables $\bar{B}$ to get K (if it were the case that $\bar{B}$ gets this message) and this convinces wiretapping law enforcement that this is a key from $\bar{A}$ to $\bar{B}$. In the message header (IP header over the internet) $\tilde{S}$ puts the value K itself in some predetermined (unencrypted) positions inside the message ID. $\tilde{R}$ has access to $\bar{B}$'s input queue, he reads the message and given K checks for the fact that it is also transmitted via the key distribution packet (this checking is possible using a filtering software such as those used in firewalls). If this is true, then $\tilde{S}$ and $\tilde{R}$ use K for their private conversation, while observers on the line (law enforcement) assume this is $\bar{A}$ and $\bar{B}$'s session which they cannot open since as far as they know and check, two honest users are following the law by using the key exchanged using the escrowed key (by the "minimization rule" of wiretapping they cannot listen to the honest users).

For example, if Fair cryptosystems are used for this key distribution, then under spoofing, they do not, in fact, enable opening of messages sent for criminals as claimed. Criminals who are good hackers and are well coordinated may enjoy total freedom as the network seems to be used only by seemingly honest users. Since compliance is provided in hardware for Clipper, first a LEAF has to be squeezed and then the above spoofing is possible. ♣

This claim shows that authentication, e.g., a digital signature protection, may be needed with key escrow system (like Fair cryptosystems) to assure the source liability to its messages and for prevention of spoofing. This is needed in any cryptosystem in such an environment, but seems crucial in escrow encryption to assure law enforcement function. Thus, one should not view popular signature providing software (like the available PGP and Privacy-Enhanced-Mail) as competition to key escrow, but rather as a necessary complementary component in environments like the Internet. Independently, [23] studied software escrow encryption systems and made similar observations about their weaknesses. One must, however, be careful to separate encryption and the signature mechanisms (which is not done in the popular systems) as the signature key should not be subject to escrow [20, 19]. We also note that with software-only escrow encryption using a public key system which simply simulates the Clipper system as in the TIS [2] one inherit the problems of both systems (and more).

To design software-based approach which has a compliance certification component and spoofing-thwarting capabilities one may rely on on-line servers and on-line proofs of compliance, and trusted servers that are not spoofable and can perform key-translation functions [20, 19]. Dealing with such compliance and possible spoofing issues, indeed, looks very much related to subliminal channel questions [31]. This is a hard channel to prevent and deal with. (It was noticed in the context of key distribution in key escrow systems first in [22]). Thus, generally dealing with the worst case of spoofing seems hard.

6 System in Operation and General Issues

The additional parties (escrow agents, the court and law enforcement officers) and their capabilities increase the complexity of operation, and present potentially new abuses, attacks and mishaps in escrow systems cryptosystems. This introduces new subtle issues, some of which we discuss.

Issue F: Threats to operation

1. How should wiretapping be made limited in time and in context?
2. How are the traditional attacks on systems: e.g., chosen messages and chosen ciphertext attacks (see [28]) apply to escrow encryption systems?

Time and context limitation can be assured using various cryptographic techniques and operational procedures depending on the technology used (e.g. limiting the life-time of keys discussed above). However, the question also applies to off-line tapping (i.e., of recorded information) where enforcement of restrictions may be harder. The Clipper program uses strict operational procedures as required by various U.S. and state laws and statutes. An interactive protocol in which the law enforcement agents are active in session key generation is used in [4]. This allows for the protection of past messages. The notion of *balanced cryptosystems* where agents open communications based on a per message granularity (rather than a per device/ session key one) is a method to limit context and time of wiretapping (consistent with the "minimization rule" policy [17]) that was introduced in [10, 12]. Using [19, 20] it is possible to achieve this with Clipper by observing that one can first decrypt inside an escrow agent's tamper-proof device without revealing the decryption key or cleartext, and then output a share of the decryption (cleartext) rather than the actual decryption. The device contains the user key and decrypts the ciphertext to obtain message m, it also has pseudorandom function shared among devices. Based on the device ID and the shared randomness rather than sending m, m's share is sent out. This technique can be further improved by externally storing users' keys and techniques from [10]. Finally note that when considering traditional cryptographic attacks, they may now be against additional parties such as the court system, escrow agents, law enforcement agents etc.

General system issues apply to key escrow systems: reliability, efficiency, computing resources and technologies, etc. We next ask about such issues:

Issue G: General System Issues:

1. Is there a possibility of a software-only solution? How is it different from the Clipper approach? What technologies are required?

2. How can we assure reliability and availability?

3. Does availability of low price key escrow devices imply overall ease of wire-tapping?

4. How does the type of application affect the technology used for escrow encryption? For example, how we make key escrow for storage channels (rather than communication channels)?

Having any system with hardware devices or purely in software is a technological tradeoff. As the vice-president has told the press, future key escrow technologies will be oriented towards software and data processing. Hardware seems to ease the design, in particular easing compliance certification (as mentioned above).

The notion of reliability and availability is an important general systems issue especially in emergency applications such as for ambulance and police services (availability vs. security in banking systems is discussed in [1]). The system must be available as denial of service may limit the usage and open new liability issues.

Claim 5 *The Clipper program puts compliance over availability.*

Discussion: As mentioned in [5, 8], a response to Blaze's attack on the 16-bit LEAF is to internally count bad LEAFs up to an upper bound $U << 2^{16}$ and then to disable the device. Now, a bad guy can flood a device with bad messages and disable it. This implies that the suggested remedy to Blaze attack may make Clipper unsuitable for sensitive high availability applications. ♣

On the issue of real-time, we observe that the FBI specifies in Requirement 2 (Real-time, Full-time Monitoring) [17]: *"Law enforcement agencies require a real-time, full-time monitoring capability for intercepts."*

Claim 6 *Cheap Clipper chips may reduce the feasibility of the Clipper program for real-time Law Enforcement.*

Discussion: The government may try to sell key-escrow systems for a low price to assure their popularity and enable simple key escrow program. This in turn, may imply that bad guys can afford a multitude of devices for relatively low investment of money. They may use them in an interleaving fashion. This will require excessive coordination/ involvement of the escrow agents and the courts. This may make real-time wiretapping a very hard task. ♣

We also note that the type of applications has always influenced how we do encryption for it. There appears to be a difference between communication channels (which are somewhat real time applications like e-mail and telephony) and storage channels where information is kept long time (e.g., a database). To the best of our knowledge no one has analyzed the government's legal need, the user's civil rights and protection mechanisms, or the cost/efficiency justifications of escrow encryption as they pertain to various applications like long term storage channels. The access-control key associated with a storage channel should have a LEAF-like element which should enable wiretapping, and the granularity of such keys depends on the definition of a proper "storage context" for wire-tapping. Finally, multicast and broadcast channels (rather than point-to-point) are another open issue.

7 Conclusions

We analyzed basic requirements of a sound key escrow system. Such a system is more complicated in parties, components, goals and exposures than traditional systems. We presented attacks and other basic concerns that apply to Clipper and recent designs, thus increasing our understanding of the nature of such systems. We believe that we are not exhaustive and that further analysis and technical discussions are needed.

Acknowledgments

Thanks to Chris Carrol for his help with the actual attack experiment. Helpful discussions on various aspects of key escrow, Clipper and this paper with Matt Blaze, Dorothy Denning, Yvo Desmedt, Matt Franklin, Michael Froomkin, Amir Herzberg, Paul Karger, David Kravitz, Hugo Krawczyk and Ron Rivest are acknowledged.

References

1. R. Anderson, *Why Cryptography Fails*, In the Proceedings of The 1st ACM Conference on Computer and Communications Security, Nov. 1993, 215–227.
2. D. M. Balenson, C. M. Ellison, S. B. Lipner and S. T. Walker *A New Approach to Software Key Escrow Encryption*, Trusted Information Systems, Inc., (also in [21]).
3. T. Beth, Zur Diskussion Gestellt, *Informatik-Spektrum* 13 (4), pp. 204-215, 1990. (An initial suggestion for Key Escrow System and Agent– a public presentation to the German Government, In German).
4. T. Beth, H.-J. Knobloch, M. Otten, G.J. Simmons and P.Wichmann, Towards Acceptable Key Escrow Systems, In the Proceedings of The 2nd ACM Conference on Computer and Communications Security, November 1994 51–58.
5. M. Blaze, *Protocol failure in the Escrowed Encryption Standard*, In the Proceedings of The 2nd ACM Conference on Computer and Communications Security, November 1994, 59–67. (also in [21]).
6. E.F. Brickel,et al. Interim Review: The SKIPJACK Algorithm, July 93. (Also in [21]).
7. D.E. Denning et al., *To Tap or Not To Tap*, CACM 93.
8. D. E. Denning and M. Smid, Key Escrowing Now, IEEE Communications Magazine, Sep. 1994, pp. 54-68.
9. Department of Justice, Letter dated Feb. 4, 1994
10. A. De Santis, Y. Desmedt, Y. Frankel, and M. Yung, *How to Share a Function Securely*, ACM STOC 94. (Initial version May 92: FOCS 92 submission).

11. Y. Desmedt, Securing Traceability of Ciphertexts: Towards a Secure Software Key Escrow Systems, Eurocrypt 95.

12. Y. Desmedt, Y. Frankel, and M. Yung, A scientific statement on the Clipper Chip technology and alternatives, (a letter to NIST as an answer to a request for comments on key escrow technology).

13. Y. Desmedt and Y. Frankel, *Threshold cryptosystems*, In G. Brassard, editor, *Advances in Cryptology, Proc. of Crypto '89 (Lecture Notes in Computer Science 435)*, pages 307–315. Springer-Verlag, 1990.

14. Y. Desmedt and J.-J. Quisquater, *Public-key systems based on the difficulty of tampering (Is there a difference between DES and RSA?)*, Advances in Cryptology– Proc. Crypto '86, Springer-Verlag LNCS 263, 1987, 111–117.

15. W. Diffie and M. Hellman, *New Directions in Cryptography*, IEEE Trans. on Information Theory 22 (6), 1976, pp. 644-654.

16. T. El Gamal, *A Public key cryptosystem and a signature scheme based on discrete logarithm*, IEEE Trans. on Information Theory 31, 465-472, 1985.

17. The FBI, Law Enforcement REQUIREMENTS for the Surveillance of Electronic Communications, June 1994. (Prepared by the Federal Bureau of Investigations (FBI) in cooperation with federal, state, and local law enforcement members of the National Technical Investigation Association).

18. FIPS PUB 185, *Escrowed Encryption Standard* February 1994. (Dept. of Commerce).

19. Y. Frankel and M. Yung, *Designs of escrow encryption systems: models, methodologies and technologies*, (Available from the authors).

20. Y. Frankel and M. Yung, Preliminary version of current paper, originally submitted to IEEE Security and Privacy 95 (Oakland), Nov. 94.

21. *Building in Big Brothers*: the cryptographic policy debate, ed. L.J. Hoffman, Springer Verlag, 1995.

22. J. Kilian and F.T. Leighton, *Failsafe Key Escrow Systems*, Crypto 95.

23. D. Kravitz, *Deficiencies of Software-based key escrow.* a letter.

24. S.M. Matyas, *Key Processing with Control Vectors*, Journal of Cryptology, 3 (2), pp 113-136, 1991

25. S. Micali, *Fair public-key cryptosystems*, Crypto '92 (also in [21]).

26. The Mosaic program office, *Mosaic: Cryptographic intertrace programmers guide for the Tessera crypto card*, Draft Revision P1.4.

27. MYK-78T Encryption/Decryption VLSI, Mykotronx Inc.

28. M. Naor and M. Yung, *Public-key cryptosystem provably secure against chosen ciphertext attack*, Proc. of the 22nd Annual Symposium on the Theory of Computing, 1990, pp. 427–437.

29. T. P. Pedersen, Distributed Provers with Applications to Undeniable Signature, Eurocrypt '91. 1991.

30. R. Rivest, A. Shamir and L. Adleman, *A Method for Obtaining Digital Signature and Public Key Cryptosystems*, Comm. of ACM, 21 (1978), pp 120-126.

31. G. Simmons, *The Subliminal Channel and Digital Signature*, Eurocrypt 84, 51–67.

32. The White House Press Release Regarding the Clipper, the White House – office of the press secretary, April 16, 93.

Robustness Principles for Public Key Protocols

Ross Anderson and Roger Needham
Cambridge University Computer Laboratory
Pembroke Street, Cambridge, England CB2 3QG

Abstract: We present a number of attacks, some new, on public key protocols. We also advance a number of principles which may help designers avoid many of the pitfalls, and help attackers spot errors which can be exploited.

1 Introduction

Cryptographic protocols are typically used to identify a user to a computer system, to authenticate a transaction, or to set up a key. They typically involve the exchange of about 2–5 messages, and they are very easy to get wrong: bugs have been found in well known protocols years after they were first published. This is quite remarkable; after all, a protocol is a kind of program, and one would expect to get any other program of this size right by staring at it for a while.

A number of remedies have been proposed. One approach is formal mathematical proof, and can range from systematic protocol verification techniques such as the BAN logic [BAN89] to the case-by-case reduction of security claims to the intractability of some problem such as factoring.

Another approach is to try to encapsulate our experience of good and bad practice into rules of thumb; these can help designers avoid many of the pitfalls, and, equally, help attackers find exploitable errors. A recent paper by Abadi and Needham undertook this exercise for cryptographic protocols in general [AN94].

The two approaches — formal proofs and structured design rules — are in many ways complementary. On the one hand, we expect that robust design principles will help us construct protocols which are more amenable to formal verification; on the other hand, protocol errors thrown up by formal analysis may lead us to new insights into the nature of robustness.

Mathematical techniques are also liable to error — quite a few protocols which had been 'proved' secure have been successfully attacked (see, e.g., [PW91] [DB93] [PW95]). These failures are typically due to assumptions which were not made explicit, and we believe that robustness principles are a large part of the solution: if they force us to consider more of the security dependencies, and from a number of aspects, then we will be less likely to produce a 'proof' which neglects a real weakness.

Our goal here is to push the state of the art a little further and deal with more complex protocols, and in particular with public key schemes. Curiously enough, although public key algorithms are based more on mathematics than secret key algorithms, and have much more compact mathematical descriptions, public key protocols have proved much harder to deal with by formal methods.

In this paper, we propose a number of principles. We present a number of new attacks, and give a (hopefully) new perspective on some old ones.

D. Coppersmith (Ed.): Advances in Cryptology - CRYPTO '95, LNCS 963, pp. 236-247, 1995.
© Springer-Verlag Berlin Heidelberg 1995

2 The Order of Encryption and Signature

We will start by expanding on the fifth principle from [AN94].

Principle 1: Sign before encrypting. If a signature is affixed to encrypted data, then one cannot assume that the signer has any knowledge of the data. A third party certainly cannot assume that the signature is authentic, so nonrepudiation is lost.

This was motivated by attacks in which the opponent could remove a signature from an encrypted message and replace it with one of his own — X.509 [CCITT88] suffered from such an attack. However, there is an even more powerful attack on several protocols which do encryption before signature, including X.509 and a number of the proposals in ISO CD 11770 [ISO94a].

Suppose that Alice wishes to use RSA [RSA78] with a modulus of 500 - 600 bits to send Bob the message M. The standard technique would be for her to first sign the message with her private key and then encrypt it with his public key. However, suppose that Alice first encrypts M under Bob's public key and then signs it with her private key. Denoting the modulus, public exponent and private exponent of party α by n_α, e_α and d_α, and ignoring hashing (as it makes no difference to our argument), the signed encrypted message would be:

$$\{M^{e_B} \pmod{n_B}\}^{d_A} \pmod{n_A} \tag{1}$$

This is vulnerable, and in a novel way. Since Bob can factor n_B and its factors are only 250–300 bits long, he can work out discrete logarithms with respect to them and then use the Chinese Remainder theorem to get discrete logs modulo n_B. So if he wants to get Alice's 'signature' on a different message, M', he can find x such that

$$[M']^x = M \pmod{n_B} \tag{2}$$

He then registers (xe_B, n_B) as a public key with a certification authority, and claims that the message signed by Alice was not M but M'.

This provides a direct attack on CCITT X.509, in which Alice signs a message of the form $\{T_A, N_A, B, X, \{Y\}^{e_B} \pmod{n_B}\}$ and sends it to Bob. Here T_A is a timestamp, N_A is a serial number, and X and Y are user data. It also breaks the draft ISO CD 11770; there, Y consists of A's name and a random challenge in key agreement mechanism 5, and A's name followed by a session key in key transport mechanisms 2, 5 and 6.

The attack is not limited to RSA: it works with ElGamal too [Elg85], provided this time that Bob can choose his own modulus. Recall that in ElGamal the message m is encrypted to (r, c) where $r = g^k \pmod{p}$, $c = y^k m \pmod{p}$, the message key is k, the recipient's private key is x, and his public key is $y = g^x$. Suppose that Bob selects a so-called 'trapdoor' modulus, under which he can work out discrete logarithms [RLS+92]. Then, for any given m', r and c, he can find a suitable y' such that $(y')^k = m'/c$.

The obvious countermeasure, of requiring all users to share the same modulus, may be politically difficult, as attempts have been made in the past to foist suspect moduli on the user community [And93a].

Key spoofing attacks are also possible on symmetric systems. For example, given a message M and a ciphertext C, the effort required to find a key such that $C = \{M\}_K$ is about 2^{55} when the algorithm used is single key DES, but thanks to the birthday problem it is only 2^{28} or so with double key DES. There are systems where a single, double DES encrypted block is used to authorise a payment, such as described in [And92b] — although that particular system is not vulnerable as one of the two keys is fixed.

With public key systems, key spoofing attacks seem easy to prevent — just always sign before encrypting. However, inverting the order of encryption and signature is a surprisingly common misfeature. A recent internet cash proposal [Oto94] also used it, and Kailar pointed out that this destroyed accountability in the invoicing system [Kai95]. Our attack goes further; it could allow invoices to be forged. It also dents a protocol for anonymous credit cards [LMP94][1].

Encryption before signature can also cause problems for formal verification techniques. The BAN logic ignores the algorithm issues, but at least it will not verify that Alice signed M in equation (1) above as she has no jurisdiction over n_B. Kailar's logic also rejects a signed encrypted message.

However, the verification tools which do try to deal with algorithm properties (such as those discussed in [KMM94]) do not seem able to deal with this attack at all. In order to fix this, the scope of their assumptions may need to be extended; one conventionally worries about the factorisation properties of RSA keys, not their discrete log properties, and the worry about trapdoor primes has been that users might be attacked by authority, rather than by each other.

In any case, it is prudent to sign before encrypting.

3   Spot the Oracle

Nonrepudiation is complicated by the fact that signature and decryption are the same operation in RSA, which many people use as their mental model of public key cryptography. They are actually quite different in their semantics: decryption can be simulated, while signature cannot. By this we mean that an opponent can exhibit a ciphertext and its decryption into a meaningful message, while he cannot exhibit a meaningful message and its signature (unless it is one he has seen previously).

Consider for example the following protocol by Woo and Lam [WL92]:

---

[1] message 2 in the extension of credit protocol can be arbitrarily manipulated by B_p

Message 1 $A \longrightarrow S$: A, B
Message 2 $S \longrightarrow A$: CB
Message 3 $A \longrightarrow B$: $\{A, N_A\}_{KB}$
Message 4 $B \longrightarrow S$: $A, B, \{N_A\}_{KS}$
Message 5 $S \longrightarrow B$: $CA, \{\{N_A, K_{AB}, A, B\}_{KS^{-1}}\}_{KB}$
Message 6 $B \longrightarrow A$: $\{\{N_A, K_{AB}, A, B\}_{KS^{-1}}\}_{KA}$
Message 7 $A \longrightarrow B$: $\{N_A\}_{K_{AB}}$

Here we are using the standard notation from [BAN89]: CX is a certificate containing the public key KX of participant X; the corresponding private key is KX^{-1}; N_X is a nonce generated by participant X; K_{AB} is a shared secret between Alice and Bob (which it is the purpose of the protocol to generate), and Sam is the key distribution centre.

There are a number of problems with this protocol, including the obvious one that Bob has no assurance of freshness (he does not check a nonce or see a timestamp). However, a subtler and more serious problem is that Alice never signs anything; the only use made of her secret is to decrypt a message sent to her by Bob. The consequence of this is that Bob can only prove Alice's presence to himself — he cannot prove anything to an outsider, as he could easily have simulated the entire protocol run. The effect that such details can have on the beliefs of third parties is one of the interesting (and difficult) features of public key protocols: few of the actual or proposed standards provide a robust nonrepudiation mechanism, and yet there is a substantial risk that many of them may be used as if they did.

We shall return to this topic later. For the meantime let us just say that we must be careful what we mean by 'Bob'. This may be 'whoever controls Bob's signing key', or it may be 'whoever controls Bob's decryption key'. Both keys are written as KB^{-1} in the standard notation, but they are actually rather different.

> **Principle 2:** Be careful how entities are distinguished. If possible avoid using the same key for two different purposes (such as signing and decryption), and be sure to distinguish different runs of the same protocol from each other.

Beaver's attack on Den Boer's oblivious transfer protocol falls into this category: when the same public key primitive is reused in the oblivious transfer context, various sneaky attacks become possible [Bea92]. Also, Landrock recently pointed out that if someone uses the same key in the ISO protocols for signature and zero knowledge proof, there is a massive security failure: the zero knowledge protocol can be used as an oracle to generate signatures [Lan95].

Woo and Lam's protocol also suffers from an oracle problem: if decryption and signature are performed using the same key, then Sam can be impersonated. This is because between messages 4 and 5, he decrypts a nonce N_A sent to him encrypted under his own public key.

Even where keys are only used for one purpose, there may still be an oracle attack; a recent example was found in the documentation for Lotus Notes Internals [Dwo94]. Oracle attacks can be fixed in various ways, such as by explicit typing of nonces, or by using different keys for different purposes (as Lotus apparently do in their current implementations). However, they can sometimes be quite subtle, and an interesting example is the attack found by Simmons on the TMN (Tatebayashi-Matsuzaki-Newmann) scheme.

Here, two users want to do a key exchange, but with a trusted server doing most of the work (we can think of the users as smartcards). If Alice and Bob are the users, and the trusted server Sam can factor N, then the protocol goes as follows:

Message 1 $A \rightarrow S : r_A^3 \pmod{N}$
Message 2 $B \rightarrow S : r_B^3 \pmod{N}$
Message 3 $S \rightarrow A : r_A \oplus r_B$

Each party chooses a random number, cubes it, and sends in to Sam. He extracts cube roots, xors the two random numbers together, and sends the result to Alice. The idea is that Alice and Bob can now use r_B as a shared secret key. However, Simmons pointed out that if Charlie and David conspire, or if just one user (David) generates predictable random numbers r_D, then Charlie can get hold of r_B in the following way [Sim94]:

Message 1 $C \rightarrow S : r_B^3 r_C^3 \pmod{n}$
Message 2 $D \rightarrow S : r_D^3 \pmod{n}$
Message 3 $S \rightarrow C : r_B r_C \oplus r_D$

We will sum all this up simply as

Principle 3: Be careful when signing or decrypting data that you never let yourself be used as an oracle by your opponent.

4   Count the Bits

We mentioned above the need to distinguish different runs of the same protocol. This means, for example, that systems based on discrete log typically need a fresh message key for each session, which brings us to the topic of subliminal channels.

The message keys in ElGamal type schemes contain various covert channels. For example, 160 of the 320 signature bits in the digital signature standard give security — apparently making the computational security $O(2^{80})$ — but the other 160 bits are available for covert communication [Sim94b].

Counting bits is not always as straightforward, as it may involve specific properties of the public key primitive. One of the earliest examples of an attack on a public key protocol is due to DeMillo and Merritt, who showed that a poker protocol leaked information through quadratic characters [DM83]. A similar attack has been reported by Lomas on a protocol of Mao.

Subliminal channels may seem rather abstract, but counting the precise amount of redundancy can bring us right down into the muddy details of particular implementations. It has been known since the earliest days of public key cryptography [DH76] that digital signatures are inherently vulnerable to attacks by forward search — an attacker applies your public key to a lot of random signatures until (with luck) she gets something which she can pass off as a message from you.

Such attacks can in principle be prevented by putting enough redundancy in each message to be signed. However, it is common to rely on naming information and counters for this purpose, and this can lead to errors — especially if neither the cryptologist nor the system designer pays attention to the other's work.

Consider ISO 11166 [ISO94b]. Here, as in X.509 and ISO CD 11770, encryption is done before signature; but as the RSA exponent is fixed, the key spoofing attack of section 2 above does not work. However, an attacker can just as easily replace the modulus and do a forward search.

How much effort will this take? In ISO 11166, the protected message consists of a key used to authenticate banking transactions, an eight bit key control vector, and a 56 bit counter. However, the standard specifies that if the user receives a count which is higher than the retained value, he should accept it and send a service message confirming the new count. Assuming that the attacker can intercept and discard this message, the counter contains only one bit of real redundancy, and so forging a message is trivial.

The effects of this are surprisingly subtle. For example, public key certificates must be checked anew with every key service message. If they are cached in the local host after checking, then programmers could forge a key service message to their own bank and could then authenticate a bogus transaction.

Another problem with ISO 11166 is that the redundancy in the key certificates is rather low. It is apparently 45 bits for a short certificate, but depends on the redundancy of the namespace for a long certificate; this means that the redundancy will steadily deteriorate. If in future there were 30,000 banks sharing the US banking namespace, then the search effort might be as little as 2^{42} modular multiplications — a large computation, but not large enough to stop a determined attacker. We conclude

Principle 4: Account for all the bits — how many provide equivocation, redundancy, computational complexity, and so on. Make sure that the redundancy you need is based on mechanisms which are robust in the application context, and that any extra bits cannot be used against you in some way.

5   Assume Nothing

We will next look at a number of related types of protocol failure, which are nicely illustrated by a pernicious attack which Burmester found on protocols by Goss, Günter, Yacobi and the EC's RIPE project team [Bur94]. These protocols try to fortify Diffie Hellman key exchange by adding authentication. As an example, we will consider the Goss protocol [Gos90], which is apparently used in the German railway system.

Here, there is a common prime p and a generator g of a high order subgroup of Z_p^*. Each user U has a secret key x_U and a public key $y_U = g^{x_U}$. At each run of the protocol, user U generates the random number r_U. Alice and Bob exchange the message key $y_A^{r_B} \oplus y_B^{r_A}$ as follows:

Message 1 $A \rightarrow B : CA, g^{r_A}$ (mod p)
Message 2 $B \rightarrow A : CB, g^{r_B}$ (mod p)

Burmester showed that there is a triangle attack: if old session keys can be obtained by an attacker, then Charlie can discover the key shared between Alice and Bob by using suitably chosen values in subsequent exchanges with them. The details can be found in [Bur94].

There is an easier way to look at this attack: Alice supplies g^{r_A} to Bob, and Bob returns to her $(g^{r_A})^{x_B}$ — so Alice can try to send him an arbitrary z and get back z^{x_B}. In fact, she gets back $y_A^{r_B} \oplus z^{x_B}$ and knows only the first of these two terms, but given the key which Bob thought he generated she can work out z^{x_B}. Thus if Bob lets old message keys leak, he will have allowed himself to be used as an oracle for his own secret operation, namely raising numbers to the exponent x_B.

Anyway, Burmester described the flaw in these protocols not as an oracle attack but as a consequence of failing to consider what might happen if a counterparty failed to keep an old message key secret. A number of other protocols fail if a message key is later revealed, and some early examples can be found in [BAN89]; Simmons' attack on the TMN protocol provides another example. We therefore propose as our next principle:

Principle 5: Do not assume the secrecy of anybody else's 'secrets' (except possibly those of a certification authority).

A related error is to make simplifying assumptions about the kind of messages which an opponent might insert in the course of an attack. The weakness in the Goss protocol which we discussed above can also be interpreted in this way: once one sees that the number received from the other party is not necessarily g^r, for some r known to either the other player or an attacker, but can be any number z, then the existence of an oracle attack becomes obvious.

However, there have been other attacks in the same mould which principle 5 does not tackle. Desmedt and Burmester broke a 'proven' secure protocol by showing that the opponent did not have to act in a nice (simulatable) way [DB93], and a number of server assisted signature schemes have also failed in this way [And92b]. We therefore state as a separate principle:

Principle 6: Do not assume that a message you receive has a particular form (such as g^r for known r) unless you can check this.

Next, we have to look at conspiracy attacks on threshold schemes and other multiparty constructions. These have caused a lot of confusion in the past, and perhaps the obvious thing to say would be something like "It is prudent to make explicit the number of conspirators against whom security is claimed".

But we need to go further. One of the present authors proposed a scheme for hiding trapdoors in RSA public keys [And93a] which turned out to be vulnerable to an attack by lattice basis reduction once a certain number of keys had been generated [Kal93]. We have also seen above that some encryption algorithms are vulnerable to key spoofing. So the next principle is a bit more general:

Principle 7: Be explicit about the security parameters of crypto primitives. A key generation routine should be claimed as good for so many keys; a threshold scheme for resistance to so many conspirators; a block cipher for so many blocks; and so on.

Of course, some of the above principles overlap, and Burmester's attack is particularly interesting as it can be construed in different ways — as the consequence of the Goss protocol's failing to observe principles 3, 5 or 6 (at least). However, we are not trying to provide a minimal set of principles; a certain redundancy of robustness concepts is unlikely to do us any harm.

Finally, there are principles which are either too algorithm specific, or too general, for the level of abstraction at which we are trying to operate. Consider for example Coppersmith's attack on NIKS-TAS, a scheme which combines discrete exponentiation with combinatorics [Cop94]. One might formulate a principle that one should shield secrets behind the public key primitive rather than the combinatorics, or one might adopt Coppersmith's own conclusion that in a discrete log scheme one should always have the secret information 'upstairs' and the public information 'downstairs'. However, it is unclear that either of these is general enough for our list.

At higher levels of abstraction, we have 'engineering commonsense' such as the KISS principle ('Keep It Simple Stupid'), which cryptographers often ignore. Many highly complex schemes are proposed for digital cash and other applications, and many of them turn out to be unsound (e.g., [TT94]). Proof is no panacea: several 'proven' secure systems fail because of unexamined assumptions [PW91] [PW95], and others omit to provide desirable properties such as unlinkability [Yac94] and arbitration [Kai95]. Of course, particular schemes may breach one or more of our principles, as they often concatenate a number of public key primitives without hashing or redundancy in order to achieve exotic effects. This is a subject of ongoing research interest.

6   The Explicitness Principle

Looking at the above seven principles, we are led to ask whether there is any overarching principle of which the others are in some sense instances. We propose the following, which one of us put forward in the computer security context in [And94a] and the other in the general protocol context in [AN94].

> **Principle 8:** Robust security is about explicitness; one must be explicit about any properties which can be used to attack a public key primitive, such as multiplicative homomorphism, as well as the usual security properties such as naming, typing, freshness, the starting assumptions and what one is trying to achieve.

With symmetric key algorithms it is often possible to treat the algorithm as a black box, as symmetric block ciphers have a certain amount of muddle in them. For example, it is not common to find that the internals of DES interact malignantly with the way you use it. However, the known asymmetric algorithms are much more structured[2], and depend on fairly straightforward mathematical operations such as integer or matrix multiplication. Thus they are much more likely to interact with other things in the protocols they are used with, and we have to be that much more careful.

Thus it is prudent to hash data before signing it, using a hash function which does not interact with the signature scheme. In an ideal world, signature schemes would be proof against adaptive chosen ciphertext attacks, but in the real world it seems that we can only achieve this by combining hashing with signature, and by being very explicit about what properties of our signature scheme we wish our hash function to mask.

A good example is correlation freedom. Until fairly recently, it was thought sufficient for a hash function to be one-way and collision-free [Dam87]. Then at Crypto 92, Okamoto defined correlation freedom to be the property that we cannot find $M \neq M'$ with $h(M)$ and $h(M')$ agreeing in more bits than we would expect to find from random chance. He conjectured that correlation freedom was strictly stronger than collision freedom, and this was proved in [And93b]. Since then, Vaudenay has shown that MD4 is not correlation free, and Knudsen has established the same for the hasing mode of SAFER K-64 [Knu95].

Here, the explicitness principle is wider than the concept of resistance to adaptive chosen ciphertext attacks. For example, we may also have to protect message keys: in the Schnorr signature scheme [Sch89], we must not have $h(g^r, m)$ equal to $f(k) + h(g^{r+k}, m)$ for any function f which our opponent is able to compute. It is also wider than any possible set of freedom properties (collision freedom, correlation freedom, multiplication freedom, ...) [And93b].

The explicitness principle can be applied to algorithms as well as protocols. We saw above, for example, that algorithm designers should be explicit about the

---

[2] It was conjectured to one of us by JWS Cassels in 1977 that this was necessarily so, and that asymmetric algorithms would in consequence always be rather fragile.

difficulty of finding key collisions on a given message-ciphertext pair. Another example is the persistence of attacks on hash functions based on modular multiplication, such as that proposed with X.509, where a failure to make the round function sufficiently multiplication free leads to attacks based on techniques such as lattice basis reduction [Cop89]. However extending the explicitness principle too far into the domain of algorithms would take us away from the subject matter of this paper.

Finally, our standard disclaimer: the weaknesses we have discussed do not necessarily imply that any given system based on a protocol criticised above is insecure, as there are many ways to implement compensating controls. However, it is prudent to avoid using standards which are questionable, and which make security depend closely on application detail. Once the application code is brought inside our security perimeter, we lose the advantages of a trusted computing base; we run the risk of unpredictable security failures as documented in [And94a]; and we acquire the legal exposures described in [And94b]. Ignoring prudent design practice can be just as expensive in cryptography as in other branches of engineering.

7 Conclusion

We have tried to extend the prudent engineering principles of Abadi and Needham to the world of public key protocols, which are even more prone than conventional ones to subtle errors, and thus may be even more in need of robustness guidelines. We do not claim that our proposed principles are either necessary or sufficient, just that they are useful; at least, we have found them to be useful both in looking for attacks and in explaining this subject to new students.

Acknowledgement: We are grateful to Martin Abadi for a number of helpful comments.

References

[And92a] RJ Anderson, "Attack on server-assisted authentication protocols", in *Electronics Letters* v **28** no 15 (16th July 1992) p 1473

[And92b] RJ Anderson, "UEPS - A Second Generation Electronic Wallet", *Computer Security — ESORICS 92*, Springer LNCS v **648** in 411–418

[And93a] RJ Anderson, "A practical RSA trapdoor", in *Electronics Letters* v **29** no 11 (27th May 1993) p 995

[And93b] RJ Anderson, "The Classification of Hash Functions", in *Codes and Ciphers* (proceedings of fourth IMA Conference on Cryptography and Coding, December 1993), published by IMA (1995) pp 83–93

[And94a] RJ Anderson, "Why Cryptosystems Fail", in *Communications of the ACM* v **37** no 11 (November 1994) pp 32–40

[And94b] RJ Anderson, "Liability and Computer Security — Nine Principles", in *Computer Security — ESORICS 94*, Springer LNCS v **875** pp 231–245

[AN94]   M Abadi, RM Needham, *'Prudent Engineering Practice for Cryptographic Protocols'*, DEC SRC Research Report **125** (June 1 1994)

[Bea92]   D Beaver, "How to Break a 'Secure' Oblivious Transfer Protocol", in *Advances in Cryptology — EUROCRYPT '92*, Springer LNCS v **658** pp 284–296

[BAN89]   M Burrows, M Abadi, RM Needham, "A Logic of Authentication", in *Proceedings of the Royal Society of London A* v **426** (1989) pp 233–271; earlier version published as DEC SRC Research Report **39**

[Bur94]   M Burmester, "On the Risk of Opening Distributed Keys", in *Advances in Cryptology — CRYPTO '94*, Springer LNCS v **839** pp 308–317

[Cop89]   D Coppersmith, "Analysis of ISO/CCITT Document X.509 Annex D", submitted to ISO

[Cop94]   D Coppersmith, "Attack on the Cryptographic Scheme NIKS-TAS", in *Advances in Cryptology — CRYPTO '94*, Springer LNCS v **839** pp 294–307

[CCITT88] CCITT X.509 and ISO 9594-8, "The Directory — Authentication Framework", CCITT Blue Book, Geneva, March 1988

[Dam87]   IB Damgård, "Collision free hash functions and public key signature schemes", in *Advances in Cryptology — EUROCRYPT '87*, Springer LNCS **304** pp 203–216

[Dwo94]   C Dwork, "Distributed Computing Column", ACM SIGACT News v 26 mo 1 (Mar 94) pp 17–19

[DB93]   Y Desmedt, M Burmester, "Towards Practical 'Proven Secure' Authenticated Key Distribution", in *1st ACM Conference on Computer and Communications Security* (ACM November 1993) pp 228–231

[DH76]   W Diffie, ME Hellman, "New Directions in Cryptography", in *IEEE Transactions on Information Theory*, **IT-22** no 6 (November 1976) p 644–654

[DM83]   R DeMillo, M Merritt, "Protocols for Data Security", in *IEEE Computer* v **16** no 2 (Feb 1983) pp 39–50

[Elg85]   T El-Gamal, "A Public-Key Cryptosystem and a Signature Scheme Based on Discrete Logarithms", in *IEEE Transactions on Information Theory* **IT-31** no 4 (July 1985) pp 469–472

[FS86]   A Fiat, A Shamir, "How To Prove Yourself: Practical Solutions to Identification and Signature Problems", in *Advances in Cryptology — CRYPTO 86*, Springer LNCS v **263** pp 186–194

[Gos90]   KC Goss, *'cryptographic method and apparatus for public key exchange with/ authentication'*, US patent no. 4,956,863 (September 11, 1990)

[ISO94a]   ISO DIS 11770, *'Information Technology — Security Techniques — Key Management — Part 3: Mechanisms using asymmetric techniques'*, ISO IST/33/-/2:94/211

[ISO94b]   ISO 11166-1:1994, *'Banking — Key management by means of asymmetric algorithms — Part 1: Principles, procedures and formats'*, and *Part 2: Approved algorithms using the RSA cryptosystem'*, 15 November 1994

[Kai95]   R Kailar, "Reasoning about Accountability in Protocols for Electronic Commerce", accepted for *Oakland 95*

[Kal93]   B Kaliski, "Anderson's RSA trapdoor can be broken", in *Electronics Letters* v **29** no 15 (22nd July 1993) pp 1387–1388

[Knu95]   L Knudsen, "A Weakness in SAFER K-64", *this volume*

[KMM94] R Kemmerer, C Meadows, J Millen, "Three Systems for Cryptographic Protocol Verification", in *Journal of Cryptology* v **7** no 2 (Spring 1994) pp 79–130

[Lan95] P Landrock, talk given at Combridge Protocols Workshop, 19–21 April 1995

[LMP94] "Anonymous Credit Cards", SH Low, NF Maxemchuk, S Paul, in *Proceedings of 2nd ACM Conference on Computer and Communications Security* (ACM, Nov 94) pp 108–117

[Oto94] K O'Toole, The Internet Billing Server — Transaction Protocol Alternatives", *Carnegie Mellon University report INI TR 1994-1* (April 26, 1994)

[PW91] B Pfitzmann, M Waidner, "How to Break and repair a 'Provable Secure' Untraceable Payment System", in *Abstracts of Crypto '91* pp 8-14 to 8-19

[PW95] B Pfitzmann, M Waidner, "How to Break Another 'Provably Secure' Payment System", *to appear in proceedings of Eurocrypt 95*

[RLS+92] RA Rueppel, AK Lenstra, ME Smid, KS McCurley, Y Desmedt, A Odlyzko, P Landrock, "The Eurocrypt '92 Controversial Issue — Trapdoor Primes and Moduli", in *Advances in cryptology — EUROCRYPT '92*, Springer LNCS v **658** pp 194–199

[RSA78] RL Rivest, A Shamir, L Adleman, "A Method for Obtaining Digital Signatures and Public-Key Cryptosystems", in *Communications of the ACM* **21** (1978) pp 120–126

[Sch89] CP Schnorr, "Efficient identification and signatures for smart cards", in *Advances in Cryptology — CRYPTO '89*, Springer LNCS **435**, pp 239–251

[Sim94a] GJ Simmons, "Cryptanalysis and Protocol Failures", in *Communications of the ACM* v **37** no 11 (November 1994) pp 56–65

[Sim94b] GJ Simmons, "Subliminal Channels; Past and Present", in *European Transactions on Telecommunications* v **5** no 4 (July/Aug 1994) pp 459–473

[TMN89] M Tatebayashi, N Matsuzaki, DB Newman, "Key distribution protocol for digital mobile communication systems", in *Advance in Cryptology — CRYPTO '89*, Springer LNCS **435** pp 324–333

[TT94] L Tang, D Tygar, "A fast off-line electronic currency protocol for smart cards", in *proceedings of the First Smart Card Research and Advanced Application Conference* (University of Lille, Oct 94) pp 89–100

[Vau94] S Vaudenay, "On the need of multipermutations - Cryptanalysis of MD4 and SAFER", in *'Fast Software Encryption'*, proceedings of KU Leuven workshop on cryptographic algorithms (Springer, to appear)

[WL92] TYC Woo, SS Lam, "Authentication for Distributed Systems", in *IEEE Computer* (January 1992) pp 39–52

[Yac94] Y Yacobi, "Efficient Electronic Money", in *Preproceedings of Asiacrypt 94* pp 131–140

Cryptanalysis of the Matsumoto and Imai Public Key Scheme of Eurocrypt'88

Jacques PATARIN

CP8 TRANSAC, 68 route de Versailles, BP 45 - 78430 Louveciennes Cedex - France

Abstract

In [1] Matsumoto and Imai have developed a new public key scheme for enciphering or signing. (This scheme is completely different and should not be mistaken with another scheme of Matsumoto and Imai developed in 1983 and broken at Eurocrypt'84).

No attacks have been published as yet for this scheme. However we will see in this paper that for almost all the keys almost each cleartext can be found from his ciphertext after only about $m^2 n^4 \log n$ computations where m is the degree of the field K chosen, and where mn is the number of bits of the text.

Moreover for absolutely all the keys that give a practical size for the messages it will be possible to find almost all the cleartexts from the corresponding ciphertexts after a feasible computation.

So the algorithm of [1] is insecure.

1   Introduction

In [1] Matsumoto and Imai have developed a new public key scheme. The aim of this paper is to see how this scheme can be attacked.

For Crypto'95 this paper must be short. Also I have written an extended version of this paper which gives more proofs, details and examples than this short paper. I will be happy to give this extended version to anybody who wants it.

2   A short description of the Matsumoto-Imai Algorithm

2.1   A Mathematical Property

Let K be a finite field of characteristic 2, and let $q = 2^m$ be the number of elements of K (for example $K = GF(2)$ the field with 2 elements). Let L_N be an extension field of degree N of K, and let θ be an integer.

Then the function

$$f : \begin{array}{ccc} L_N & \to & L_N \\ x & \mapsto & x^{1+2^{m\theta}} \end{array}$$

is a bijection if $1 + 2^{m\theta}$ is coprime with $2^{mN} - 1$.

More precisely when f is a bijection f is easily invertible, and its inverse function f^{-1} is such that $f^{-1}(x) = x^{\hbar}$, where $\hbar$ is the multiplicative inverse of $1 + 2^{m\theta}$ modulo $2^{mN} - 1$.

D. Coppersmith (Ed.): Advances in Cryptology - CRYPTO '95, LNCS 963, pp. 248-261, 1995.

Let B be a basis of L_N, then the expression of f in the basis B is :

$$f(x_1, \ldots, x_N) = (p_1(x_1, \ldots, x_N), \ldots, p_N(x_1, \ldots, x_N))$$

where $p_1, \ldots, p_N$ are N polynomials in N variables of degree 2. The reason for this is that $x \mapsto x$ and $x \mapsto x^{2^{m\theta}}$ are both linear functions, so $f(x) = x.x^{2^{m\theta}}$ is a quadratic function and its components in the basis B have quadratic expressions.

The polynomials $p_1, \ldots, p_N$ are found by choosing a "representation" of L_N. Such a "representation" is typically given by the choice of an irreducible polynomial $i_N(X)$ over K, of degree N, so we can identify L_N with $K[X]/(i_N(X))$. It is then easy to find the polynomials $p_1, \ldots, p_N$.

2.2 Description of the Matsumoto-Imai Algorithm

The field K, with 2^m elements, is public. Each message will have $n.m$ bits, where n is another public integer.

n is split in $n = n_1 + \ldots + n_d$ with d integers $n_1, \ldots, n_d$. Then with these integers we will need d extensions of $K, L_{n_1}, \ldots, L_{n_d}$ of degree respectively $n_1, \ldots, n_d$. We will call "word" a value represented by some components of K. For example an element of L_{n_e}, $1 \leq e \leq d$, can be seen as a word of length n_e. Some quadratic functions $f_1, \ldots, f_d$ will be used, giving d words. These d words will then be recombined in a word of length n.

The secrets items will be :
1. Two affine bijections s and t from $K^n \to K^n$ (these affine bijections can be represented in a basis by polynomials of total degree 1 and with coefficients of the polynomials in K).
2. The separation of n in d integers : $n = n_1 + \ldots + n_d$.
3. The "representation" of the fields $L_{n_1}, \ldots, L_{n_d}$. These "representations" are given with the choice of d irreducible polynomials. We will denote by ψ_e the isomorphism from K^{n_e} to L_{n_e} given by these representations, $1 \leq e \leq d$.
4. Some integers $\theta_1, \ldots, \theta_d$ such that $1 \leq \theta_e < n_e$ and $GCD(2^{\theta_e} + 1, 2^{mn_e} - 1) = 1, 1 \leq e \leq d$. These integers θ_e give the quadratic functions $f_1, \ldots, f_d$ as we have seen in paragraph 2.1. (with $N = n_e$). (GCD is the Greatest Common Divisor function).

The enciphering is described in figure 1. (This figure must be read from the top to the bottom). The functions $\mu_1, \ldots, \mu_d$ are the function projections from K^n to K^{n_e}, and μ is the concatenation function.

The important point is that the composition of all these operations is still a quadratic function in its components in a basis. So this function can be given by n polynomials over K, $(p_1, \ldots, p_n)$ (these polynomials give the ciphertext y from the cleartext x).

The public items are :

1. The field K of length 2^m, and the length n of the messages.

2. The n polynomials $(p_1, \ldots, p_n)$ in n variables over K.

So anyone can encipher a message.

Moreover to decipher is easy if the secret items are known : all the operations given in figure 1 will then be easily inverted. For example the quadratic functions f_e will be inverted by exponentiation $x \mapsto x^{h_e}$.

Note For various reasons (we explain these reasons in the extended version of this paper) Matsumoto and Imai limit themselves to the choice of integers θ_e and n_e such that there are some integers ℓ_e, r_e and b_e such that $n_e = (2\ell_e + 1).2^{r_e}$ and $\theta_e = b_e.2^{r_e}$, where $1 \le b_e \le \ell_e$.

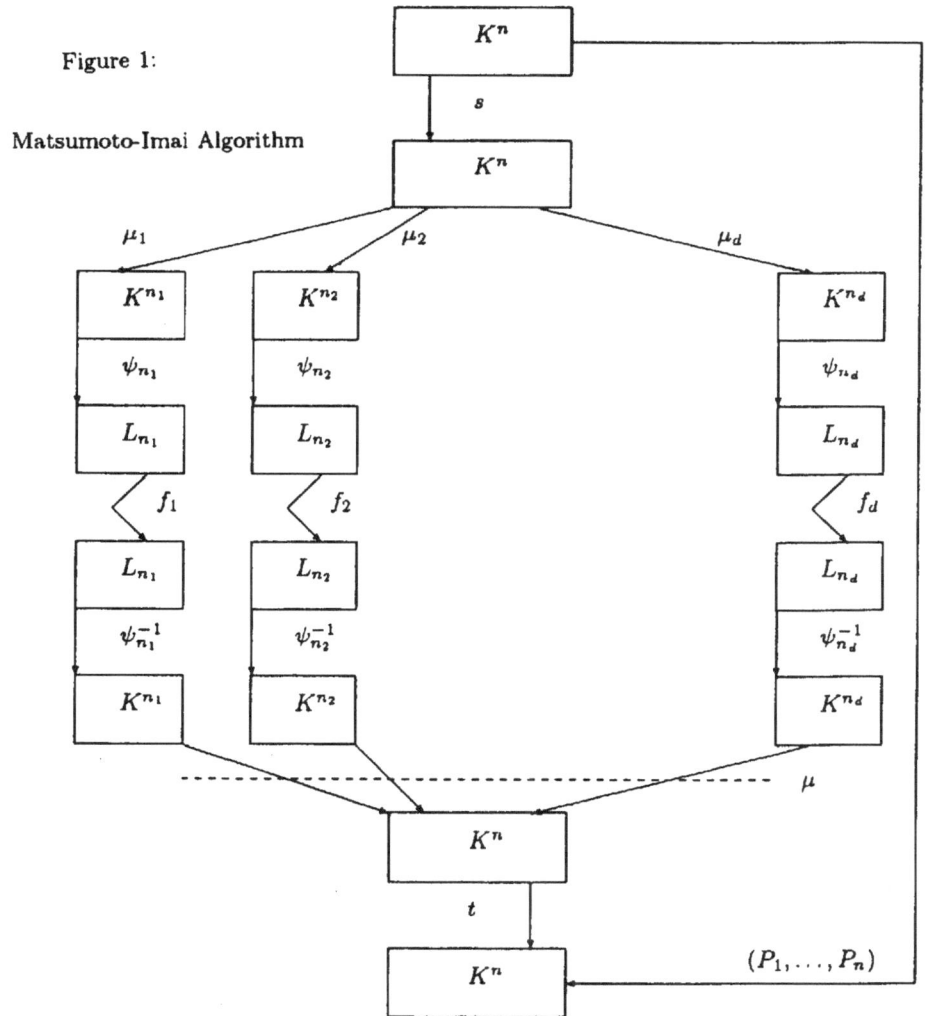

Figure 1:

Matsumoto-Imai Algorithm

3   Complexity

The complexity of the public transformation is in $O(m^2 n^3)$ and the complexity of the secret transformation is in $O((mn)^2(m + \log n))$, (cf.[1]).

Remark   The m^2 coefficient in this expression comes from the fact that a multiplication of two elements of K requires $O(m^2)$ basic computations, and an addition of two elements of K requires $O(m)$ basic computations. However when m is not too big ($m \leq 8$ for example) we can store the table of the multiplication of two elements of K (and also the addition table if we want, but it is easy to compute addition in K without the table) and so computing the Matsumoto-Imai Algorithm will be about m^2 times quicker.

4   Our notations

Thoughout this paper (as in [1]), m is the degree of the field K chosen, n is the number of components of K that we have in each message, and d is the number of integers in the secret split of $n : n = n_1 + \ldots + n_d$.

Let e be an index, $1 \leq e \leq d$, and let x be the plaintext and y the ciphertext. In this paper we will also denote by L_{n_e} the extension field of degree n_e over K, by a_e the element of L_{n_e} affine in x, by b_e the element of L_{n_e} affine in y, and by θ_e the secret parameter such that $b_e = a_e^{1+2^{m\theta_e}}$. (So, with the notations of Figure 1 that we will not use any more, we have $a_e = \psi_{n_e}(\mu_e(s(x)))$). Moreover in order to simplify the notations throughout this paper we will denote most of the time θ_e by θ, a_e by a, and b_e by b.

Figure 1 becomes Figure 2 with these notations :

x

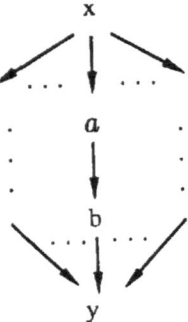

y

Figure 2: Our notations for the Matsumoto-Imai Algorithm. $b = a^{1+2^{m\theta_e}}$ is quadratic in a, a is affine in x and b is affine in y.

5   A family of weak keys

In this paragraph we will show that there are some weak keys in the Matsumoto-Imai algorithm. It does not seem that Matsumoto and Imai where aware of these weak choices. However it is very easy to avoid these weak keys, so it's not a very serious problem for the Matsumoto-Imai Algorithm. (In the following paragraphs we will see much more serious problems).

Let us again use the notations of paragraph 4, with x the plaintext, y the ciphertext, a affine in x, b affine in y, and $b = a^{1+2^{m\theta_e}}$ (1) (in a field L_{n_e} of degree n_e over K).

The equation (1) can also be written : $a = b^{\hbar_e}$, where $\hbar_e$ is the inverse multiplicative of $(1 + 2^{m\theta_e})$ modulo $2^{mn_e} - 1$.
If α is an integer, let us denote by $HW(\alpha)$ the number of 1 in the expression of α in base 2. (HW stands for "Hamming Weight in base 2").
Let x_i be the bits of x and y_i be the bits of y, $1 \leq i \leq mn$.
Each value $y_j, 1 \leq j \leq n$, has a quadratic expression in the values $x_i, 1 \leq i \leq n$. Similary each value x_j has an expression as a polynomial of degree $\sup_{e,1\leq e\leq d} HW(\hbar_e)$ in the values y_i. So, in order to make this expression of the $x_j, 1 \leq j \leq n$, as polynomials in the $y_i, 1 \leq i \leq n$, impractible, it is necessary for at least one variable e, that $HW(\hbar_e)$ be not too small (for example $HW(\hbar_e) \geq 6$ for at least one variable e). This fact was very clear from [1], but we will see now that in fact it is much better if for all variables e $HW(\hbar_e)$ is not too small (for example $HW(\hbar_e) \geq 6$ for all variables e).

Let us assume that this is not the case, i.e. that for one variable $e, 1 \leq e \leq d$, $HW(\hbar_e)$ is very small and : $a = b^{\hbar e}$ (2).
Let a_i be the bits of a and b_i be bits of b in a basis, $1 \leq i \leq n_e m$.
Since a is affine in x, there are some values $\alpha_{1i}, 0 \leq i \leq nm$, such that
$$a_1 = \alpha_{10} + \sum_{i=1}^{nm} \alpha_{1i} x_i.$$
From (2) we know that a_1 has a polynomial expression of total degree $HW(\hbar_e)$ in $b_1, \ldots, b_{n_e m}$. Since all these values $b_1, \ldots, b_{n_e m}$ are affine in $y_1, \ldots, y_{nm}$, a_1 has a polynomial expression of total degree $HW(\hbar_e)$ in $y_1, \ldots, y_{nm}$.
So there is a polynomial P of total degree $HW(\hbar_e)$ such that :

$$\alpha_{10} + \sum_{i=1}^{nm} \alpha_{1i} x_i = P(y_1, \ldots, y_{nm}) \quad (3).$$

And it's the same for $a_2, a_3, \ldots, a_{n_e m}$. So there is at least $n_e m$ equations similar to (3), of degree 1 in the x_i, and of total degree $HW(\hbar_e)$ in the y_i. So if for a particular e, $HW(\hbar_e)$ is very small, (say $HW(\hbar_e) \leq 4$ for example) we will be able to find these $n_e m$ equations similar to (3) (and this even if there is another index f such that $HW(\hbar_f)$ is very big). In order to find these equations, we will simply write the most general form of such an equation of degree $HW(\hbar_e)$, and by generating some values for x and y from the public form, we will obtain some equations of degree 1 in the coefficients of the polynomials.
After collecting sufficiently such equations by Gaussian reduction on these equa-

tions we will find the vectorial space of solution for the coefficients of the polynomials. So we will find at least $n_e m$ independent equations similar to (3).

Now from these equations, when y is given, we have immediatly $n_e m$ equations of degree one on the bits x_i of the cleartext. This may be a very dangerous thing in some applications, and at least it allows the cryptanalyst to do the exhaustive search in $2^{(n-n_e)m}$ instead of 2^{nm} on the cleartext.

Conclusion : All the values n_e and θ_e must be chosen in order that $HW(\hbar_e)$ is not too small (say ≥ 6 for example).

6   Our first general attack on all the keys

6.1   An example

Let us use again the notations of paragraph 4, with x the plaintext, y the ciphertext, a affine in x, b affine in y, and $b = a^{1+2^{m\theta}}$. (We are looking at what happening in $L_{n_e}, 1 \leq e \leq d$).
In this example, we will assume that $m = 1$ and $\theta = 1$ (the general case will be considered in paragraph 6.2). So

$$b = a^3. \quad (1).$$

Let $(b_1, \ldots, b_{n_e})$ be the representation of b in L_{n_e}, and let $(a_1, \ldots, a_{n_e})$ be the representation of a in L_{n_e}.
From (1) we see that all the $b_j, 1 \leq j \leq n_e$, have a quadratic expression in $(a_1, \ldots, a_{n_e})$, because $b = a.a^2$, and a^2 is linear (because here $m = 1$).
However we would like to find a useful expression which gives the a_j values from the b_j values (instead of the b_j values from a_j values).
The first idea is of course to write

$$a = b^\hbar \quad (2),$$

but in most cases $HW(\hbar)$ is big so (2) gives an intractable expression for the a_j values. So, what should be done ?
Let us start again from (1) and multiply both terms of (1) by a.
We obtain :

$$b.a = a^4. \quad (3).$$

The equation (3) gives n_e equations of degree one on the b_j values **and** of degree one on the a_j values !! (because a^4 is linear in a, because $m = 1$ here).

Moreover for any $b \neq 0$, there are exactly two solutions of (3) : the solution $a = 0$, and the solution $a = b^\hbar$. So in a way with (3) we will loose one equation over $GF(2)$, but the important point is that this equation (3) will be useful even if the equation (2) is intractable.

a is affine in x, b is affine in y, and a^4 is linear in a.

So from equation (3), we know that there are some equations of the form :

$$\sum_{i=1}^{n}\sum_{j=1}^{n}\gamma_{ij}x_iy_j + \sum_{i=1}^{n}\alpha_ix_i + \sum_{i=1}^{n}\beta_iy_i + \delta_0 = 0. \quad (4)$$

These equations are true for all the x, y, when x is the plaintext of y. Moreover if $b = 0$, there is only one solution for a of (3). So we must necessary have at least n_e "formally" independent equations like (4). ("Formally" means here that the vectorial space of the solutions for the values $\gamma_{ij}, \alpha_i, \beta_i$ and δ_0 is at least n_e). However, for a given value y we can just say that we will have at least $n_e - 1$ independent equations (4) (and not n_e) because (3) has two solutions in a when $b \neq 0$.

By choosing some values for x and computing the value y from x and the public form, and then by replacing these values x_i and $y_i, 1 \leq i \leq n$ in (4), we will obtain some equations of degree one in the $n^2 + n + n + 1 = (n+1)^2$ variables $\gamma_{ij}, \alpha_i, \beta_i$ and δ_0.

In this way we are able to find quickly all the equations (4) (by Gaussian reductions). Then, from a given y for which we want to find x, these equations give us some equations (at least $n_e - 1$ independent equations) of degree one on the values $x_1, \ldots, x_n$. So by Gaussian reduction, we are able to find $n_e - 1$ variables $x_1, \ldots, x_m$ from the others.

Let us now see the general case.

6.2   The general case

Let us again use the notations of paragraph 4. We have

$$b = a^{1+2^{m\theta}}. \quad (5).$$

By composition on each side of this equation with $g : x \mapsto x^{2^{m\theta}-1}$, we obtain :

$$b^{2^{m\theta}-1} = a^{2^{2m\theta}-1}.$$

Now let us multiply each side by $a.b$. We obtain :

$$a.b^{2^{m\theta}} = b.a^{2^{2m\theta}}. \quad (6).$$

Let $(a_1, \ldots, a_{n_e})$ be the representation of a in L_{n_e}, and $(b_1, \ldots, b_{n_e})$ be the representation of b in L_{n_e}. (So all the a_i and b_i, $1 \leq i \leq n_e$, are elements of K). This equation (6) gives n_e equations (not necessarily independent) of degree one on the b_j values **and** of degree one on the a_j values ! (because $b \mapsto b^{2^{m\theta}}$ is linear, and $a \mapsto a^{2^{2m\theta}}$ is linear, i.e. in a basis the n_e components of $b^{2^{m\theta}}$ can be written as a polynomial of degree one in the components of b, with coefficients of the polynomial in $K = GF(2^m)$).

Moreover a is affine in x, and b is affine in y. So these n_e equations ((6) in a basis) when we write them in the components $(x_1, \ldots, x_n)$ and $(y_1, \ldots, y_n)$ of x and y give n_e equations of the form :

$$\sum_{i=1}^{n}\sum_{j=1}^{n}\gamma_{ij}x_iy_j + \sum_{i=1}^{n}\alpha_i x_i + \sum_{i=1}^{n}\beta_i y_i + \delta_0 = 0. \quad (7).$$

These equations are true for all x, y when x is the plaintext of y. So by choosing some values for x and computing the value y from x and the public form, and then by replacing these values x_i and y_i, $1 \leq i \leq n$, in (7), we will obtain some equations of degree one in the $(n+1)^2$ variables of $GF(2^m)$: $\gamma_{ij}, \alpha_i, \beta_i$ and δ_0.

By this way we are able to find quickly, by Gaussian reductions, all the equations (7). This is the Part 1 of our Attack. Maybe we will find some equations (7) which do not come from the equation (6) (because we found all the equations which have the general form of (7)), but the important point is that at least we will found all the equations (7) which come from (6).

Part 2 of our attack : then, from a given y for which we want to find x, these equations will give us some equations of degree one on the values $x_1, \ldots, x_n$. By Gaussian reduction these equations will allow us to find λ variables $x_1, \ldots, x_n$ from the others, where λ is the number of independent equations (7) in $x_1, \ldots, x_n$ when $y_1, \ldots, y_n$ are replaced by the value we want. So, in order to evaluate the power of this attack, we have to evaluate λ. This is what we will do now.

Remark If $m = 1$ and $\theta_e = 1$ (and whatever the value of n_e) when we will find all the equations (7) we will find all the equations which come from $b^2.a = b.a^4$ and all the equations which come from $b.a = a^4$, and these equations are not formally the same (because if $b = 0$ the first ones vanish, but not the second ones). For example, when $m = 1$, $\theta = 1$, and $n_e = 5$ we have explicitely found all the equations (7). In this case we have found a vectorial space of solutions for the coefficients $\gamma_{ij}, \alpha_i, \beta_i$ and δ_0 of dimension exactly 10. In that case $b^2.a = b.a^4$ and $b.a = a^4$ gives 10 equations. When we choose for y a given value, these 10 equations will of course give us at most 5 independent equations, and if for this y we have $b \neq 0$, when they will give us exactly 4 independent equations (because we have exactly two solutions for a).

6.3   Evaluation of λ

Theorem 6.3   *For all the practical keys and for most of the ciphertexts y, the number λ of independent equations of degree one in $x_1, \ldots, x_n$ that we will obtain from the equations (7) for this given y is $\lambda \geq \sum_{e=1}^{d}(n_e - GCD(\theta_e, n_e)) \geq \dfrac{2n}{3}$. Moreover this shows that for a lot of secret keys for most of ciphertext we will have $\lambda \geq n - d$.*

For the proof of this Theorem, we will need three lemmas.

Lemma 1 *Let L be a finite field with q elements. Let p be an integer, and let y be an element of L. Then the equation $x^p = y$ has at most $GCD(p, q-1)$ solutions x.*

Proof of Lemma 1

If $y = 0$, then $x = 0$ is the only solution, and a lemma 1 is true.
So we can assume that $y \neq 0$. Then $x = 0$ is not solution, so we can assume also that $x \neq 0$, so $x^{q-1} = 1$.
Let $\mu = GCD(p, q-1)$.
From Bezout Theorem we know that there are two integers α and β such that $\alpha p - \beta(q-1) = \mu$. So

$$\begin{aligned} x^p = y &\Rightarrow x^{\alpha p} = y^\alpha \\ &\Rightarrow x^\mu . (x^{q-1})^\beta = y^\alpha \\ &\Rightarrow x^\mu = y^\alpha . \end{aligned}$$

In a field (commutative by definition) each equation of degree k has at most k solutions. So there is at most μ solutions of $x^\mu = y^\alpha$, and so there is at most μ solutions of $x^p = y$, as claimed.

Lemma 2 *For all integers m, α and β we have :*

$$GCD(2^{m\alpha} - 1, 2^{m\beta} - 1) = 2^{mGCD(\alpha, \beta)} - 1.$$

Proof of lemma 2

- Clearly $GCD(2^{m\alpha} - 1, 2^{m\beta} - 1) \geq 2^{mGCD(\alpha, \beta)} - 1$, because in $2^{m\alpha} - 1$ and $2^{m\beta} - 1$ we can put $2^{mGCD(\alpha, \beta)} - 1$ in factor (use the formula $x^k - 1 = (x-1)(x^{k-1} + x^{k-2} + \ldots + x + 1)$).

- Clearly also, we can assume that $\alpha > \beta$ by the symmetry of the hypothesis in lemma 2 and since when $\alpha = \beta$ lemma 2 is obvious.

- Now if x and y are two integers, and if μ is an integer such that $x - y.2^\mu > 0$, we have :
$$GCD(x, y) = GCD(y, x - y.2^\mu).$$

So with $x = 2^{m\alpha} - 1, y = 2^{m\beta} - 1$ and $2^\mu = 2^{m(\alpha - \beta)}$ we have :

$$GCD(2^{m\alpha} - 1, 2^{m\beta} - 1) = GCD(2^{m\beta} - 1, 2^{m(\alpha - \beta)} - 1). \quad (8).$$

By iterating this technique, $GCD(\alpha, \beta)$ will appear in a way similar to the computation of $GCD(\alpha, \beta)$ by the euclidian algorithm so we will obtain: $GCD(2^{m\alpha} - 1, 2^{m\beta} - 1) \leq 2^{mGCD(\alpha, \beta)} - 1$.

Lemma 3 *In L_{n_e} (the field with 2^{mn_e} elements) the equation (6) that we have written before :*
$a.b^{2^{m\theta}} = b.a^{2^{2m\theta}}$ *(6), has at most $2^{mGCD(\theta, n_e)}$ solutions in a, for each given $b \neq 0$.*

Proof of lemma 3

When $b \neq 0$ this equation (6) has two families of solution a :
1) $a = 0$.
2) a such that : $(a^{2^{m\theta}}-1)^{2^{m\theta}+1} = b^{2^{m\theta}-1}$.   (9).

We know that the function $g : z \mapsto z^{2^{m\theta}+1}$ is a bijection in L_{n_e} (because by construction of the Matsumoto-Imai Algorithm θ and n_e are chosen in order to have this property).
So a is solution of (9) if and only if :

$$a^{2^{m\theta}-1} = g^{-1}(b^{2^{m\theta}-1}).   (10).$$

Now from lemma 1 we know that this equation (10), for a given b has at most $GCD(2^{m\theta}-1, 2^{mn_e}-1)$ solutions a.
So with lemma 2 we obtain that (9) has at most $2^{mGCD(\theta,n_e)} - 1$ solutions in a.
So, by adding the solution $a = 0$, we obtain that : when $b \neq 0$, (6) has at most $2^{mGCD(\theta,n_e)}$ solutions in a, as claimed.

Corollary of lemma 3

For a given $b \neq 0$, if we write the equation (6) in a basis on the components of a (i.e. with a representation of L_{n_e} as an extension of degree n_e of $GF(2^m)$), then we will obtain at least $n_e - GCD(\theta, n_e)$ independent equations of degree one in the components of a.

Proof of the corollary

We have seen that the equations that we obtain are of degree one in the components of a. Moreover these equations have at least one solution : $a = 0$, so there is no contradiction in these equations. If λ_e is the number of independent equations, we will have exactly $2^{m(n_e-\lambda_e)}$ solutions. However from Lemma 3 we know that we have at most $2^{mGCD(\theta,n_e)}$ solutions. So $\lambda_e \geq n_e - GCD(\theta, n_e)$, as claimed.

Proof of theorem 6.3

(Here exceptionally we will use the notations a_e, b_e, and θ_e for a, b and θ because we will need the d values e, $1 \leq e \leq d$).
Let y be a ciphertext such that for this y we have :
$\forall e, 1 \leq e \leq d, b_e \neq 0$. (So $a_e \neq 0$ as well, since $b_e = a_e^{1+2^{m\theta_e}}$).

Note. For a given e, the probability that $b_e = 0$ is $1/2^{mn_e}$. So if mn_e is very small this probability may not be negligible. However if mn_e is very small (for example if $m = 1$ and $n_e = 3$) then $a_e = b_e^{\hbar_e}$ with a very small $\hbar_e$, so with a very small $HW(\hbar_e)$, and we have seen in paragraph 5 that this gives very weak keys. (We have seen that in this case it is easy to "eliminate" the variables of the branch number e).

So we can assume that mn_e is not so small, and so that "most" of the cipher-texts y will be such that $\forall e, 1 \le e \le d, b_e \ne 0$. (For very big d this may not be the case, but if n is of a reasonable size, then d cannot be too big and for most ciphertext y, $\forall e, 1 \le e \le d, b_e \ne 0$ as claimed).

We have seen in paragraph 6.2 that in the equations (7) we will find at least all the equations which come from all the equations (6) when we write these equations with the components of x and y instead of a and b, and this for all the values of $e, 1 \le e \le d$.

So from the Corollary of Lemma 3, we know that we will obtain at least $\sum_{e=1}^{d}(n_e - GCD(\theta_e, n_e))$ independent equations of degree one in the components of x (for each given y such that $\forall e, b_e \ne 0$).
So we have proved the first part of Theorem 6.3.

We will now need another lemma.

Lemma 4 $\forall e, 1 \le e \le d$, let $\delta_e = GCD(\theta_e, n_e)$, and let $k_e = n_e/\delta_e$ (k_e is an integer because δ_e divides n_e). Then k_e is always odd, and $k_e \ge 3$.

Proof of lemma 4

We know from the values chosen in [1] that we can write $n_e = \alpha.2^r$ and $\theta_e = \beta.2^s$ with α and β odd and $s \ge r$. (This is also explain in the extended version of this paper).
So $\delta_e = 2^r GCD(\alpha, \beta)$, and $k_e = \alpha/GCD(\alpha, \beta)$, with α and β odd.
So k_e is odd. Moreover $\theta_e < n_e$, so $\delta_e < n_e$, so $k_e > 1$, so $k_e \ge 3$.

We can now finish the proof of Theorem 6.3.
We have proved that $\lambda \ge \sum_{e=1}^{d}(n_e - GCD(\theta_e, n_e))$.
So from Lemma 4, $\lambda \ge \sum_{e=1}^{d}(n_e - \frac{n_e}{3}) = \frac{2}{3}\sum_{e=1}^{d}n_e = \frac{2n}{3}$.
Moreover for a lot of secret keys $GCD(\theta_e, n_e) = 1$, and if that occurs we have $\sum_{e=1}^{d}(n_e - 1) = n - d$, so $\lambda \ge n - d$.

6.4 Improved Gaussian elimination

Our attack, as we have described it, proceeds in two parts :

Part 1. We find all the equations (7), and this has to be done once and for all.

Part 2. For a specific ciphertext y, we try to find x with the help of equations (7). So this has to be done each time we have another x to find from another y.

In Part 1 we have to do a Gaussian reduction on $(n+1)^2$ variables and about $(n+1)^2$ equations of $GF(2^m)$. So in the most general case our Part 1 will have a complexity of at most $m^2.n^6$.

In most practical cases Part 1 will be dominant in time from Part 2. So it is worth improving Part 1. This is what we do in the extended version of this paper where we show that there is an Algorithm in $O(m^2n^4 \log n)$ for Part 1 instead $O(m^2n^6)$. (The idea is to choose some values of x in order to have a more easy Gaussian reduction).

7 Our second general attack

7.1 Description of the second attack

In paragraph 6 our attack was based on the idea that if $b = a^{1+2^{m\theta}}$, then $a.b^{2^{m\theta}} = b.a^{2^{2m\theta}}$ (6).

In equation (6) a and b are on both sides.

In this paragraph 7 we will now try to find some equations of the general form :

$$a^2.b^u = b^v$$

with $HW(u)$small and $HW(v)$ small. (Here a is only on one side of the equation).

For this purpose we will not start from $b = a^{1+2^{m\theta}}$, but we will start from $a = b^{\hbar_e}$. So we will have to evaluate the value of $\hbar_e$.

Theorem 7.1 *Let $\delta = mGCD(\theta_e, n_e)$, and let α and k be integers such that $\alpha\delta = m\theta_e$, and $k\delta = mn_e$. Let $\hbar_e$ be, as usual, the multiplicative inverse of $1 + 2^{m\theta_e}$ modulo $2^{mn_e} - 1$.*

Then : 1. k is odd and $k \geq 3$.

and : 2. $\hbar_e = 2^{k\delta-1} + \sum_{i=1}^{k-1}(-1)^i.2^{\alpha\delta i-1}$.

Proof

$k = n_e/GCD(\theta_e, n_e)$ and from Lemma 4 of paragraph 6 we know that this value is odd and ≥ 3. So k is odd and $k \geq 3$.

Now $1 - 2^{\alpha\delta} + 2^{2\alpha\delta} + \ldots + (-2^{\alpha\delta})^{(k-1)} = (1 - (-2^{\alpha\delta})^k)/(1 + 2^{\alpha\delta})$.

(Because this is a well know sum of a geometric series).

So we have :

$$(1 + 2^{\alpha\delta})\left(1 + \sum_{i=1}^{k-1}(-1)^i.2^{\alpha\delta i}\right) = 1 - (-1)^k(2^{k\delta})^{\alpha}.$$

So since k is odd, we have :

$$(1 + 2^{\alpha\delta}) \left(2^{k\delta} + \sum_{i=1}^{k-1} (-1)^i . 2^{\alpha\delta i} \right) = 2 \quad \text{modulo} \quad (2^{k\delta} - 1).$$

So $2^{k\delta - 1} + \sum_{i=1}^{k-1} (-1)^i . 2^{\alpha\delta i - 1}$ is the multiplicative inverse of $1 + 2^{m\theta_e}$ modulo $2^{mn_e} - 1$, as claimed.

From $a^2 = b^{2^{\hbar_e}}$ and Theorem 7.1 we have :

$$a^2 . b^u = b^v, \quad \text{with } u = \sum_{i=1}^{(k-1)/2} 2^{m\theta_e(2i-1)} \text{ and } v = 1 + \sum_{i=1}^{(k-1)/2} 2^{2m\theta_e i}.$$

So $HW_m(u) = (k-1)/2$ and $HW_m(v) = (k+1)/2$, where $HW_m(z)$ denotes the minimum number of terms when we write z as powers of 2^m.
So the equation $a^2 . b^u = b^v$, when we write it as n_e equations in the components x_i^2 and y_i gives n_e equations of degree one in the x_i^2, and of total degree $(k+1)/2$ in the y_i.

Remark In paragraph 6 the most difficult keys were the keys with very small k. Strangely theses keys are the easiest keys in this paragraph 7.

For example when $k = 3$ we will obtain n_e equations of the general form :

$$\sum_{i=1}^{n} \sum_{j=1}^{n} \gamma_{ij} x_i^2 y_j + \sum_{i=1}^{n} \sum_{j=i+1}^{n} \eta_{ij} y_i y_j + \sum_{i=1}^{n} \alpha_i x_i + \sum_{i=1}^{n} \beta_i y_i + \delta_0 = 0 \quad (9).$$

Our attack will then still be in two Parts.

Part 1. We will find all the equations (7) of paragraph 6 and also all the equations (9) (as usual by Gaussian reduction after computing some couples (x, y) from the public form).

Part 2. Then for a given y we will put to the power 2^{m-1} the equations (9) found. Since in K we have $(\alpha + \beta)^{2^{m-1}} = \alpha^{2^{m-1}} + \beta^{2^{m-1}}$ and since $x_i^{2^m} = x_i$, the equations (9) will give us like this equations of degree one in the x_i. Therefore we will use both equations (7) and (9).

Note 1. Here, when $k = 3$, the Gaussian reduction for Part 1 is in $m^2 n^6$ (instead of $m^2 n^4 \log n$), so Part 1 takes more time with this attack. However Part 2 will be quicker because for all $b \neq 0$, there is only one solution in a to $a^2 . b^u = b^v$. Since Part 1 has to be done only once we see that $k = 3$ is not a good choice.

Note 2. Another solution would have been to write all the equations (9) on the bits. However the complexity for Part 1 when $k = 3$ would then have been in $O(m^6 n^6)$ instead of $O(m^2 n^6)$.

7.2 An example of attack

In [1] p. 435 we know that an implementation with $m = 8$ and $n = 32$ has been done. We know nothing about the secret keys chosen. Whatever the value of the secret key we will be able to find a cleartext from a ciphertext of this implementation after at most 2^{32} computations, and much less most of the time, with our attacks of paragraph 6 and 7. (More details are given about this example in the extended version of this paper).

8 Further Improvements

If for some index e, mn_e is not too big (for example if $mn_e \le 32$), then we have found that it is possible to further improve our attacks. We will do more computations to be done once and for all but less to be done for each message. We have no room here to explain these improvements in details but the idea is to use the fact that for some values of the x_i the number of independent equations (7) in the y_i decreases suddenly. This comes from the fact that when $a = 0$ our equations (6) of paragraph 6 becomes $0 = 0$. So these values of the x_i give useful information that allow us when mn_e is not too big to "separate" the variables. (More details are given about this in the extended version of this paper).

9 Conclusion

In this paper we have seen that the scheme of [1] is insecure. Whatever the value of the keys we are able to find easily most of the cleartexts from their ciphertexts, at least for texts of reasonable size. Some different schemes based on similar ideas may be much more difficult to break. However the choice of the secret keys and the choice of these variations will have to be done with great care.

Acknowledgements

I want to thank Paul Camion, Daniel Augot and Sandrine Grellier for their descriptions of the Matsumoto-Imai Algorithm.

References

[1] T. Matsumoto and H. Imai, *Public Quadratic Polynomial-tuples for efficient signature-verification and message-encryption*, EUROCRYP'88, Springer-Verlag 1988, pp. 419-453.

Cryptanalysis Based on 2-Adic Rational Approximation

Andrew Klapper

Dept. of Computer Science
University of Kentucky
Lexington, KY 40506-0046 USA
klapper@cs.engr.uky.edu

Mark Goresky

Dept. of Mathematics
Northeastern University
Boston, MA 03115 USA
goresky@nuhub.neu.edu

Abstract. This paper presents a new algorithm for cryptanalytically attacking stream ciphers. There is an associated measure of security, the *2-adic span*. In order for a stream cipher to be secure, its 2-adic span must be large. This attack exposes a weakness of Rueppel and Massey's summation combiner. The algorithm, based on De Weger and Mahler's rational approximation theory for 2-adic numbers, synthesizes a shortest *feedback with carry shift register* that outputs a particular key stream, given a small number of bits of the key stream. It is adaptive in that it does not neeed to know the number of available bits beforehand.

Index Terms – **Binary sequences, feedback with carry shift registers, cryptanalysis, rational approximation, 2-adic numbers.**

1 Introduction

In this paper a new general purpose attack on stream ciphers is presented. This attack can be used successfully, for example, against Rueppel and Massey's summation combiner [16, 19]. All future stream cipher designers will need to consider resistance to this attack.

The development of cryptosystems tends to alternate between the design of new systems that resist known attacks, and the design of new attacks against systems. Often such attacks are highly specialized and only work against particular systems. Occasionally a very general attack is found that can potentially be used against a large class of cryptosystems. Examples include differential cryptanalysis, which can be used against many round based block ciphers such as DES, [1], and the Berlekamp-Massey algorithm [15], which can be used against stream ciphers. Sometimes we can numerically measure the extent to which cryptosystems resist a particular attack. This is the case with the Berlekamp-Massey algorithm: the *linear span* of a binary sequence S is the size of the smallest linear feedback shift register (or LFSR) that generates S. The higher the linear span of a sequence, the greater the resistance to the Berlekamp-Massey algorithm when S is used as the key in a stream cipher. Thus when a cryptographer designs a new stream cipher, it is imperative that she show that the key stream has large linear span.

D. Coppersmith (Ed.): Advances in Cryptology - CRYPTO '95, LNCS 963, pp. 262-273, 1995.

As with the Berlekamp-Massey algorithm, the method of cryptanalysis described in this paper has an associated measure of security, the *2-adic span*. There is a precise way in which the Berlekamp-Massey algorithm can be thought of as a rational approximation algorithm in the ring of power series with integer coefficients. Similarly, the new attack can be thought of as a rational approximation algorithm in the ring of 2-adic integers. The Berlekamp-Massey algorithm is closely related to continued fractions [3, 4, 13, 17, 22]. Over the 2-adic numbers, however, continued fractions fail to converge in general. The algorithm presented here is based on Mahler and De Weger's lattice theoretic substitute for 2-adic continued fractions [12, 21]. Unlike Mahler and De Weger's approach, our algorithm is adaptive. The number of known bits of key stream does not need to be determined before the algorithm can be executed. Another nonadaptive 2-adic rational approximation algorithm was discovered by Gregory and Krishnamurthy [6].

Recall that a stream cipher is a private key system in which the message, a binary sequence, is encrypted by adding it bit by bit modulo two to a key stream, which is another binary sequence. The key stream is known to the legitimate receiver of the message who can thus recover the message by adding the key to the ciphertext. Individual stream ciphers are defined by the manner in which the key is generated. A cryptosystem (or the associated key) is secure if the message cannot be determined by an eavesdropper who does not know the key. A cryptosystem is generally only considered secure if it is secure against a known plaintext attack. In the case of stream ciphers, this means that an eavesdropper should be unable to determine the key even if part of the key is known.

For an attack to be practical, it must also be possible for the eavesdropper to generate the key at a rate commensurate with the rate at which it is generated by the sender of the message. The power of the Berlekamp-Massey algorithm is that it produces a description of a fast device – a short LFSR – that generates the key, if such a device exists. Recently, a new class of feedback registers, *feedback with carry shift registers* or FCSRs was developed [7, 9]. These registers are designed to quickly output the 2-adic representation of a rational number. Moreover, they have associated with them a number of algebraic structures that are analogs of the algebraic structures used to analyze linear feedback shift registers.

When FCSRs were originally described by Klapper and Goresky, no adaptive, provably convergent algorithm was known for synthesizing FCSRs from short prefixes of eventually periodic sequences. The algorithm described in this paper fills this gap. It proceeds in two stages. The eavesdropper is assumed to have access to the first few bits of the key stream $a_0, a_1, \cdots$. A rational approximation algorithm (described in Section 3) is used to find the best rational approximation p/q to the 2-adic number $\alpha = \sum_{i=0}^{\infty} a_i 2^i$. This approximation will be best in the sense that, for all approximations that are accurate modulo a particular power of 2, the maximum of $|p|$ and $|q|$ is as small as possible. We show that if in fact α is rational, $\alpha = p'/q'$, then our rational approximation algorithm finds the reduced rational representation of α if $\lceil 2\log(\max(|p'|, |q'|)) \rceil + 2$ bits are known (see Theorem 6). This is approximately twice the size of the smallest

FCSR that generates the key (see Corollary 8). Thus the rate of convergence of our algorithm is analogous to that of the Berlekamp-Massey algorithm. The complexity of our algorithm is $\mathcal{O}(T^2 \log T \log \log T)$, where T is the number of known bits. This exceeds the complexity of the Berlekamp-Massey algorithm only by the log factors. Once the rational representation of α is found, there is a fast algorithm for finding an initial loading of a FCSR that outputs α. This is described in Section 4.

It follows that any key stream that can be generated by a small FCSR (or euivalently, whose associated 2-adic number is a rational number p/q, with $|p|$ and $|q|$ small) leads to an insecure stream cipher. This fact gives rise to a security measure.

Definition 1. The *2-adic span* of an eventually periodic sequence $\mathbf{a} = a_0, a_1, \cdots$ is the number of bits of storage used by the smallest FCSR that generates $\mathbf{a}$. It is denoted by $\lambda_2(\mathbf{a})$.

The idea of cryptanalysis using FCSRs is that if a key stream can be generated by a short FCSR, and we can determine this FCSR efficiently from a small subsequence of the key stream, then we can construct an efficient generator for the key stream. It may not be the same device that was originally used to generate the key stream, but this is immaterial to the cryptanalyst. Thus in order for a stream cipher to be secure, its key stream must have large 2-adic span. In the conclusions we describe the effect on the security of a well known key stream generator, the summation combiner [16, 19].

2 Review of Feedback with Carry Shift Registers

In this section we describe FCSRs and some of their basic properties. Details of the construction and results in this section may be found in [7, 9].

FCSRs are based on division in the ring of 2-adic integers. Recall that a 2-adic integer is a series $\sum_{i=0}^{\infty} a_i 2^i$, $a_i \in \{0, 1\}$, where we have replaced the indeterminate x by the integer 2. Addition and multiplication are defined term by term as for ordinary integers. The set of 2-adic integers forms a complete valued ring $\mathbf{Z}_2$. The ordinary rational numbers intersect $\mathbf{Z}_2$ in the set of rationals with odd denominator or, equivalently, in the set of 2-adic integers whose sequence of coefficients is eventually periodic. See [11] for background on 2-adic numbers.

Definition 2. ([7, 9]) The FCSR with connection integer $q = -1 + \sum_{i=1}^{r} q_i 2^i$, $q_i \in \{0, 1\}$, is a device with r bits of storage plus an auxiliary memory containing an integer. If the auxiliary memory is m, and the contents of the register consists of the r bits $(a_{r-1}, a_{r-2}, \ldots, a_1, a_0)$, then the operation of the shift register is defined as follows:

A1. Form the integer sum $\sigma = \sum_{k=1}^{r} q_k a_{r-k} + m$.

A2. Shift the contents one step to the right, outputting the rightmost bit a_0.

A3. Place $a_r = \sigma \bmod 2$ into the leftmost cell of the shift register.

A4. Replace the memory integer with $m = (\sigma - a_r)/2 = \lfloor \sigma/2 \rfloor$.

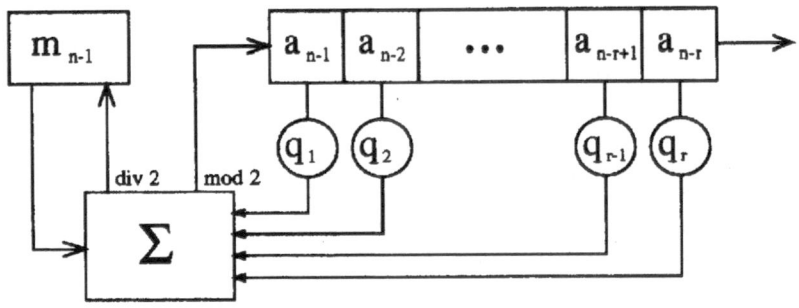

Fig. 1. Feedback with Carry Shift Register

The integer q is referred to as the connection integer because it is the analog of the connection polynomial associated with a LFSR. Notice that $q_0 = -1$ does not correspond to a feedback tap. A FCSR is depicted in Figure 1.

We have the following facts about FCSRs

Theorem 3.

1. *The output from a FCSR is eventually periodic. If q is the connection integer, then the 2-adic number associated with the output sequence is a rational number of the form p/q.*

2. *Conversely, if p is any integer, then the 2-adic expansion of the rational number p/q can be realized as the output of a FCSR with connection integer q. At times we identify p/q with its associated 2-adic number and its associated sequence of coefficients and simply refer to p/q as the output of the register.*

3. *The rational number p/q is strictly periodic if and only if $-q < p \leq 0$.*

4. *Consider a FCSR with connection integer $q = \sum_{i=0}^{r} q_i 2^i$ and initial memory m. Suppose $0 \leq m \leq |\{i : q_i = 1\}|$. Then this condition holds throughout the operation of the register. Also, this condition holds for strictly periodic p/q. Thus for these sequences the space required for the auxiliary memory is fewer than $\log r$ bits.*

5. *Adding b to the memory of a FCSR with connection integer q changes the 2-adic value of the output by $-b2^r/q$.*

3 The Rational Approximation Algorithm

In this section we present an algorithm for synthesizing FCSRs. The goal is to find the rational approximation p/q to a given 2-adic number α with $\Phi(p,q) \overset{def}{=} \max(|p|, |q|)$ minimal. For eventually periodic sequences, this quantity is related to the 2-adic span by Theorem 7. If α is eventually periodic, the algorithm finds the precise rational representation p/q of α if $\lceil 2 \log(\Phi(p,q)) \rceil + 2$ consecutive bits of a are known. It follows that the number of bits of a needed to find such a register is at most $2\lambda_2(\mathbf{a}) + 2 \lceil \log(\lambda_2(\mathbf{a})) \rceil + 4$.

Assume we have consecutive terms $a_0, a_1, \ldots$ of a binary sequence a which is the 2-adic expansion of a number α. We wish to determine integers p and q such that $\alpha = p/q$. The algorithm proceeds by computing successive approximations g_1/g_2 to α such that at stage k

$$\alpha \cdot g_2 - g_1 \equiv 0 \ (\bmod \ 2^k). \tag{1}$$

The algorithm is given in Figure 2. It is based on Mahler and De Weger's lattice theoretic approach to p-adic rational approximation [12, 21], but has the advantage that the number of known bits of α need not be predetermined. Each new bit that is found with certainty leads to a better approximation. The symbols $f = (f_1, f_2)$ and $g = (g_1, g_2)$ denote pairs of integers. Note that the minimization steps can be performed by a pair of integer divisions. This is explained in more detail in the last paragraph of this section.

To understand the algorithm some notation is useful. For any $h = (h_1, h_2) \in \mathbf{Z} \times \mathbf{Z}$, let $h(\alpha) = \alpha \cdot h_2 - h_1$. Let $L_k = \{h : h(\alpha) \equiv 0 \bmod 2^k\} \subseteq \mathbf{Z} \times \mathbf{Z}$. Thus the goal of the algorithm is to find the element of L_k with h_2 odd and the minimum value of Φ over all elements of L_k with h_2 odd. The set L_k is a Z-lattice – a finitely generated module over Z. De Weger characterized bases for L_k as follows.

Lemma 4. The pair f, g is a basis for L_k if and only if $|f_1 g_2 - f_2 g_1| = 2^k$.

Correctness of the algorithm is proved by the following lemma.

Lemma 5. At each stage we have

1. f and g are in L_k;
2. $\langle f, g \rangle$ is a basis for L_k;
3. $f \in 2(\mathbf{Z} \times \mathbf{Z}) - L_{k+1}$;
4. g minimizes Φ over all elements of L_k with g_2 odd.
5. f minimizes Φ over all elements of L_k with f_1 and f_2 even.

Proof: We prove this by induction. It is straightforward to check that the conditions hold initially.

Let us suppose that the conditions hold at stage k. If $g(\alpha) \equiv 0 \bmod 2^{k+1}$, or equivalently, $g \in L_{k+1}$, then it is again straightforward to check the conditions. Therefore, assume $g \notin L_{k+1}$. We treat the case when $\Phi(g) < \Phi(f)$. The other case is similar.

Let f' and g' be the new values after updating.

Rational_Approximation()

begin
Input $a_0, a_1, \cdots, a_{k-1} = 0$, where $a_0 = \cdots = a_{k-2}$, and $a_{k-1} = 1$
$\alpha = a_0 + a_1 \cdot 2 + \cdots + a_{k-1} \cdot 2^{k-1}$
$f = (0, 2)$
$g = (2^{k-1}, 1)$
while there are more known bits **do**
    Input a_k
    $\alpha = \alpha + a_k 2^k$
    if $g(\alpha) \equiv 0 \bmod 2^{k+1}$ **then**
        $f = 2f$
    else if $\Phi(g) < \Phi(f)$ **then**
        Let d be odd and minimize $\Phi(f + dg)$
        $\langle g, f \rangle = \langle f + dg, 2g \rangle$
    else
        Let d be odd and minimize $\Phi(g + df)$
        $\langle g, f \rangle = \langle g + df, 2f \rangle$
    fi fi
    $k = k + 1$
od
return g
end

Fig. 2. The 2-Adic Rational Approximation Algorithm.

1. We have

$$
\begin{aligned}
g'(\alpha) &= \alpha \cdot g_2' - g_1' \\
&= \alpha \cdot (f_2 + dg_2) - (f_1 + dg_1) \\
&= f(\alpha) + dg(\alpha) \\
&\equiv 2^k + d2^k \bmod 2^{k+1} \\
&\equiv 0 \bmod 2^{k+1},
\end{aligned}
$$

since f and g are in $L_k - L_{k+1}$ and d is odd. It follows that $g' \in L_{k+1}$. Also, g is in L_k, so $f' = 2g$ is in L_{k+1}.

2. By De Weger's lemma, we have $|f_1 g_2 - f_2 g_1| = 2^k$. Therefore $|f_1' g_2' - f_2' g_1'| = |2g_1(f_2 + dg_2) - 2g_2(f_1 + dg_1)| = |2(f_1 g_2 - f_2 g_1)| = 2^{k+1}$. Again by De Weger's lemma, g', f' is a basis for L_{k+1}.

3. We have $g \in \mathbf{Z} \times \mathbf{Z} - L_{k+1}$, so $f' = 2g \in 2(\mathbf{Z} \times \mathbf{Z}) - L_{k+2}$.

4. Suppose that minimality fails. By the fact that f', g' form a basis for L_{k+1}, this means that there are integers a and b so that

$$
\Phi(ag' + bf') < \Phi(g') \tag{2}
$$

and $ag'_2 + bf'_2$ is odd. The latter condition is equivalent to a being odd since f'_2 is even and g'_2 is odd. By possibly negating both a and b, we can assume a is nonnegative. Further, if $a = 1$, then $ag' + bf' = f + (d + 2b)g$ and this contradicts the choice of d in the algorithm. Thus we can assume that $a > 1$. Equation (2) can be rewritten

$$\Phi(af + (ad + 2b)g) < \Phi(f + dg).$$

Let c be the odd integer closest to $d + 2b/a$. Then $|c - (d + 2b/a)| \le (a-1)/a$. It follows that

$$\Phi(f + cg) = \Phi\left(f + \left(d + \frac{2b}{a}\right)g + \left(c - \left(d + \frac{2b}{a}\right)\right)g\right)$$

$$\le \Phi\left(f + \left(d + \frac{2b}{a}\right)g\right) + \Phi\left(\left(c - \left(d + \frac{2b}{a}\right)\right)g\right)$$

$$\le \frac{1}{a}\Phi(af + (d + 2b)g) + \frac{a-1}{a}\Phi(g)$$

$$< \frac{1}{a}\Phi(f + dg) + \frac{a-1}{a}\Phi(g)$$

$$\le \frac{1}{a}\Phi(f + dg) + \frac{a-1}{a}\Phi(f + dg)$$

$$= \Phi(f + dg),$$

where we have used the triangle inequality for Φ, and inductively used the minimality condition on g. This contradicts the choice of d.

5. Suppose there is an element $h' \in L_{k+1}$ with h'_1 and h'_2 even, such that $\Phi(h') < \Phi(f') = 2\Phi(g)$. We can write $h' = 2h$ for some $h \in L_k$. If both h_1 and h_2 are even, then $\Phi(h) < \Phi(g) < \Phi(f)$ by the inequality in the algorithm leading to this case. This contradicts the minimality of f. If h_2 is odd, then $\Phi(h) < \Phi(g)$ contradicts the minimality of g. It is impossible that h_1 is odd and h_2 is even for $k \ge 1$ since $h(\alpha) \equiv 0 \bmod 2^k$. □

We observe that point (5) of the lemma is not strictly necessary for convergence of the algorithm. In fact, the algorithm runs correctly if we always update g and f by the first method of update, independent of the relation between $\Phi(f)$ and $\Phi(g)$. However, point (5) ensures that the size of f is small and leads to better bounds on the complexity.

Suppose the input sequence to the algorithm is, in fact, an eventually periodic sequence giving a rational 2-adic $\alpha = p/q$. We want to know how many iterations of the rational approximation algorithm are required to output (p, q).

Theorem 6. *Suppose* $\mathbf{a} = a_0, a_1, a_2, \cdots$ *is an eventually periodic sequence with associated 2-adic number* $\alpha = \sum a_i 2^i = p/q$, $p, q \in \mathbf{Z}$, *and* $\gcd(p, q) = 1$. *Then the 2-adic rational approximation algorithm outputs* (p, q) *if* $T \ge \lceil 2\log\Phi(p, q)\rceil + 2$ *bits are used.*

Proof: The output from the algorithm is a pair (g_1, g_2) satisfying $g_1 - \alpha g_2 \equiv 0 \bmod 2^T$. Hence $g_1 q \equiv p g_2 \bmod 2^T$. By the Φ-minimality, we have $\Phi(g_1, g_2) \leq \Phi(p, q)$. It follows that $|g_1 q| \leq \Phi(p, q)^2 \leq 2^{T-2}$ and $|p g_2| \leq \Phi(p, q)^2 \leq 2^{T-2}$. Therefore, $g_1 / g_2 = p/q$. Again by the Φ-minimality, we must have $g_1 = p$ and $g_2 = q$. $\square$

It is preferable to bound the number of bits needed by the 2-adic span (the size of the smallest FCSR that outputs **a**). The 2-adic span is related to Φ by the following.

Theorem 7. *If $\alpha = \sum_{i=0}^{\infty} a_i 2^i = p/q$ is the rational number corresponding to **a**, reduced to lowest terms, then the 2-adic span is bounded by*

$$\lceil \log \Phi(p, q) \rceil - \lceil \log \log \Phi(p, q) \rceil \leq \lambda_2(\mathbf{a}) \leq \lceil \log \Phi(p, q) \rceil + \lceil \log \log \Phi(p, q) \rceil .$$

It follows that

$$\lambda_2(\mathbf{a}) - \lceil \log(\lambda_2(\mathbf{a})) \rceil - 1 \leq \lceil \log \Phi(p, q) \rceil \leq \lambda_2(\mathbf{a}) + \lceil \log(\lambda_2(\mathbf{a})) \rceil + 1.$$

This gives the desired bound.

Corollary 8. *Suppose **a** is an eventually periodic sequence with associated 2-adic number $\alpha = p/q$, $\gcd(p, q) = 1$. Then the 2-adic rational approximation algorithm outputs (p, q) if $T \geq 2\lambda_2(\mathbf{a}) + 2 \lceil \log(\lambda_2(\mathbf{a})) \rceil + 4$ bits are used.*

In implementing the algorithm, the value of d can be found by division. For example, suppose we are in the case where $g(\alpha) \not\equiv 0 \bmod 2^{k+1}$ and $\Phi(g) < \Phi(f)$. If $g_1 \neq \pm g_2$, then d is an odd integer among the four odd integers immediately less than or greater than $(f_2 - f_1)/(g_1 - g_2)$ and $-(f_1 + f_2)/(g_1 + g_2)$. Thus it suffices to consider the value of $\Phi(f + dg)$ for these four values of d. When $g_1 = \pm g_2$, one or the other of these quotients is not considered, and in the second case of the algorithm, the roles of f and g are switched.

4 Initial Loading of a FCSR

Now let us show how to construct a FCSR which generates the bit sequence for a given rational number p/q. We assume q is a positive odd integer and let $r = \lfloor \log_2(q + 1) \rfloor$. Write $q = \sum_{i=0}^{r} q_i 2^i$ with $q_0 = -1$ and $q_i \in \{0, 1\}$ for $i > 0$. We want to determine the initial setting (including the extra memory) of the FCSR with connection integer q that outputs the 2-adic expansion of p/q. The number of nonzero taps in such a FCSR is $t = wt(q + 1)$, the Hamming weight of the binary expansion of $q + 1$. The initial setting is related to p and q by the following proposition.

Proposition 9. *Suppose a FCSR with connection integer $q = \sum_{i=0}^{r} q_i 2^i$ is set up with initial setting $a_0, \cdots, a_{r-1}$ and initial memory m. Then its output is the 2-adic expansion of the fraction p/q where*

$$p = \sum_{k=0}^{r-1} \sum_{i=0}^{k} q_i a_{k-i} 2^k - m 2^r. \tag{3}$$

Furthermore, if we let $x = \sum_{j=0}^{r-1} a_j 2^j$, then the double sum in equation (3) is the product qx with all products of terms whose indices sum to r or more omitted. Such a double sum can be computed essentially as quickly as the full product qx. It follows that, for a given fraction p/q, the initial loading can be derived by the following steps.

B1. Compute $a_0 + a_1 2 + \cdots + a_{r-1} 2^{r-1} = p/q \bmod 2^r$. In general computing modular quotients is apparently hard. However, when the modulus is a power of a prime they can be computed efficiently. It is straightforward to do so in time $\mathcal{O}(r^2)$.

B2. Compute $y = \sum_{k=0}^{r-1} \sum_{i=0}^{k} q_i a_{k-i} 2^k$, say by a modified multiplication algorithm.

B3. Compute $m = (y - p)/2^r$ in time $\mathcal{O}(r)$.

We can then use $a_0, \cdots, a_{r-1}$ as the initial loading and m as the initial memory in a FCSR with connection integer q. This FCSR will output the 2-adic expansion of p/q. Of course if p and q have been determined by the rational approximation algorithm, then we already have the initial loading $a_0, \cdots, a_{r-1}$ and need only determine the initial memory m from steps (**B2**) and (**B3**).

5 Complexity Issues

Suppose the rational approximation algorithm is executed with a sequence **a** which is eventually periodic, with rational associated 2-adic number $\alpha = p/q$. Let λ be the 2-adic span of **a**. Then the rational approximation algorithm takes $T = 2\lambda + 2 \lceil \log(\lambda) \rceil + 4$ steps.

Consider the kth step. If $g(\alpha) \not\equiv 0 \bmod 2^{k+1}$, then we say that a discrepancy has occurred. The complexity of the algorithm depends on the number of discrepancies. To simplify the computation of αg_2, we maintain αf_2 as well. When no discrepancy occurs, these values and the value of f can be updated with k bit operations.

Suppose a discrepancy occurs. The minimization step can be done with two divisions of k bit integers. The remaining steps take time $\mathcal{O}(k)$. Then αg_2 and αf_2 can be updated with $\mathcal{O}(k)$ bit operations and two multiplications of k bit integers by d.

Let D be the number of discrepancies, and let M be the maximum time taken by a multiplication or division of T bit integers. The Schönhage-Strassen algorithm [20], gives $M = \mathcal{O}(T \log T \log \log T)$. This can be improved to $M \sim r \log r$ using Pollard's nonasymptotic algorithm and Newton interpolation for $T < 2^{37}$ on a 32 bit machine or $T < 2^{70}$ on a 64 bit machine [18]. These are ranges that are typical in current usage.

When T bits of input are used, this form of the algorithm guarantees that both f and g have at most T bits. This follows for g because $(\alpha \bmod 2^k, 1)$ is in L_k, and by induction for f since the f at stage k requires at most one nore bit than the maximum required by f and g at stage $k-1$. The complexity of the algorithm is thus $4DM + \mathcal{O}(T^2)$. Strictly in terms of T, this is $\mathcal{O}(T^2 \log T \log \log T)$.

However, if the sequence is chosen so the number of discrepancies is small, the complexity is lower. In particular a cryptographer designing a stream cipher should try to choose sequences for which many discrepancies occur.

6   Conclusions

We have exhibited an algorithm for synthesizing a minimal size feedback with carry shift register that outputs a sequence given a relatively small number of its bits. This is a general approach to attacking stream ciphers, and 2-adic span, the associated security measure, must now be considered whenever cryptologists design stream cipher.

This approach can be used to attack Massey and Rueppel's summation combiner [16, 19]. In their setup, the outputs from several short maximal period LFSRs (the outputs of such sequences are called *m-sequences*) with pairwise relatively prime periods are combined using addition with carry. It has been shown that the linear span of the resulting sequence tends to be close to the period of the sequence. This period is the product of the periods of the m-sequences, and is exponentially larger than the sizes of the original LFSRs. For this reason, summation combiners have been suggested for use in stream ciphers.

However, addition with carry is precisely addition in the 2-adic numbers. (In fact, it was this observation that motivated this work.) If a and b are two sequences, and c is the result of combining them with a summation combiner, then $\lambda_2(c)$ is at most $\lambda_2(a) + \lambda_2(b) + 2\lceil \log(\lambda_2(a)) \rceil + 2\lceil \log(\lambda_2(b)) \rceil + 6$. Even if these 2-adic spans are maximal (and ensuring this is problematic), we will be able to determine the resulting sequence from far fewer bits than previously thought. For example, if we combine m-sequences of period $2^n - 1$ for $n = 7, 11, 13, 15, 16, 17$, then the resulting sequence has linear span nearly 2^{79}, but the 2-adic span is less than 2^{18}. Thus 2^{19} bits suffice to determine this sequence – *and far fewer unless care is taken in the choice of the m-sequences.*

How can we build summation combiners that are secure against this sort of attack? It is necessary to choose underlying LFSRs whose output sequences have large 2-adic span. Suppose, for example, we hope to have 2^{60} secure bits. We might build a summation combiner based on maximal period LFSRs with length about 60. The resulting sequence is certainly secure against the classical Berlekamp-Massey attack. However, it is only secure against the attack described in this paper if we choose the individual output sequences have 2-adic spans close to their periods.

A slightly different analysis arises if we consider security based on complexity. If we consider a system secure if it takes 2^{60} word operations to crack it, then it must have 2-adic span at least 2^{27}. If we allow 2^{70} word operations, the 2-adic span must be 2^{32}. However, the analysis we have given is a worst case analysis. The actual speed of the rational approximation algorithm depends on the number of updates that must be performed. Thus the sequences chosen must not only have large 2-adic span, they must guarantee that many updates occur in the rational approximation algorithm.

Many other questions remain concerning FCSRs and 2-adic span. To understand the average behavior of the algorithm, we are led to the question of how the 2-adic span varies as a random sequence is lengthened. This is closely related to the question of what the 2-adic span of an average sequence is. Experimental evidence suggests that on average the 2-adic span is about half the length of the sequence. This would be consistent with what happens in the case of linear span, and would imply that on average many updates occur in the rational approximation algorithm.

The question of how to generate sequences with large 2-adic span is now an important one. We cannot have secure stream ciphers without such sequences. In fact we need to be able to generate sequence that simultaneously have large 2-adic and large linear span,

Finally, a number of generalizations of FCSRs have been suggested, based on other complete valued fields over number fields [8, 10]. Generalization of the ideas in this paper to p-adic shift registers (p a prime) is straightforward. Generalizations to registers over ramified and unramified extensions of the 2-adic (or p-adic) numbers are more difficult. With care, the registers can be constructed and have the appropriate algebraic structures. However, the rational approximation algorithm only appears to generalize under very special conditions. For example, we must have an algebraic number field whose ring of integers is a Euclidean domain. Furthermore, it seems that the convergence rate can only be guaranteed if the norm function on this field has certain properties.

7 Acknowledgements

The authors thank J. Goldsmith for several helpful suggestions for improving the manuscript.

References

1. E. Biham and A. Shamir: Differential Cryptanalysis of DES-like Cryptosystems, *Journal of Cryptology*, vol. 4, 1991, pp.3-72.
2. L. Blum, M. Blum, and M. Shub: A simple unpredictable pseudo-random number generator, *Siam J. Comput.* vol. 15, pp. 364-383 (1986).
3. U. Cheng: On the continued fraction and Berlekamp's algorithm. *IEEE Trans. Info. Theory* vol. 30, 1984 pp. 541-544.
4. Z. D. Dai and K. C. Zeng: Continued fractions and the Berlekamp-Massey algorithm. *Auscrypt '90*, Springer Lecture Notes in Comp. Sci. vol. 453, Springer Verlag, N. Y., 1990.
5. S. Golomb: *Shift Register Sequences.* Aegean Park Press,
6. R. T. Gregory and E. V. Krishnamurthy: *Methods and Applications of Error-Free Computation*, Springer Verlag, N. Y., 1984.
7. A. Klapper and M. Goresky: 2-Adic Shift Registers, *Fast Software Encryption: Proceedings of 1993 Cambridge Security Workshop, Springer-Verlag LNCS*, vol. 809, 1994, pp. 174-178.

8. A. Klapper, and M. Goresky: Feedback Registers Based on Ramified Extensions of the 2-Adic Numbers, *Proceedings, Eurocrypt 1994*, Perugia, Italy,

9. A. Klapper and M. Goresky: Feedback Shift Registers, Combiners with Memory, and Arithmetic Codes, *University of Kentucky, Deptartment of Computer Science Technical Report No. 239-93*.

10. A. Klapper: Feedback with Carry Shift Registers over Finite Fields, *Proceedings of Leuven Algorithms Workshop*, Leuven, Belgium, December, 1994.

11. N. Koblitz: *p-Adic Numbers, p-Adic Analysis, and Zeta Functions*. Graduate Texts in Mathematics Vol. 58, Springer Verlag, N. Y. 1984.

12. K. Mahler: On a geometrical representation of p-adic numbers, *Ann. of Math.* vol. 41, 1940 pp. 8-56.

13. D. Mandelbaum: An approach to an arithmetic analog of Berlekamp's algorithm. *IEEE Trans. Info. Theory*, vol. IT-30, 1984 pp. 758-762.

14. G. Marsaglia and A. Zaman: A new class of random number generators, *Annals of Applied Probability*. vol. 1, 1991 pp. 462-480.

15. J.L. Massey: Shift register sequences and BCH decoding, *IEEE Transactions on Infoormation Theory*, vol. IT-15, pp. 122-127, 1969.

16. J. Massey and R. Rueppel: Method of, and Apparatus for, Transforming a Digital Data Sequence into an Encoded Form, U.S. Patent No. 4,797,922, 1989.

17. W. H. Mills: Continued fractions and linear recurrences, *Math. Comp.* vol. 29, 1975, pp. 173-180

18. J. Pollard: The Fast Fourier Transform in a Finite Field, *Math. Comp.*, vol. 25, 1971, pp. 365-374.

19. R. Rueppel: *Analysis and Design of Stream Ciphers*. Springer Verlag, New York, 1986.

20. A. Schönhage and V. Strassen: Schnelle Multiplikation Grosser Zahlen, *Computing*, vol. 7, 1971, pp. 281-292.

21. B. M. M. de Weger: Approximation lattices of p-adic numbers, *J. Num. Th.* vol. 24, 1986, pp. 70-88.

22. L. R. Welch and R. A. Scholtz: Continued fractions and Berlekamp's algorithm, *IEEE Trans. Info. Theory*, vol. 25, 1979 pp. 19-27.

A Key-schedule Weakness in SAFER K-64

Lars R. Knudsen*

Laboratoire d'Informatique
École Normale Supérieure
Paris, France

Abstract. In this paper we analyse SAFER K-64 and show a weakness in the key schedule. It has the effect that for almost every key K, there exists at least one different key K^*, such that for many plaintexts the outputs after 6 rounds of encryption are equal. The output transformation causes the ciphertexts to differ in one of the 8 bytes. Also, the same types of keys encrypt even more pairs of plaintexts different in one byte to ciphertexts different only in the same byte. This enables us to do a related-key chosen plaintext attack on SAFER K-64, which finds 8 bits of the key requiring from 2^{44} to about 2^{47} chosen plaintexts. While our observations may have no greater impact on the security of SAFER K-64 when used for encryption in practice, it greatly reduces the security of the algorithm when used in hashing modes, which is illustrated. We give collisions for the well-known secure hash modes using a block cipher. Also we give a suggestion of how to improve the key schedule, such that our attacks are no longer possible.

1 Introduction

In [6] a new encryption algorithm, SAFER K-64, hereafter denoted SAFER, was proposed. Both the block and the key size is 64. The algorithm is an iterated cipher, such that encryption is done by iteratively applying the same function to the plaintext in a number of rounds. Finally an output transformation is applied to produce the ciphertext. For SAFER the suggested number of rounds is 6. Strong evidence has been given that SAFER is secure against differential cryptanalysis [7] and against linear cryptanalysis [2]. In [11] it was shown that by replacing the S-boxes in SAFER by random permutations, about 6% of the resulting ciphers can be broken faster than by exhaustive search.

In this paper we analyse SAFER and show a weakness in the key schedule. It has the effect that for virtually every key K, there exists at least one different key K^*, such that for a non-negligible fraction of all plaintexts the outputs after 6 rounds of encryption are equal. The output transformation causes the ciphertexts to differ in one of the 8 bytes. These pairs of plaintexts and ciphertexts can be found in time from about 2^{22} to 2^{28} encryptions. All estimates of complexity in this paper are the number of 6 round SAFER encryptions. Two keys encrypting a plaintext into the same ciphertext is called a "key-collision" in the literature

---

* Postdoctoral researcher sponsored by the Danish Technical Research Council.

D. Coppersmith (Ed.): Advances in Cryptology - CRYPTO '95, LNCS 963, pp. 274-286, 1995.
© Springer-Verlag Berlin Heidelberg 1995

and in [10] a brute-force key-collision attack on the DES was given, which can be applied to any block cipher. Given a plaintext P the method finds two keys for which the two encryptions of P are equal and requires about 2^{32} operations for a 64 bit block cipher.

What we have found for SAFER is much stronger. For (almost every) given key K there exists (at least) one other key K^*, different from K only in one byte, such that the encryption functions induced by the two keys encrypt from 2^{22} to 1.7×2^{28} plaintexts the same way in the 6 rounds of encryptions. The output transformation makes the ciphertexts differ in one byte, the same byte in which the keys differ. For some keys, K, there are up to 9 other keys encrypting a non-negligible fraction of all plaintexts in the same way as K. Also, for the same types of keys, K and K^*, the encryption functions induced by the two keys encrypt from 2^{29} to 2^{35} pairs of plaintexts, P and P^*, different only in one byte, the same way in the 6 rounds of encryptions. The output transformation makes the ciphertexts differ in the same byte. Interestingly, the keys, the plaintexts and the ciphertexts differ in the same byte.

We use our observations to establish related-key chosen plaintext attacks, which using from 2^{44} to 2^{47} chosen plaintexts finds 8 bits of the secret key with probabilities from 1 to 2^{-59} depending on certain circumstances of the attacks. Related-key attacks are not the most realistic attacks, and our results may have no greater impact on the security of SAFER in practice when used for encryption. However, first of all, it can be avoided by re-constructing the key schedule, secondly it greatly reduces the security of the algorithm when used in hashing modes.

In hashing modes using a block cipher algorithm as building block the plaintext (and/or the key) is exclusive-or'ed to the ciphertext to produce a kind of one-wayness in the hash algorithm. We found collisions of such hash functions in estimated time about 2^{23} encryptions when SAFER is used as the underlying block cipher. This should be compared with a brute force collision attack, which requires about 2^{32} operations. The keys we used were well-chosen, but with our method collisions can be found faster than a brute force attack for most keys.

This paper is organised as follows. First we give a short description of SAFER. In Sect. 3 we describe the weakness in the key schedule and give examples of the above mentioned (pseudo)-collisions. Next we use our observations to establish a related-key chosen plaintext attack on SAFER. In Sect. 4 we describe attacks on hash modes using SAFER and give examples of collisions. In Sect. 5 we give different methods of how to improve SAFER to avoid the problems described in the preceding sections.

2 Description of SAFER

SAFER is an r round iterated cipher with both block and key size of 64 bits and with all operations done on bytes. The key is expanded to $2r + 1$ round keys each of 64 bits, described later. The designer's recommendation for r is 6 [6]. Each round takes 8 bytes of text input and two round keys each of 8 bytes.

The input and the round keys are divided into 8 bytes and the first round key is xor'ed, respectively added modulo 256, according to Fig. 1. The bytes are then processed using 2 permutations or S-boxes, $X(a) = (45^a \bmod 257) \bmod 256$, and the inverse of X, $L(a) = log_{45}(a) \bmod 257$ for $a \neq 0$ and where $L(0) = 128$. After the S-boxes each byte of the second round key is added modulo 256, respectively xor'ed, and finally the so-called *Pseudo-Hadamard Transformation (PHT)* is applied to produce the output of the round. PHT is defined by three layers of the 2-PHT, which is defined by

$$2\text{-}PHT(x,y) = (2*x+y, x+y)$$

where each coordinate is taken modulo 256. Between two layers of 2-PHT's a permutation of the bytes is done, see Fig. 1. After the last round an output transformation is applied, which consists of xoring, respectively adding modulo 256, the last-round key.

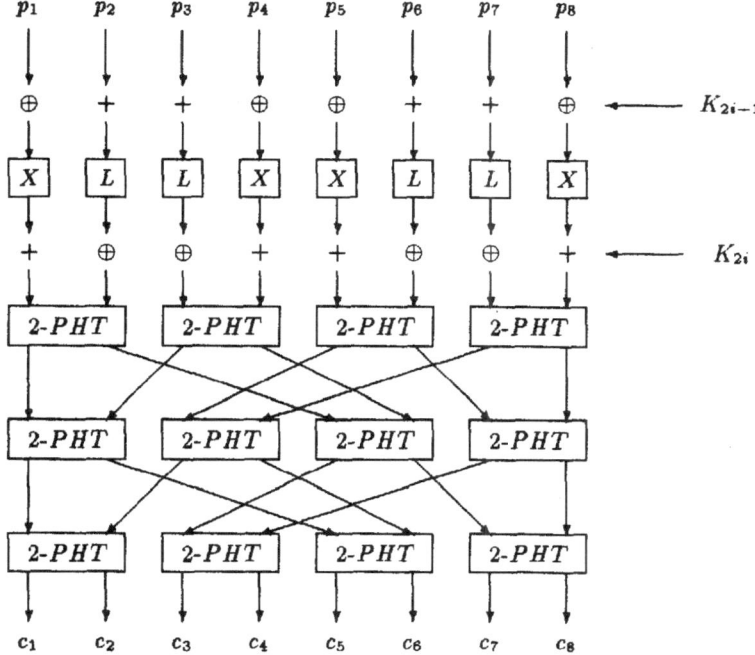

Fig. 1. One round of SAFER.

The key of 64 bits is expanded to $2r + 1$ round keys each of 64 bits in the following way. Let $K = (k_{1,1}, ..., k_{1,8})$ be an 8 byte key. The round key byte j in round i is denoted $K_{i,j}$. The round key bytes are derived as follows: $K_{1,j} = k_{1,j}$ for $j = 1, ..., 8$ and

$$k_{i,j} = k_{i-1,j} << 3 \tag{1}$$

$$K_{i,j} = k_{i,j} + bias[i,j] \bmod 256 \tag{2}$$

for $i = 2, ..., 2r + 1$ and $j = 1, ..., 8$. '<< 3' is a bitwise rotation 3 positions to the left and $bias[i, j] = X(X(9i + j))$, where X is the exponentiation function described above.

2.1 Some Properties of SAFER

The following lemma is used in this paper.

Lemma 1. *Let X be the exponentiation function of SAFER and let a be any byte value. Then it holds that*

$$X(a) + X(a + 128) = 1 \bmod 256$$

Proof: The statement is proved as follows.

$$\begin{aligned} X(a) + X(a + 128) \bmod 256 &= (45^a + 45^{a+128} \bmod 257) \bmod 256 \\ &= (45^a \times (1 + 45^{128}) \bmod 257) \bmod 256 \\ &= (0 \bmod 257) \bmod 256 \end{aligned}$$

since $45^{128} = -1 \bmod 257$. And since both $X(a)$ and $X(a+128)$ are in the range $[0, 256]$ and their sum is not zero, the statement follows. $\qquad\square$

The mixed use of addition modulo 256 and exclusive-or operations in SAFER was introduced to give the cipher *confusion* [6]. There is a simple and useful connection between the two operations when used on bytes, namely

Lemma 2. *Let a be a byte value. Then $a \oplus 128 = a + 128 \bmod 256$.*

Proof: Follows easily from the fact that the only possible carry bit of $a + 128$ disappears. $\qquad\square$
A result similar to Lemma 2 is shown in [7].

3 Weakness in the Key Schedule

From the previous section it is seen that key byte j affects only S-box j directly in every round. Let $K = (k_1, ..., k_8)$ be an 8 byte key. Consider the first byte in the first round. A key byte is first xor'ed to the plaintext byte, the result is exponentiated and another key byte is added modulo 256, the ciphertext byte after one round is $X(y \oplus K_{1,1}) + K_{2,1}$, where $K_{1,1}, K_{2,1}$ are derived from k_1. While it is true that this is a permutation of the plaintext byte to the ciphertext byte for a fixed key, it is not a permutation of the key byte to the ciphertext byte for a fixed plaintext. Let $K^* = (k_1^*, ..., k_8^*)$ be an 8 byte key different from K in only one byte, say byte no. 1. Then if k_1 and k_1^* encrypt some of the 256 possible inputs to S-box 1 in every round the same way, obviously K and K^* encrypt some 64 bit plaintexts over 6 rounds the same way.

If, say, n inputs to an S-box in the s'th round are encrypted the same way by two such keys we will say that the keys encrypt equally with probability $p_s = \frac{n}{256}$. Also we will call two such keys *related*. Consider S-box 1, K and K^* again. If a byte y is evaluated the same way with the two keys in S-box 1, i.e.

$$X(y \oplus K_{1,1}) + K_{2,1} = X(y \oplus K^*_{1,1}) + K^*_{2,1}$$

then so is the byte $\tilde{y} = y \oplus K_{1,1} \oplus K^*_{1,1} \oplus 128$. This follows from Lemma 1 and 2. Since L is the inverse of X, a similar property holds for the logarithmic S-boxes. Therefore n is always a multiple of 2. The probability that a 64 bit plaintext encrypts into the same ciphertext using such two keys is

$$\prod_{s=1}^{6} p_s \geq 2^6/2^{48} = 2^{-42}, \tag{3}$$

and the number of plaintexts is $Pl = 2^{64} \times \prod_{s=1}^{6} p_s \geq 2^{22}$. Here we have tacitly assumed that the p_i's are independent. This is not the case, however our experimental results have shown that the product (3) of the round probabilities is a good approximation for SAFER with 6 rounds. Since this phenomenon is isolated to one S-box we could easily do an exhaustive search for all such pairs of keys. We found that for two keys different only in the sixth byte with the values 132 and 173 respectively, $\prod_{s=1}^{6} p_s = \frac{6912}{2^{48}} \simeq 2^{-35}$ and $Pl \simeq 1.7 \times 2^{28}$. Note that since the only requirement we make is that the two keys have certain values in the eighth bytes, $Pl \simeq 1.7 \times 2^{28}$ for 2^{56} pair of keys. For another 3×2^{56} pairs of keys $Pl \simeq 1.13 \times 2^{28}$. How do we determine for how many keys there exist another key which encrypts from 2^{22} to about 2^{28} plaintexts the same way? Take a key K. Consider all $2^8 - 1$ keys K^* different from K only in byte 1. If none of them are related to K, choose keys K^* different from K only in byte 2 and so on. Again we can do an exhaustive search for all S-boxes isolated. The total number of keys for which there are no such other keys different in only one byte is about 2^{40}. For many keys K there exists more than one related key, on average about 2 related keys, and in some cases there are as many as 9 keys related to K.

In the search for the plaintext/ciphertext pairs that coincide for two keys it is not necessary to do two full 6 rounds of encryptions. One can start the encryptions in the second round with the inputs to this round such that the ciphertexts after the first two rounds of encryption are the same. This can be done easily by precomputing two small tables. Assume that the two keys differ in the first byte only. For the 256 possible values of the text output of the first S-box in the first round, store in a table the values for which the two keys decrypt to equal plaintexts. For the 256 possible values of the text input to the first S-box in the second round, store in a table the values for which the two keys encrypt to equal values. By pairing the values in the two tables and determining which PHT inputs whose first byte equals the first byte of a pair give a PHT output whose first byte equals the second byte of this pair, one can compute all the 64 bit inputs to the second round, such that the two keys encrypt equally in both the first and the second round.

Then after every round of encryption one checks whether the encryptions are equal. In most trials only 1 round of encryption is needed for every plaintext in a pair. Therefore one needs only to do about $\frac{1}{6} \times 2/ \prod_{i=3}^{6} p_i$ encryptions, which is 2^{22} in the optimal cases. Again we note that the output transformation, which consists of xoring, respectively adding modulo 256, the key K_{2r+1} makes the above ciphertexts differ in one byte, exactly the byte for which the keys differ. As illustrations we list in Fig. 1 two such examples. The first collision was found in time 2^{22}, the second in time $2^{22.1}$. We summarize our results.

Plaintext	Keys	Ciphertexts
8a 2c 62 a2 a2 81 c1 8c	e0 81 01 85 eb 3b 48 76	ca dd fc f6 30 ac 71 38
8a 2c 62 a2 a2 81 c1 8c	e0 81 01 85 eb 3b 48 bc	ca dd fc f6 30 ac 71 5c
50 1c 7a 44 39 63 f7 8c	e0 81 01 85 eb 3b 48 76	6a 7d db 51 44 89 5a f7
50 1c 7a 44 39 63 f7 8c	e0 81 01 85 eb 3b 48 bc	6a 7d db 51 44 89 5a 93

Table 1. Pseudo key-collisions for SAFER (hex notation).

Theorem 3. *For all but 2^{40} keys K in SAFER, there exists at least one and on average two keys, K^*, different from K in one byte, say byte b_k, such that K and K^* encrypt from 2^{22} to about 2^{29} plaintexts the same way in 6 rounds. The output transformation of SAFER makes the ciphertexts differ in one byte, byte b_k. The related keys can be found easily by exhaustive search over a single 8 bit S-box in 6 rounds. Given two related keys one such plaintext (and the two ciphertexts) can be found in time from about 2^{22} to 2^{28} encryptions.*

From the above also the following result follows.

Theorem 4. *For all but 2^{17} keys K in SAFER, there exists at least one and on average 3.5 keys, K^*, different from K in one byte, say byte b_k, such that K and K^* encrypt from 2^{29} to about 2^{35} pairs of plaintexts, P, P^*, different in only byte b_k the same way in 6 rounds. The output transformation of SAFER makes the ciphertexts differ in one byte, byte b_k.*

To find such "collisions", one can use the same method as described above for the result of Theorem 3, but this time start the search in the third rounds, such that the encryption in the second and third rounds are equal. Once two ciphertexts different in only byte b_k are found, the ciphertexts after one round are decrypted into two plaintexts different in only byte b_k. Examples of collisions from Theorem 4 are given in the section about collisions of hash functions. We can use Theorem 4 to establish a related-key attack on SAFER.

3.1 A Related-key Chosen Plaintext Attack

In [3, 4, 1] new attacks based on related keys were introduced. In this section we apply the principles of these attacks and introduce a chosen plaintext attack on

SAFER. Assume we have access to two oracles, one encrypting plaintexts with a key K, the other encrypting plaintexts with a key K^*, such that K and K^* are related, i.e. encrypt a non-negligible fraction of all plaintexts the same way. Assume without loss of generality that the keys differ only in byte b_1. Consider the following attack

- Choose the values of the bytes b_2 to b_8 at random.
- Get the 256 encryptions $\{C_i\}$ of the plaintexts $b_1, b_2, ..., b_8$ for all values of b_1 encrypted under the first key.
- Get the 256 encryptions $\{C_j\}$ of the plaintexts $b_1, b_2, ..., b_8$ for all values of b_1 encrypted under the second key.
- Sort the ciphertexts just received and check, if any ciphertext in $\{C_i\}$ differs from any ciphertext in $\{C_j\}$ only in byte b_1. If a match is found the two ciphertexts are output.

If ciphertexts are output in the last step of the above attack, we search exhaustively for two 8 bit keys k and $k*$ for which the encryptions of the bytes b_1 for the two corresponding plaintexts yields equal outputs after one round. For these key bytes we check if the xor of the byte b_1 for the two ciphertexts is the value of the xor of the last-round key bytes induced by k and $k*$. If this is the case we have found 8 bits of the secret key with a high probability. It could happen by accident that two ciphertext blocks are different only in one byte without the property that the encryptions after each of the 6 rounds are equal. But clearly that would happen only with negligible probability.

The attack is repeated until the last step of the algorithm outputs two ciphertexts. Note that since we choose all 256 plaintexts different in one byte, we can consider 2^{16} pairs of plaintexts, consisting of one plaintext encrypted under one key and another plaintext encrypted under the second key. It follows that there are 256 pairs of plaintexts encrypted the same way in the first round. According to Theorem 4 the above algorithm succeeds with probability at least $2^{29} \times 2^{-64} = 2^{-35}$ and therefore needs to be repeated at most about 2^{35} times, in the optimal cases only 2^{29} times. The number of chosen plaintexts needed in the worst cases is about $2 \times 2^8 \times 2^{35} = 2^{44}$. The probability of success is about 0.63. The attack can be extended to the case where the attacker has no knowledge of the byte for which the keys differ. The above attack is simply repeated for all 8 bytes requiring a total of 2^{47} chosen plaintexts. If the two keys are chosen at random different in only one byte, the attack succeeds with a probability of $\frac{3.5}{256}$, according to Theorem 4. Two randomly chosen 8 byte keys will be different in only one byte with probability $8 \times \frac{255}{256} \times 2^{-56} \simeq 2^{-53}$. Therefore, if all of the 8 bytes of the two keys are chosen at random, the attack succeeds with a probability of $2^{-53} \times \frac{3.5}{256} \simeq 2^{-59}$. We summarize our results in Table 2 for SAFER with the recommended 6 rounds. We note that the complexities given are worst case considerations. The factor 0.63 in the probabilities can be increased by using more chosen plaintexts. In Table 3 we give the complexities for similar related-key attacks on SAFER reduced to (the first) 4 and 5 rounds.

Our attacks may seem unrealistic. But imagine Alice and Bob are sending many messages to each other every day. Alice and Bob have been acting in many

Chosen plaintexts	Probability	Conditions
2^{44}	0.63×1	Two related keys
2^{44}	$0.63 \times 1/73$	The two keys differ in one known byte position.
2^{47}	$0.63 \times 1/73$	The two keys differ in one unknown byte position.
2^{47}	0.63×2^{-59}	The two keys are randomly chosen.

Table 2. Related-key chosen plaintext attacks on SAFER finding one byte of the key. (Worst case considerations.)

4 rounds		5 rounds		
Ch. pl.texts	Prob.	Ch. pl.texts	Prob.	Conditions
2^{30}	0.63×1	2^{37}	0.63×1	Two related keys.
2^{30}	$0.63 \times 1/14$	2^{37}	$0.63 \times 1/35$	The two keys differ in one known byte position.
2^{33}	$0.63 \times 1/14$	2^{40}	$0.63 \times 1/35$	The two keys differ in one unknown byte position.
2^{33}	0.63×2^{-57}	2^{40}	0.63×2^{-58}	The two keys are randomly chosen.

Table 3. Related-key chosen plaintext attacks on SAFER reduced to four and five rounds finding one byte of the key. (Worst case considerations.)

cryptographic papers, so they know that the key should be changed often. So, they change the key every day, but to save computations only in one byte, so that all the bytes in the key are changed after eight days. Nowhere in the literature have they found evidence that this should be dangerous. Using SAFER it will be. Eve hasn't appeared in as many papers as Alice and Bob, but is smart enough to trick one of the parties into encrypting many chosen plaintexts every day. Eve finds 8 bits of the secret key with probability $\frac{3.5}{256}$ every day, except the first day, using at most 2^{47} chosen plaintexts. We assume here that the time to sort and compare ciphertexts is negligible compared to the time of getting the many encryptions. After 73 days Eve has used about 2^{53} chosen plaintexts and with a probability 0.63 found at least 8 key bits. The number of chosen plaintexts can be reduced to 2^{50}, if Eve can predict which byte of the secret key is changed from day to day. Similar attacks on SAFER with a reduced number of rounds will have much lower complexities.

3.2 The Rotations and Bias Additions

In this section we consider the rotations and bias additions used in the key schedule of SAFER. In [6] it is argued that the bias additions prevent weak keys. Moreover, by letting out the key biases, for any key K there exists another key K^*, such that the first 5 rounds of the encryption function induced by K are the same as the last 5 rounds of the encryption function induced by K^*. This is not a desirable property as illustrated in [3, 4, 1]. We have found a reason to have byte rotations as well.

Lemma 5. *PHT has 256 fixed points.*

This result can be found by using Gauss-eliminations on the 8×8 matrix of PHT etc. In each fixed point every byte value is a multiple of 64. There are 16 fixed points where every byte value is either 0 or 128. They are given in Table 4. If

```
(   0   0   0   0 0   0   0   0 ) (   0   0   0   0 128 128   0   0 )
(   0   0 128   0 0   0 128   0 ) (   0   0 128   0 128 128 128   0 )
(   0 128   0 128 0   0   0   0 ) (   0 128   0 128 128 128   0   0 )
(   0 128 128 128 0   0 128   0 ) (   0 128 128 128 128 128 128   0 )
( 128   0   0 128 0 128 128 128 ) ( 128   0   0 128 128   0 128 128 )
( 128   0 128 128 0 128   0 128 ) ( 128   0 128 128 128   0   0 128 )
( 128 128   0   0 0 128 128 128 ) ( 128 128   0   0 128   0 128 128 )
( 128 128 128   0 0 128   0 128 ) ( 128 128 128   0 128   0   0 128 )
```

Table 4. The 16 fixed points for the PHT with only entries 0 and 128.

one leaves out the key rotations, but keeps the addition of the biases then these 16 fixed points for PHT are "linear structures" for SAFER with any number of rounds in the following way. Let $a_1, \ldots, a_{16}$ be the fixed points from Table 4. Let $E(K, P) = C$ be the encrypted value of plaintext P using key K, then

$$E(K, P) = C \implies E(K + a_i, P + a_i) = C.$$

where '+' is bytewise addition modulo 256. Thus, an exhaustive search for the key could be reduced by a factor of 16 using 16 chosen plaintexts. The 16 fixed points are the only linear structures. Fixed points with entries of values 64 or 192 are affected/destroyed by the group operation changes exclusive-or/addition mod 256, but the values 0 and 128 are not, which follows from Lemma 2. The above illustrates that SAFER needs both key rotations and bias additions in the key schedule.

4 Collision of Hash Functions

Often a block cipher is used as building block in hash functions. A hash function for which the hash code is of the same size as the block cipher is called a *single block length hash function*. In these hash functions the message blocks are hashed in a number of rounds, each round requiring one encryption of the underlying block cipher. There are essentially 12 secure single block length hash functions, which by a linear transformation of the inputs to one round of the hash function can be transformed into only 2 different schemes [8, 9]:

$$H_i = E_{M_i}(H_{i-1}) \oplus H_{i-1} \tag{4}$$

$$H_i = E_{M_i}(H_{i-1}) \oplus H_{i-1} \oplus M_i \tag{5}$$

The first scheme is known as the Davies-Meyer scheme. These schemes are believed to be secure, in the sense that, if the underlying block cipher has no weaknesses, free-start preimage attacks and free-start collision attacks have time complexities 2^m and $2^{m/2}$ encryptions, respectively, of the underlying m-bit block cipher [5, 8]. In a free-start attack the attacker is free to choose the initial values. Using SAFER as the underlying block cipher it is possible to find both free-start and fixed-start collisions with a complexity of much less than the brute force method of 2^{32} operations.

Also, we note that the attacks to follow will be applicable to many double block length hash functions based on a block cipher, since in free-start attacks it is possible to attack the two blocks independently. In the next section we show how to find free-start collision for the schemes (4) and (5).

4.1 Free-start Collisions

In this section we exploit the phenomenon of Theorem 4. In the attacks to follow we choose two plaintexts different only in the byte for which both the keys and ciphertexts differ. We hope in this way to obtain plain- and ciphertexts and keys, such that

$$E_{K_1}(P_1) \oplus P_1 = E_{K_2}(P_2) \oplus P_2 \ or$$
$$E_{K_1}(P_1) \oplus P_1 \oplus K_1 = E_{K_2}(P_2) \oplus P_2 \oplus K_2$$

We can speed up this search by choosing the inputs of SAFER to the third round, such that the keys encrypt equally in the second and third rounds. When we find two ciphertexts different in only one byte, we calculate the plaintexts and check for a collision. In the optimal cases these collisions can be found in estimated time about $2^{22.8}$ encryptions of SAFER. In Table 5 we give examples of such collisions for hash functions of type (4). The first collision was found in time $2^{20.6}$ encryptions, the second collision in time $2^{19.3}$ encryptions.

Initial value (pl. text)	Message (key)	Hash code
6e 32 68 46 c8 fd f1 a9	6f 2d 73 46 e1 2f 62 45	e5 12 8b 4d 3d 58 c2 18
6e 32 68 46 c8 fd f1 9c	6f 2d 73 46 e1 2f 62 f7	e5 12 8b 4d 3d 58 c2 18
f4 b1 a3 27 0b ed 78 a9	57 f5 9b 4e 49 77 0a 45	54 43 57 c4 be f9 88 c9
f4 b1 a3 27 0b ed 78 9c	57 f5 9b 4e 49 77 0a f7	54 43 57 c4 be f9 88 c9

Table 5. Free-start collisions for hash functions of type (4) with SAFER.

It is possible to find free-start collisions for hash functions of type (5) also. We found such collisions in time about 2^{22}. In the next section we give examples of collisions for hash functions of type (5) with a fixed start.

Initial value (pl. text)	Message (key)	Hash code
ff 4e 79 3f c3 4f 52 5b	6d e6 02 f2 54 f0 59 a8	a7 a9 3e 8c 23 30 c3 b4
ff 4e 79 3f c3 4f 52 5b	e5 e6 02 f2 54 f0 59 a8	a7 a9 3e 8c 23 30 c3 b4
ff 9d e5 f5 c1 bc eb 71	6d 9b 13 2f 4d f5 7a b5	11 47 f9 f4 53 c8 e3 17
ff 9d e5 f5 c1 bc eb 71	e5 9b 13 2f 4d f5 7a b5	11 47 f9 f4 53 c8 e3 17

Table 6. Fixed-start collisions for hash functions of type (5) with SAFER.

4.2 Fixed-start Collisions

Although the collisions found in the last section are considered hard to find, if the underlying block cipher has no weaknesses, it is interesting to find collisions also for a fixed start. Using our observations about SAFER this cannot be done for the hash round function (4), since if the plaintexts are equal for two related keys the hash value of (4) will always be different. However, it is possible to find collisions if we consider two rounds of the hash function. Assume H_0 is a fixed initial value. Using the related key properties described earlier in this paper one finds M_1 and M_1', such that $H_1 = E_{M_1}(H_0) \oplus H_0$ and $H_1' = E_{M_1'}(H_0) \oplus H_0$ differ in one byte. Then use the related key properties once again in the second round and find M_2 and M_2', such that $H_2 = E_{M_2}(H_1) \oplus H_1$ equals $H_2' = E_{M_2'}(H_1') \oplus H_1'$. We did not implement this attack. For the hash functions (5) it is possible to find fixed start collisions for the hash round function. For our pseudo-collisions for SAFER, see Table 1, the ciphertexts and keys differ in the same byte. Therefore when both the plaintexts and the keys are fed forward in the hash mode, we can obtain collisions. The difference in the ciphertexts of Table 1 is equal to the difference in the last-round keys, which is not necessarily the difference in the keys themselves. Therefore for this attack to work we must use pairs of keys for which the byte differences in the keys are equal to the byte differences in the last-round keys of the keys. An exhaustive search reveals many pairs of keys with this property. Two keys different only in the fifth byte with values 9 and 129 respectively, encrypt about 2^{28} plaintexts in the same way. By using similar techniques as for free-start collisions one can show that a collision can be found in expected time about 2^{22} encryptions. In Table 6 we list such collisions. The first collision was found in time $2^{22.3}$ encryptions, the second collision in time $2^{20.0}$ encryptions. Many of our collision implementations ran faster than expected, which may be due to the fact that probabilities in (3) are not independent as assumed.

5 Improvements of SAFER

In this section we suggest modifications of SAFER, such that the above attacks cannot be effected. An obvious and immediate way is to increase the number of rounds.

5.1 An Increased Number of Rounds

In SAFER with 8 rounds there are still many pairs of keys encrypting some plaintexts the same way. In the optimal case a pair of keys encrypt 1.5×2^{14} plaintexts into the same ciphertexts after 8 rounds of encryption using our method. The output transformation makes those ciphertexts differ in one byte. But in contrast to SAFER with 6 rounds collisions cannot be found faster than the time of 2^{32} encryptions. Still, it must be an undesirable property for a block cipher.

In the optimal case for SAFER with 10 rounds a pair of keys encrypt equally for all 10 rounds with probability of only 2^{-66} using our method. Since there are only 2^{64} different plaintexts there are no keys with the above phenomenon.

5.2 New Key Schedule for SAFER

Another and in our taste better solution is to change the key schedule. The discoveries in this paper come from the fact that a key is applied to the text input before and just after the S-box, thus enabling collisions considering one byte isolated in every round. One way to hinder this is, paradoxically, to remove the second xor/addition of the key in every round or just in one of the middle rounds. To find collisions similar to the ones we've found would now require an incorporation of the PHT-transformation. That seems very unlikely to succeed. But, the fact that a one byte key is connected to the same S-box in every round seems dangerous and unnecessary. We give a modified key schedule for SAFER with any number of rounds. Let $K = (k_{1,1}, ..., k_{1,8})$ be an 8 byte key and let

$$k_{1,9} = \bigoplus_{i=1}^{8} k_{1,i}$$

The round keys are defined as follows. $K_{1,j} = k_{1,j}$ for $j = 1, ..., 8$ and

$$k_{i,j} = k_{i-1,j} << 3$$
$$K_{i,j} = k_{i,(i+j-2 \bmod 9)+1} + bias[i,j] \bmod 256$$

for $i = 2, ..., 2r+1$, $j = 1, ..., 8$. There is a circular shift of the nine key bytes. In that way the 8 key bytes $k_1, ..., k_8$ are connected to different S-boxes from round to round. The parity byte is introduced to provide an avalanche effect in the key schedule. The new key schedule ensures that the round keys of two different keys are always different in two bytes in some rounds and in one byte in the remaining rounds. For instance, in SAFER with 6 rounds, two keys will be different in two bytes in 9 out of the 13 round keys. In SAFER with 8 rounds, this will be the case in 13 out of the 17 round keys. Thus, our method of finding key-collisions is no longer applicable. Also, note that if the key is chosen uniformly at random, any round key is uniformly random.

6 Conclusion

We described a weakness in the key schedule of SAFER K-64 and exploited it to establish a related-key attack much faster than a brute force attack, and showed by examples that collisions for the standard hashing modes based on a block cipher using SAFER K-64 are easy to find. A new key schedule was suggested, so that the resulting cipher is invulnerable to our attacks. To conclude, we believe that the results presented in this paper show that a change in the key schedule of SAFER K-64 is necessary.

7 Acknowledgments

I would like to thank Jim Massey and Serge Vaudenay for fruitful discussions and Carlo Harpes, Xuejia Lai and Torben Pedersen for valuable comments.

References

1. E. Biham. New types of cryptanalytic attacks using related keys. *Journal of Cryptology*, 7(4):229–246, 1994.
2. C. Harpes, G.G. Kramer, and J.L. Massey. A generalization of linear cryptanalysis and the applicability of Matsui's piling-up lemma. In L. Guillou and J.-J. Quisquater, editors, *Advances in Cryptology - Eurocrypt'95, LNCS 921*, pages 24–38. Springer Verlag, 1995.
3. L.R. Knudsen. Cryptanalysis of LOKI'91. In J. Seberry and Y. Zheng, editors, *Advances in Cryptology, AusCrypt 92, LNCS 718*, pages 196–208. Springer Verlag, 1993.
4. L.R. Knudsen. *Block Ciphers – Analysis, Design and Applications*. PhD thesis, Aarhus University, Denmark, 1994.
5. X. Lai. *On the Design and Security of Block Ciphers*. PhD thesis, ETH, Zürich, Switzerland, 1992.
6. J.L. Massey. SAFER K-64: A byte-oriented block-ciphering algorithm. In *Fast Software Encryption - Proc. Cambridge Security Workshop, Cambridge, U.K., LNCS 809*, pages 1–17. Springer Verlag, 1994.
7. J.L. Massey. SAFER K-64: One year later. In *Proc. - The Leuven Algorithms Workshop*. Springer Verlag, 1995. To appear.
8. B. Preneel. *Analysis and Design of Cryptographic Hash Functions*. PhD thesis, Katholieke Universiteit Leuven, January 1993.
9. B. Preneel. Hash functions based on block ciphers: A synthetic approach. In D.R. Stinson, editor, *Advances in Cryptology - Proc. Crypto'93, LNCS 773*, pages 368–378. Springer Verlag, 1993.
10. J.-J. Quisquater and J.-P. Delescaille. How easy is collision search. Applications to DES. In J.-J. Quisquater and J. Vandewalle, editors, *Advances in Cryptology - Eurocrypt '89, LNCS 434*, pages 429–433. Springer Verlag, 1990.
11. S. Vaudenay. On the need for multipermutations: Cryptanalysis of MD4 and SAFER. In *Proc. - The Leuven Algorithms Workshop*. Springer Verlag, 1994. To appear.

Cryptanalysis of the Immunized LL Public Key Systems

Yair Frankel[1] * and Moti Yung[2]

[1] Sandia National Laboratories, Albuquerque, NM
[2] IBM T. J. Watson Research Center, Yorktown Heights, NY

Abstract. In CRYPTO '93 Lim and Lee provided a valuable investigation of public key encryption systems secure against *adaptive chosen ciphertext attacks*. In this paper we identify several insecurities of both their RSA and El Gamal based schemes. We first demonstrate that the RSA based scheme is insecure under an adaptive chosen ciphertext attack. We also point weaknesses in the design of both their RSA and El Gamal based schemes regarding the use of pseudorandom-generators, and in particular show that their choice of pseudorandom-generators for the RSA based scheme may be insecure even with respect to a *known ciphertext only attack*.

They further claim that their schemes are particularly useful in the context of group-oriented cryptosystems due to the unique verification method used. (In fact their scheme is the only group-oriented practical encryption claimed to be secure against chosen ciphertext attacks). Group oriented cryptosystems distribute the decryption process amongst a multiple of individuals in order to provide a mechanism in which no single person is trusted. We further demonstrate that both their schemes are completely insecure in this setting.

---

* Research was performed while the author was at GTE Laboratories Incorporated.

D. Coppersmith (Ed.): Advances in Cryptology - CRYPTO '95, LNCS 963, pp. 287-296, 1995.
© Springer-Verlag Berlin Heidelberg 1995

1 Introduction

We analyze the Lim and Lee public key cryptosystems (which are built upon the RSA [RSA78] and El-Gamal [ElGamal85] cryptosystems) presented at CRYPTO '93 [LimLee93]. Their systems introduced new insights and were designed (1) to strengthen the basic public-key systems against adaptive chosen ciphertext attacks, (2) to improve problems with previous such schemes, and (3) to be adaptable for group-oriented systems. In this paper we:

- demonstrate that the Lim-Lee RSA based encryption scheme is insecure against an adaptive chosen ciphertext attack;
- show that the Lim-Lee El Gamal based encryption scheme is insecure against an adaptive chosen ciphertext attack when used in a group oriented setting;
- find weaknesses with regards to the usage of pseudorandom generators in the Lim-Lee schemes (in the RSA based scheme, one implementation is even insecure w.r.t. known ciphertext only attack); and
- make heuristic recommendations on potential improvements of the Lim-Lee El Gamal based encryption scheme, strengthening the bindings between ciphertext portions of the scheme.

1.1 Chosen ciphertext security

For many years, no public key system [DiffieHellman76] was shown to be secure under a chosen ciphertext attack. Indeed, Rabin's scheme [Rabin79] which was proven to be secure under chosen message attack and a dense message space was shown, in fact, to be strongly insecure with respect to chosen ciphertext attacks. Thus, the question of how to design or prove security against such attacks was open for a while.

Provably secure chosen-ciphertext secure systems :
The first solution for such systems was given in [NaorYung90] which presented a public key cryptosystem secure against chosen ciphertext attack based on zero-knowledge non-interactive proof systems and probabilistic encryption. Chosen ciphertext attack formalizes the situation where the adversary (say, an **operator** of the decryption equipment in a company) as part of the attack obtains the decryption equipment and is allowed to sequentially query it as a black box (an input-output oracle). The system is said to be secure under a chosen ciphertext attack if the attacker cannot decrypt a new message. This is also known as the "lunch break attack", "midnight-attack" or "querying-only attack". The names visualize the situation where the supervisor is out of the office (for lunch) and the attacker is using the opportunity to play with the equipment over the break but it has no meaningful ciphertext in its possession (as the supervisor has put relevant documents in a safe and suspended communication).

Strengthening this attack is possible: an "adaptive attack" is one where the attacker gets to query the decryption equipment also after receiving the meaningful ciphertext that the attacker wishes to open. However, the attacker is not allowed to query the machine with the target ciphertext. This can be visualized as an "equipment testing available-ciphertext attack" in which the attacker (a **technician**) who now can have access to a meaningful ciphertext it tries to understand. The technician is allowed to test the equipment in front of the supervisor, yet the attacker should query the machine with other messages (which look seemingly harmless to the supervisor overlooking the testing). That is, in this attack the adversary may ask queries based on the ciphertext it wants to decrypt but not the ciphertext itself. In [DoDwNa91] a system secure against adaptive chosen ciphertext attack was given as part of "non-malleable cryptography". Also [RackoffSimon91] gave a solution to such an attack assuming a stronger setting where the sender (the technician) who queries the machine also has a personal public key known to the decryption machine.

Practical chosen-ciphertext secure systems :
The above theoretical methods employed the impractical non-interactive zero-knowledge proof systems. This motivated the study of practical immunization methods by somewhat simulating what is achieved by the theoretical methods. First [Damgard91] suggested a practical approach (against the operator adversary), then [ZhengSeberry92] claimed a system that is secure even against adaptive chosen ciphertext attacks (the technician adversary). They also showed that the scheme of [Damgard91] does not withstand such attacks. Under a very strong assumption about a cryptographic hash function being as good as a random oracle a method like theirs has been validated to be secure [BellareRogaway93]. Finally, [LimLee93] suggested public key schemes which they claimed to be secure against the (technician) adaptive attack as well. They also demonstrated some problems with the methods of [ZhengSeberry92]. Then they further suggested that their immunization technique works in the context of group-oriented cryptography [DesmedtFrankel89] where the cryptographic power to decrypt is shared amongst a multiple of agents.

Immunization methods against chosen ciphertext attacks attempt to entangle cryptographic operations and introduce dependencies among them so that it may be easy to be convinced that the party which has generated the ciphertext, indeed must have known the cleartext and thus could not have abused the system as it must have known the result of its query to begin with. This is the spirit of the theoretical work.

We would like to remark that the work by [LimLee93] certainly teaches us desirable properties of a practical chosen ciphertext public key. They do point out certain problems with previous schemes and describe how a scheme for the group-oriented case should work. In fact for this purpose they advocate a validity check which is based on the ciphertext and not on the recovered plaintext; which is a true statement. In short, we feel that their analysis is valuable. In their arguments they follow the above direction of thinking by simulating the theoretical approach. It would have seemed that it is plausible to argue that due to the computational relationships within the ciphertext gram, the attacker cannot generate something new based on the available information without actually starting from a known cleartext message. Its information being past answers to queries and, in the adaptive attack – the available ciphertext challenge. In fact this is what [LimLee93] argue, and they further give additional claims about "semantic security" (i.e., all bits are secure based on a hard problem [GoldwMicali84]) and that certain approaches to breaking the method do not work. Nevertheless, we show that such general arguments should be taken with a grain of salt, and there are ways to abuse the system even when the queries must have portions which look quite dependent on each other and the plaintext message, and even if in addition weaker attacks seem not to be successful.

1.2 Organization of paper

In Section 2 we discuss the necessary background for the paper. The Lim-Lee RSA based scheme is cryptanalyzed in Section 3, and the El Gamal based scheme is broken in Section 4 under the group-oriented cryptography setting. In Section 5 we discuss inherent weaknesses in their use of pseudo-random generators. Potential improvements and conclusion are given in Section 6.

2 Background

Notation:
Throughout this paper we denote $|x|$ as the length of the string x and $x_1||x_2$ as the concatenation of the string x_1 with string x_2.

Let $G(x_1, x_2)$ denotes an x_1 bit output sequence produced by cryptographically strong pseudorandom generator on a seed x_2 and let $h(x)$ denote the output of a cryptographically secure hash function on a string x.

Attacks:
We consider "adaptive chosen ciphertext attacks" when the adversary is able to make queries to a decryption oracle based on but different from a target ciphertext. We say that an attack is successful if, for a large enough fraction of the cases, the attacker is able to recover the ciphertext after the attack. Namely, for a large portion and appropriate choice of system's parameters (as specified by the designer), the attacker can be successful. A strong attack, is one that every ciphertext (not only a fraction of them) can be recovered.

Group-Oriented Cryptography:
Let us give a brief description of group-oriented cryptosystems in our public key decryption setting. Group oriented cryptosystems, also called threshold cryptosystems, were developed as a method to distribute the decryption process amongst a multiple of individuals in order to provide a decryption mechanism in which no single person is trusted [Desmedt87, DesmedtFrankel89]. In such a scheme, the decryption key is initially distributed to a number of agents and each agent is given a partial key. When an encryption is received by the group, a quorum of agents is available to apply its partial keys. The results of the partial keys applied to the ciphertext are called partial results. These partial results are then combined by a polynomial time algorithm into the final decryption result.

No security is placed in the combining function (it can be a public server or any available agent).

As long as a less than a threshold of agents is attacked by an intruder that reads their memory the system remains secure (the intruder does not have access to a threshold of the partial keys). A successful chosen ciphertext attack on a group oriented cryptosystem allows the adversary access to the combining function (as it is a public function with no security requirements put on it) and access to less than a threshold of the partial keys. Then, the attacker is given the allowed information, he is allowed queries, and it attempts to break the system. Lim and Lee provide important insight on why group oriented cryptosystems are particularly prone to chosen ciphertext attacks [LimLee93].

3 Breaking the Lim-Lee RSA based scheme

First let us review the details of their RSA based scheme.

For user A let e_A be the public exponent, N_A be the public modulus and d_A be the private exponent of the RSA encryption scheme [RSA78].

Encryption:
The ciphertext for a message m is $C = (c_0, c_1, c_2)$ where:

- $c_1 = s^{3 \cdot e_A} \bmod N_A$ where $s \in_R \{1, \ldots N_A - 1\}$
- $c_2 = z \oplus m$ where $z = G(|m|, s)$
- $c_0 = s^{3 \cdot H} \bmod N_A$ where $H = h(c_1 \| c_2)$

Decryption:
To decrypt $C = (c_0, c_1, c_2)$.

- Verify that $c_0^{3 \cdot e_A} = c_1^{H'}$ where $H' = h(c_1 \| c_2)$; return NULL if false
- Compute $z = G(|c_2|, s)$ where $s \equiv c_1^{(1/3) \cdot d_A} \bmod N_A$
- Return $z \oplus c_2$

Note that the ciphertext was generated by encrypting the seed s, employing the seed for encryption (stream cipher) and encrypting a one-way integrity certificate of the two fields. This seems to be a strong binding of the seed and the cleartext.

We now break the Lim-Lee RSA based encryption scheme.

Theorem 1. *The Lim-Lee RSA based encryption scheme can be broken using an adaptive chosen ciphertext attack.*

Proof. We first describe the attack and then show that the adversary is successful with sufficient probability.

Observe using the Extended Euclidean Algorithm that there exist integers u, v such that $e_A \cdot u + H \cdot v = \gcd(e_A, H)$. Since e_A is public and H can be computed from publicly known values c_1 and c_2, the values of u and v can be generated in polynomial time. Then if $\gcd(e_A, H) = 1$, the attacker is successful and it can compute $s^3 \equiv c_1^u \cdot c_0^v \bmod N_A$. The attacker, when successful, can simply send the oracle the ciphertext $C' = (s^{3 \cdot H''}, c_1, 0)$ where $H'' = h(c_1 || 0)$ and 0 is a string of all zero bits of length $|c_2|$. To complete the attack notice that the oracle will pass the verification step and return $z = G(|c_2|, s)$.

It is now shown that the adversary succeeds with a large enough fraction of the ciphertexts. Using [Apostol76, Thm 3.9] with probability about $\frac{6}{\pi^2}$ the $\gcd(e_A, H) = 1 \equiv 1 \bmod \phi(N_A)$. In this estimate we assumed that the public exponent has been chosen at random and that H is almost random, since h as a strong cryptographic hash function – has an almost-random output distribution. In case when e_A is chosen as a constant (e.g., 3 is a popular choice for a globally chosen exponent) then the probability of a random number in the range being relatively prime can be estimated as the constant $\frac{\phi(e_A)}{e_A}$ (e.g., $\frac{2}{3}$ for exponent 3) as the range of the function h is quite large. $\qquad\square$

4 Attacks on the El Gamal based scheme

First let us review the details of the Lim-Lee El Gamal based scheme.

Let α be a generator for $GF(p)$ where p is a large prime. For user A let $y_A \equiv \alpha^{x_A}$ be the public exponent, and x_A be the private key of the El Gamal encryption scheme [ElGamal85].

Encryption:
 The ciphertext for a message m is $C = (c_1, c_2, c_3, c_4)$ where:

- $c_0 \equiv \alpha^{r_0}$ where $r_0 \in_R \{1, \ldots, p-1\}$
- $c_1 \equiv \alpha^{r_1}$ where $r_1 \in_R \{1, \ldots, p-1\}$
- $c_2 = z \oplus m$ where $z = G(|m|, y_A^{r_1} c_0)$
- $c_3 = h(c_0 || c_2)$
- $c_4 \equiv r_0 + c_3 r_1 \bmod p - 1$.

Decryption:
 To decrypt $C = (c_1, c_2, c_3, c_4)$.

- Verify that $c_3 = h(c_0' || c_2)$ where $c_0' \equiv \alpha^{c_4} c_1^{-c_3}$; return NULL if false
- Compute $z = G(|c_2|, s)$ where $s = c_1^{x_A} c_0' \bmod p$
- Return $z \oplus c_2$

4.1 Cryptanalyzing the group-oriented setting

Lim and Lee discuss the case of using this scheme for group-oriented cryptography. In this case, we secretly share the private exponent and we need a secret sharing scheme to do it. Thus, what is generally needed is to make sure that the generator works in a field (or module), so assume $p - 1 = wq$ where w is small and q is a prime and choose the generator α above to be in $GF(q)$, similarly the choice of exponents r_0 and r_1 (see [DesmedtFrankel89]). Otherwise, the description stays the same as above.

With the group-oriented setting the decryption agents verify the hash value (the first step in the decryption process) individually, before they jointly compute s. This is an important property for group-oriented decryption as [LimLee93] suggest. To compute s, an agent uses its share of the private key x_A and c_1 to get its partial result. A threshold of partial results can be interpolated to get the result $c_1^{x_A}$. Now s can be computed given c_0'. If the verification process is successful with a different ciphertext, then a random seed unrelated to s should be generated. We demonstrate that this is not the case and in fact one can attack the system.

Theorem 2. *The Lim-Lee El Gamal based encryption scheme in the group-oriented setting can be strongly broken using an adaptive chosen ciphertext attack.*

Proof. We only need to prove that an attacker can generate an s' related to the s of a given ciphertext and recover s itself.

The attacker can generate a ciphertext $C' = (c_1' \equiv c_1 \cdot \alpha \bmod p, c_2, c_3, c_4' \equiv r_0 + c_3 r_1 + c_3 \bmod q)$ and provide it to the oracle. Observe that the verify step will be accepted since c_2 remains the same and $c_0' \equiv c_0 \bmod p$. The seed generated in the decryption step will be $s' = (\alpha^{r_1+1})^{x_A} c_0' \equiv (\alpha^{r_1+1})^{x_A} c_0$. Thus $s \equiv s'(\alpha^{x_A})^{-1} \bmod p$.

To finish the proof we note that one agent (or combiner server) will receive the information needed in order to calculate s' and therefore can generate s. That person (server) can generate C' and be successful in performing the adaptive chosen ciphertext attack. □

5 Cautions on the use of pseudorandom generators

Pseudorandom number generators do not necessarily exhibit the strength to withstand attacks, when correlated seeds can be generated or when the seed is poorly encrypted. Lim and Lee suggest to use generators like [AleChGoSch88] as possible pseudorandom generators. They specifically say that this is a good choice since these generators rely on the same (RSA) assumption as the cryptosystems and there is no need for further tools. They claim their system is semantically secure (that is, all encrypted bits are secure). However we show that the opposite is true and, in fact, this choice may be quite bad due to an algebraic interplay between the generator and the insecurity of the seed encryption.

Claim 1 *The Lim-Lee RSA based scheme when used with the RSA based pseudorandom generator of [AleChGoSch88] with exponent being 3, is insecure with respect to a "known ciphertext" only attack.*

Proof. From Theorem 1 an attacker can determine s^{3g} where $g = \gcd(e_A, H)$. Therefore, as argued, the attacker is able to determine, s^3 with good probability. Now using [AleChGoSch88] with the RSA function with exponent 3, the pad used in the encryption comprised of the concatenation of least significant bits of s, $s^3 \bmod n$, $(s^3)^3 = s^9 \bmod n$, $(s^9)^3 = s^{27} \bmod n$, and so on. Knowing s^3, in turn, gives all the bits but the first one, since exponentiation is easy! $\qquad\square$

Note again that the attack above did not employ any access to the device as an oracle, it was a pure known ciphertext attack. Encryption exponent being 3 (as in the generator) is a popular choice, since it provides the fastest modular exponentiation for RSA.

In addition we can also point out that Theorem 2 tells us that an attacker can generate outputs from the decryption functions based on seeds related to the original seed used to encrypt the meaningful ciphertext (i.e., the original C being attacked). Therefore we deduce that:

Corollary 3. *The Lim-Lee El Gamal based encryption scheme has a weakness when attacked by an adaptive chosen ciphertext attack – the attacker can correlate the seeds of the various ciphertexts.*

6 Conclusion and Discussion

The work of [LimLee93] discusses immunized public key encryption systems. It shows problems with previous such systems, and carefully discusses original requirements and designs of the first group-oriented such systems. We found their design goals, criticism of previous systems, discussions, methodologies, and some of their efficient techniques highly valuable.

On the other hand, we presented cryptanalysis of and security problems with the actual schemes of [LimLee93]. We showed that one can generate relations in a meaningful way even if it does not seem so or even if the ciphertext gram is partially authenticated by a hash function. We showed that in the case of group-oriented cryptography, the availability of partial results in the combiner may enable strong attacks (in the original schemes no trust was put in any single point and surely not in the combining stage). We also showed potential problems with the usage of a single key and the same algebraic problem for more than one task in a cryptosystem.

6.1 Potential improvements to the scheme

It seems difficult to modify the Lim-Lee El Gamal based scheme so that it is strong as the Diffie-Hellman problem due to the value c_4 in the ciphertext. That is, due to the way c_4 is generated, we do not see how opening up ciphertext C reduces to breaking the Diffie-Hellman problem.

We make the following preliminary suggestions to improve their El Gamal scheme.

(1) The function G should be a pseudorandom function rather than a pseudorandom number generator where s comprises of two values: a function index and an input value for that function. (In practice it is a block cipher operation). This, heuristically, will reduce the relation between ciphertext grams, but it may make the computations less efficient, though. In fact a random function based on generators built on a totally different algebraic problem than the seed encryption is preferable (e.g., one based on DES or triple-DES) to disentangle potential algebraic dependencies that may ease cryptanalysis.

(2) The value c_3 should be $h(c_0||c_1||c_2)$ instead of $h(c_0||c_2)$; it is preferable that again this hash function should not be based on the same algebraic problem as the other encryption mechanisms. The type of bindings in the resulting "ciphertext gram" above has been recently formalized and a suggestion for immunized RSA based system has been shown as well in [FranklinReiter95].

(3) At some point in their work, Lim and Lee suggest to also include in the preimage of h in c_3 information like date and other related information under the hash function. We remark that, in doing so, one has to be careful not to introduce unstructured redundancy which may enable potential "birthday" attack on this ciphertext validation component by playing with the additional information.

(4) Finally, when claiming security against chosen ciphertext attacks, the claims that the system is semantically secure and that it seems that the sender knows the cleartext from the ciphertext structure are not enough. Also, having a key perform different functions may be dangerous. We feel that the following two things better be done carefully when designing a chosen-ciphertext system based on heuristics. First, it may be a good idea to attempt to prove chosen ciphertext security under certain (strong, if not necessary known or widely assumed) properties of the tools used. Second, it is also useful to extensively characterize how the attacker may have produced the ciphertext or what may be easily computed from it, and given a ciphertext what ciphertexts can be computed adaptively from it (as the quality of the system against adaptive chosen ciphertext attacks relates directly to how the designers capture this fact in their arguments).

A natural open problem is designing a chosen ciphertext system which is practical and proven secure (e.g., as secure as El Gamal or RSA).

References

[AleChGoSch88]   W. Alexi, B. Chor, O. Goldreich and C. P. Schnorr, *RSA and Rabin functions: certain parts are as secure as the whole*, SIAM J. Computing vol. 17 (2). 1988.

[Apostol76]   T. M. Apostol *Introduction to analytic number theory*, Springer-Verlag, New York, 1976.

[BellareRogaway93]   M. Bellare and P. Rogaway, *Random Oracles are Practical: a paradigm for designing efficient protocols*, ACM, 1-st Comp. and Com. Sec. 1993.

[Damgard91]	I. Damgård, *Towards practical public key cryptosystems secure against chosen ciphertext attacks*, Advances in Cryptology–Proc. of Crypto '91.
[Desmedt87]	Y. Desmedt, *Society and group oriented cryptography: a new concept*, Advances in Cryptology, Proc. of Crypto '87, Springer-Verlag, 1988.
[DesmedtFrankel89]	Y. Desmedt and Y. Frankel, *Threshold cryptosystems*, Advances in Cryptology, Proc. of Crypto '89 Springer-Verlag, 1990.
[DiffieHellman76]	W. Diffie and M. Hellman, *New Directions in Cryptography* , IEEE Trans. on Information Theory 22 (6), 1976, pp. 644-654.
[DoDwNa91]	D. Dolev, C. Dwork and M. Naor, *Non-Malleable Cryptography*, Proc. of the 23rd Annual ACM Symposium on the Theory of Computing, 1991, pp. 542–560.
[ElGamal85]	T. El Gamal, *A Public key cryptosystem and a signature scheme based on discrete logarithm*, IEEE Trans. on Information Theory 31, 465-472, 1985.
[FranklinReiter95]	M.K. Franklin and M.K. Reiter, *Adaptive Chosen Ciphertext Security for RSA from G-Q Signatures*, Preliminary manuscript.
[GoGoMi86]	O. Goldreich S. Goldwasser and S. Micali, *How to Construct Random Functions* , J. of the ACM 33 (1986), pp. 792-807.
[GoldwMicali84]	S. Goldwasser and S. Micali, *Probabilistic Encryption*, J. Com. Sys. Sci. 28 (1984), pp 270-299.
[LimLee93]	C. H. Lim and P. J. Lee, *Another method for attaining security against adaptive chosen ciphertext attacks*, Advances in Cryptology–Proc. of Crypto '93.
[NaorYung90]	M. Naor and M. Yung, *Public-key cryptosystem provably secure against chosen ciphertext attack*, Proc. of the 22nd Annual Symposium on the Theory of Computing, 1990, pp. 427–437.
[Rabin79]	M. O. Rabin, *Digital Signatures and Public Key Functions as Intractable as Factoring*, Technical Memo TM-212, Lab. for Computer Science, MIT, 1979.
[RackoffSimon91]	C. Rackoff, and D. Simon, *Non-Interactive Zero-Knowledge Proof of Knowledge and Chosen Ciphertext Attacks*, Advances in Cryptology– Proc. of Crypto '91.
[RSA78]	R. Rivest, A. Shamir and L. Adleman, *A Method for Obtaining Digital Signature and Public Key Cryptosystems*, Comm. of ACM, 21 (1978), pp 120-126.
[ZhengSeberry92]	Y. Zheng and J. Seberry, *Immunizing public key cryptosystems against chosen ciphertext attacks*, IEEE Jour. on Selected Areas in Communications, 11(5), 1993, pp. 715-724. (Also in: Advances in Cryptology–Proc. of Crypto '92).

Secure Signature Schemes based on Interactive Protocols

Ronald Cramer (CWI, Amsterdam, cramer@cwi.nl),
Ivan Damgård (BRICS *, Aarhus University, ivan@daimi.aau.dk)

Abstract. Given only an interactive protocol of a certain type as a primitive, we can build a (non-interactive) signature scheme that is secure in the strongest sense of Goldwasser, Micali and Rivest (see [11]): not existentially forgeable under adaptively chosen message attacks. There are numerous examples of primitives that satisfy our conditions, e.g. Feige-Fiat-Shamir, Schnorr, Guillou-Quisquater, Okamoto and Brickell-Mc.Curley ([9], [17], [12], [15], [3]).
A main consequence is that efficient and secure signature schemes can now also be based on computationally difficult problems other than factoring (see [11]), such as the discrete logarithm problem.
In fact, the existence of one-way group homomorphisms is a sufficient assumption to support our construction. As we also demonstrate that our construction can be based on claw-free pairs of trapdoor permutations, our results can be viewed as a generalization of [11].

1 Introduction

This paper deals with the construction of *secure* signature schemes. By "secure", we mean that some well-defined computational assumption can be shown to be sufficient for the scheme not to be existentially forgeable, even under an adaptive chosen message attack. This notion of optimal security was introduced in [11]. Most, if not all, signature schemes used in practice such as ISO9796/RSA or DSA are based on a computational assumption that is certainly *necessary* for this kind of security, but not known to be *sufficient*.

Goldwasser, Micali and Rivest [11] were the first to find a provably secure signature scheme, based on the existence of claw-free pairs of trapdoor permutations. Merkle [13] showed essentially that existence of collision intractable hash functions is a sufficient assumption. Naor and Yung showed that any one-way permutation is also enough [14], and finally this was reduced to any one-way function (which is also a necessary assumption) by Rompel [16].

Although secure signature schemes are generally less efficient than the ones used in practice, the efficiency of the GMR scheme is not too bad when based on factoring. Measured in signature length and time for signature generation, GMR is worse than plain RSA by a factor of 5-20, depending on the number of messages to be signed. By relying on the (perhaps) stronger assumption that

---

* Basic Research in Computer Science, Centre of the Danish National Research Foundation

D. Coppersmith (Ed.): Advances in Cryptology - CRYPTO '95, LNCS 963, pp. 297-310, 1995.
© Springer-Verlag Berlin Heidelberg 1995

RSA is hard to invert, Bos and Chaum [2] have been able to build an even more efficient secure scheme. Dwork and Naor [7] have exhibited an efficient and secure signature scheme whose security is also equivalent to the difficulty of RSA-inversion. In contrast with other schemes that use authentication trees, such as [11], they are able to re-use the authenticating nodes many times. As a result of this and further exploitations of the specific properties of the RSA functions, the length of their signatures can be made quite small, although a price has to be paid in the form of a large public file. In [6], similar trade-offs between shared random strings and the size of secure signatures have been achieved for certain families of claw-free pairs of trapdoor permutations, in particular a family based on the dificulty of factoring integers.

On the theoretical side, the reduction in the necessary assumptions by [13], [14] and [16] have come at the price of dramatically reduced efficiency. In particular, signatures have become larger. Where a GMR signature is of length $O(k \log i)$ bits, where k is the security parameter and i indicates the number of signatures made, a Naor-Yung signature would typically be of length $O(k^2 \log i)$ bits, because a full preimage under a one-way function is required to authenticate 1 bit.

Thus it has been an open question whether secure signatures with efficiency comparable to or better than that of GMR could be based on more general assumptions than claw-free pairs of trapdoor permutations.

In this paper, we show that secure signature schemes with signatures as short as those of GMR can be built if so called signature protocols exist. In particular, our schemes have the same property as GMR that the length of signatures grow logarithmically with the number of messages signed. Note, however, that Goldreich [10] has shown that the GMR scheme can be modified so that *all* signatures have length $O(k \log k)$ bits. This same modification applies to our scheme as well. Note also that Goldreich's modification makes the signature scheme memoryless, which implies that a signature will not reveal the number of signatures made so far.

Dropping some technical details, a signature protocol is an interactive protocol for a hard problem that uses three messages, where the prover speaks first and the verifier sends a random challenge as the second message. The essential properties are

- The protocol must be secure (zero-knowledge) against the honest verifier.
- The challenge must be longer than the prover's first message.
- It must be infeasible for a cheating prover to answer more than one challenge in a given protocol execution.

Furthermore, we show that it is sufficient for the existence of signature protocols (and hence for the existence of secure signatures) that one-way group homomorphisms exist. This has a nice theoretical consequence, because it shows that, compared to GMR, the trapdoor property can be traded for the homomorphism property without getting longer signatures. Moreover, our construction allows us, in both signature generation and verification, to minimize the number of evaluations of the one-way function and replace them by evaluations of the group operation in the groups involved. This means that we can use the discrete logarithm assumption as a basis for secure signatures in a much more efficient way than known before. Where earlier methods would, with security parameter k, require $O(k^2)$ exponentiations per basic authentication step and give signatures of length $O(k^2 \log i)$ bits, our method requires $O(1)$ exponentiations and gives signatures of length $O(k \log i)$.

We also show that existence of a three pass public coin proof of knowledge for any hard problem (A hard random self-reducible problem would be enough for this) and a collision intractable hash function implies existence of signature protocols. Although the hash function alone would be enough to construct secure signatures, using our method may lead to shorter signatures ($O(k \log i)$ compared to $O(k^2 \log i)$), depending on the protocol used.

2 Signature Protocols

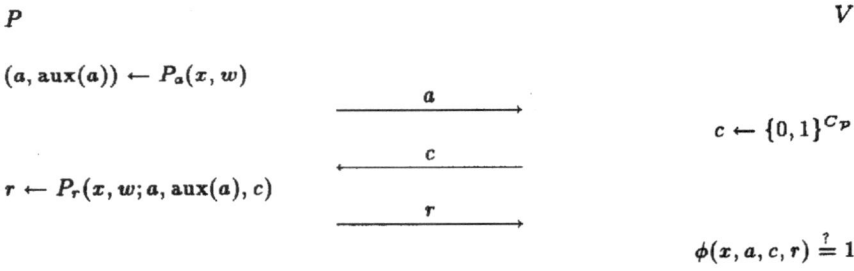

Fig. 1. Protocol $\mathcal{P}$, common input x, private input for P is w

This section is devoted to defining the basic building block, a *signature protocol*, that is used in our construction for secure signatures. Let $\mathcal{P}$ be a three round public coin protocol where the prover speaks first. Figure 1 depicts the kind of protocol we will look at. It resembles a proof of knowlege for a binary relation R (see for instance [8] for details), in that the prover can always make the verifier accept on common input x, if the prover knows w such that $(x, w) \in R$.

Indeed, by running (probabilistic) polynomial time algorithm P_a on x and his secret witness w, the prover P computes his initial message a, and some (secret) auxiliary information $\mathrm{aux}(a)$. The length of this first message a is denoted $A_{\mathcal{P}}$, the *authentication length*, which only depends on x. After having received a, the verifier V chooses a challenge c uniformly at random, and sends it to P. The length of admissible challenges in $\mathcal{P}$ is called the *challenge length* $C_{\mathcal{P}}$ (we will sometimes abuse this notation to refer to the *set* of possible challenges). Also here, it is assumed to depend only on x. The prover P completes the conversation by running (probabilistic) polynomial time algorithm P_r on x, w, a, c, and, the auxiliary information aux(a) for a. The resulting response r is submitted to the verifier V. We will assume that the procedure ϕ that the verifier V invokes to test the validity of the conversation, is a polynomial time algorithm. The collection of all possible accepting conversations with respect to x will be denoted $Acc(x)$. For the rest of this paper, $\mathcal{P}$ will denote a protocol as described above.

For the purpose of constructing secure signature schemes, we require the following from $\mathcal{P}$ in stead of the ordinary soundness condition:

Definition 1 *Let k be a security parameter for protocol $\mathcal{P}$. Suppose we are given a probabilistic polynomial time generator G for relation R that on input 1^k produces $(x, w) \in R$, such that no probabilistic polynomial time algorithm, given x as input, can generate two accepting conversations (with respect to x) (a, c, r), (a, c', r') from $Acc(x)$, with $c \neq c'$, except with negligible probability of success. Then $\mathcal{P}$ is called* collision intractible *over G.*

Next, we need the protocol $\mathcal{P}$ to be honest verifier zero-knowledge, that is, we only demand that conversations with a verifier who follows protocol $\mathcal{P}$ can be simulated. Additionally, we require that the simulator outputs accepting conversations where the challenge can be chosen in advance, i.e., the simulator can take any value c as input, and will output an accepting conversation where the challenge is equal to c. A protocol $\mathcal{P}$ satisfying these conditions will be called *special honest verifier zero-knowledge*.

More precisely, let $(x, w) \in R$ and let a prover P and a verifier V with common input x be given. The prover has w as private input. Then $\mathcal{P}(x, w)$ denotes the probability distribution on $Acc(x)$ induced by conversations between P and V, provided that they both follow protocol $\mathcal{P}$ honestly. We require the following:

Definition 2 *Let $(x, w) \in R$. Suppose we are given a probabilistic polynomial time algorithm S with the following properties.*

1. *On input x and any $c \in C_\mathcal{P}$, S outputs an accepting conversation from $Acc(x)$, where c is the challenge.*
2. *The distribution of $S(x, c)$, where c is chosen uniformly at random from $C_\mathcal{P}$, is equal to $\mathcal{P}(x, w)$.*

Then $\mathcal{P}$ is called special honest verifier zero knowledge, *and S its* special simulator.

In the following we will demonstrate that a protocol $\mathcal{P}$ that is special honest verifier zero-knowledge, is in fact secure against a slightly more general verifier. It follows immediately from Definition 2 that, for each fixed $c \in C_\mathcal{P}$, $S(x, c)$ outputs conversations $(a, c, r) \in Acc(x)$ with exactly the same distribution as $(a \leftarrow P_a(x, w), c, r \leftarrow P_r(x, w, a, \text{aux}(a), c))$, i.e., according to the honest prover who has access to (x, w). Therefore, in order for the conversations to be simulatible, it is sufficient that the verifier chooses challenges c independently from the first message in any given execution of $\mathcal{P}$. This proves the following theorem:

Theorem 1 *If $\tilde{V}$ is any probabilistic polynomial time verifier who, in any given execution of protocol $\mathcal{P}$, chooses the challenge c independently from the prover's first message a, then the conversation between prover P and verifier $\tilde{V}$ can be simulated by means of the special simulator $\mathcal{S}$.*

Summarizing, we require the following of our protocol $\mathcal{P}$ in order for it to support our construction of (non-interactive) secure signature schemes.

Definition 3 *Suppose $\mathcal{P}$ satisfies the following conditions.*

1. $C_{\mathcal{P}} > A_{\mathcal{P}}$.
2. $\mathcal{P}$ is collision-intractible over G.
3. $\mathcal{P}$ is special honest verifier zero-knowledge.

Then $\mathcal{P}$ is called a signature protocol. *If $\mathcal{P}$ satisfies the second condition and is honest verifier zero-knowledge (so it does not necessarily have a special simulator), $\mathcal{P}$ is called a* quasi signature protocol.

The following theorem shows that any given signature protocol $\mathcal{P}$ can be transformed into a new signature protocol $\mathcal{P}^*$ where the challenge length $C_{\mathcal{P}^*}$ can be of any size polynomial in the security parameter k.

Theorem 2 *Suppose there exists a signature protocol $\mathcal{P}$ for relation R and generator G, then there is a signature protocol $\mathcal{P}^*$ for R and G, satisfying that $C_{\mathcal{P}^*} = t$, for any t polynomial in the security parameter k.*

Proof. Without loss of generality, we may assume that $A_{\mathcal{P}} + 1 = C_{\mathcal{P}}$ The protocol $\mathcal{P}^*$ goes as follows:

1. The prover sends a first message a to the verifier, where a is computed as in $\mathcal{P}$.
2. The verifier sends t random bits $b_1, \ldots, b_t$.
3. The prover sends t conversations in $\mathcal{P}$, $(a_i, c_i, r_i), i = 1, \ldots, t$, where $c_i = b_i || a_{i+1}$ for $i = 1, \ldots, t-1$ and $c_t = b_t || 0 || \cdots || 0$.
4. The verifier checks that $a = a_1$, that all conversations are accepting conversations, and that $c_i = b_i || a_{i+1}$ for $i = 1, \ldots, t-1$, and that $c_t = b_t || 0 || \cdots || 0$.

By construction, the challenge length t for $\mathcal{P}^*$ can be chosen what we want it to be, provided $t = \text{poly}(k)$. Suppose now that we are given two accepting conversations in $\mathcal{P}^*$ for some public string x with the same first message a, but with different challenges $(b_1, \ldots, b_t)$ and $(b_1', \ldots, b_t')$. Let, for $j = 1 \ldots t$, (a_j, c_j, r_j) and (a_j', c_j', r_j') be the respective replies in those conversations in $\mathcal{P}^*$, and let i be an index such that $b_i \neq b_i'$. Clearly, this implies that $c_i \neq c_i'$. Take i to be the smallest index such that $c_i \neq c_i'$. If $i = 1$, we have a collision in $\mathcal{P}$ with respect to x, as by definition of $\mathcal{P}^*$, we must have $a_1 = a_1' = a$. On the other hand, if $i > 1$, c_{i-1} must be equal to c_{i-1}', i.e., $b_{i-1} || a_i = b_{i-1}' || a_i'$. But then $a_i = a_i'$ and we have a collision (a_i, c_i, r_i), (a_i', c_i', r_i') in $\mathcal{P}$ with respect to x. Therefore, $\mathcal{P}^*$ is collision-intractible over R and G.

As for special honest verifier zero-knowledge of $\mathcal{P}^*$, we now exhibit a special simulator S^* for $\mathcal{P}^*$, that runs S as a subroutine. S^* starts by receiving a public string x and a challenge $(b_1, \ldots, b_t)$ as input. It proceeds by putting $c_t = b_t||0||\cdots||0$, and feeding x and c_t to S. After S has output an accepting conversation (a_t, c_t, r_t) in $\mathcal{P}$ with respect to x, S^* repeats the following for $i = t-1 \ldots 1$. Put $c_i = b_i||a_{i+1}$, feed x and c_i to S and receive an accepting conversation (a_i, c_i, r_i) from S. By invoking Theorem 1, it is clear that S^* generates accepting conversations in $\mathcal{P}^*$ with respect to x, with exactly the same distribution as the conversations with the honest verifier in $\mathcal{P}^*$.

Thus, in the constructions to follow, whenever we have a signature protocol, we may assume that the challenge length is whatever we need it to be. Before investigating under which general assumptions signature protocols can be shown to exist, we mention Guillou-Quisquater [12], Okamoto [15] (both the factoring and the RSA-versions) and Fiat-Shamir [9] (if the number of secret roots is chosen sufficiently large) as examples of proofs of knowledge that can be viewed as signature protocols. Schnorr's discrete log protocol [17] does not directly satisfy the conditions, but can be modified to do so since it is based on a one-way group homomorphism (see below).

3 Sufficient Assumptions

The most general computational assumptions we have been able to find, sufficient for existence of signature protocols, is the existence of one-way group homomorphisms, and the existence of claw-free pairs of trapdoor permutations. No implication is known in either direction between these two assumptions.

One-Way Group Homomorphisms

Definition 4 *A family of one-way group homomorphisms is a family of group homomorphisms $\mathcal{F} = \{f : G \to H\}$. In the following, we let $k_f = \log_2(|H|)$, i.e. the number of bits needed to represent an element in H. We will sometimes drop subscript f, if it is clear which f we refer to. The family has to satisfy the following properties:*

1. *There is a polynomial time algorithm which given f and $w \in G$, computes $f(w)$ in time polynomial in k.*
2. *There is a probabilistic polynomial time algorithm which on input 1^k outputs an element $f : G \to H$ chosen uniformly from $\mathcal{F}$, subject to $k = k_f$.*
3. *The elements $f : G \to H \in \mathcal{F}$ satisfy that there is a probabilistic algorithm which given G outputs an element chosen uniformly from G, in time polynomial in k.*
4. *The one-way property: Let A be any probabilistic polynomial time algorithm which receives input f and $f(w)$, where f, w are chosen as in points 2 and 3. Then the probability that A outputs y such that $f(y) = f(w)$ is superpolynomially small in k.*

5. *The elements $f : G \to H \in \mathcal{F}$ satisfy that group operation and inversion in G and H can be computed in time polynomial in k.*

Examples of possible one-way group homomorphisms are the RSA functions, squaring modulo a composite number, or discrete exponentiation modulo a prime, or on an elliptic curve. Given a family as in this definition, we can make the following binary relation and generator for it:

Definition 5 *Let $\mathcal{F}$ be as in Definition 4. Then $R_{\mathcal{F}}$ is the binary relation consisting of pairs $((f, x_1, \ldots, x_{k_f+1}), (w_1, \ldots, w_{k_f+1}))$, where $f \in \mathcal{F}$ and $f(w_i) = x_i$. $G_{\mathcal{F}}$ is the generator that on input 1^k generates f using property 2 of Definition 4, generates $w_1, \ldots, w_{k_f+1}$ using property 3 and finally computes $x_i = f(w_i)$.*

Theorem 3 *Suppose $\mathcal{F}$ is a family of one-way group homomorphisms. Then there exists a signature protocol for $R_{\mathcal{F}}$ and $G_{\mathcal{F}}$.*

Proof. The protocol claimed takes $f, x_1, \ldots, x_{k+1}$ as common input, while the witnesses $w_1, \ldots, w_{k+1}$ are private input to the prover. The protocol is now a straightforward generalization of Feige-Fiat-Shamir [9] and goes as follows:

1. The prover chooses a random $r \in G$ and sends $f(r)$ to the verifier.
2. The verifier chooses bits $e_1, \ldots, e_{k+1}$ at random and sends them to the prover.
3. The prover returns $z = r \cdot w_1^{e_1} \cdots w_{k+1}^{e_{k+1}}$. The verifier checks that $f(z) = f(r) \cdot x_1^{e_1} \cdots x_{k+1}^{e_{k+1}}$

This protocol is clearly complete with probability 1. Honest verifier zero knowledge is clear by standard arguments: first choose z and $e_1, \ldots, e_{k+1}$ at random, then use this to compute an $f(r)$-value. It is also clear that the challenge is one bit longer than the first message from the prover. Thus, only the collision intractable property remains to be argued. Assume by contradiction that some enemy A can produce z, z' and $(e_1, \ldots, e_{k+1}) \neq (e'_1, \ldots, e'_{k+1})$ such that $f(z) = f(r) \cdot x_1^{e_1} \cdots x_{k+1}^{e_{k+1}}$ and $f(z') = f(r) \cdot x_1^{e'_1} \cdots x_{k+1}^{e'_{k+1}}$. This means that

$$f(z \cdot z'^{-1}) = x_1^{d_1} \cdots x_{k+1}^{d_{k+1}},$$

where all d_i are 1, -1 or 0, and at least one of them is non-zero.

We can then build the following algorithm which will invert f with the help of A: given a random f-image x, generate an output seemingly coming from $G_{\mathcal{F}}$ as follows: choose $w_1, \ldots, w_{k+1}$ and $1 \leq j \leq k+1$ at random. Put $x_i = f(w_i)$ for $i \neq j$, and $x_j = f(w_j) \cdot x$. Now run A's algorithm with f and the x_i's as input. Clearly the set of x_i is distributed exactly as output from $G_{\mathcal{F}}$, whence A's success probability is the same as in real life. Note that if A has success, we can write x^{d_j} as

$$x^{d_j} = f(z \cdot z'^{-1} \cdot \prod_i w_i^{-d_i})$$

Now note that the set of x_i's contains no information about j, whence the probability that $d_j \neq 0$, given that A has success, is at least equal to $1/(k+1)$.

Remark 1 *It is clear that the protocol constructed in the proof above can be modified to have any challenge length desired by having more x_i-values. Enlarging the challenge length in this way will be more efficient than using Theorem 2.*

We briefly state that, in the definition of a signature protocol, we can exchange the assumption on the challenge length and the existence of a *special* simulator for the assumption that we are given a family of collision-intractable hash-functions, as can be seen from the following theorems. The proofs are given in [5].

Theorem 4 *Let $\mathcal{P}$ be honest verifier zero-knowledge and collision-intractable over R and G. Then $\mathcal{P}$ can be compiled into a protocol $\mathcal{P}^*$ (for relation R and generator G), that is also collision-intractable over R and G but that additionally satisfies special honest verifier zero-knowledge.*

Theorem 5 *Suppose there exists a quasi signature protocol $\mathcal{P}$ for relation R and generator G and that a family $\mathcal{H}$ of collision intractable hash functions exists. Then there exists a signature protocol $\mathcal{P}^*$ for $R_{\mathcal{H}}$ and $G_{\mathcal{H}}$. Here $R_{\mathcal{H}}$ consists of pairs $((x, h), w)$ where $(x, w) \in R$, w is of length k bits and $h \in \mathcal{H}$ has output length k. The generator $G_{\mathcal{H}}$ runs G to generate (x, w) and then selects $h \in \mathcal{H}$ with the desired output length.*

Claw-Free Pairs of Trapdoor Permutations

In [11], a secure signature scheme is exhibited, based on (a family of) claw-free pairs of trapdoor permutations. The following theorem is proved in [5].

Theorem 6 *Suppose that a family of claw-free pairs of trapdoor permutations exists. Then there exists a signature protocol.*

4 Main Result

We will now present the new signature scheme $\Sigma_{\mathcal{P}}$, based on a signature protocol $\mathcal{P}$. In Section 5, we prove that $\Sigma_{\mathcal{P}}$ is not existentially forgeable under adaptively chosen message attacks. A concrete example is given in Section 6.

Theorem 7 *Let $\mathcal{P}$ be a signature protocol for relation R and generator G. Then the signature scheme $\Sigma_{\mathcal{P}}$ is not existentially forgeable under adaptively chosen message attacks.*

It is assumed that we are given a signature protocol $\mathcal{P}$ for relation R and generator G. By Theorem 2, we may assume that for each security parameter k and for each instance (x, w) as output by running $G(1^k)$, the (non-constant) polynomial $t(k)$ satisfies $t = C_{\mathcal{P}} \geq 3 \cdot A_{\mathcal{P}}$. The construction of $\Sigma_{\mathcal{P}}$ from $\mathcal{P}$ works as follows.

Initialization Phase

Given a security parameter k, the signer uses the generator G to generate two solved instances x_0 and x_1, with respective witnesses w_0 and w_1. He also computes $(a_1^1, \text{aux}(a_1^1)) \leftarrow P_a(x_1, w_1)$ and puts (x_0, x_1, a_1^1) in his public directory.

Signing Phase

Let $m \in \{0,1\}^t$ be the message to be signed and let $i \geq 1$. The i-th signature, on a message $m \in \{0,1\}^t$, is computed as follows. First, the signer computes

1. $(a_0^i, \text{aux}(a_0^i)) \leftarrow P_a(x_0, w_0)$,
2. $r_0^i \leftarrow P_r(x_0, w_0; a_0^i, \text{aux}(a_0^i), m)$,
3. $(a_1^{2i}, \text{aux}(a_1^{2i})) \leftarrow P_a(x_1, w_1)$, $(a_1^{2i+1}, \text{aux}(a_1^{2i+1})) \leftarrow P_a(x_1, w_1)$,
4. $r_1^i \leftarrow P_r(x_1, w_1; a_1^i, \text{aux}(a_1^i), a_1^{2i} \| a_1^{2i+1} \| a_0^i)$.

The signer stores a_1^{2i}, $\text{aux}(a_1^{2i})$, a_1^{2i+1}, $\text{aux}(a_1^{2i+1})$, a_0^i, r_1^i. Let $\text{Auth}(a_0^i)$ be an authentication path for a_0^i, i.e., $\text{Auth}(a_0^i)$ consists of all tuples of the form $(a_1^j, a_1^{2j}, a_1^{2j+1}, a_0^j, r_1^j)$, with $1 \leq j \leq i$, such that a_1^j is an ancestor of a_1^i. We assume that the tuples in $\text{Auth}(a_0^i)$ are ordered in decreasing ancestry from left to right. The signature $\sigma(m)$ on m consists of $(\text{Auth}(a_0^i), r_0^i)$.

Verification Phase

The receiver puts $\sigma(m) \equiv (\text{Auth}(a_0^{j_r}), r_0^{j_r})$, where r is the number of tuples in $\text{Auth}(a_0^i)$ and $(a_1^{j_l}, a_1^{2j_l}, a_1^{2j_l+1}, a_0^{j_l}, r_1^{j_l})$ is the l-th tuple in $\text{Auth}(a_0^{j_r})$. After having checked whether $a_1^{j_1} \overset{?}{=} a_1^1$, the receiver has to perform the following verifications, for $j = 2, \ldots, r$.

1. $a_1^{j_l} \overset{?}{\in} \{a_1^{2j_{l-1}}, a_1^{2j_{l-1}+1}\}$
2. $\phi(x_1, a_1^{j_l}, a_1^{2j_l} \| a_1^{2j_l+1} \| a_0^{j_l}, r_1^{j_l}) \overset{?}{=} 1$.

Finally, he checks whether $\phi(x_0, a_0^{j_r}, m, r_0^{j_r}) = 1$. If all verifications hold, the signature is accepted.

Note that, by assumption on the challenge length $t(k)$, $2 \cdot A_P(x_1) + A_P(x_0) \leq t$, so the challenges are long enough to encode the strings $a_1^{2i} \| a_1^{2i+1} \| a_0^i$. These strings can be padded up to t bits, if necessary, using standard techniques. As we have also assumed that all occurring values have fixed length descriptions (depending only on the corresponding public string), parsing these concatenations is easy.

5 Proof of Security

Our notion of security for signature schemes is that of [11]. In this section we show that no polynomially bounded adversary can construct a forgery on a message that hasn't been signed by the real signer, even if he is allowed to get polynomially many signatures on messages that he has chosen in an adaptive fashion. It will be shown that the existence of such a successful forger contradicts the assumption that the protocol $\mathcal{P}$ is collision intractable over the generator G. To this end, we compile this successful forger into an attacker that breaks that assumption.

In the following theorem, it is assumed that we are given a signature protocol $\mathcal{P}$ for generator G and relation R. By Theorem 2, we may assume that for each security parameter k and for each instance (x, w) as output by running $G(1^k)$, the (non-constant) polynomial $t(k)$ satisifies $t = C_{\mathcal{P}} \geq 3 \cdot A_{\mathcal{P}}$.

Theorem 8 *Any probabilistic polynomial time cracking algorithm $\mathcal{A}$ that forges a signature on a new message with probability $\epsilon(k)$, after at most polynomially many calls to a signer, can be compiled into probabilistic polynomial time procedure $\mathcal{A}^*$ that breaks the collision intractability of $\mathcal{P}$ over G with probability of the order of $\epsilon(k)$. The running time of $\mathcal{A}^*$ is of the same order as the running time of $\mathcal{A}$.*

Proof. Let a security parameter k be given, and let x be an instance of $\mathcal{P}$ generated by G on input 1^k.

We now describe how $\mathcal{A}^*$ cracks the collision intractability of $\mathcal{P}$ by using the forger $\mathcal{A}$ and the following simulation of $\Sigma_{\mathcal{P}}$. $\mathcal{A}^*$ receives x as input.

$\mathcal{A}^*$ first runs G on input 1^k in order to obtain a solved instance (x', w'). Then a bit b is chosen at random. Put $(x_b, w_b) = (x', w')$, and $x_{1-b} = x$.

For the simulation, we distinguish between two cases.

Case $b = 0$: We create an authentication tree with $P(k)$ internal nodes, starting at the leaves. The leaves a_1^j are generated as follows.

1. $c^j \leftarrow \{0, 1\}^t$
2. $(a_1^j, c^j, r_1^j) \leftarrow \mathcal{S}(x_1, c^j)$.

For children a_1^{2i} and a_1^{2i+1}, generate $a_0^i \leftarrow P_a(x_0, w_0)$. Then the parent a_1^i will be generated as

$$(a_1^i, a_1^{2i}||a_1^{2i+1}||a_0^i, r_1^i) \leftarrow \mathcal{S}(x_1, a_1^{2i}||a_1^{2i+1}||a_0^i).$$

The resulting instance (x_0, x_1, a_1^1) of $\Sigma_{\mathcal{P}}$ is sent to the forger $\mathcal{A}$. After this, the cracking algorithm can start making its (at most $P(k)$) calls.

The above takes care of $\text{Auth}(a_0^i)$, for $i = 1, \ldots, P(k)$. Note that this simulation can now deal with any signature request, as the i-th signature, on a message m^i, can be completed by computing $r_0^i \leftarrow P_r(x_0, w_0; a_0^i, \text{aux}(a_0^i), m^i)$.

Case $b = 1$:

1. Generate $(a_1^1, \text{aux}(a_1^1)) \leftarrow P_a(x_1, w_1)$, and send the instance (x_0, x_1, a_1^1) to the forger $\mathcal{A}$.
2. Let $m^i \in \{0, 1\}^t$ be the i-th message to be signed. Generate $a_0^i \leftarrow \mathcal{S}(x_0, m^i)$. Proceed as in Step 3 of the signing phase of $\Sigma_{\mathcal{P}}$.

Note that in both cases the simulation can deal with any signature request, by the properties of the special simulator $\mathcal{S}$. Furthermore, the distribution of the a_0^i, r_0^i, a_1^i and r_1^i is always according to the honest signer who has access to both w_0 and w_1, by Theorem 1. Thus the simulation is perfect, and we may now assume that the cracking algorithm outputs a forgery on a new message (i.e, a message that has not been signed by the simulator) $\tilde{m}$. Without loss of generality, we assume that this happens after exactly $P(k)$ calls, with probability $\epsilon(k)$.

Let $(\text{Auth}(a_0), r_0)$ be the forgery, on a new message $\tilde{m}$. Suppose that $a_0 = a_0^j$ for some $1 \leq j \leq P(k)$, with probability $\epsilon_1(k)$. As $\tilde{m}$ has not been signed by the simulation, we must have $\tilde{m} \neq m^j$, so $\mathcal{A}^*$ can get a collision for $\mathcal{P}$ from $(a_0, \tilde{m}, r_0)$ and (a_0^j, m^j, r_0^j).

If, on the contrary, $a_0 \neq a_0^j$ for all $1 \leq j \leq P(k)$, then there clearly exist a tuple $(a_1', a_1'', a_1''', a_0', r_1')$ in $\text{Auth}(a_0)$ and a node a_1^i in the tree, with $a_1' = a_1^i$, such that a_1^i is a leaf or a_1^i is an internal node with $a_1''||a_1'''||a_0' \neq a_1^{2i}||a_1^{2i+1}||a_0^i$.

In case a_1^i is an internal node, say with probability $\epsilon_2(k)$, we immediately get a collision. If a_1^i is a leaf, with probability $\epsilon_3(k)$, however, the probability that $a_1''||a_1'''||a_0' \neq c^i$ is $1 - \frac{1}{2^t}$, as the distribution of a_1^i is independent of the distribution of c^j (by the properties of the special simulator), and c^j was chosen uniformly at random. Thus in this case we get a collision with probability $1 - \frac{1}{2^t}$. From the perfectness of the simulation it follows that the distribution of everything sent to $\mathcal{A}$ is independent of b. Therefore the probability that $\mathcal{A}^*$ can compute a collision for the instance $x_{1-b} = x$ is

$$\frac{1}{2}\epsilon_1(k) + \frac{1}{2}\epsilon_2(k) + \frac{1}{2}(1 - \frac{1}{2^t})\epsilon_3(k) \geq \frac{1}{2}\epsilon(k) - \frac{1}{2^{t+1}}\epsilon_3(k),$$

which is clearly of the same order as $\epsilon(k)$. Thus we have shown that any forger of the signature scheme $\Sigma_{\mathcal{P}}$ can be turned very efficiently into a cracker of the collision intractibility of $\mathcal{P}$, with essentially the same probability of success.

6 Concrete Examples

We now describe a signature scheme whose security is equivalent to the difficulty of computing discrete logarithms, by applying our main construction to a suitable transformation of the discrete log based protocol of Schnorr [17]. In its basic form, this is a protocol for proving knowledge of a discrete log in a group $\mathcal{G}$ of prime order q. Such a group can be realized, for example as a subgroup of $\mathbf{Z}_p^*$, where p is a prime, and q divides $p - 1$.

This protocol is a quasi-signature protocol by standard arguments. With some modifications, it can be turned into a signature protocol: we will have as input to the protocol d instances instead of one, $(x_1, w_1), \ldots, (x_d, w_d)$, where $x_i = g^{w_i}$. Then the new protocol $\mathcal{P}$ goes as follows:

1. The prover chooses z at random in $[0, \ldots, q)$, and sends $a = g^z$ to V.
2. The verifier chooses $c_1, \ldots, c_d$ at random in $[0, \ldots, 2^l)$, and sends them to P.
3. P sends $r = (z + c_1 w_1 + \cdots + c_d w_d) \bmod q$ to V, and V checks that $g^r = a \cdot x_1^{c_1} \cdots x_d^{c_d}$,.

where $l = \lfloor \log_2(q) \rfloor$. Completeness and special honest verifier zero-knowledge are clear by the same arguments as above. Collision intractability can be shown by essentially the same proof as for Theorem 3. Finally, it is clear that by choosing

d large enough, we can get a large enough challenge length, and therefore a signature protocol.

We can now carry out our construction of $\Sigma_\mathcal{P}$ (see also Section 4). To set up an instance of $\Sigma_\mathcal{P}$, the signer generates two independent instances of $\mathcal{P}$, $(x, w) \equiv ((x_1, w_1), \ldots, (x_d, w_d))$ and $(\overline{x}, \overline{w}) \equiv ((\overline{x}_1, \overline{w}_1), \ldots, (\overline{x}_d, \overline{w}_d))$, with $x_i = g^{w_i}$ and $\overline{x}_i = g^{\overline{w}_i}$ for $i = 1, \ldots, d$. The w_i and $\overline{w}_i$ are chosen at random from $\mathbf{Z}_q$. Note that both these instances use the same pair $(g, \mathcal{G})$. The root of the authentication tree, a_1^1, is computed as $a_1^1 = g^{z_1^1}$, where z_1^1 is chosen at random from $\mathbf{Z}_q$. The initialization phase of $\Sigma_\mathcal{P}$ is completed when the public key of the signer, $(x, \overline{x}, a_1^1)$, is placed in the public directory.

We will now show how the signer computes the first signature on a message $m \in \{0, 1\}^{d \cdot l}$, where $m = m_1 || \ldots || m_d$ and the m_i are l-bitstrings, to be interpreted as members of $[0 \ldots 2^l)$.

First, he computes a_0^1 as $a_0^1 = g^{z_0^1}$, with z_0^1 chosen at random from $\mathbf{Z}_q$, and r_0^1 as $r_0^1 = z_0^1 + m_1 w_1 + \cdots + m_d w_d$. Before establishing an authentication for a_0^1, he computes a_1^2 and a_1^3 (in the same way as a_1^1). Next, a_0^1 is authenticated, together with a_1^2 and a_1^3, by computing r_1^1 as $r_1^1 = z_1^1 + \mu_1 \overline{w}_1 \cdots + \mu_d \overline{w}_d$, where $\mu_1 || \cdots || \mu_d = a_1^2 || a_1^3 || a_0^1$. The μ_i are l-bitstrings, to be interpreted as members of $[0, \ldots, 2^l)$.

The values r_0^1, r_1^1, a_0^1, a_1^2 and a_1^3 are forwarded to the receiver, who checks whether

1. $g^{r_0^1} \stackrel{?}{=} a_0^1 \cdot x_1^{m_1} \cdots x_d^{m_d}$, and
2. $g^{r_1^1} \stackrel{?}{=} a_1^1 \cdot \overline{x}_1^{\mu_1} \cdots \overline{x}_d^{\mu_d}$.

Note that the values a_1^2 and a_1^3 are ready to play the role of a_1^1 in the second and third execution of $\Sigma_\mathcal{P}$, i.e., to authenticate a_0^2, a_1^4, a_1^5, and a_0^3, a_1^6, a_1^7, respectively.

The i-th signature $(i > 1)$ consists of a_0^i, r_0^i and an authentication path for a_0^i, which is a list of all tuples $(a_1^j, a_1^{2j}, a_1^{2j+1}, a_0^j, r_1^j)$ such that a_1^j is an ancestor of a_1^i.

For example, such a list for a_0^{11} would effectively consist of a_1^1, a_1^2, a_1^3, a_1^4, a_1^5, a_1^{10}, a_1^{11}, a_0^1, a_0^2, a_0^5, r_0^1, r_0^2 and r_0^5. Note that verification of the path requires only three exponentiations in $\mathcal{G}$.

We get signatures of length $O(k \log i)$ bits, where k is the number of bits needed to represent an element in $\mathcal{G}$, and i indicates the number of signatures made. Moreover, one authentication step requires a constant number of exponentiations in $\mathcal{G}$, both for signing and verification. Note that the 1 exponentiation needed from the signer uses input independent from the bits authenticated $(c_1, \ldots, c_d)$. Therefore we can use the idea suggested by Schnorr of having the signer precompute this exponentiation if some idle time is available on his computer. This way the on-line time to generate a signature becomes almost negligible.

Previously, the only known way to get a signature scheme provably secure based on discrete log was to use the method from [4] to build a collision intractable hash function and then use Merkle's construction. This would require an exponentiation for each bit processed in the hashing, and moreover we would need as a part of the signature a full preimage under the hash function to authenticate 1 bit. Therefore we would get signatures of length $O(k^2 \log i)$ bits and would need $O(k^2 \log i)$ exponentiations to make a signature.

7 Conclusion

We have shown that the existence of *signature protocols* is a sufficient condition for the existence of signature schemes that are not existentially forgeable under adaptively chosen message attacks, which is the strongest notion of security for signature schemes (see [11]). The length of the signatures in our schemes grows logarithmically in the number of signatures. In addition to the existence of claw-free pairs of trapdoor permutations, on which the scheme from [11] is based, the most general computational assumption we have been able to find, sufficient for the existence of signature protocols, is the existence of one-way group homomorphisms. As an example, we have presented a signature scheme whose security is equivalent to the difficulty of computing discrete logarithms.

8 Acknowledgements

It's a pleasure to thank Berry Schoenmakers for many valuable discussions and comments. Also thanks to Matt Franklin for commenting on an earlier version of this paper.

References

1. M. Abadi, E. Allender, A. Broder, J. Feigenbaum and L. Hemachandra: *On Generating Solved Instances of Computational Problems*, Proc. of Crypto 88, Springer Verlag LNCS series.
2. J.Bos and D.Chaum: *Provably Unforgeable Signatures*, Proc. of Crypto 92, Springer Verlag LNCS series.
3. E. F. Brickell and K. S. McCurley: *An Interactive Identification Scheme Based on Discrete Logarithms and Factoring*, Journal of Cryptology, 5(1), pp.29–39, 1992.
4. I. Damgård: *Collision Free Hash Functions and Public-Key Signature Schemes*, Proc. of EuroCrypt 87, Springer Verlag LNCS series.
5. R. Cramer, I. Damgård: *Secure Signature Schemes based on Interactive Protocols*, BRICS report series, RS–94–29, September 1994, Aarhus University.
6. R. Cramer: *On Shared Randomness and the Size of Secure Signatures*, CWI technical report CS–R9530, April 1995.
7. C. Dwork, M. Naor: *An Efficient Existentially Unforgeable Signature Scheme and its Applications*, Proceedings of Crypto'94, Santa Barbara, August 1994, Springer Verlag LNCS series, pp. 234–246.
8. U.Feige, A.Shamir: *Witness Indistinguishable and Witness Hiding Protocols*, Proc. of STOC 90.
9. U. Feige, A. Fiat and A. Shamir: *Zero-Knowledge Proofs of Identity*, Journal of Cryptology 1 (1988) 77–94.

10. O. Goldreich: *Two Remarks concerning the GMR Signature Scheme*, Proc. of Crypto 86, Springer Verlag LNCS series.

11. S. Goldwasser, S. Micali and R. Rivest: *A Digital Signature Scheme Secure Against Chosen Message Attacks*, SIAM Journal on Computing, 17(2): 281-308, 1988.

12. L. Guillou and J.-J. Quisquater: *A Practical Zero-Knowledge Protocol fitted to Security Microprocessor Minimizing both Transmission and Memory*, Proc. of EuroCrypt '88, Springer Verlag LNCS series.

13. R.C.Merkle: *A Digital Signature Based on a Conventional Encryption Function*, Proc. of Crypto 87, Springer Verlag LNCS series.

14. M.Naor and M.Yung: *Universal One-Way Hash Functions and their Cryptographic Applications*, Proc. of STOC 89.

15. T. Okamoto: *Provably Secure and Practical Identification Schemes and Corresponding Signature Schemes*, Proc. of Crypto '92, pp.31–53, Santa Barbara, August 1992.

16. J.Rompel: *One-Way Functions are Necessary and Sufficient for Secure Signatures*, Proc. of STOC 90.

17. C.P. Schnorr: *Efficient Signature Generation by Smart Cards*, Journal of Cryptology, 4(3):161–174, 1991.

18. M.Tompa and H.Woll: *Random Self-Reducibility and Zero-Knowledge Proof of Information Possession*, Proc. of FOCS 87.

Improved Efficient Arguments

(Preliminary version)

Joe Kilian[1]

NEC Research Institute, 4 Independence Way, Princeton, NJ 08540.
joe@research.nj.nec.com

Abstract. We consider complexity of perfect zero-knowledge arguments [4]. Let T denote the time needed to (deterministically) check a proof and let L denote an appropriate security parameter. We introduce new techniques for implementing very efficient zero-knowledge arguments. The resulting argument has the following features:

- The arguer can, if provided with the proof that can be deterministically checked in $O(T)$ time, run in time $O(TL^{O(1)})$. The best previous bound was $O(T^{1+\epsilon}L^{O(1)})$.
- The protocol can be simulated in time $O(L^{O(1)})$. The best previous bound was $O(T^{1+\epsilon}L^{O(1)})$.
- A communication complexity of $O(L \lg L)$, where L is the security parameter against the prover. The best previous known bound was $O(L \lg T)$.

This can be based on fairly general algebraic assumptions, such as the hardness of discrete logarithms.

Aside from the quantitative improvements, our results become qualitatively different when considering arguers that can run for some super-polynomial but bounded amount of time. In this scenario, we give the first arguments zero-knowledge arguments and the first "constructive" arguments in which the complexity of arguing a proof is tightly bounded by the complexity of verifying the proof.

We obtain our results by a hybrid construction that combines the best features of different PCPs. This allows us to obtain better bounds than the previous technique, which only used a single PCP. In our proof of soundness we exploit the error correction capabilities as well as the soundness of the known PCPs.

1 Introduction

One of the great achievements in the study of interactive proof systems has been the discovery of transparent/probabilistically checkable proofs [6, 17]. While most of this research has been aimed at proving complexity results, it is interesting to consider the original application, proving theorems. By requiring the verifier to look at a vanishing section of a proof, one might hope to use them to speed the verification of large, unwieldy proofs. For example, [6] discusses an application to checking the executions of long computations, saying, "In this setup, a single reliable PC can monitor the operation of a herd of supercomputers

D. Coppersmith (Ed.): Advances in Cryptology - CRYPTO '95, LNCS 963, pp. 311-324, 1995.
© Springer-Verlag Berlin Heidelberg 1995

working with possibly extremely powerful but unreliable software and untested hardware."

Beyond the practical difficulties of this scenario are problems of a more fundamental nature. First, the prover must work quite hard to produce the transparent proof. The most easy to verify PCPs ([3] and its descendants) can be checked in $O(1)$ probes, but require the prover to spend time superquadratic in the size of the execution trace. Polishchuk and Spielman [27] construct a constant probe PCP using only $O(T^{1+\epsilon})$ work, where T is the time to deterministically verify the original proof. Furthermore, they show that for any positive constant c_2 there exists a constant c_1 such that proofs of size T are converted into transparent proofs in $O(T \lg^{c_1} T)$ time, which may be verified using $O(T^{c_2})$ work. Thus, at present, there is a continuum of achievable proof complexity/verification complexity tradeoffs, but there does not exist any "optimal" PCP that dominates the others. One can either force the prover to work hard or force the verifier to work hard.

Another difficulty is that in order for the proof to work, the verifier must have possession of the entire proof, or at least a guarantee that the prover cannot change any bits of the proof. Thus, it is not clear how to verify a PCP over a network. The work needed to receive such a proof would be much more than the work required to receive and check a standard proof.

By a standard transformation, results for PCPs carry over into two-prover proofs with essentially optimal (logarithmic in the size of the PCP) communication requirements. Furthermore, these proofs may be made to be zero-knowledge [10]. However, it is open how to surmount this last difficulty within the more realistic framework of single-prover interactive proof systems.

1.1 Efficient zero-knowledge arguments

Brassard, Chaum and Crépeau introduce the notion of arguments [4]. Unlike interactive proofs, which place no assumptions the power of potentially malicious provers, the argument framework puts some bound on the capabilities of the prover, weakening the ordinary soundness condition to one of *computational soundness*. This more realistic assumption leads to dramatically improved properties over interactive proof systems. For example, there exist constant-round perfect zero-knowledge arguments for NP based on reasonable number-theoretic complexity assumptions [4]; [26] shows how to base such proofs on one-way functions at the expense of greater round complexity.

Fiat and Shamir [19] introduce a technique whereby interactive arguments of a fairly general form may be converted into noninteractive arguments. Their basic idea is to replace random questions from the verifier by the results of a random-behaving hash function. They observe that this transformation is rigorously analyzable given a truly random hash function as a black box. Damgård also uses similar ideas for developing practical noninteractive arguments [9]. More recently, Bellare and Rogaway have developed a much more extensive treatment black-box hash functions [8]. Unfortunately, there is no known clean computational assumption under which the soundness of the resulting noninteractive argument can be established.

In [21], it is shown (under suitable complexity assumptions) how to

1. Convert PCPs into a type of "perfect zero-knowledge" PCPs, and
2. Convert { "perfect zero-knowledge"} PCPs into {perfect zero-knowledge} arguments.

Asymptotically, this construction requires much less communication than any previous construction. For example, by using the PCPs constructed in [3], the total communication is only $O(L \lg T)$, where T is the number of steps needed to check the original proof and L is the security parameter for the prover (informally, L specifies the size of the problems the prover is assumed unable to solve).

More recently, Micali has put forth the notion of "CS Proofs" [25]. A stronger result may be obtained by a straightforward application of the Fiat-Shamir transformation and the method of [21]. We strongly recommend a careful reading of [4, 19, 9, 21, 8] prior to reading [25].

1.2 Limitations of previous techniques

The arguments of [21] inherit their work/verification time tradeoffs from the work/verification time tradeoffs in the original PCPs. So to obtain the lowest communication costs advertised, one must use proofs that are very difficult to construct.

The time required to simulate the argument is polynomial in the size of the original proof. While in line with previous proof systems and arguments, one can hope to do much better. The verifier only communicates $O(L \lg T)$ bits and performs $O(L^{O(1)} \lg T)$ computations. The intuition behind our notions of zero-knowledge is that what one obtains by participating in a proof should not be more than what one could have obtained using the same resources but not participating in the proof. In the program-checking application of [6], one can by oneself reconstruct the original "proof" in $O(T \lg^{O(1)} T)$ time. Hence, to say that the simulation can be performed in $O(T^{O(1)})$ time doesn't preclude being able to obtain information about entire execution, with a computational investment of only $O(L^{O(1)} \lg T)$. In such a scenario, the standard notion of zero-knowledge is too weak to be meaningful.

Finally, even when optimized for communication, there remains a significant gap between the $O(L \lg T)$ communication required by this protocol what one could reasonable hope for. Intuitively, one might achieve communication of $O(L)$ bits (it would be amazing if one could achieve $o(L)$ communication, without assuming the existence of problems of size $o(L)$ that the prover cannot solve), and since T might conceivably be nearly exponential in L, $O(L \lg T)$ is nearly a quadratic factor off from what one can hope for.

1.3 Our main result

We make progress on the above mentioned difficulties. We show that one can use substantially less communication, even while using the computationally cheap proofs from [27]. As before, this protocol requires the existence of secure perfect zero-knowledge bit-commitment and collision-intractable hash functions, with security parameter L. We will also assume in this paper that the size of ones proofs is much larger than the statement of what is to be proven.

Theorem 1. Under the above assumptions, a proof $\mathcal{P}$ deterministically verifiable in T steps, can be implemented as a perfect zero-knowledge argument for the correctness of $\mathcal{P}$ with the following properties:

(communication efficiency) Only $O(L \lg L)$ bits of communication are required.

(computational efficiency) The prover only has to perform only $O(L^{O(1)}T)$ computational steps

(completeness) If $\mathcal{P}$ is correct and P follows the protocol, then V will always accept.

(soundness) If $\mathcal{P}$ is false, then either V will reject with probability at least $\frac{1}{2}$ or there exists a program which will break the bit-commitment or collision-intractability assumption (with security parameter L) with nonnegligible probability, in $T^{O(1)}$ time and using $T^{O(1)}$ calls to an oracle for P.

(strong perfect zero-knowledge) There exists a simulator that given an oracle for a possibly malicious verifier $\hat{V}$ will perfectly simulate $\hat{V}$'s view of the proof using expected $L^{O(1)}$ computation and $L^{O(1)}$ calls to the oracle for $\hat{V}$.

Note that we are implicitly using a black-box notion of soundness and zero-knowledge in the statement of our theorem, which has ample precedent in the literature. We prefer this approach because is allows one to make meaningful statements about arguments of specific theorems. The older formalisms strictly make sense only in the context of infinite languages L.

1.4 How our protocol scales for large T

The improved efficiency of our protocol is particularly striking if one considers large T. For example, it is not unreasonably to posit a *super-arguer* that can run for superpolynomially many steps (e.g. $T = O(n^{\lg n})$), but cannot perform exponential-time computations. The original arguments required communication at least T, and hence the verifier would also have to run in superpolynomial time. The arguments of [21] don't have this problem (as noted in [25]) but they still are very problematic. Suppose that an arguer works very hard to generate a proof whose verification takes as long as the time to construct the proof. To use the [21] protocol would require him to run for $T' = \Omega(T^{1+\epsilon})$ steps. However, T'/T is also superpolynomial. We contend that this is not in the spirit of superpolynomial time. If one believes that T^2 or $T^{1+\epsilon}$ is "of the same order" as T, then one is really treating T as polynomial time.

In contrast, our techniques yield a polynomial *multiplicative* blowup in the running time of the super-arguer as opposed to the polynomial *compositional* blowup of previous techniques. Thus, the transformation from a proof to an argument is much more robust than the previous one.

We also note that if one uses the [21] protocol the simulation time is at least T, and thus fails to be zero-knowledge for superpolynomial T. Our protocol continues to be zero-knowledge for as long as a polynomially large security parameter L is appropriate. It is quite possible that the cryptographic primitives we require can be based on problems which have no subexponential time solutions, in which case exponential-sized T may be accommodated.

1.5 Techniques Used

For our new protocol, we use the techniques employed by [21], namely zero-knowledge proofs on committed bits [28, 14], transparent proofs [6, 17] and Merkle's hash-tree commitments [24],[1] and introduce three new techniques.

To improve the communication complexity of our protocols while using computationally inexpensive transparent proofs, we add a further recursive step to our protocol. Interestingly, in these recursive proofs both parties know that the statement is true (with probability extremely close to 1) ahead of time. Rather, the prover convinces the verifier that he knows a particular way of proving this statement. The use of recursive proofs in the transparent proof context is not new; more sophisticated noncryptographic examples of this technique can be found in, for example, [1, 3, 7]. Here, our use of recursion is intermingled with the manner in which individual bits of the transparent proof are revealed, allowing us to obtain much stronger bounds than if we simply restricted ourselves to these techniques.

To achieve the improved zero-knowledge result, we augment the basic hash tree construction with a randomization step. It is difficult to simulate the interior nodes of the hash tree, in particular the root node, in our desired time bound. To get around this problem we use the recursive proof technique to hide the values of all but the root node of the tree. We then show how to randomize this root node so that it may be very efficiently simulated.

Finally, we show how to save random bits by using cryptographically secure pseudorandom generators. This last technique allow us to efficiently exploit transparent proof system in which the verifier flips many more coins than is allowed for in our communication bound. Interestingly, whereas the previous techniques don't say anything about the standard transparent proof/PCP model, this result gives strong evidence for PCPs with a better size/bits of randomness tradeoff than our current techniques can establish without relying on computational assumptions. Note that we do not however obtain fewer random bits than the most efficient systems [1, 3], but merely show how to shrink the number of random bits used in proofs that obtain a very small size in exchange for a large

---

[1] [21] missed this reference - our apologies to Merkle.

amount of required randomness. It is an interesting open question to achieve these tradeoffs using noncryptographic techniques.

1.6 Outline of the Abstract

In Section 2 we introduce some preliminary definitions and background. In Section 3 we show how to combine PCPs to obtain greater efficiencies than are obtainable using a single PCP. In Section 4 we show how to obtain a protocol with strong simulatability properties. In Section 5 we note that one use complexity assumptions to save random bits. Finally, in Section 6 we give a brief idea of how soundness and zero-knowledge are proven for our interactive arguments.

2 Preliminaries

2.1 Cryptographic Primitives

The original construction of [21] makes use of Merkle's hash-tree technique, which in turn relied on families of *collision-intractable hash functions* and perfect zero-knowledge bit-commitment functions. Let $\phi(L)$ denote ones notion of an intractable amount of time. A family $\{H_{k,r}\}$ of collision-intractable hash functions has the following property: with high probability, if one chooses a function $h : \{0,1\}^{2L} \to \{0,1\}^L$ uniformly from $\{H_{L,*}\}$, then one cannot in expected time $\phi(L)$ compute x and y such that $h(x) = h(y)$.

The perfect zero-knowledge commitment scheme has functions $\{E_{L,*}\}$, such that if $E_{L,q}$ is appropriately chosen, then the distribution on $E_{L,q}(0, r)$ will be identical to that of $E_{L,r}(1, r)$, where r is a uniformly distributed infinite sequence of random bits, of which an expected $L^{O(1)}$ bits are actually read. However, one cannot in expected time $\phi(L)$ compute prefixes r_0 and r_1 such that $E_{L,q}(0, r_0) = E_{L,q}(1, r_1)$.

By using some global randomness, these schemes can be decided on before any proofs take place, and the same functions can be used by everyone.

2.2 Review of the [21] Protocol

The [21] protocol uses two very separate techniques. The first technique is a method for adding "zero-knowledge" to transparent proof systems. We will leave this basic trick unchanged, though we will alter our method of making our commitments in the next section. The second technique is to use Merkle's hash tree technique to allow a time-bounded prover P to commit to a large number of bits to a verifier V with little communication, and then efficiently reveal these individual bits. This procedure works as follows. Let $b_1, \ldots, b_n$ be the sequence of bits to be committed by the prover, let $h : \{0,1\}^{2L} \to \{0,1\}^L$ be uniformly chosen (by the verifier) from a family of collision-intractable hash functions. For ease of exposition, assume that $n = 2^k L$ for some integer k (otherwise, pad the sequence to make it the right size).

To commit to $b_1, \ldots, b_n$, P breaks $b_1, \ldots, b_n$ into $m = 2^k$ blocks,

$$C_{k+1,1}, \ldots, C_{k+1,m},$$

of L bits each. For $1 \leq i \leq k$ and $1 \leq j \leq 2^{i-1}$, P computes

$$C_{i,j} = h(C_{i+1,2j-1}, C_{i+1,2j}).$$

We can think of the array $[C_{i,j}]$ as specifying a hash tree in which each parent is equal to the hashed values of its children. Finally, P sends $C_{1,1}$, the root node, to V.

To reveal block $C_{k+1,l}$, P defines $j_1, \ldots, j_{k+1}$ by $j_{k+1} = l$ and $j_{i-1} = \lceil j_i/2 \rceil$ for $1 \leq i \leq k$. P then sends V

$$(C_{2,2j_1-1}, C_{2,2j_1}), \ldots, (C_{k+1,2j_k-1}, C_{2,2j_k}).$$

V checks that

$$C_{i,j_i} = h(C_{i+1,2j_i-1}, C_{i+1,2j_i})$$

for $1 \leq i \leq k$. The collision intractability of h ensures that P can expand a path in the tree in only one way. We refer to this sequence of pairs as a "witness" for block $C_{k+1,l}$, and denote it as W_l.

In the original protocol, the leaves of the tree were encrypted as perfect zero-knowledge blobs, but in the next section we achieve zero-knowledge in a more efficient way that (in its first step) works on the root of the tree. For the moment, we ignore this issue, and implement transparent proofs as follows:

1. P constructs the transparent proof and commits to it using the commit protocol described above. This step requires $O(L)$ bits of communication and that P compute for time within an $O(L^c)$ multiplicative factor of the time needed to construct the transparent proof.
2. V generates its sequence r of random bits, and sends r to P.
3. P and V compute which bits, and hence which L-bit blocks V would have accessed from the original transparent proof. P reveals these blocks using the above protocol. This requires computing time proportional to that used by in evaluating the original transparent proof, with an $O(L^c \lg n)$ multiplicative overhead. The total communication required by this step is $O(L \lg n)$ times the total number of revealed blocks.
4. V decides whether to accept or reject based on the values of the bits revealed by P. Again, this requires relatively little computation.

3 Greater Efficiencies Using Hybrid Schemes

Assume without loss of generality that V uses relatively few ($O(L)$) random bits. If not, then we can use the straightforward technique in Section 5 to shrink the number of bits required. Most of the communication cost is incurred in Step 3. To reveal a single bit requires $O(L \lg n)$ communication, which already exceeds our desired bounds.

To get around this problem, we employ the "proof within a proof" technique that was used to achieve zero-knowledge. In the zero-knowledge version of the above protocol, the prover doesn't reveal the actual bits, but rather convinces the verifier that had he revealed these bits the verifier would have accepted. Similarly, the prover does not really need to send the witness for each block the verifier wishes to see. It suffices that he convince the verifier that he could indeed have sent such a message and that the verifier would have accepted. This involves a zero-knowledge proof of knowledge that can be recursively solved using current techniques.

3.1 Using "witnesses" for answers

Given input x, random string r and the root node $C_{1,1}$, we denote a witness $\mathcal{W}$ for $(x, r, C_{1,1})$ as a sequence

$$\mathcal{W} = \left((l_1, W_{l_1}), \ldots, (l_p, W_{l_p})\right),$$

where

- $l_1, \ldots, l_p$ denote those sections of the original transparent proof that V would have looked at on input x and with random string r,
- W_{l_i} is a valid witness for a block C_{k+1,l_i}, and
- On seeing the bits given in blocks

$$C_{k+1,l_1}, \ldots, C_{k+1,l_p},$$

V will accept.

We note that the existence of a witness for $(x, r, C_{1,1})$ is not sufficient to guarantee that the original theorem is true. However, under the collision intractability assumption, the fact that P *knows* such a witness is strong evidence for the validity of the theorem.

Consider a machine that in the pointer machine model (used by [6] and [27]) nondeterministically guesses $\mathcal{W}$ and then deterministically verifies that $\mathcal{W}$ is a witness for $(x, r, C_{1,1})$. We denote the transcript of this verification by $\mathcal{T}_\mathcal{W}$. We bound the length of $\mathcal{T}_\mathcal{W}$ up to polylogarithmic factors. First, suppose that the original verifier for the transparent proof used total time T_V. Note that $p \leq T_V$. Guessing $\mathcal{W}$ involves $O(L \lg n T_V)$ operations. Verifying that each witness is consistent requires $O(L^c \lg n)$ operations for some constant c, for a total cost of $O(L^c \lg n T_V)$. Finally, verifying that seeing the given portion of the tableau will cause the verifier to accept requires at most $O(LT_V)$ operations. Thus, the total transcript will be of size $O(L^c \lg n T_V)$.

For illustrative purposes, we first show how to obtain near optimal communication costs, without regard for the computational costs involved. Suppose that the original proof $\mathcal{P}$ was of size n and that we used the PCPs from [3]. Then the size of the resulting transparent proof $\mathcal{P}'$ will be $O(n^{O(1)})$ and the time used by the verifier to check this proof will be $O(\lg n)$ (not counting the initial cost for putting T in error-correcting code format). Thus, the size of the resulting proof of knowledge will be (up to polylogarithmic factors) $O(L^{c_1} \lg^{c_2} n)$, which for reasonable sized n is $O(L^{c_3})$ for some constant c_3.

For n large compared to L, $O(L^{c_3})$ is small compared to n. P can therefore recursively prove that it knows a valid $\mathcal{W}$ by using the [21] zero-knowledge construction, optimizing for low communication by again using the transparent proof of [3]. The resulting communication for this step will be

$$O(L \lg(O(L^{c_3}))) = O(L \lg L).$$

Since the [3] type proof uses superquadratic time complexity, the above proof is very inefficient. To obtain simultaneously low-communication and low-computation proofs, we use a three-step recursion. On the top level, we use a Polishchuk-Spielman proof which requires $O(n \lg^{c_1} n)$ time and verified using $O(n^{c_2})$ work, where c_2 is sufficiently small ($< \frac{1}{4}$ suffices). The resulting proof of knowledge will therefore be of size $O(L^{O(1)} n^{c_2})$. Since c_2 is so small, we can then use the computationally intensive protocol described above.

Since malicious prover can try to cheat in the proofs at each level of the recursion, we amplify the error probabilities of these proofs so that in each one he can escape detection with probability at most, for example, $1/100$. More precisely, our proof of soundness requires that whenever a PCP causes the verifier to accept with probability $\geq 1/1000$ an accepting computation path can be reconstructed (technically, this condition is not needed for the first proof).

The number of rounds of communication required when use of recursive proofs is of course greater than that of the original protocol. We can offset this somewhat by noting a simple optimization to the protocol of [21]. Instead of using proofs on committed bits, which requires multiple rounds, it suffices to use the zero-knowledge PCPs of [10]. These PCPs have the same qualitative properties as those of [3] (though with worse constants) but with the property that the queries are easy to simulate. This eliminates the round cost incurred by the proofs on committed bits.

We note that the techniques in [10] can be applied to the the protocol of [27] to obtain a zero-knowledge PCP in which the prover only performs $T^{1+\epsilon}$ work. However, it is open whether there exists zero-knowledge PCPs in which the prover only performs $T \log^{O(1)} T$ work. Fortunately, the elimination of the proof on committed bits steps occur in the last proof invoked, in which the work required is small in any case.

4 Achieving Strong Zero-Knowledge

In this section, we further modify the construction of [21] in order to make a more efficiently simulatable zero-knowledge protocol. In terms of a verifier's $\hat{V}$ view, the argument from the previous section can be summarized as follows:

1. P sends $\hat{V}$ a string $C_{1,1}$ (the root of the hash tree).
2. $\hat{V}$ sends P a string r (which may not be random).
3. P and $\hat{V}$ engage in a low-communication zero-knowledge proof of knowledge.

Since the transparent proofs we use have perfect completeness, it is straightforward to simulate the last two steps of the protocol. Regardless of the distribution on r, P will always know a witness for $(x, r, C_{1,1})$, and this proof can be simulated in using the original simulator for this protocol. Furthermore, the simulator for this last proof will work in time polynomial in L, since the entire transparent proof is bounded by a polynomial in L.

However, as we have written our protocol, it is not at all clear how to simulate the distribution on $C_{1,1}$, even in time polynomial in n. $C_{1,1}$ is a hashed down version of a transparent proof that S does not know. In the original [21] argument, the root node was a hashed version of a large number (polynomial in n) of L-bit perfect zero-knowledge blobs. Since these blobs were easy to simulate, the root node could be simulated in time polynomial in n. However, it is not clear how to speed up this simulation for arbitrary collision-intractable hash functions.

We get around this problem by using zero-knowledge blobs and a further use of hash trees. We modify the naive, stripped down protocol as follows: Instead of sending $\hat{V}$ the L-bit value of $C_{1,1}$, P generates a sequence of L perfect zero-knowledge L-bit blobs, denoted $C'_1, \ldots, C'_L$. Then, P computes a $\lceil \lg L \rceil$-depth hash-tree for $C'_1, \ldots, C'_L$, generating a second root node, C'. P sends C' to $\hat{V}$.

Naively, P can run the basic (nonzero-knowledge) argument as follows: On receiving r, P first sends the entire hash tree for $C'_1, \ldots, C'_L$, and then opens these blobs to reveal $C_{1,1}$. V will later check this part of the proof by verifying that the hash tree is self consistent and that the blob openings were correct. Now, P can then behave just as in the basic protocol without zero-knowledge. We call this concatenation of a valid hash tree rooted at C', the valid opening of these blobs to reveal $C_{1,1}$ and a valid witness for $(x, r, C_{1,1})$ a witness for (x, r, C').

Naively, this extra step involves sending $O(L^2)$ bits just to reveal the secondary hash-tree, which is prohibitive. However, we can use the same recursion trick to demonstrate knowledge of a witness $\mathcal{W}$ for (x, r, C'). The size of $\mathcal{T}_{\mathcal{W}}$ will be $O(L^c)$ bigger than the transcript for a witness for $(x, r, C_{1,1})$ (for some constant c), but this will at most add an $O(L \lg L)$ factor to the communication complexity of the resulting protocol.

5 Saving random bits

The transparent proofs and PCPs in the literature in general use relatively few bits. However, as one optimize ones PCPs in order to minimize the prover's overhead the number of random bits may conceivably become problematic. However, we can use pseudorandom generators to conserve bits in a straightforward manner. Suppose that V would normally send m random bits to P, and let $g : \{0,1\}^n \rightarrow \{0,1\}^m$ be a pseudorandom generator. Then if suffices for V to send a random n-bit string x to P, who then behaves as if sent the m-bit string $y = g(x)$. We claim that if g is sufficiently strong, then the modified PCP will remain a PCP, even against infinitely powerful provers.

Suppose that for some supposed theorem T (not necessarily true), and supposed transparent proof/PCP $\mathcal{P}'$ (not necessarily correctly generated), there is a nonnegligible difference between the probability that V accepts using random coin tosses y and the probability that V accepts using pseudorandom coin tosses $y = g(x)$. Consider the circuit $C_{T,\mathcal{P}'}$ which on input y, outputs 1 iff V would accept $\mathcal{P}'$ after generating its queries according to y. It is straightforward to generate $C_{T,\mathcal{P}'}$ given $\mathcal{P}'$, T and V, and its size will typically be $O(\mathcal{P}')$. Furthermore, $C_{T,\mathcal{P}'}$ can then be used to distinguish a random $g(x)$ from a random y.

Thus, by the contrapositive, if one believes that g is resistant against all sufficiently large circuits, then in particular it may be used for all transparent proofs of a given size.

The above argument relies on the nonuniform complexity of g; we don't know how to prove the analogous result result based on uniform complexity. We also note that thus far there has been no advantage to using cryptography based randomness reduction techniques instead of those currently used for PCPs.

6 Establishing Soundness and Zero-knowledge

In this abstract, we omit the proof of soundness and completeness, and give only a brief sketch of how they are proven. how they work.

Theorem 2. For the protocol described above, there exists a breaker B with the following property. Let T be a false statement, and $\hat{P}$ be a malicious prover who claims to have a proof of length n for this theorem. Suppose that $\hat{P}$ can convince V to accept T with probability greater than $\frac{1}{100}$ using the protocol outlined above. Then given black-box access to $\hat{P}$, B can break either the commitment assumption or the sibling-intractable hash function assumption, running in time $L^{c_1} n^{c_2}$, where c_1 and c_2 are explicitly computable constants. $\square$

Thus, if we believe that breaking the assumption with security parameter L requires more than $L^{c_1} n^{c_2}$ times the number of steps that could plausible taken by $\hat{P}$, then it is reasonable to believe such an argument. Note that it is not so crucial to prove an optimal for c_2, since one can modestly increase the security parameter L.

Here is the very basic idea - numerous trivial details are omitted. For simplicity of exposition, we concentrate on a single proof; the analysis of a cascade of $O(1)$ recursive proofs behaves similarly. Thus, we consider the simplified protocol where $\hat{P}$ commits to a root C and then expands its leaves in response to V's challenges. In each challenge, $\hat{P}$ either,

1. Opens up paths in the hash-tree, revealing PCP values that cause V to accept, or,
2. Opens up legitimate paths in the hash-tree, but reveals values that cause V to reject, or,
3. Otherwise causes V to reject. We call this a garbage response.

While in the actual protocol only a single challenge is given, B can roll $\hat{P}$ back and ask him several challenges. He can then combine the branches for non-garbage responses in the natural way to recover entries of a partial PCP. If at any time these non-garbage answers cannot reconciled, i.e., the values of the hash-tree (or the root-node commitments and their hash-tree) are inconsistent, then B trivially has either,

1. Two strings that hash to the same value, or
2. Identical bit commitment strings for a 0 or a 1.

If this ever happens then we are done. Otherwise, we can show that if B runs $\hat{P}$ sufficiently many times and receives a Type 1 response with sufficiently high probability on each random run, then he will reconstruct a PCP that would cause the original verifier to accept with sufficiently high probability.

It remains to low-bound how many times B must run $\hat{P}$ in order to obtain a sufficiently good PCP. Suppose that B ran through all of V's coin tosses, which are identified with those used by the initial PCP verifier. Then, provided that no "breaks" were found, the resulting PCP would cause the verifier to accept with at least the probability that $\hat{P}$ makes a Type 1 response to the next question by V. Here we implicitly assume (at least this assumption trivializes the claim) that V's coin tosses specifies what bits of the PCP he will look at - all known PCPs have this property.

We suspect that the above bound on how often B needs to invoke $\hat{P}$ to construct a PCP is sufficient for known PCP's. That is, the verifiers require randomness that is typically within a constant of optimal. But to make sure, we note that it also suffices for B to run $\hat{P}$ $O(n^c)$ times, where c is a sufficiently large constant (2 may suffice) and n is the size of the PCP. Here, no effort has been made to reduce the $O(n^c)$ - sharper bounds are possible but are not needed. The intuition for why this is true is that any portion of the PCP that is not revealed after so many trials cannot have been needed to make the verifier accept with the given probability. Thus, setting these entries to 0 will not significantly damage the resulting PCPs acceptance probability.

Recall that in each recursive proof $\hat{P}$ that he could have given a good response to the preceding question. Using the PCP construction procedure given above and the the self-correcting properties of [3]-style PCPs (and by extension, those of [10]) the breaker can, whenever the verifier still has a sufficiently large chance of accepting, actually produce a good response. The breaking can then proceed recursively.

Of course, one must be a little more careful than described above. The PCP reconstruction step will only work when it holds that for a random challenge $\hat{P}$ will cause V to accept the current proof. One must perform a simple averaging argument to verify that if the acceptance probability for the entire protocol is sufficiently high than for the most of the time it will hold that most of the time $\hat{P}$ will deal with the next challenge so as to make V accept with high probability. Further details are omitted.

Theorem 3. For the protocol described above, there exists a simulator S with the following property. Let $\mathcal{P}$ be a correct proof for T, and let $\hat{V}$ be a malicious verifier. Then given T, $|\mathcal{P}|$ and black-box access to $\hat{V}$, S can generate the view obtained by $\hat{V}$ in time L^c for some explicitly computable constant c. □

Finally, we observe that a simulator can trivially simulate the distribution on C', simply by generating L random blobs (0-blobs and 1-blobs are identically distributed) and creating the hash-tree for them. We can no longer use the simulator for the proof system given in [21], since it implicitly relied on the multiphase nature of the standard protocol for zero-knowledge proofs on committed bits. However, if one uses the [3] protocol, then $\hat{V}$ is only allowed to look at a constant number of bits. The simulator exploits this fact by committing to a sequence of random bits. With constant probability, the query made by $\hat{V}$ will be satisfiable by the proof, at which point it the simulation of the zero-knowledge proof will be perfect. Here the proof is simplified by the fact that the bit commitments are information theoretically secure. Note that to argue that the concatenation of the simulation phases is zero-knowledge, we can use the fact that our protocols are auxiliary input zero-knowledge, a weaker property than black-box zero-knowledge.

7 Acknowledgments

We would like to acknowledge Lance Fortnow and Carsten Lund for valuable information about the self correction properties of PCPs. Dan Spielman provided early and invaluable information on his work with Polishchuk, which greatly improved the results of an earlier version of this manuscript.

References

1. S. Arora and S. Safra. Probabilistic Checking of Proofs, *Proceedings of STOC 1992.*
2. S. Arora and T. Leighton and B. Maggs. On-line algorithms for path selection in a nonblocking network. *Proceedings of STOC 1990*, pp. 149–158
3. S. Arora, C. Lund, R. Motwani, M. Sudan and M. Szegedy. Proof Verification and Hardness of Approximation Problems, *Proceedings of STOC 1992.*
4. G. Brassard, D. Chaum, and C. Crépeau. Minimum Disclosure Proofs of Knowledge, *J. Comput. System Sci.* **37** (1988), 156–189.
5. L. Babai, L. Fortnow, and C. Lund. Non-Deterministic Exponential Time has Two-Prover Interactive Proofs, *Proceedings of FOCS 1990*
6. L. Babai, L. Fortnow, L. Levin and M. Szegedy. Checking computation in polylogarithmic time. *Proceedings of STOC 1991.*
7. M. Bellare, S. Goldwasser, C. Lund, A. Russell, "Efficient probabilistic checkable proofs and applications to approximation," *Proc. 25th STOC, 1993, pp. 294-304.*
8. M. Bellare, P. Rogaway. Random Oracles are Practical: A paradigm for Designing Efficient Protocols, *Proc. First ACM Conference on Computer and Communications Security,* ACM, November 1993.

9. I. Damgård, Non-interactive Circuit Based Proofs and Non-Interactive Perfect Zero-Knowledge with Preprocessing, Advances in Cryptology - EUROCRYPT 92, pp. 341–355.

10. C. Dwork, U. Feige, J. Kilian, M. Naor and S. Safra. Low communication 2-Prover Zero-Knowledge Proofs for NP. *Advances in Cryptology - Crypto '92*, pp. 215–227.

11. U. Feige, A. Fiat and A. Shamir. Zero knowledge proofs of identity, *Proceedings of 19nd Annual Symposium on the Theory of Computation, 1987, pp. 210-217.*

12. U. Feige, S. Goldwasser, L. Lovasz, M. Safra and M. Szegedy. Approximating Clique is Almost NP-Complete, *Proceedings of 32nd Annual Symposium on Foundations of Computer Science, 1991, pp. 2-12.*

13. U. Feige, D. Lapidot and A. Shamir. *Multiple Non-Interactive Zero-Knowledge Proofs Based on a Single Random String*, Proceedings of the 22th Annual Symposium on the Theory of Computation, 1990, pp. 308–317

14. C. Bennett. personal communication via Gilles Brassard.

15. M. Ben-Or, S. Goldwasser, J. Kilian, and A. Wigderson. Multi-Prover Interactive Proofs: How to Remove Intractability, *Proceedings of STOC 1988.*

16. De Santis, A., S. Micali and G. Persiano, "Bounded-Interaction Zero-Knowledge proofs," *Advances in Cryptology - Crypto '88*

17. U. Feige, S. Goldwasser, L. Lovász, S. Safra and M. Szegedy. Approximating clique is almost NP-Complete. *Proceedings of FOCS 1991*, pp. 2-12.

18. L. Fortnow, J. Rompel, and M. Sipser. On the Power of Multi-Prover Interactive Protocols, *Proceedings of Structure 1988.*

19. A. Fiat and A. Shamir. How to Prove Yourself: Practical Solution to Identification and Signature Problems. *Advances in Cryptology - Crypto '86*, pp. 186–189.

20. S. Goldwasser, S. Micali, and C. Rackoff. The Knowledge Complexity of Interactive Proof Systems, *SIAM J. Comput.* 18 (1989), 186–208.

21. J. Kilian. A note on efficient zero-knowledge proofs and arguments, *Proceedings of STOC 1992.*

22. J. Kilian On the complexity of bounded interaction and noninteractive proofs. *Proceedings of FOCS 1994.*

23. C. Lund, L. Fortnow, H. Karloff, and N. Nisan. The polynomial-time hierarchy has interactive proofs, *Proceedings of STOC 1990*, pp. 2-10.

24. R. Merkle. A Certified Digital Signature. *Proceedings of Crypto '89*, pp. 218–238.

25. S. Micali. Computationally Sound Proofs, *Proceedings of FOCS 1994.*

26. M. Naor, R. Ostrovsky, R. Venkatesan and M. Yung. Perfect Zero-Knowledge Arguments for NP can be Based on General Complexity Assumptions. *Advances in Cryptology - Crypto '92*, pp. 196–214.

27. A. Polishchuk and D. Spielman. Nearly-linear Size Holographic Proofs. *Proceedings of STOC 1994.*

28. S. Rudich, Personal communication via Gilles Brassard.

29. A. Shamir. IP = PSPACE, *Proceedings of FOCS 1990*, IEEE.

Honest Verifier vs Dishonest Verifier in Public Coin Zero-Knowledge Proofs

Ivan Damgård[*]   Oded Goldreich[†]   Tatsuaki Okamoto[‡]

Avi Wigderson[§]

June 7, 1995

Abstract

This paper presents two transformations of public-coin/Arthur-Merlin proof systems which are zero-knowledge *with respect to the honest verifier* into (public-coin/Arthur-Merlin) proof systems which are zero-knowledge *with respect to any verifier.*

The first transformation applies only to constant-round proof systems. It builds on Damgård's transformation (see *Crypto93*), using ordinary hashing functions instead of the interactive hashing protocol (of Naor, Ostrovsky, Venkatesan and Yung – see *Crypto92*) which was used by Damgård. Consequently, the protocols resulting from our transformation have much lower round-complexity than those derived by Damgård's transformation. As in Damgård's transformation, our transformation preserves statistical/perfect zero-knowledge and does not rely on any computational assumptions. However, unlike Damgård's transformation, the new transformation is not applicable to argument systems or to proofs of knowledge.

The second transformation can be applied to proof systems of arbitrary number of rounds, but it only preserves statistical zero-knowledge. It assumes the existence of secure commitment schemes and transforms any public-coin proof which is statistical zero-knowledge with respect to the honest into one which is statistical zero-knowledge (in general). It follows, by a result of Ostrovsky and Wigderson (1993), that any language which is "hard on the average" and has a public-coin proof system which is statistical zero-knowledge with respect to the honest verifier, has a proof system which is statistical zero-knowledge (with respect to any verifier).

---

[*]Dept. of Computer Science, Aarhus Univesity, Denmark and BRICS, Basic Research In Computer Science, center of the Danish National Research Foundation.

[†]Dept. of Computer Science and Applied Math., Weizmann Institute of Science, Rehovot, Israel. Work done while visiting BRICS, Basic Research In Computer Science, center of the Danish National Research Foundation. Supported in part by grant No. 92-00226 from the United States – Israel Binational Reseach Foundation (BSF), Jerusalem, Israel.

[‡]NTT Laboratories, Yokosuka-shi, 238-03 Japan. Work done while visiting AT&T Bell Laboratories, Murray Hill, NJ, USA

[§]Institute for Computer Science, Hebrew University, Jerusalem, Israel. Work done while visiting BRICS, Basic Research In Computer Science, center of the Danish National Research Foundation. This research was partially supported by a grant from the Wolfson Research Awards, administered by the Israeli Academy of Sciences and Humanities.

D. Coppersmith (Ed.): Advances in Cryptology - CRYPTO '95, LNCS 963, pp. 325-338, 1995.
© Springer-Verlag Berlin Heidelberg 1995

Part I

Hashing Functions Can Simplify Zero-Knowledge Protocol Designs (too)[1]

1 Introduction to Part I

Zero-knowledge proof systems, introduced by Goldwasser, Micali and Rackoff [16], are a key tool in the design of cryptographic protocols. The results of Goldreich, Micali and Wigderson [14] guarantee that such proof systems can be constructed for any NP-statement, provided that one-way functions exist. However, the general construction presented in [14] and subsequent works may yield quite inefficient proof systems for particular applications of interest. Thus, developing methodoligies for the design of zero-knowledge proofs is still of interest.

Designing proof systems which are merely zero-knowledge with respect to the honest verifier (i.e., the verifier specified for the system) is much easier than constructing proof systems which are zero-knowledge in general (i.e., with respect to any efficient strategy of trying to extract knowledge from the specified prover). For example, the simple 1-round interactive proof for Graph Non-Isomorphism [2] is zero-knowledge with respect to the honest verifier. Yet, cheating verifiers may extract knowledge from this system and a non-trivial modification, which utilizes proofs of knowledge and increases the number of rounds, is required to make it zero-knowledge in general. Likewise, assuming the existence of one-way function, there exist constant-round interactive proofs for any NP-language which are zero-knowledge with respect to the honest verifier. Yet, constant-round interactive proofs for NP which are zero-knowledge in general are known only under seemingly stronger assumptions and are also more complex (cf., [11]).

In view of the relative simplicity of designing protocols which are zero-knowledge with respect to the honest verifier, a transformation of such protocols into protocols which are zero-knowledge in general (i.e., w.r.t. any verifier) may be very valuable. Assuming various intractability assumptions, such transformations have been presented by Bellare et. al. [2], and Ostrovsky et. al. [23]. A transformation which does not rely on any intractability assumptions has been presented by Damgård in *Crypto93*. His transformation (of honest-verifier zero-knowledge into general zero-knowledge) has two shortcomings. Firstly, it can be applied only to constant-round protocols of the Arthur-Merlin type (i.e., in which the verifier's messages are uniformly distributed in the set of strings of specified length). Secondly, the transformation produces protocols of very high round complexity; specifically, the round complexity of the resulting protocol is linear in the randomness complexity of the original one.

In this part of paper, we improve the round complexity of Damgård's transformation, while preserving the class of interactive proofs to which it can be applied. Our transformation only increases the number of rounds by a factor of two. However, it also increases the error probability of the proof system by a non-negligible amount which can be made arbitrarily small. This increase is inevitble in view of a result of Goldreich and Krawcyzk [12], see discussion in subsection 3.4. Thus, to get proof systems with negligible error probability, one may repeat the protocols resulting from our transfor-

---

[1] by Ivan Damgård, Oded Goldreich and Avi Wigderson.
[2] To be convinced that G_0 and G_1 are not isomorphic, the verifier randomly selects n random isomorphic copies of each graph, randomly shuffles all these copies together, and asks the prover to specify the origin of each copy.

mation a non-constant number of times. Still, the resulting proof systems will have much lower round complexity than those resulting from Damgård's transformation.

We preserve some of the positive properties of Damgård's transformation. In particular, our transformation does not rely on any computational assumptions and preserves perfect and almost-perfect (statistical) zero-knowledge. However, unlike Damgård's transformation, the new transformation is not applicable to argument systems (i.e., the BCC model [4]) or to proofs of knowledge.

Our transformation builds on Damgård's work [6]. In his transformation, the random messages sent by the verifier (in each round) are replaced by a multi-round interactive hashing protocol, which in turn originates in the work of Ostrovsky, Venkatesan and Yung [22]. Instead, in our transformation, the random messages sent by the verifier are replaced by a $\frac{3}{2}$-round protocol, called Random Selection. The Random Selection protocol uses a family of ordinary hashing functions; specifically, we use a family of t-wise indepedent functions, for some parameter t (which is polynomial in the input length).

We believe that the Random Selection protocol may be of independent interest. Thus, a few words are in place. The goal of this protocol is to allow two parties to select a "random" n-bit string. There is a parameter ε which governs the quality of this selection and the requirement is asymmetric with respect to the two parties. Firstly, it is required that if the first party follows the protocol then, no matter how the second player plays, the output of the protocol will be at most ε away (in norm-1) from uniform. Secondly, it is required that if the second party follows the protocol then, no matter how the first player plays, no string will appear as output of the protocol with probability greater than $\mathrm{poly}(n/\varepsilon) \cdot 2^{-n}$. Our Random Selection protocol has the additional property of being simulatable in the sense that, given a possible outcome, it is easy to generate a (random) transcript of the protocol which ends with this outcome.

Other Related Work

The idea of transforming honest verifier zero-knowledge into zero-knowledge in general was first studied by Bellare, Micali and Ostrovsky [2]. Their transformation needed a computational assumption of a specific algebraic type. Since then several constructions have reduced the computational assumptions needed. The latest in this line of work is by Ostrovsky, Venkatesan and Yung [23], who give a transformation which is based on interactive hashing and preserved statistical zero-knowledge. Their transformation relies on existence of a one-way permutation. The transformation works for any protocol, provided that the verifier is probabilistic polynomial-time.

In the other part of this paper, a secure commitment scheme[3] is used to transform honest-verifier zero-knowledge Arthur-Merlin proofs (with unbounded number of rounds) into (general) zero-knowledge Arthur-Merlin proofs. This transformation increases the round-complexity of the proof system by an additive term which is linear in the number of coin tosses used in the original proof system.

An indirect way of converting protocols which are zero-knowledge with respect to the honest verifier into ones which are zero-knowledge in general, is available through a recent result of Ostrovsky and Wigderson [24]. They have proved that the existence of honest verifier zero-knowledge proof system for a language which is "hard on the average" implies the existence of one-way functions. Combined with the results of [14] and [19, 3], this yields a (computational and general) zero-knowledge proof for the same language. Thus, computational honest-verifier zero-knowledge interactive proofs, for "hard on the average" languages, get transformed into computational zero-knowledge interactive proofs for these languages. However, perfect honest-verifier zero-knowledge proofs (for such languages) do not get transformed into *perfect* zero-knowledge proofs.

---

[3] Secure commitment schemes exist provided that one-way functions exist [18, 20] and the latter exist if some languages which is hard on the average have proof systems which are zero-knowledge with respect to the honest verifier [24].

A two-party protocol for random selection, with unrelated properties, has been presented in [10]. This protocol guarantees that, as long as one party plays honestly, the outcome of the protocol hits any set $S \subset \{0,1\}^n$ with probability at most $\tilde{O}(\sqrt{|S|/2^n})$, where $\tilde{O}(\varepsilon) \stackrel{\text{def}}{=} \varepsilon \cdot \text{polylog}(1/\varepsilon)$.

Another two-party protocol for random selection, with other unrelated properties, has been presented in [13]. Loosely speaking, this protocol allows a computationally restricted party, interacting with a powerful and yet untrustful party, to uniformly select an element in an easily recognizable set $S \subset \{0,1\}^n$.

Remarks Concerning this Part of the Paper

We use the standard definitions of interactive proofs and zero-knowledge, except for the following minor modification. We require the simulator (in the definition of zero-knowledge) to to run in strictly polynomial-time (rather than in expected polynomial-time), but we allow it to produce output only with some non-negligible probability (rather than always). Clearly, this definition implies the standard one, but the converse is not known to hold – see [9]. This definition is more convenient for establishing zero-knowledge claims and in particular for our purposes, but our results do not depend on it (and can be derived under the standard definitions).

Due to space limitations the proofs of all propositions have been omitted. The complete proofs appear in our technical report [7].

2 Random Selection

We consider a randomized two-party protocol for selecting strings. The two parties to the protocol are called the *challenger* and the *responder*. These names are supposed to reflect the asymmetric requirements (presented below) as well as the usage of the protocol in our zero-knowledge transformation. Loosely speaking, we require that

- if the challenger follows the protocol then, no matter which strategy is used by the responder, the output of the protocol is almost uniformly distributed;
- if the responder follows the protocol then, no string may appear with probability much greater than its probability under the uniform distribution. Furthermore, for any string which may appear as output, when an arbitrary challenger strategy is used, one can efficiently generate a random transcript of that protocol ending with this output.

We postpone the formal specification of these properties to the analysis of the protocol presented below. Actually, we present two version of the protocol.

Construction 1 (Random Selection Protocol – two versions): *Let n and $m < n$ be integers[4], and $H_{n,m}$ be a family of functions, each mapping the set of n-bit long strings onto[5] the set of m-bit long strings.*

C1: *the challenger uniformly selects $h \in H_{n,m}$ and sends it to the responder;*

R1:
- *(version 1): the responder uniformly selects $x \in \{0,1\}^n$, computes $\alpha = h(x)$ and sends α to the challenger;*
- *(version 2): the responder uniformly selects $\alpha \in \{0,1\}^m$ and sends it to the challenger;*

C2: *the challenger uniformly selects a preimage of α under h and outputs it.*

---

[4] In particular, we will use $m \stackrel{\text{def}}{=} n - 4\log_2(n/\varepsilon)$, where ε is an error-bound parameter.

[5] We stress that each function in $H_{n,m}$ rages over all $\{0,1\}^m$. Thus, the challenger may always respond in step C2 even if the responder deviates from the protocol or version 2 is used.

We remark that if version 1 is used and both parties follow the protocol then the output is uniformly distributed in $\{0,1\}^n$. However, the interesting case is when one of the parties deviates from the protocol. In this case, the protocol can be guaranteed to produce "good" output, provided that "good" families of hash functions are being used as $H_{n,m}$. These functions must have relatively succient representation as well as strong random properties. Furthermore, given a function h, it should be easy to evaluate h on a given image and to generate a random preimage (of a given range element) under h. Using the algorithmic properties of $H_{n,m}$ it follows that the instructions specified in the above protocol can be implemented in probabilistic poly(n/ε)-time, which for $\varepsilon = 1/\text{poly}(n)$ means poly(n)-time.

Construction 2 (Preferred family $H_{n,m}^t$): *Let n, $m < n$ and $t = \text{poly}(n)$ be integers. We associate $\{0,1\}^n$ with the finite field $GF(2^n)$ and consider the set of $(t-1)$-degree polynomials over this field. For each such polynomial f, we consider the function h so that, for every $x \in \{0,1\}^n$, $h(x)$ is the m most significant bits of $f(x)$. The family $H_{n,m}^t$ consists of all such functions h. The canonical description of a function $h \in H_{n,m}^t$ is merely the sequence of t smallest coefficients of the corresponding polynomial. Finaly, we modify the functions in $H_{n,m}^t$ so that for each $h \in H_{n,m}^t$ and every $x' \in \{0,1\}^m$ it holds $h(x'0^{n-m}) \stackrel{\text{def}}{=} x'$.*

In the sequel, we will use the family $H_{n,m} \stackrel{\text{def}}{=} H_{n,m}^n$. We now list the following, easy to verify, properties of the above family.

P1 There is a poly(n)-time algorithm that, on input a function $h \in H_{n,m}^t$ and a string $x \in \{0,1\}^n$, outputs $h(x)$.

P2 The number of preimages of an image y under $h \in H_{n,m}^t$ is bounded above by $2^{n-m} \cdot t$; furthermore, there exists a poly$(2^{n-m}t)$-time algorithm that, on input y and h, outputs the set $h^{-1}(y) \stackrel{\text{def}}{=} \{x : h(x)=y\}$. (The algorithm works by trying all possible extensions of y to an element of $GF(2^t)$; for each such extension it remains to find the roots of a degree $t-1$ polynomial over the field.)

P3 $H_{n,m}^t$ is a family of **almost t-wise independent** hashing functions in the following sense: for every t distinct images, $x_1,...,x_t \in (\{0,1\}^n - \{0,1\}^m0^{n-m})$, for a uniformly chosen $h \in H_{n,m}^t$, the random variables $h(x_1),...,h(x_t)$ are indepedently and uniformly distributed in $\{0,1\}^m$.

2.1 The output distribution for honest challeger

We now turn to analyze the output distribution of the above protocol, assuming that the challenger plays according to the protocol. In the analysis we allow the responder to deviate arbitrarily from the protocol and thus as far as this analysis goes the two versions in Construction 1 are equivalent. The analysis is done using the "random" properties of the family $H_{n,m}^t$. Recall that the statistical difference between two random variable X and Y is
$$\frac{1}{2} \sum_\alpha |\text{Prob}(X=\alpha) - \text{Prob}(Y=\alpha)|$$
We say that X is ε-away from Y if the statistical difference between them is ε.

Proposition 1 *Let n be an integer, $\varepsilon \in [0,1]$ and $m \stackrel{\text{def}}{=} n - 4\log_2(n/\varepsilon)$. Suppose that $H_{n,m}$ is a family of almost n-wise independent hashing functions. Then, no matter which strategy is used by the responder, provided that the challenger follows the protocol, the output of the protocol is at most $(2\varepsilon + 2^{-n})$-away from uniform distribution.*

2.2   The output distribution for honest responder

We now show that no matter what strategy is used by the challenger, if the responder follows the protocol then the set of possible outputs of the protocol must constitute a non-negligible fraction of the set of n-bit long strings. This claim holds for both versions of Construction 1. Furthermore, we show that no single string may appear with probability which is much more than 2^{-n} (i.e., its probability weight under the uniform distribution).

Proposition 2 *Suppose that $H_{n,m} = H_{n,m}^t$ is a family of hashing functions satisfying property (P2), for some $t = \text{poly}(n)$. Let C^* be an arbitrary challenger strategy. Then, for every $x \in \{0,1\}^n$, the probability that an execution of version 1 of the protocol with challenger strategy C^* ends with output x is at most $(t \cdot 2^{n-m}) \cdot 2^{-n}$.*

Proposition 3 *Let C^* be an arbitrary challenger strategy. Then, for every $x \in \{0,1\}^n$, the probability that an execution of version 2 of the protocol with challenger strategy C^* ends with output x is at most 2^{-m}. Furthermore, for every deterministic challenger strategy c, exactly 2^m strings may appear as output, each with probability exactly 2^{-m}.*

2.3   Simultability property of the protocol

We conclude the analysis of the above protocol by showing that, one can efficiently generate random transcripts of the protocol having a given outcome. Throughout this analysis, we assume that the responder follows the instruction specified by the protocol. As in the proof of the last two propositions, it suffices to consider an arbitrary deterministic challenger strategy, denoted c.

Now, suppose that $H_{n,m} = H_{n,m}^t$ is a family of hashing functions satisfying property (P1), for some $t = \text{poly}(n)$. Then, on input x and access to a function $c : \{0,1\}^* \mapsto \{0,1\}^*$, we can easily test if $c(h(x)) = x$, where $h \stackrel{\text{def}}{=} c(\lambda)$. In case the above condition holds, the triple $(h, h(x), x)$ is the only transcript of the execution of the protocol, with challenger strategy c, which ends with output x. Otherwise, there is no execution of the protocol, with challenger strategy c, which ends with output x. Thus,

Proposition 4 *Consider executions of the Random Selection protocol in which the challenger strategy, denoted c, is an arbitrary function and the responder plays according to the protocol. There exists a polynomial-time oracle machine that, on input $x \in \{0,1\}^n$ and $h \in H_{n,m}$ and oracle access to a function c, either generates the unique transcript of a c-execution which outputs x or indicates that no such execution exists.*

2.4   Setting the Parameters

Proposition 1 motivates us to set ε (the parameter governing the approximation of the output in case of honest challenger) as small as possible. On the other hand, Propositions 2 and 3 motivates us to maintain the difference $n - m$ small and in paricular logarithmic (in n). Recalling that $n - m = 4\log_2(n/\varepsilon)$, this suggests setting $\varepsilon = 1/p(n)$ for some fixed positive polynomial p.

3   The Zero-Knowledge Transformation

Our transformation is restricted to interactive proofs in which the verifier sends the outcome of every coin it tosses. Such interactive proofs are called *Arthur-Merlin games* [1] or *public-coins interactive proofs* (cf., [17]). Note that in such interactive proofs the verifier moves, save the last, may consist merely of tossing coins and sending their outcome. (In its last move the verifier decides, based on the entire history of the communication, whether to accept the input or not.) Without loss of generality, we may assume that in every round of such an interactive proof the verifier tosses at least $4\log(|x|/\varepsilon)$ coins, where x is the common input to the interactive proof and ε specifies the desired bound on the statistical distance (between one round in the resulting interactive proof and the original one). Furthermore, assume for sake of simplicity that at each round the verifier tosses the same number of coins, denoted n.

3.1 The Transformation

In the following description, we use the second version of the Random Selection protocol presented in Construction 1. This simplifies the construction of the simulator for the transformed interactive proof. The first version can be used as well, at the expense of some modification in the simulator construction.

The protocol transformation consists of replacing each verifier move (except the last, decision move) by an execution of the Random Selection protocol, in which the verifier plays the role of the challenger and the prover plays the role of the responder.

Construction 3 (transformation of round i in (P, V) interaction): *Let (P, V) be an interactive proof system in which the verifier V only uses public coins, let $\varepsilon(n) = 1/\text{poly}(n)$ be the desired error in the Random Selection protocol, $m \stackrel{\text{def}}{=} m(n) \stackrel{\text{def}}{=} n - 4\log_2(n/\varepsilon(n))$ and $H_{n,m}$ be as specified in Construction 2 (for $t = n$). The ith round of the (P, V) interaction, on common input x, is replaced by the following two rounds of the resulting interactive proof (P', V'). Let $(h_1, \alpha_1, r_1, \beta_1, ..., h_{i-1}, a_{i-1}, r_{i-1}, \beta_{i-1})$ be the history so far of the interaction between prover P' and verifier V'. Then, the next two rounds consist of an execution of the (second version of the) Random Selection protocol follows by P' mimicking the response of P. Namely, in the first round, the verifier V' uniformly selects $h_i \in H_{n,m}$ and sends it to the prover P' who replies with a_i uniformly selected in $\{0, 1\}^m$. In the second round, the verifier V' uniformly selects $r_i \in h_i^{-1}(a_i)$ and sends it to the prover P' who replies with $\beta_i \stackrel{\text{def}}{=} P(x, r_1, ..., r_i)$.*

The final decision of the new verifier V' mimics the one of the original verifier V; namely,

$$V'(h_1, \alpha_1, r_1, \beta_1, ..., h_t, a_t, r_t, \beta_t) = V(r_1, \beta_1, ..., r_t, \beta_t)$$

3.2 Preservation of Completeness and Soundness

In this subsection, we may assume that V' follows the interactive proof. Thus, if for some $x \in \{0, 1\}^*$, prover P always convinces V on common input x then P' always convinces V' on this common input. We stress that both V' and P' run in polynomial-time when given oracle access to V and P, respectively. Thus, the new verifier is a legitimate one. Furthermore, if the original prover P, working in polynomial-time with help of a suitable auxiliary input, could convince the original verifier to accept some common input, then the resulting prover P' could do the same (i.e., can convince V' to accept this common input, while working in polynomial-time with help of the same auxiliary input).

We have just seen that the completeness properties of the original interactive proof is preserved, by the transformation, in a strong sense. Soundness properties are preserved as well, but with some slackness which results from the imperfectness of the Random Selection protocol. In particular,

Proposition 5 *Let $\mu: \{0, 1\}^* \mapsto [0, 1]$ be a function bounding the probability that verifier V accepts inputs when interacting with any (possibly cheating) prover. Namely,*

$\mu(x)$ is a bound on the probability that V accepts x. Suppose that on input x, the interactive proof (P,V) runs for $t(|x|)$ rounds. Then, $\mu'(x) \stackrel{\text{def}}{=} \mu(x) + O(t(|x|) \cdot \varepsilon(|x|))$ is a function bounding the probability that verifier V' accepts inputs when interacting with any (possibly cheating) prover.

proof: Recall that V' plays the role of the challenger in the Random Selection protocol. Thus, the proposition follows quite immediately from Proposition 1. ∎

We stress that the above proposition remains valid no matter which of the two version of Random Selection is used. The same holds with respect to the comments regarding completeness (made above).

3.3 Zero-Knowledge

In this subsection, we may assume that P' follows the interactive proof. Assuming that P is zero-knowledge with respect to the verifier V, we prove that P' is zero-knowledge with respect to any probabilistic polynomial-time verifier strategy. This statement holds for the three versions of zero-knowledge; specifically, perfect, almost-perfect (statistical), and computational zero-knowledge.

Proposition 6 Let (P,V) be a constant-round Arthur-Merlin interactive proof. Suppose that P is perfect (resp. almost-perfect) [resp. computational] zero-knowledge with respect to the honest verifier V over the set $L \subseteq \{0,1\}^*$. Then P' is perfect (resp. almost-perfect) [resp. computational] zero-knowledge (with respect to any probabilistic polynomail-time verifier) over the set $L \subseteq \{0,1\}^*$.

A few comments regarding the proof: Let M be a simulator witnessing the hypothesis of the proposition. Then, for every $x \in L$, with non-negligible probability $M(x)$ halts with output, and given that this happens the output has distributed indistinguishable from that of $(P,V)(x)$. For every verifier strategy V^* interacting with P', we construct a simulator M^*, which uses M and V^* as black-boxes, as follows. By uniformly selecting and fixing coin tosses for V^*, we may assume that V^* is deterministic.

On input x, the simulator M^* invokes M and assuming $M(x)$ halts with output, sets $(r_1, \beta_1, ..., r_t, \beta_t) \stackrel{\text{def}}{=} M(x)$; otherwise M^* also halts with no output. The simulator M^* now tries to form transcripts of the Random Selection protocol which end with output r_1, r_2 through r_t, respectively. (Here we use the simulatability of the Random Selection protocol.) A transcript with output r_1 is formed as follows. M^* feeds V^* with input x and obtains h_1, which can be assumed as in Propositions 2 and 3 to be in $H_{n,m}$. Next, M^* computes $a_1 = h_1(r_1)$ and feeds V^* with (x, a_1). If V^* replies with r_1, we've succeeded in forming a transcript for the first invocation of Random Selection and we proceed to the next. (This happens with non-negligible probability.) Otherwise, M^* halts with no output. We note that for the next invocations of Random Selection, V^* is fed with the entire history so far; for example, to obtain h_2 we feed V^* with (x, a_1, β_1) and next we feed it with (x, a_1, β_1, a_2), where $a_2 = h_2(r_2)$. If all t rounds were completed successfully[6], M^* halts with output $(h_1, a_1, r_1, \beta_1, ..., h_t, a_t, r_t, \beta_t)$.

To complete the proof we prove six claims. Firstly, we show that in each of the three cases (perfect, almost-perfect, or computational zero-knowledge), the simulator M^* produces output with non-negligible probability. Secondly, for each of the three cases, we establish the required relationship between the transcript of the real interaction

---

[6]This happens with probability $p(|x|)^t$, where $p(\cdot)$ is the non-negligible probability that we've completed successfuly a single round. This is the reason we can handle any constant number of rounds.

and the output of the simulator. As expected, the proofs become more involved as we move from perfect to computational zero-knowledge.

The above proposition remains valid even if one uses the first version of the Random Selection protocol. However, a slightly more complex simulator will have to be used. The reason being that in the first version (of the Random Selection protocol) the a_i's are not selected uniformly but are rather weighted by the number of their preimages under the corresponding h_i's. Thus, r_i's which are mapped to a_i's with small preimage may be less likely in the real interactions. To compensate for this phenomenon, one may modify the simulator so that it skews the probabilities in the same manner. Namely, when producing a transcript with less likely r_i's, the simulator will discard it with some probability. The required probability (with which to discard transcripts) can be easily computed.

3.4   Conclusions

Combining Propositions 5 and 6, we get

Theorem 1 *Let $\mu : \mathbb{N} \mapsto [0,1]$. Suppose L has a constant-round Arthur-Merlin proof system, with error bound μ, which is perfect (resp. almost-perfect) [resp. computational] zero-knowledge with respect to the honest verifier. Then, for every positive polynomial $p(\cdot)$, L has a constant-round Arthur-Merlin proof system, with error bound $\mu'(n) \stackrel{\text{def}}{=} \mu(n) + \frac{1}{p(n)}$, which is perfect (resp. almost-perfect) [resp. computational] zero-knowledge (with respect to any probabilistic polynomial-time verifier). Furthermore, the zero-knowledge property can be demonstrated using a black-box simulation. Also, if the original system had no error on inputs in L then the same holds for the new system.*

Theorem 1 does not preserve the error probability of the original system. This seems inevitible, in light of [12]. Recall that there are languages believed not to be in $\mathcal{BPP}$ which have constant-round Arthur-Merlin proof systems, with exponentially small error probability, which are zero-knowledge with respect to the honest verifier. For example, Graph Isomorphism has such a system (for perfect zero-knowledge), and assuming the existence of one-way functions, every language in $\mathcal{NP}$ has such a system (for computational zero-knowledge) [14]. Now, a stronger version of Theorem 1, say one in which $\mu'(n) - \mu(n)$ is a negligible function of n, would imply that these languages have constant-round Arthur-Merlin (balck-box) zero-knowledge proof systems (with negligible error probability). But, according to [12], languages having constant-round Arthur-Merlin (balck-box) zero-knowledge proof systems lie in $\mathcal{BPP}$. Needless to say that $\mathcal{NP}$ and even Graph Non-Isomorphism are believed not to lie in $\mathcal{BPP}$.

We now compare the round complexity of the protocols resulting from our transformation to those resulting from Damgård's transformation of [6]. Suppose we start with a c-round proof system which uses $r(n)$ random coins and has error $\mu(n)$. Clearly, $\mu(n) \geq 2^{-r(n)}$ and $r(n) > \log_2 n$ (otherwise the language is in BPP [15]). Now, the proof system resulting from Damgård's transformation will have $c + r(n)$ rounds and maintain the error bound of the original proof system. On the other hand, the protocol resulting from our transformation will have $2c$ rounds and error $\mu(n) + \frac{1}{\text{poly}(n)}$. In case $\mu(n)$ is non-negligible, we have a clear advatage. Otherwise, to make the comparison fair, we use sequentail repetitions to reduce the error in the protocols resulting from our transformation to the bound $\mu(n)$. This requires $\log_{\text{poly}(n)}(1/\mu(n))$ repetitions yielding round complexity bounded by $\frac{\log_2(1/\mu(n))}{\log_2 n} \leq \frac{r(n)}{\log_2 n}$. (Typically, $\mu(n)$ is much larger than $2^{-r(n)}$.)

Part II

Using Commitment Schemes to Simplify Zero-Knowledge Protocol Design [7]

4 Introduction to Part II

In this part, we will show another transformation, which can be applied to *arbitrary number of round* statistical zero-knowledge proofs, assuming the existence of secure commitment schemes (i.e., one-way functions [18, 20]). This assumption can be replaced by the restriction on the applicable languages, that they are "hard on the average" (not in $\mathcal{AVBPP}$) [24].

This result can be considered to improve the two previous results partially: one is the result by Ostrovsky, Venkatesan and Yung [23] and the other is by Damgård[6] (see Introduction of Part I). That is, our result generalizes the assumption of [23], from one-way permutations to *one-way functions*, although our transformation is only applicable to *public coin* proof systems. On the other hand, this result relaxes the round complexity restriction for applicable proof systems, from constant number of rounds to *arbitrary number of rounds*, although our transformation does not preserve perfect zero-knowledge, and the applicable languages should not be in $\mathcal{AVBPP}$.

The technique of using the bit-commitment for the transformation can be also applied to the argument model [4]. In this transformation, the roles of the committer and receiver are reversed (i.e., the verifier is the committer.)

5 The Zero-Knowledge Transformation

Theorem 2 *If language L has a statistical zero-knowledge public-coin proof against a "honest verifier", then L has a statistical zero-knowledge public-coin proof against "any verifier", assuming the existence of secure bit-commitment schemes (i.e., one-way functions).*

Proof

Let (M, A) be a statistical zero-knowledge public-coin proof against a "honest verifier", A, for language L. Then we will construct a statistical zero-knowledge public-coin proof, (M^*, A^*), against any verifier, A^*, for L.

We assume

1. If $x \in L$, then $\text{Prob}[(M, A)(x) \text{ accepts }] \geq 1 - 1/2^n$

2. If $x \notin L$, then for any $\widetilde{M}$, $\text{Prob}[(\widetilde{M}, A)(x) \text{ accepts }] \leq 1/2^n$,

where n is the size of x.

Suppose that the conversation of $(M, A)(x)$ is $(\alpha_1, \beta_1, \ldots, \alpha_k, \beta_k)$, where α_i ($i = 1, \ldots, k$) is the i-th public coin message by A, and β_i is the i-th message by M. Let l_i be the (bit) size of α_i.

Let BC be Naor's bit-commitment function based on a pseudo-random generator, G, [20]. That is, Naor's bit-commitment protocol is as follows:

---

[7]by Tatsuaki Okamoto.

1. [Commit stage:]
   Receiver (R) sends a $(3n$ bits) random string, t, to Committer (C).

   C randomly selects a $(n$ bits) seed, s, of a pseudo-random generator, G, and calculates $BC(s, t, b) = G^{(3n)}(s) \oplus bt$, where $b \in \{0, 1\}$ is the bit C is committed to, bt is t (if $b = 1$) or 0^{3n} (if $b = 0$), and $G^{(3n)}(s)$ is the first $3n$ bits output of $G(s)$. P sends $BC(s, t, b)$ to R.

2. [Reveal stage:]
   C sends s and b to R, and R checks the validity.

A pseudo-random generator exits if and only if a one-way function exists [18].

Next, we show the protocol of (M^*, A^*) using Naor's bit-commitment protocol.

Protocol (M^*, A^*)

Common input: x

What to prove: $x \in L$.

Repeat the following protocol for i from 1 to k sequentially. Here, when $i = j$, we suppose that (M^*, A^*) has already executed the protocol for i from 1 through $j - 1$. (When $i = 1$, suppose that no protocol has been executed before.)

1. Repeat the following protocol for I from 1 to l_i sequentially.

   (a) A^* sends a $(3n$ bits) random string, $t_I^{(i)}$, to M^*.

   (b) M^* randomly selects a $(n$ bits) seed, $s_I^{(i)}$, of a pseudo-random generator, and a random bit, $b_I^{(i)} \in \{0, 1\}$. M^* calculates $BC(s_I^{(i)}, t_I^{(i)}, b_I^{(i)})$, and sends it to A^*.

   (c) A^* sends a random bit, $c_I^{(i)} \in \{0, 1\}$, to M^*.

   (d) M^* sends $s_I^{(i)}$ and $b_I^{(i)}$ to A^*.

   (e) A^* checks the validity of $s_I^{(i)}$ and $b_I^{(i)}$, and if it is invalid A^* halts. Otherwise, go to the next step.

2. M^* sets
$$\alpha_i \leftarrow (b_1^{(i)} \oplus c_1^{(i)}, b_2^{(i)} \oplus c_2^{(i)}, \ldots, b_{l_i}^{(i)} \oplus c_{l_i}^{(i)}).$$

   M^* runs M with α_i as the i-th message by A and gets the i-th message by B, β_i. Here, we suppose that M, given $(\alpha_1, \ldots, \alpha_{i-1})$, has already outputs $(\beta_1, \ldots, \beta_{i-1})$ sequentially. M^* sends β_i to A^*.

Finally, for $i = 1, \ldots, k$, A^* sets

$$\alpha_i \leftarrow (b_1^{(i)} \oplus c_1^{(i)}, b_2^{(i)} \oplus c_2^{(i)}, \ldots, b_{l_i}^{(i)} \oplus c_{l_i}^{(i)}).$$

Then, A^* runs A with $(\alpha_1, \ldots, \alpha_k)$ as A's random string, and $(\beta_1, \ldots, \beta_k)$ as messages from M. If A accepts, then A^* accepts.

$$\text{[End of Protocol } (M^*, A^*)]$$

[Completeness]

If $x \in L$ and M^* and A^* are honest, then, clearly, (M^*, A^*) accepts x with the same probability by (M, A), where M and A are also honest.

[Soundness]

If $x \notin L$, we will show that for any prover, $\widetilde{M^*}$, $(\widetilde{M^*}, A^*)(x)$ accepts with probability less than $\epsilon(n)$.

First, we assume that there exist $\widetilde{M^*}$ and a constant a such that $(\widetilde{M^*}, A^*)(x)$ accepts with probability greater than $1/n^a$. Here, we suppose that $\widetilde{M^*}$ is deterministic, by selecting the optimum coin flips of $\widetilde{M}^*$ which maximize the accept probability of $(\widetilde{M}^*, A^*)(x)$.

Then we will show that $\widetilde{M}^*$ must break the condition of Naor's bit-commitment. For any M, $\text{Prob}[(M, A)(x) \text{ accepts }] \leq 1/2^n$, and (M^*, A^*) is the same as (M, A) except the procedure of determining $\{\alpha_i\}$. Hence, if $(\widetilde{M}^*, A^*)(x)$ accepts with probability greater than $1/n^a$ for a constant a, then $(\alpha_1, \ldots, \alpha_k)$, which is input to A by A^* to decide the acceptance, must be in a negligible $(< 1/2^n)$ fraction, Γ, of $\{(\alpha_1, \ldots, \alpha_k)\}$ with probability greater than $1/n^a$ for a constant a. Here, Γ is fixed when $\widetilde{M}^*$ is fixed.

On the other hand, from the condition of Naor's bit-commitment, the committer $(\widetilde{M}^*)$ can change the committed value with probability at most $1/2^n$. Since A^* sends a true random bits c_I (for $I = 1, \ldots, l_i$; $i = 1, \ldots, k$), $e_I^{(i)}$ is uniformly distributed with probability greater than $1 - 1/2^n$. Hence, $(\alpha_1, \ldots, \alpha_k) = (e_1^{(1)}, \ldots, e_{l_k}^{(k)})$ is uniformly distributed with probability greater than $(1 - 1/2^n)^{\sum_{i=1}^{k} l_i} > 1 - \epsilon(n)$. Therefore, the probability that $(\alpha_1, \ldots, \alpha_k) \in \Gamma$ is at most $(1/2^n)(1 - \epsilon(n)) + \epsilon(n) < \epsilon(n)$.

Thus, if $(\widetilde{M}^*, A^*)(x)$ accepts with probability greater than $1/n^a$ for a constant a, then $\widetilde{M}^*$ must break the condition of Naor's bit-commitment.

[Zero-knowledgeness (Black-box simulation zero-knowledgeness]

When $x \in L$, for any verifier A^*, simulator $\widetilde{S}$ for (M^*, A^*), which utilizes A^* as a black-box, can be constructed as follows:

[Simulator $\widetilde{S}$]

1. For $x \in L$, $\widetilde{S}$ runs Simulator S for (M, A), then gets the simulated conversation, $(\alpha_1, \beta_1, \ldots, \alpha_k)$ of $(M, A)(x)$. Let

$$(e_1^{(i)}, e_2^{(i)}, \ldots, e_{l_i}^{(i)}) = \alpha_i,$$

for $i = 1, \ldots, k$.

2. Repeat the following procedure for i from 1 to k, and for I from 1 to l_i, sequentially. (So, totally, $(\sum_{i=1}^{k} l_i)$ procedures are repeated sequentially.) We denote each procedure by $[i, I]$. Here, when $i = j$ and $I = J$, we suppose that $\widetilde{S}$ has already executed the procedures for i from 1 through $j - 1$ and the procedures for I from 1 through $J - 1$ in the procedure for $i = j$. (i.e., $[1, 1], \ldots, [1, l_1]$, $\ldots, [j - 1, 1], \ldots, [j - 1, l_{j-1}], [j, 1], \ldots, [j, J - 1]$.)

So, the initial status of A^* in the following procedure is the final status of A^* just before the procedure. Let $Init_{[i, I]}$ be the initial status of A^* in procedure $[i, I]$,

During the following procedure $[i, I]$, $\widetilde{S}$ can make A^* to $Init_{[i, I]}$ from the first initial status of A^* (i.e., $Init_{[1, 1]}$). Since a simulated conversation from $[1, 1]$ through $[i, I - 1]$ has been fixed, $\widetilde{S}$ can make A^* to $Init_{[i, I]}$ just by simulating the fixed simulated conversation from $[1, 1]$ through $[i, I - 1]$ again. (Then the execution is straightforward and no trial and error.) (Note: $[i, 0]$ means $[i - 1, l_{i-1}]$.) When $i = 1$, suppose that no procedure has been executed before.

(a) $\widetilde{S}$ runs A^* and gets a (3n bits) string, $t_I^{(i)}$ from A^*.

(b) $\widetilde{S}$ randomly selects a (n bits) seed, $s_I^{(i)}$, of a pseudo-random generator, and a random bit, $b_I^{(i)} \in \{0, 1\}$. $\widetilde{S}$ calculates $BC(s_I^{(i)}, t_I^{(i)}, b_I^{(i)})$, and gives it to A^*.

(c) $\widetilde{S}$ runs A^* and gets a bit, $c_I^{(i)} \in \{0, 1\}$, from A^*.

(d) $\widetilde{S}$ checks whether the following equation holds or not:

$$b_I^{(i)} \oplus c_I^{(i)} = e_I^{(i)}.$$

If it holds, then $\widetilde{S}$ goes to the next procedure, $[i, I+1]$. (Note: $[i, l_i+1]$ means $[i+1, 1]$.) Otherwise, $\widetilde{S}$ makes A^* to $Init_{[i,I]}$ and returns to the first step of this procedure, $[i, I]$.

3. Finally $\widetilde{S}$ arranges these values in the order of (M^*, A^*) protocol, and outputs them.

Next, we will show that $\widetilde{S}$ terminates in expected polynomial-time.

Since A^* is a polynomial time bounded Turing machine, from the property of the bit-commitment protocol,

$$|\text{Prob}[c_I^{(i)} \mid b_I^{(i)} = 0] - \text{Prob}[c_I^{(i)} \mid b_I^{(i)} = 1]| < \epsilon(n).$$

Therefore, if $b_I^{(i)}$ is randomly selected,

$$\text{Prob}[b_I^{(i)} \oplus c_I^{(i)} = e_I^{(i)}] > 1/2 - \epsilon(n).$$

Thus, in each procedure, the expected repetition number is less than $1/(1/2 - \epsilon(n)) < 2 + 4\epsilon(n)$. Clearly, after procedure $[i, I]$ is completed, the simulated conversation from $[1, 1]$ to $[i, I]$ is not affected by the following procedures. (i.e., there is no back-track.) Hence, totally, $\widetilde{S}$ terminates in expected time of polynomial (i.e., $O(2(\sum_{i=1}^{k} l_i) \times T)$; where T is the running time of each procedure described above).

Next, we will show that the simulated conversation is statistically close to the real conversation.

Since this is a black-box simulation, if the simulated messages of M^* is statistically close to the real messages, then the total simulation is also statistically close to the real conversation.

To prove this, it is sufficient to show that the simulated α_i is statistically close to the real one. Since (M, A) is a statistical zero-knowledge proof, the distribution of the simulated $\alpha_i = (e_1^{(i)}, \ldots, e_{l_i}^{(i)})$ (output of simulator S for (M, A)) is statistically close to the uniform distribution. On the other hand, the real α_i is also statistically close to the uniform distribution. This is because: (same as the related part of the proof that $\widetilde{S}$ terminates in expected polynomial-time)

$$|\text{Prob}[c_I^{(i)} \mid b_I^{(i)} = 0] - \text{Prob}[c_I^{(i)} \mid b_I^{(i)} = 1]| < \epsilon(n),$$

and $b_I^{(i)}$ is truly random in the real conversation. Hence,

$$\text{Prob}[e_I^{(i)} = b_I^{(i)} \oplus c_I^{(i)} = 0] > 1/2 - \epsilon(n).$$

Thus, the simulated α_i is statistically close to the real one.

$\square$

We can immediately obtain the following corollary from Theorem 2 and [24].

Corollary 1 *If language L has a statistical zero-knowledge public-coin proof and L is not in $\mathcal{AVBPP}$, then L has a statistical zero-knowledge public-coin proof against "any verifier".*

By using the commitment scheme reversely, we can obtain the following:

Corollary 2 *If language L has a statistical zero-knowledge public-coin argument against a "honest verifier", then L has a statistical zero-knowledge public-coin argument against "any verifier", assuming the existence of secure bit-commitment schemes (i.e., one-way functions).*

References

[1] L. Babai. *Trading Group Theory for Randomness*, **Proc. of 17th STOC, pages 421–420, 1985.**

[2] M. Bellare, S. Micali and R. Ostrovsky: *The (true) Complexity of Statistical Zero-Knowledge*, **Proc. of STOC 90.**

[3] M. Ben-Or, O. Goldreich, S. Goldwasser, J. Håstad, J. Killian, S. Micali and P. Rogaway: *Everything Provable is Provable in Zero-Knowledge*, **Proc. of Crypto 88.**

[4] G. Brassard, D. Chaum and C. Crépeau: *Minimum Disclosure Proofs of Knowledge*, **JCSS.**

[5] G. Brassard, C. Crépeau and M. Yung: *Everything in NP can be Argued in Perfect Zero-Knowledge in a Constant Number of Rounds*, **16th ICALP, pp. 123–136, 1989.**

[6] I. Damgård: *Interactive Hashing can Simplify Zero-Knowledge Protocol Design Without Computational Assumptions*, **Proc. of Crypto 93.**

[7] I. Damgård, O. Goldreich, and A. Wigderson: *Hashing Functions can Simplify Zero-Knowledge Protocol Design (too)*, BRICS Technical Rerport RS-94-39, Nov. 1994.

[8] U. Feige and A. Shamir: *Zero-Knowledge Proofs of Knowledge in Two Rounds*, **Advances in Cryptology – Crypto89** (proceedings), pp. 526–544, 1990.

[9] O. Goldreich: *Foundation of Cryptography – Fragments of a Book*, February 1995. Available from the **Electronic Colloquium on Computational Complexity (ECCC),** http://www.eccc.uni-trier.de/eccc/.

[10] O. Goldreich, S. Goldwasser and N. Linial: *Fault-Tolerant Computation without Assumptions: the Two-Party Case*, **32nd FOCS, pp. 447–457, 1991.**

[11] O. Goldreich and A. Kahan: *How to Construct Constant-Round Zero-Knowledge Proof Systems for NP*, to appear in **Journal of Crypology,**

[12] O. Goldreich and H. Krawcyzk: *On the Composition of Zero-Knowledge Proof Systems*, **17th ICALP, pp. 268–282, 1990.**

[13] O. Goldreich, Y. Mansour and M. Sipser: *Proofs that Never Fail and Random Selection*, **Proc. of FOCS 87.**

[14] O. Goldreich, S. Micali and A. Wigderson: *Proofs that yield Nothing but their Validity and a Methodology of Cryptographic Protocol Design*, **Proc. of FOCS 86.**

[15] O. Goldreich and Y. Oren: *Definitions and Properties of Zero-Knowledge Proof Systems.* **Jour. of Crypto., Vol. 7, pp. 1-32, 1994.**

[16] S. Goldwasser, S. Micali and C. Rackoff: *The Knowledge Complexity of Interactive Proof Systems*, **SIAM J. Computing, Vol. 18, pp. 186–208, 1989.**

[17] S. Goldwasser and M. Sipser. *Private Coins versus Public Coins in Interactive Proof Systems*, **Proc. of 18th STOC, pages 59–68, 1986.**

[18] J. Hastad, R. Impagliazzo, L.A. Levin and M. Luby: *Construction of Pseudorandom Generator from any One-Way Function*, manuscript, 1993. See preliminary versions by Impagliazzo et. al. in **21st STOC** and Hastad in **22nd STOC.**

[19] R. Impagliazzo and M. Yung, *Direct Minimum-Knowledge Computations*, **Advances in Cryptology - Crypto87 (proceedings),** 1987, pp. 40–51.

[20] M. Naor: *Bit Commitments from Pseudorandomness*, **Proc. of Crypto 89.**

[21] M. Naor, R. Ostrovsky, R. Venkatesan and M. Yung: *Zero-Knowledge Arguments for NP can be Based on General Complexity Assumptions*, **Proc. of Crypto 92.**

[22] R. Ostrovsky, R. Venkatesan and M. Yung: *Fair Games Against an All-Powerful Adversary*, presented at DIMACS Complexity and Cryptography Workshop, October 1990, Princeton.

[23] R. Ostrovsky, R. Venkatesan and M. Yung: *Interactive Hashing Simplifies Zero-Knowledge Protocol Design*, **Proc. of EuroCrypt 93.**

[24] R. Ostrovsky and A. Wigderson: *One-Way Functions are Essential for Non-Trivial Zero-Knowledge*, **Proc. 2nd Israel Symp. on Theory of Computing and Systems, 1993.**

Proactive Secret Sharing
Or:
How to Cope With Perpetual Leakage

Amir Herzberg   Stanisław Jarecki*   Hugo Krawczyk   Moti Yung

IBM T.J. Watson Research Center
Yorktown Heights, NY 10598
{amir,stasio,hugo,moti}@watson.ibm.com

Abstract. Secret sharing schemes protect secrets by distributing them over different locations (share holders). In particular, in k out of n threshold schemes, security is assured if *throughout the entire life-time of the secret* the adversary is restricted to compromise less than k of the n locations. For long-lived and sensitive secrets this protection may be insufficient.

We propose an efficient *proactive* secret sharing scheme, where shares are periodically renewed (*without changing the secret*) in such a way that information gained by the adversary in one time period is useless for attacking the secret after the shares are renewed. Hence, the adversary willing to learn the secret needs to break to all k locations *during the same time period* (e.g., one day, a week, etc.). Furthermore, in order to guarantee the availability and integrity of the secret, we provide mechanisms to detect maliciously (or accidentally) corrupted shares, as well as mechanisms to secretly recover the correct shares when modification is detected.

1 Introduction

Secret sharing schemes protect the secrecy and integrity of information by distributing the information over different locations. For sensitive data these schemes constitute a fundamental protection tool, forcing the adversary to attack multiple locations in order to learn or destroy the information. In particular, in a $(k + 1, n)$-threshold scheme, an adversary needs to compromise more than k locations in order to learn the secret, and corrupt at least $n - k$ shares in order to destroy the information. However, the adversary has the *entire life-time of the secret* to mount these attacks. Gradual and instantaneous break-ins into a subset of locations over a long period of time may be feasible for the adversary. Therefore for long-lived and sensitive secrets the protection provided by traditional secret sharing may be insufficient.

A natural defense is to periodically refresh the secrets; however, this is not always possible. That is the case of inherently long-lived information, such as cryptographic master keys (e.g., signature/certification keys), data files (e.g., medical records), legal documents (e.g., a will or a contract), proprietary trade-secret information (e.g., Coca-Cola's formula), and more.

---

* Laboratory of Computer Science, Massachusetts Institute of Technology

D. Coppersmith (Ed.): Advances in Cryptology - CRYPTO '95, LNCS 963, pp. 339-352, 1995.
© Springer-Verlag Berlin Heidelberg 1995

Thus, what is actually required to protect the secrecy of the information is to be able to *periodically renew the shares without changing the secret*, in such a way that any information learned by the adversary about individual shares becomes obsolete after the the shares are renewed. Similarly, to avoid the gradual destruction of the information by corruption of shares it is necessary to *periodically recover lost or corrupted shares* without compromising the secrecy of the recovered shares.

These are the core properties of *proactive secret sharing* as presented here. In the proactive approach, the lifetime of the secret is divided into *periods of time* (e.g., a day, one week, etc.). At the beginning of each time period the share holders engage in an interactive *update protocol*, after which they hold completely *new* shares of the *same* secret. Previous shares become obsolete and should be safely erased. As a consequence, in the case of a $(k + 1, n)$ proactive threshold scheme, the adversary trying to learn the secret is required to compromise $k + 1$ locations during a *single* time period, as opposed to incrementally compromising $k + 1$ locations over the *entire* secret life-time. (As an example consider a secret that lives for five years; a weekly refreshment of shares will reduce the time available for the adversary to break the $k + 1$ necessary locations from five years to one week.) Similarly, the destruction of the secret requires the adversary to corrupt $n - k$ shares in a *single* time period.

Note that in this setting the adversary is *mobile* and may break into each server multiple times. It nevertheless cannot compromise the secret if at any time period it does not break into more than k locations.

Our solution to the proactive secret sharing problem can support up to $k = n/2 - 1$ corrupted parties at any time period. It assumes the existence of secure encryption and signature functions, as well as the security of the verifiable secret sharing scheme (VSS) based on homomorphic functions [11, 17]. At the system level, we assume a broadcast channel and synchrony (as in VSS). The exact model and assumptions are described in section 2.1.

The mobile adversary setting was originally presented in the context of secure systems by Ostrovsky and Yung [18] with the focus on a theoretical setting of "general distributed function evaluation". That solution allowed large (polynomial) redundancy in the system (redundancy is the ratio of total servers n to the threshold k of simultaneously faulty servers), and used the availability of huge majority of honest servers to achieve the very general task of secure computation in the information theoretic sense. That model was then used in a more practical setting by Canetti and Herzberg [3] who proactively maintained a distributed pseudorandom generator.

Applications: Proactive secret sharing has numerous applications, primarily maintaining data which is long-lived in scenarios where availability and secrecy are crucial. Recently, we employed it to implement "proactive function sharing" where the shares are never combined to reconstruct a single secret, but rather used to collectively compute a function many times and in different inputs. (This follows the *function sharing* model of [7] based on threshold encryption [8].) A particularly attractive application (presented in a forthcoming paper [14]) provides *proactive digital signatures* which achieve the benefits of threshold signatures, with the additional property that the scheme is broken only if the adversary corrupts more than a threshold of the servers in a *single* time-period. For signature keys (e.g., a certification authority) that live for long time and require very strong security, this solution is of particular importance.

Organization: Section 2 presents in some detail the proactive secret sharing model, including the adversary model and the basic definitions of security. It also describes the basic cryptographic tools used in our solution. Section 3 describes the share renewal protocol, and Section 4 describes the share recovery protocol (proofs are omitted from this extended abstract). Section 5 deals with secret reconstruction, and Section 6 shows how to maintain inter-server authentication/decryption keys securely in the proactive setting. Section 7 summarizes the result.

2 Preliminaries

2.1 Model and Assumptions

We assume a system of n servers $\mathcal{A} = \{P_1, P_2, \ldots, P_n\}$ that will (proactively) share a secret value x through a $(k + 1, n)$-threshold scheme (i.e., k shares provide no information on the secret, while $k + 1$ suffice for the reconstruction of the secret). We assume that the system is securely and properly initialized. The goal of the scheme is to prevent the adversary from learning the secret x, or from destroying it (In particular, any group of $k + 1$ non-faulty servers should be able to reconstruct the secret whenever it is necessary).

SERVERS AND COMMUNICATION MODEL. Each server in $\mathcal{A}$ is connected to a common broadcast medium C, called communication channel, with the property that messages sent on C instantly reach every party connected to it. We assume that the system is synchronized, i.e., the servers can access a common global clock, and that each server in $\mathcal{A}$ has a local source of randomness.

TIME PERIODS AND UPDATE PHASES. Time is divided into *time periods* which are determined by the common global clock (e.g., a day, a week, etc.). At the beginning of each time period the servers engage in an interactive *update protocol* (also called *update phase*). At the end of an update phase the servers hold new shares of the secret x.

THE MOBILE ADVERSARY MODEL. The adversary can corrupt servers at any moment during a time period. If a server is corrupted during an update phase, we consider the server as corrupted during both periods adjacent to that update phase. We assume that the adversary corrupts *no more* than k out of n servers in each time period, where k must be smaller than $n/2$ (this guarantees the existence of $k + 1$ honest servers at each time).

Corrupting a server means any combination of learning the secret information in the server, modifying its data, changing the intended behavior of the server, disconnecting it, and so on. For the sake of simplicity, we do not differentiate between malicious faults and "normal" server failures (e.g., crashes, power failures etc.). We assume also that the adversary is connected to the broadcast channel C, which means she can hear *all* the messages and inject her own. She cannot, however, modify messages sent to C by a server that she does not control, nor prevent a non-corrupted server from receiving a message sent on C. Additionally, the adversary always knows the non-secret data and the algorithm that each machine performs.

We assume the adversary to be *computationally bounded*, such that it cannot break the underlying cryptographic primitives on which we base our design (this includes public-key encryption and signatures, and verifiable secret sharing – see Section 2.2).

SECURITY OF A PROACTIVE SECRET SHARING SCHEME. We state the security properties of our proactive secret sharing algorithm, relative to the adversary defined above . We only sketch the definition of security, following the notion of *semantic security* introduced in [12]. In the formal definition, to be presented in the complete version of this paper, the adversary is modeled as a computationally bounded interactive probabilistic Turing machine which is fed with all the publicly available information on the secret (e.g., its length, a particular subspace from which the secret is chosen, the value of the secret under a one-way function, and so on), and with the information learned by the adversary during (one or more) runs of the update protocol (this includes all the public communication between servers, secret information of the servers that were corrupted in each of these periods, etc.).

Let κ be a function applicable to the space of secrets x. Let $p_0^{(\kappa)}$ be the probability that the adversary correctly computes the value $\kappa(x)$ when fed with the a-priori (public) information on the secret, and let $p_1^{(\kappa)}$ be the analogous probability but after the adversary is fed with the additional information gathered during the run of the protocol. (The above probabilities depend on the random coins used by the adversary and the servers.) Intuitively, the function $\kappa(x)$ models some knowledge about x, while the difference $p_1^{(\kappa)} - p_0^{(\kappa)}$ quantifies the amount of that knowledge "learned" by the adversary by watching the execution of the protocol and actively intruding the servers.

Definition 1. (sketch) We say that a proactive secret sharing scheme is semantically secure if for any function κ computable on the secret, the difference between the probabilities $p_0^{(\kappa)}$ and $p_1^{(\kappa)}$ is negligible.

The exact notion of "negligible" in the above definition depends on the exact model of the adversary. In the traditional complexity-theoretic setting of a polynomial time adversary, one considers these probabilities as functions of the secret length, and "negligible" stands for any function that decreases faster than any inverse polynomial. A more careful model would bound the difference between $p_0^{(\kappa)}$ and $p_1^{(\kappa)}$ as an explicit function of the (small) probabilities with which the adversary can break the underlying cryptographic primitives.

In some cases, in order to stress the existence of an a-priori public information $\pi(x)$ on the secret x, we will say that the proactive secret sharing scheme is semantically secure relative to $\pi(x)$.

Not only we are interested to preserve the secrecy of x, but also to guarantee its availability and recoverability. This means that we need to prevent the adversary from destroying the secret or impeding its reconstruction by, for example, destroying or modifying shares.

Definition 2. (sketch) A proactive secret sharing scheme that guarantees the correct reconstructibility of the secret at any time is called robust.

Notice that for a proactive secret sharing scheme to be robust, one needs to ensure that in any time period the honest servers (which could have been corrupted during previous time periods) have correct shares (i.e., ones that combine to the correct secret x), and that this correctness can be verified by the other servers. This requires that honest servers be able to verify whether each of them stores a correct share. Also, those who do hold correct shares must be able to cooperate in order to recover the shares of the ones that lost them (without exposing the recovered share to anybody except its intended holder).

The focus of this paper is to construct *semantically secure and robust proactive secret sharing scheme* based on the existence of secure public-key encryption [12] and signatures [13], as well as on the existence of verifiable secret schemes [11, 16]. The theorems in this paper are stated relative to these security notions and the above adversary model.

A NOTE ABOUT THE REMOVAL OF AN ADVERSARY FROM A SERVER. We assume that the adversary intruding the servers $\mathcal{A}$ is "removable" (e.g., through a reboot procedure) when it is detected. The responsibility for triggering the reboot operation (or other measures to guarantee the normal operation of a server) relies on the system management which gets input from the servers in the network. In addition to regular detection mechanisms (e.g., anti-virus scanners) available to the system management, our protocols provide explicit mechanisms by which a majority of (honest) servers can detect and alert about a misbehaving server. We assume for simplicity that the reboot operation is performed immediately when attacks or deviations from the protocol are detected. We remark that the initialization of servers and reboot operations require a minimal level of *trust* in the system management, restricted to installation of correct programs and of public keys used for server-to-server communication. Specifically, no secret information is exposed.

2.2 Cryptographic Tools

SHAMIR'S SECRET SHARING. Our secret sharing scheme is based on Shamir's scheme [19]. Let q be a prime number, $x \in Z_q$ be the secret to be shared, n the number of participants (or share holding servers), and $k + 1$ the reconstructibility threshold. The dealer D of the secret chooses a random polynomial f of degree k over Z_q subject to the condition $f(0) = x$. Each share x_i is then computed by D as $f(i)$ and then transmitted *secretly* to participant P_i. The reconstruction of the secret can be done by having $k + 1$ participants providing their shares and using polynomial interpolation to compute x.

VERIFIABLE SECRET SHARING – VSS. In Shamir's scheme a misbehaving dealer can deal inconsistent shares to the participants, from which they will not be able to reconstruct a secret. To prevent such malicious behavior of the dealer one needs to implement a procedure or protocol through which a consistent dealing can be verified by the recipients of shares. Such a scheme is called *verifiable secret sharing* (VSS) [5] . Our work uses these schemes in an essential way. We implement our solution using specific schemes due to Feldman [11] and Pedersen [17]. These schemes are based on hard to invert homomorphic functions and, in particular, on the hardness of computing discrete logarithms over Z_p, for prime p. The first scheme protects the secrecy of the secret in a computational sense, while the second provides information theoretic secrecy. Our solution works with either of these schemes. We choose to present here the solution using Feldman's scheme since it is somewhat simpler. However, the use of Pedersen's scheme strengthens and simplifies the proofs of security.

FELDMAN'S VSS. For completeness we briefly describe Feldman's scheme. Let p be a prime number with $p = mq + 1$, where q is also prime, and m an integer (possibly, a small one like $2, 3, 4$). Let g be an element of Z_p of order q. The dealer chooses the polynomial f over Z_q with coefficients $f_0, f_1, \ldots, f_k$ and broadcasts the corresponding values $g^{f_0}, g^{f_1}, \ldots, g^{f_k}$. Then it secretly transmits the value $x_i = f(i) \pmod{q}$ to P_i. Each server P_i verifies its own share by checking the following equation:

$$g^{x_i} \stackrel{?}{=} (g^{f_0})(g^{f_1})^i (g^{f_2})^{i^2} \ldots (g^{f_k})^{i^k} \pmod{p} \tag{1}$$

If the equation holds, P_i broadcasts a message accepting its share as proper. If all servers find their shares correct then the dealing phase is completed successfully. Indeed, by the homomorphic properties of the exponentiation function (i.e., $g^{a+b} = g^a g^b$) the above equation holds for all $i \in \{1 \ldots n\}$ if and only if the shares were dealt correctly. It is worth noticing that besides allowing the verification of correct dealing of shares, the public values g^{x_i} can be used at time of secret reconstruction to verify that the participating shares are correct (see Section 5).

Notice that Feldman's scheme reveals the value $y = g^x \pmod{p}$, where $x = f_0$ is the secret being shared. Although the entire value of x cannot be derived from y (assuming the hardness of discrete logarithm), still there is partial information on x (e.g., its least significant bit) which is exposed by y. Therefore the semantic security of our solution (when based on Feldman's scheme) can be only stated *relative to the knowledge of $g^x \pmod{p}$*. This is acceptable in cases where g^x is known anyway (see [14]); however, in general, one would like to prevent the leakage of partial information. In this case, one encodes the actual secret, say s, into a longer string x and performs the proactive secret sharing using x. The encoding should have the property that given x, the value of s is easy to recover, but given $g^x \pmod{p}$ then no information on s can be efficiently computed. Such schemes exist based on *hard core bits* of the exponentiation function (e.g., if s is represented by the $O(\log(|p|))$ most significant bits of x [15]; a more efficient construction can be based on [1]). For more information see [11, 16]. *This issue can be completely avoided in our solution by replacing Feldman's VSS with the information theoretic scheme of [17].*

PUBLIC-KEY ENCRYPTION AND SIGNATURES. Our solution requires semantically secure encryption [12] and existentially unforgeable signatures [13]. We do not specify or assume any particular implementation of these functions. For a pair of *sender* S and *receiver* R, we denote by $ENC_R(data)$ the probabilistic encryption of *data* under R's public key; and by $SIG_S(data)$ the signature of *data* under S's private key.

3 Periodic Share Renewal Scheme

Here we present a basic component of our solution, namely, the protocols for periodic renewal of shares which preserve the secret, and at the same time make past knowledge obsolete for the adversary. Beyond guaranteeing the secrecy of the shared secret, our scheme is robust in the sense of guaranteeing integrity and availability of the secret in the presence of up to k misbehaving servers.

3.1 Initial Setting: Black-box Public Key Assumptions

Cryptographic solutions in a distributed environment typically require the ability to maintain private and authenticated communication between the servers. This is achieved by the servers having pairs of private and public keys corresponding to public-key cryptosystems with encryption and signature capabilities (e.g., RSA, ElGamal, etc.). However, an adversary that breaks into a server and learns its private key can then impersonate that server for the whole life of that private key. Therefore, to ensure proactive security, it is necessary to maintain these private keys proactively, namely, to renew them periodically.

We will show in section 6 how this can be done in our context (a more general treatment of proactive authentication can be found in [4]). However, for clarity of presentation, we start by making the strong assumption that servers are equipped with a pair of private and public keys in a way that the private key cannot be learned or modified by the adversary, even if this adversary manages to break into the server. While such an intruder will be able to generate legal signatures and decrypt messages using the private key (as a "black-box"), it will not be able to learn or modify the key itself. We will remove this assumption and deal with the proactive maintenance of the private keys in section 6.

3.2 Initialization of Secret Sharing

We assume an initial stage where a secret $x \in Z_q$ (for prime q) is encoded into n pieces $x_1, \ldots, x_n \in Z_q$ using a k-threshold Shamir's secret sharing: Each $P_i, i \in \{1 \ldots n\}$ holds its share x_i, where $x_i = f(i)$ for some k-degree polynomial $f(\cdot)$ over Z_q s.t. $x = f(0)$.

After the initialization, at the beginning of every time period, all honest servers trigger an *update phase* in which the servers perform a *share renewal protocol*. The shares computed in period t are denoted by using the superscript (t), i.e., $x_i^{(t)}, t = 0, 1, \ldots$. The polynomial corresponding to these shares is denoted $f^{(t)}(\cdot)$.

3.3 Share renewal

To renew the shares at period $t, t = 1, 2, \ldots$, we adapt a simplified version of the update protocol presented by Ostrovsky and Yung in [18]. When the secret x is (distributively) stored as a value $f^{(t-1)}(0) = x$ of a k degree polynomial $f^{(t-1)}(\cdot)$ in Z_q, we can update this polynomial by adding it to a k degree random polynomial

$\delta(\cdot)$, where $\delta(0) = 0$, so that $f^{(t)}(0) = f^{(t-1)}(0) + \delta(0) = x + 0 = x$. By the linearity of the polynomial evaluation operation we get the renewal of the shares $x_i^{(t)} = f^{(t)}(i)$ according to:

$$f^{(t)}(\cdot) \leftarrow f^{(t-1)}(\cdot) + \delta(\cdot) \ (mod \ q) \iff \forall_i \ f^{(t)}(i) = f^{(t-1)}(i) + \delta(i) \ (mod \ q)$$

In our system we will have $\delta(\cdot) = (\delta_1(\cdot) + \delta_2(\cdot) + \ldots + \delta_n(\cdot)) \ (mod \ q)$, where each polynomial $\delta_i(\cdot)$, $i \in \{1 \ldots n\}$ is of degree k and is picked independently and at random by the ith server subject to the condition $\delta_i(0) = 0$. The share renewal protocol for each server P_i, $i \in \{1 \ldots n\}$, at time period t is as follows:

1. P_i picks k random numbers $\{\delta_{im}\}_{m \in \{1 \ldots k\}}$ from Z_q. These numbers define a polynomial $\delta_i(z) = \delta_{i1}z^1 + \delta_{i2}z^2 + \ldots + \delta_{ik}z^k$ in Z_q, whose free coefficient is zero and hence, $\delta_i(0) = 0$.

2. For all other servers P_j, P_i *secretly* sends $u_{ij} = \delta_i(j) \ (mod \ q)$ to P_j.

3. After decrypting $u_{ji}, \forall j \in \{1 \ldots n\}$, P_i computes its new share $x_i^{(t)} \leftarrow x_i^{(t-1)} + (u_{1i} + u_{2i} + \ldots + u_{ni}) \ (mod \ q)$ and *erases* all the variables it used except of its current secret key $x_i^{(t)}$.

This protocol solves the share renewal problem against a ("passive") adversary that may learn the secret information available to corrupted servers but where all servers follow the predetermined protocol. This is proven in the next theorem. Notice that we assume here that the shares are transmitted to the corresponding holders with perfect secrecy. This allows us to prove the information-theoretic secrecy of this scheme. In the next sections we use explicit encryption for the transmission of these shares and then the secrecy of the scheme is reduced to the strength of the encryption.

Theorem 3. *If all servers follow the above share renewal protocol then:*

Robustness: The new shares computed at the end of the update phase correspond to the secret x (i.e., any subset of $k + 1$ of the new shares can reconstruct the secret x).

Secrecy: An adversary that at any time period knows no more than k shares (possibly a different set of shares in each period) learns nothing about the secret.

3.4 Share Renewal Protocol in the Presence of Active Attackers

In the above basic share renewal protocol an *active* adversary controlling a server can cause the destruction of the secret by dealing inconsistent share updates or just by choosing a polynomial δ_i with $\delta_i(0) \neq 0$. In order to assure the detection of wrongly dealt shares we add to the above basic protocol a *verifiability* feature. Namely, we adapt to our scenario Feldman's verifiable secret sharing scheme as described in section 2.2. In traditional applications of verifiable secret sharing, the fact that all the share-holders find their shares to be consistent is used as a proof for a correct dealing of the secret. In our case, this is used as a proof for correct dealing of update shares by the servers.

The verifiable share renewal protocol for each server P_i at period t is as follows:

1. P_i picks k random numbers $\{\delta_{im}\}_{m\in\{1...k\}}$ from Z_q to define the polynomial $\delta_i(z) = \delta_{i1}z^1 + \delta_{i2}z^2 + \ldots + \delta_{ik}z^k$. It also computes values $\epsilon_{im} = g^{\delta_{im}} \pmod{p}$, $m \in \{1\ldots k\}$.

2. P_i computes $u_{ij} = \delta_i(j) \pmod{q}$, $j \in \{1\ldots n\}$, and computes $e_{ij} = ENC_j(u_{ij})$, $\forall j \neq i$.

3. P_i broadcast the message $VSS_i^{(t)} = (i, t, \{\epsilon_{im}\}_{m\in\{1...k\}}, \{e_{ij}\}_{j\in\{1...k\}\setminus\{i\}})$, and the signature $SIG_i(VSS_i^{(t)})$.

4. For all messages broadcasted in the previous step by other servers, P_i decrypts the shares intended for P_i (i.e., computes u_{ji} out of e_{ji}, $\forall j \neq i$), and verifies the correctness of shares using the equivalent of the verifiability equation 1 from section 2.2, namely, for all $j \neq i$ it verifies:

$$g^{u_{ji}} \stackrel{?}{=} (\epsilon_{j1})^i (\epsilon_{j2})^{i^2} \ldots (\epsilon_{jk})^{i^k} \pmod{p}. \tag{2}$$

(Notice that this equation accounts for the condition $\delta_j(0) = 0$.)

5. If P_i finds all the messages sent in the previous step by other servers to be correct (e.g., all have correct signatures, time period numbers, etc.), and all the above equations to hold, then it broadcasts a signed acceptance message announcing that all checks were successful.

6. If all servers sent acceptance messages then P_i proceeds to update its own share by performing: $x_i^{(t)} \leftarrow x_i^{(t-1)} + (u_{1i} + u_{2i} + \ldots + u_{ni}) \pmod{q}$ and erases all the variables it used except for its current share $x_i^{(t)}$.

7. If in the above step 5, P_i finds any irregularities in the behavior of other servers during step 4 then it broadcasts a signed accusation against the misbehaving server(s). When to send accusations, and how to resolve them is discussed in the next subsection.

Resolving accusations In step 5 of the above protocol, each server checks the correct behavior and dealing of other servers. If a misbehaving server is found then there are two kinds of actions to take. One is not to use the polynomial $\delta_i(\cdot)$ dealt by this server in the renewal of shares in step 6. The second is to alert the system management so that it could take measures to rectify the misbehaving server (e.g., it may be required to reboot the server in order to "expel" the adversary). However, an accusation against a server by another server requires verification, since a misbehaving server could falsely accuse others. For a consistent update of shares, the honest servers need to agree on who the "bad" servers are. We explain below how each server P_i decides on its list B_i of bad processors.

We say that a message from server P_i at period t is correct if it complies with the specifications of the above protocol, including all the specified fields and information (e.g., the correct time period number) as well as a correct signature.

We distinguish between three classes of irregularities in the protocol: (a) Formally incorrect messages: wrong time period numbers, numbers out of bounds etc.; (b) Two or more correct messages from the same server containing a valid signature, or no message at all from some server; and (c) a mismatch in equation 2.

Notice that irregularities of the first two types are discovered using public information only, and then are found by all (honest) servers. Any such misbehaving processor is marked as "bad" by each of the other processors. The faults of the third kind are more problematic since they are discovered only locally by a server that receives a share causing a mismatch in equation 2.

When a server P_i finds that equation 2 corresponding to the information sent by P_j does not hold, it needs to broadcast an accusation against P_j. Then the servers decide whether it is P_i or P_j who is cheating. A way to do this is by having P_j "defend" itself. If P_j sent a correct u_{ji}, namely, one that passes equation 2, then it can expose this value and prove that it corresponds to the publicly available encryption value e_{ji} sent by P_j to P_i. To prove this P_j may need to reveal additional information used to compute the encryption (like the random vector used in probabilistic encryption). However, P_j does not need to reveal any private information of itself. Then everybody can check whether u_{ji} and the additional information published by P_j encrypts under P_i's public key to e_{ji} as broadcasted by P_j in step 2. Second, everybody can check whether this u_{ji} matches equation 2. If P_j defends itself correctly then all servers mark P_i as bad, otherwise P_j is marked as bad. Notice that in some encryption schemes (like RSA), the information published by the accuser is sufficient for public verification, which simplifies the above general protocol.

Once all accusations are resolved, every honest server P_i holds the same list of bad servers $\mathcal{B}_i$ (i.e., for each pair (P_i, P_j) of non-faulty servers, $\mathcal{B}_i = \mathcal{B}_j$). Now the computation of the new shares is done by replacing step 6 of the share renewal protocol by:

$$x_i^{(t)} \leftarrow x_i^{(t-1)} + \sum_{j \notin \mathcal{B}_i} u_{ji} \ (mod \ q)$$

3.5 Security Properties

In Theorem 3 we dealt with the share renewal assuming the servers are curious but honest. Here we claim the analogous result for the case that up to k servers are arbitrarily misbehaving in the renewal protocol.

Theorem 4. *If the adversary controls up to k servers during the protocol, then:*

Robustness: The new shares computed at the end of the update phase by honest servers correspond to the secret x (i.e., any subset of $k + 1$ of the new shares of honest servers interpolate to the secret x).

Secrecy: The above secret sharing scheme is semantically secure. [2]

4 Share Recovery Scheme

A proactive secret sharing system must be able to check whether a share of each participating server has been corrupted (or lost), and restore the correct share if

---

[2] In the protocol above we achieve semantic security relative to the a priori knowledge of the exponent $g^x \ (mod \ p)$ of the secret x. This extra knowledge is avoided by using Pedersen's *VSS* scheme [17] (see discussion on Feldman's *VSS* in section 2.2).

necessary. Otherwise, an adversary could cause the loss of the secret by gradually destroying $n - k$ shares. Below we present the necessary mechanisms for detection and recovery of corrupted shares.

The share $x_i^{(t)}$ held by processor P_i in period t is called *correct* if $x_i^{(t)} = f^{(t)}(i) \pmod{q}$, where $f^{(t)}$ is the current secret sharing polynomial. Otherwise, we say that the share is *incorrect*. A server can have an incorrect share because it was controlled by the adversary during the share renewal protocol (and hence it was prevented to update its share correctly), or because the adversary attacked the server after the update phase and modified the server's secret share. A secret share can also be lost because a server was rebooted or replaced by a new server.

4.1 Detection of Corrupted Shares

How are corrupted shares detected? In some cases it is easy to detect that a server requires to recover its correct share. This is the case of servers that do not participate of an update phase (e.g., due to a crash), or servers that misbehave during that phase. However, if the share of some server is ("silently") modified by the adversary (e.g., after an update phase) then this modification may go undetected. Hence, in the spirit of proactiveness, the system must periodically test the correctness of the local states of the participating servers, detecting in this way lost or modified shares.

To implement the distributed verifiability of shares, we add an invariant that in each time period t, each server P_i stores a set $\{y_j^{(t)}\}_{j \in \{1...n\}}$ of exponents $y_j^{(t)} = g^{x_j^{(t)}} \pmod{p}$ of current shares of all servers in $\mathcal{A}$. This is achieved as follows. First, we augment section 3.2 with the requisite that each server stores the values $y_j^{(0)}$ corresponding to the initial shares $x_j^{(0)}, j \in \{1 \ldots n\}$ (this can be achieved by performing Feldman's VSS at initialization). Also, using the homomorphism of the exponentiation function, we supplement step 6 of the update protocol in section 3.4 so that each server P_i updates its set $\{y_j\}_{j \in \{1...n\}}$ by computing for every j:

$$y_j^{(t)} \leftarrow y_j^{(t-1)} * (g^{u_{1j}} * g^{u_{2j}} * \cdots * g^{u_{nj}}) \pmod{p}$$

(in the general case, the above product is computed using only update shares u_{mj} corresponding to servers that did not misbehave in the update phase, i.e., $m \notin \mathcal{B}_i$).

We extend the *update phase* between time periods to include a *share recovery* protocol. Its first part is the *lost share detection protocol* which works as follows: Every server broadcasts the values $\{y_j^{(t)}\}_{j \in \{1...n\}}$ stored in that server, together with a signature on these values. After collecting these messages from all servers and checking their signatures, each server decides by majority on the current proper set of share exponents. Now each server P_i can decide on a set $\mathcal{B}_i$ of servers which presented an incorrect (i.e., different from majority) exponent of their own share. These are the servers that P_i believes to need a share recovery. (In particular, it can be the case that for some i, $P_i \in \mathcal{B}_i$, which means that server P_i decided that its own share is not correct). It is clear that every pair of non-faulty servers (P_i, P_j) has the same view about who has an incorrect share, i.e., $\mathcal{B}_i = \mathcal{B}_j = \mathcal{B}$, where $|\mathcal{B}| \leq k$.

4.2 Basic Share Recovery Protocol

A straightforward way to reconstruct the shares $x_r = f^{(t)}(r)$ for $r \in \mathcal{B}$, is to let each server in $\mathcal{D} = \mathcal{A} \setminus \mathcal{B}$ send its own share to P_r. However, this would expose the secret x to P_r. Instead, for each $r \in \mathcal{B}$, the servers in the set $\mathcal{D}$ will collectively generate a random secret sharing of x_r in a way analogous to that used to re-randomize the secret sharing of the main secret x in the share renewal protocol: Every server P_i in $\mathcal{D}$ deals a random k-degree polynomial $\delta_i(\cdot)$, such that $\delta_i(r) = 0 \pmod{q}$. In this way, a new, random secret sharing $\{x'_i\}_{i \in \mathcal{D}}$ of x_r is obtained. The servers can now send these new shares to P_r, to allow it to compute x_r without letting P_r learn anything about the original shares $\{x_i\}_{i \in \mathcal{D}}$. Also, any coalition of k or less servers, not including P_r, will learn nothing about the value of x_r.

We first present the share recovery protocol stripped of verifiability. It is secure only against an adversary that eavesdrops into k or less servers, but does not change the servers behavior. For each P_r that requires share recovery, the following protocol is performed:

1. Each $P_i, i \in \mathcal{D}$, picks a random k-degree polynomial $\delta_i(\cdot)$ over Z_q such that $\delta_i(r) = 0$, i.e. it picks random coefficients $\{\delta_{ij}\}_{j \in \{1...k\}} \subset Z_q$ and then computes $\delta_{i0} = -\sum_{j \in \{1...k\}} \delta_{ij} r^j \pmod{q}$.
2. Each $P_i, i \in \mathcal{D}$, broadcasts $\{ENC_j(\delta_i(j))\}_{j \in \mathcal{D}}$.
3. Each $P_i, i \in \mathcal{D}$, creates its new share of x_r, $x'_i = x_i + \sum_{j \in \mathcal{D}} \delta_j(i)$ and sends it to P_r by broadcasting $ENC_r(x'_i)$.
4. P_r decrypts these shares and interpolates them to recover x_r.

4.3 Full Share Recovery Protocol

In the general case, the adversary not only can eavesdrop into the servers but also cause the corrupted servers to deviate from their intended protocol. To cope with these cases, we add to the above protocol (section 4.2) the necessary "verifiability" properties for the dealing of polynomials $\delta_i(\cdot)$ in step 2 and for reconstruction of x_r from x_i's in step 3 and 4. Due to space constraints we omit the description of this protocol in these Proceedings, but we sketch the properties of this protocol in the following theorem:

Theorem 5. *If the adversary compromises no more than k servers in any time period, then the full share recovery protocol has the following two properties:*

Robustness: Each recovering server that follows the protocol recovers its correct share $x_r = f^{(t)}(r) \pmod{q}$.
Secrecy: The semantic security of the secret x is preserved.

5 Secret Reconstruction

In the above section we have shown how to renew shares consistent with the secret, and preserve their integrity such that at any time any $k+1$ parties could reconstruct the secret if desired. However, when coming to actually reconstruct the secret, the participants in the reconstruction protocol (namely, polynomial interpolation as in Shamir's scheme) must be able to detect servers that provide corrupted shares to the reconstruction. This detection is easily accomplished by verification of the submitted shares against values $y_i^{(t)}$ held by the majority of servers.

6 Dynamically Secure Private Keys

We now show how to remove the protected key assumption of section 3.1. We extend the *update phase* between time periods to include a third component, the *private key renewal protocol*, which will be triggered before share recovery and share renewal. As a result of the private key renewal, an adversary that breaks into a server in period t, but which does not control the server at period $t + 1$, cannot learn this server's new key.

The private key renewal protocol at the beginning of each update phase works as follows: Each server P_i chooses a new pair of private and public keys $a_i^{(t)}, b_i^{(t)}$. The server then broadcasts its new public key $b_i^{(t)}$ authenticated by its signature using its *previous* private key $a_i^{(t-1)}$. The other servers, having $b_i^{(t-1)}$ from the previous period, can verify this signature. Clearly, an adversary that controlled the server at time period $t - 1$, or before, but not during the update phase between periods $t - 1$ and t, cannot learn the new private key chosen by the server. However, if the adversary knows $a_i^{(t-1)}$ then, even if she is not controlling P_i during the private key renewal protocol of period t, she can choose her own private key and inject its public counterpart into the broadcast channel, authenticated as if it originated from P_i. But since P_i is not actively controlled by the adversary anymore, it will send its own authenticated public key to the communication channel as well. This will result in two different messages legally authenticated as coming from P_i. This will constitute a public proof of P_i's compromise, after which a reboot procedure must be triggered.

When a server is rebooted, it internally chooses its new private key, publishing only the corresponding public key, which must be then installed on all servers in A.

7 Proactive Secret Sharing: the Combined Protocol

Combining all the above pieces we get our full protocol for proactive secret sharing: At the beginning of every time period, the update phase is trigerred, which consist of three stages: the private key renewal protocol, the share recovery protocol (including lost share detection) and the share renewal protocol. The following theorem summarizes our main result:

Theorem 6. *If there are no more than k corrupted servers in each period (where servers compromised at a renewal phase are considered as compromised in both adjacent periods), then the above protocol constitutes a secure and robust proactive secret sharing scheme.*

References

1. M. Bellare and P. Rogaway, *Optimal Asymmetric Encryption*, Eurocrypt '94.
2. G.R. Blakley, *Safeguarding Cryptographic Keys*, AFIPS Con. Proc (v. 48), 1979, pp 313–317.
3. R. Canetti and A. Herzberg, *Maintaining Security in the Presence of Transient Faults*, Proc. Crypto'94 (LNCS 839), pp. 425-438.
4. R. Canetti and A. Herzberg, *Proactive Maintenance of Authenticated Communication*, manuscript, 1995.
5. B. Chor, S. Goldwasser, S. Micali and B. Awerbuch, *Verifiable Secret Sharing and Achieving Simultaneous Broadcast*, Proc. of IEEE Focs 1985, pp. 335-344.
6. M. Blum and S. Micali, How to Construct Cryptographically Strong Sequences of Pseudorandom Bits. SIAM J. Comp. 13, 1984.
7. A. De Santis, Y. Desmedt, Y. Frankel, and M. Yung, *How to Share a Function Securely*, ACM STOC 94.
8. Y. Desmedt and Y. Frankel, *Threshold cryptosystems*, In G. Brassard, editor, Advances in Cryptology, Proc. of Crypto '89 (Lecture Notes in Computer Science 435), pages 307–315. Springer-Verlag, 1990.
9. D. Dolev, C. Dwork, O. Waarts, and M. Yung, *Perfectly Secure Message Transmission*, Proc. 31st Annual Symposium on the Foundations of Computer Science, 1990, (also JACM).
10. T. El Gamal, *A Public key cryptosystem and a signature scheme based on discrete logarithm*, IEEE Trans. on Information Theory 31, 465-472, 1985.
11. P. Feldman, *A Practical Scheme for Non-Interactive Verifiable Secret Sharing*, Proc. of the 28th IEEE Symposium on the Foundations of Computer Science, 1987, 427-437
12. S. Goldwasser and S. Micali, *Probabilistic Encryption*, J. Com. Sys. Sci. 28 (1984), pp 270-299.
13. S. Goldwasser, S. Micali and R. Rivest, *A Secure Digital Signature Scheme*, Siam Journal on Computing, Vol. 17, 2 (1988), pp. 281-308.
14. A. Herzberg, M. Jakobsson, S. Jarecki, H. Krawczyk and M. Yung, *Proactive Public Key and Signatures Systems*, draft.
15. D.L. Long and A. Wigderson, *The Discrete Log. Problem Hides O(Log N) Bits*. SIAM J. Comp. 17, 1988, pp. 363-372.
16. T. P. Pedersen, *Distributed Provers with Applications to Undeniable Signature*, Proc. Eurocrypt '91 (LNCS 547), pp. 221-242.
17. T. P. Pedersen, *Non-Interactive and Information-Theoretic Secure Verifiable Secret Sharing*, Proc. Crypto'91 (LNCS 576), pp. 129-140.
18. R. Ostrovsky and M Yung, *How to withstand mobile virus attacks*, Proc. of the 10th ACM Symp. on the Princ. of Distr. Comp., 1991, pp. 51-61.
19. A. Shamir. *How to share a secret*, Commun. ACM, 22 (1979), pp 612-613.

Secret Sharing with Public Reconstruction (extended abstract)

Amos Beimel* and Benny Chor **

Department of Computer Science
Technion, Haifa 32000, Israel

Abstract. All known constructions of information theoretic t-out-of-n secret sharing schemes require *secure, private* communication channels among the parties for the reconstruction of the secret. In this work we investigate the cost of performing the reconstruction over *public* communication channels. A naive implementation of this task distributes $O(n)$ one times pads to each party. This results in shares whose size is $O(n)$ times the secret size. We present three implementations of such schemes that are substantially more efficient:

- A scheme enabling multiple reconstructions of the secret by different subsets of parties, with factor $O(n/t)$ increase in the shares' size.
- A one-time scheme, enabling a single reconstruction of the secret, with $O(\log(n/t))$ increase in the shares' size.
- A one-time scheme, enabling a single reconstruction by a set of size *exactly t*, with factor $O(1)$ increase in the shares' size.

We prove that the first implementation is optimal (up to constant factors) by showing a tight $\Omega(n/t)$ lower bound for the increase in the shares' size.

1 Introduction

Secret sharing schemes were introduced by Blakley [7] and Shamir [19], and were the subject of a considerable amount of work, e.g. [18, 15, 16, 4, 20]. In these schemes, a dealer holds a secret piece of information. Upon system initialization, the dealer gives one share of the secret to each of n parties. These shares are distributed privately, and are kept by each party in a secure way. Later on, any authorized subset (a subset containing at least t parties) of the parties collect their shares, and use them to reconstruct the secret. All known schemes that guarantee information theoretic secrecy require the use of secure, private communication channels between the parties that participate in the reconstruction.

The question we raise in this work is whether reconstruction can be done without assuming that the channels are secure, while maintaining the security of the schemes. We consider the worst case scenario: The "bad" parties can overhear any communication, so from their point of view the channels are public. On the

---

* email: beimel@cs.technion.ac.il, URL: http://www.cs.technion.ac.il/~beimel
** email: benny@cs.technion.ac.il, URL: http://www.cs.technion.ac.il/~benny

D. Coppersmith (Ed.): Advances in Cryptology - CRYPTO '95, LNCS 963, pp. 353-366, 1995.
© Springer-Verlag Berlin Heidelberg 1995

other hand, "good" parties hear only messages sent to them. (In particular, from the point of view of the "good guys", the channels do not carry any of the potential advantages of a broadcast channel.)

The simplest way to implement such public reconstruction securely is to hand to each party upon system initialization, in addition to his original share, $2(n-1)$ one time pads. These pads are used in order to simulate a private channel on a public one. In the private channel scenario, reconstruction is typically done by exchanging shares among parties. To enable such exchange with every other participant, each party will need two pads per participant: one for receiving a share, and one for sending the share. Thus the simple implementation results in $O(n)$ multiplicative factor increase in the size of each share.

We design substantially more efficient schemes of three typees. The first type is *unrestricted schemes*. In these schemes, any number of authorized sets (each containing at least t parties) may reconstruct the secret, after communicating on the public channel. Any disjoint coalition of at most $t - 1$ parties, does not gain any partial information on the secret, given the coalition's shares and the the communication of the sets that reconstructed the secret. We describe unrestricted schemes in which the size of the shares in $O(n/t)$ times the size of the original secret. We complement this result by proving a tight $\Omega(n/t)$ lower bound on the increase in the shares' size for any unrestricted scheme.

In order to participate in more than one reconstruction, every party that has already reconstructed the secret must store the secret. This is problematic in applications where an adversary might break into the computer of the secret holder. (One of the advantages of traditional secret sharing is that breaking into the computer of a "share holder" does not compromize the secret.) The unrestricted non-reactive schemes of Section 5 solve this problem, but the share size there is n times the secret size.

The second type is *one time schemes*, in which only a single authorized set (containing at least t parties) will reconstruct the secret. It is not known during system initialization which set will reconstruct the secret, and the dealer has to accommodate any possible set. For example, these schemes can be used to enable one time activities like the firing of a ballistic missile or opening of a sealed safe. We describe one-time schemes in which the size of the shares in $O(\log(n/t))$ times the original secret size. It is an open problem if this bound is tight for one-time schemes. Finally, we consider one time schemes where one authorized set of size *exactly* t will reconstruct the secret. Additional parties in supersets with more than t parties may not reconstruct the secret, because communicating it from members of the authorized set over the public channel is not possible in a secure way. This means that the authorized sets that can securely reconstruct the secret do not form a *monotone* access structure. We design such schemes with just $O(1)$ multiplicative increase in the share size (for any threshold t).

In light of our results, one may wonder if the initial distribution of shares can also be done over public channels. By the properties of "regular" schemes, each participant requires a share whose conditional mutual information with the secret (given $t - 1$ shares) is at least the entropy of the secrets [15]. This conditional entropy cannot be increased by communicating over public channels [17, 1]. Thus in our model, it is necessary to have secure initial distribution of shares from the dealer to the participants.

Some bibliographical remarks: A similar setting of public interaction was considered for interactive key distribution schemes (e.g. [10, 2]).Our schemes employ key distribution schemes, though not interactive ones. Another solution to eliminating the use of secure private channels assumes that the parties have limited computing power. A common assumption is that the parties are probabilistic polynomial-time Turing machines, and the security of the channels is achieved by means of public key cryptography [12, 13]. Public channels have been used in secret sharing (in addition to private channels) in dynamic sharing of secrets. These are schemes where the dealer enables parties to reconstruct different secrets in different time instants (e.g. [20, 6, 9]). A different scenario in which a public broadcast channel is used (in addition to private channels) is to protect against Byzantine parties [3]. Unlike our scenario, in that work the broadcast channel is heard by *all* parties.

The rest of this paper is organized as follows: In Section 2 we define the of the model, secret sharing schemes, and key distribution schemes. Section 3 describes the unrestricted schemes, and Section 4 the one time schemes. In Section 5 we introduce non-reactive, unrestricted schemes. Finally, Section 6 provides lower bounds for unrestricted schemes.

2 Definitions

In this section we define our model, secret sharing schemes (traditional and public channels), and key distribution schemes. We consider a system with n parties denoted by $\{P_1, P_2, \ldots, P_n\}$. In addition to the parties, there is a dealer in the system, who has a secret input s. A *scheme* is a probabilistic mapping, which the dealer applies to the input, and generates n pieces of information. These pieces of information are called shares, and the i-th pieces is called the share of P_i. For every i, the dealer gives the i-th share to P_i. The dealer is only active in this initial stage. After the initial stage, the parties can communicate, according to some pre-defined, possibly randomized, protocol. The parties are honest, that is, they follow their protocols. However, they are curious and after the protocol has ended some of them can collude and try to gain some partial information on the secret.

Definition 1. Let B be a (bad) coalition (set of parties). The *view* of B, denoted by VIEW$_B$, after an execution of a protocol is all the information it has, i.e. the shares of the parties in the coalition, and the messages exchanged by *all* parties over the communication channels. (Here the insecurity of the communication is manifested.) The coalition B *has no information* on a random variable X if for every two possible values x_1, x_2 of X:

$$\Pr[\text{ VIEW}_B \mid X = x_1] = \Pr[\text{ VIEW}_B \mid X = x_2],$$

where the probability is taken over the random inputs of the dealer, and the random inputs of the members outside the coalition. Notice that we do not make any assumptions on the distribution of X.

We define both traditional secret sharing scheme, i.e with private channels, and secret sharing schemes with public reconstruction.

Definition 2. Let S be a finite set of secrets. A *t-out-of-n secret sharing scheme* is a scheme, in which the input is a secret taken from S, and which satisfies the following two conditions:

Reconstructability: Any set of parties whose size is at least t can reconstruct the value of the secret after communicating among them self. Any party in the reconstructing set gets the value of the secret with certainty.

Security: Every disjoint coalition B of size at most $t - 1$ has no information on the secret as defined in Definition 1. There are three variates we consider:

1. *Traditional* secret sharing schemes in which the reconstruction takes place via secure, private channels. In this case the view of a disjoint coalition is its shares.

2. *Unrestricted* secret sharing scheme *with public reconstruction* in which a coalition B can hear all communications that took place. The security is guaranteed even if any collection of sets (maybe even all) will reconstruct the secret using the public channel. In this case the view of a disjoint coalition is its shares and all the communications that took place.

3. *One-time* scheme in which the security is guaranteed only if one set will reconstruct the secret. In this case the view of a disjoint coalition is its shares and a communication of one reconstructing set.

The security should hold for any coalition of at most $t - 1$ parties. As a special case ($B = \emptyset$), a listener who heard all communications but has no shares should gain no partial information about the secret.

Shamir [19] presented a traditional scheme in which the size of the shares is the same as the size of the secrets (for domains of secrets which contain at least $n + 1$ secrets). The domain of shares in Shamir's scheme is the smallest possible, since the size of the share has to be at least as large as the size of the secrets [15]. In traditional secret sharing schemes, while one set reconstructs the secret, no information is leaked to disjoint coalitions (due to the security of the channels). Hence, these schemes are always unrestricted. Furthermore, in traditional schemes, if a set can reconstruct the secret, then every superset of the set can reconstruct the secret. However, secret sharing schemes with public reconstruction do not necessarily have this monotone property. We require that every party of the superset should know the reconstructed secret. However, it is not necessarily possible to "distribute" the secret to members of a superset without leaking information on it to other parties.

We describe unrestricted, non-interactive key distribution schemes. (Formal definition can be found in [10, 2].) These schemes are used in the constructions of the schemes with public reconstruction.

Let b be a positive integers such that $b \leq n - 2$, and K be a set of keys. A $(2, b)$ *key distribution scheme* with n users and domain of keys K is a scheme in which a dealer (who has no input) generates n shares such that the following requirements hold:

Every pair of parties has a key, which is uniformly distributed. Each member of the pair can deterministically reconstruct G's key from his share. Any "bad" coalition "B" of cardinality at most b gets no information on the key of any disjoint pair. In this case their view is the collection of their pieces.

Consider a $(2, 2t-3)$ key distribution scheme, a coalition B of $t-1$ parties, and a disjoint set G of t parties. It holds that from the point of view of the coalition, the $\binom{t}{2}$ keys of pairs of parties in G are distributed uniformly and independently (for proof see [2]). Blom [8] constructed efficient $(2, b)$-key distribution schemes. For every prime-power q (where $q \geq n$) he presented a scheme in which the keys are taken from $\mathrm{GF}(q)$ and the shares are taken from $\mathrm{GF}(q)^{b+1}$.

3 Unrestricted Schemes

In this section we construct unrestricted secret sharing schemes with public reconstruction in which the size of the share of every party is $O(n/t)$ times the size of the secret. We first describe a simple scheme in which the size of the shares is $O(n)$ times the size of the secret. In this scheme, the dealer shares the secret using Shamir's secret sharing scheme [19]. The dealer also deals to every pair of parties two random strings whose size is the same as the size of the secret. These two random strings, which we call keys, are given to the two parties of the pair, and will be used as one-time pads. Overall, every party receives $2(n-1)$ keys, each one with the same size as the secret. When the parties of a set of size at least t wish to reconstruct the secret, all the parties "send" their shares to the "leader" of the set, say the party with minimal index in the set. The leader gets at least t shares, which enable him to reconstruct the secret. Then, the leader "sends" the secret to the other parties. The parties use their keys as one time pads to simulate the private channels. Specifically, let P_{i_0} be the party with smallest identity in the set. Every party P_i, holding the share s_i from Shamir's scheme, adds s_i and the first key of the pair $\langle P_{i_0}, P_i \rangle$ and sends this sum on the public channel. The party P_{i_0} can reconstruct all the shares from these messages, and therefore reconstruct the secret. Now, P_{i_0} sends messages, one message to every party in the reconstruction set. For every party P_i, he sums the secret and the second key of the pair $\langle P_{i_0}, P_i \rangle$ and sends this sum on the public channel. Since the one-time pads are independent, coalitions of parties disjoint to the reconstructing set do not gain any information on the shares or the secret. Furthermore, even if many reconstructions took place, this will not leak any information to a disjoint set.

Suppose P_{i_0} is the leader in a set of size at least t. In the previous scheme, during the reconstruction for this set, only the keys that were given to P_{i_0} were used. To improve the space efficiency we will use all the keys of the parties in the reconstructing set. Following [2], we partition the secret into t sub-secrets, and share each sub-secret using Shamir's scheme. Now we choose t parties of the reconstructing set, and each one will be responsible for reconstructing one sub-secret. Each party will act as the leader in the previous scheme. That is, every leader receives shares only from the other $t-1$ leaders (this is enough!), but sends his sub-secret (after reconstruction) to every member of the reconstructing set. This way we can handle t sub-secrets "at the price of one". The domain of the secrets in the scheme is $\mathrm{GF}(q)^t$, where q is a prime-power such that $q > n$. In the scheme we view the secret as t secrets from $\mathrm{GF}(q)$.

Unrestricted Secret Sharing Scheme

Distribution stage:
Input: t secrets $s_1, s_2, \ldots, s_t \in \mathrm{GF}(q)$
Shares:
Share each s_i using Shamir's scheme for every i, where $1 \leq i \leq t$.
Denote the n shares of secret s_i by $s_{i,1}, s_{i,2}, \ldots, s_{i,n}$.
For every pair of parties generate 4 independent keys from $\mathrm{GF}(q)$.
Denote the keys of $\langle P_i, P_j \rangle$ by $k_{i,j}^1, k_{i,j}^2, k_{i,j}^3, k_{i,j}^4$.
The share of P_i is $s_{1,i}, \ldots, s_{t,i}$ and $k_{i,j}^1, k_{i,j}^2, k_{i,j}^3, k_{i,j}^4$ for $1 \leq j \leq n$.

Reconstruction stage:
A set $G = \{P_{i_1}, \ldots, P_{i_\ell}\}$ that wants to reconstruct the secret ($\ell \geq t$).
Every $P_{i_j} \in G$ announces if he has previously reconstructed the secret.
Let P_{i_j} for $1 \leq j \leq t$ be the leaders of G.
Each leader P_{i_j} ($1 \leq j \leq t$) sends (at most) $t-1$ messages to other leaders that have not previously reconstructed the secret:
$$s_{i_j, j'} + k_{i_j, i_{j'}}^1 \text{ to } P_{i_{j'}} \text{ for } 1 \leq j' < j$$
$$s_{i_j, j'} + k_{i_j, i_{j'}}^2 \text{ to } P_{i_{j'}} \text{ for } j < j' \leq t$$
Each leader P_{i_j} ($1 \leq j \leq t$) computes s_j from $s_{j,i_1}, \ldots, s_{j,i_t}$.
Each leader P_{i_j} sends a message to every $P_{i_{j'}} \in G$ that has not previously reconstructed the secret:
$$s_j + k_{i_j, i_{j'}}^3 \text{ to } P_{i_{j'}} \text{ for } 1 \leq j' < j$$
$$s_j + k_{i_j, i_{j'}}^4 \text{ to } P_{i_{j'}} \text{ for } j < j' \leq \ell$$
Each party concatenates the sub-secrets $s_1, \ldots, s_t$ to obtain the secret.

Figure 1: Unrestricted scheme.

As described, the scheme has two technical points we should elaborate. The first is the fact that in one reconstruction two parties P_i and P_j might need to exchange 4 different messages. This is the reason for giving them 4 common keys. The second difficulty is that in different reconstructions the same party can be responsible for different sub-secrets. This means that P_i will have to send to P_j two different messages, using the same key as a one time pad. This might leak information to disjoint coalitions. Therefore, every party that participated in one reconstruction will remember the secret, and in latter reconstructions will inform other parties (in the clear) that he need not receive new messages. It does not prevent a leader from sending all messages that he has to send according to the scheme, since these message either depend on his share, or on the secret. (The party need only remember the secret, and *not* the messages that he heard.) Thus, every key is used as a one-time pad at most once (in the first reconstruction that the pair participates together). Therefore, the scheme satisfies the unrestricted security requirement. A detailed code of the scheme appears in Figure 1.

Let us calculate the size of the share of every party in the unrestricted scheme. Each party is given t shares generated by Shamir's scheme for secrets taken from $GF(q)$. The dealer also distributes to each party $4(n-1)$ keys taken from $GF(q)$. Hence, each share contains $(4n+t-4)$ elements from $GF(q)$, compared to t elements from $GF(q)$ for the secret. We summarize these results in the the next theorem.

Theorem 3. *Let q be a prime-power such that $q > n$. The above mentioned scheme is an unrestricted t-out-of-n secret sharing scheme with public reconstruction for secrets taken from $GF(q)^t$. The share of each party is an element of $GF(q)^{4n+t-4}$. So the size of each share is $1 + 4(n-1)/t$ times the size of the secrets.*

4   One-Time Schemes

In the unrestricted scheme, we need totally independent keys in order to guarantee the security of the scheme during repeated reconstructions. In this section we deal with the scenario where the secret is going to be reconstructed only once. For example, to enable the firing of a ballistic missile or opening of a sealed safe. In this case, total independence among the keys is not needed, and weaker independence requirements suffice. Shares can therefore be taken from a smaller sample space, which translates into smaller size shares. Specifically, we use Blom's key distribution scheme [8] for this purpose.

The first scheme we present enables one-time reconstruction of the secret by sets of size *exactly t*. The size of the shares is a constant (less than 10) times the size of the secret, namely only $O(1)$ increase in shares' size. We employ this "exactly t" scheme as a building block for "at least t" schemes. We use $\log(n/t)$ independent instances of "exact schemes" for thresholds $t, 2t, 4t, \ldots$ up to n, and an additional instance of size t. Now, given any set G with ℓ parties ($\ell \geq t$), we represent it as a union of subsets (not necessary disjoint) with cardinalities $t, 2t, 4t, \ldots$ – at most two sets subsets of cardinality t and at most one subset of cardinality $2^i t$ for each $i \geq 1$. The secret is now separately reconstructed by each subset. Any member of G takes part in at least one of these reconstructions, and thus learns the secret. On the other hand, any disjoint coalition containing at most $t-1$ parties gets no partial information on the secret from any single instance. Due to the independence of the instances, this remain valid with respect to the joint reconstructions. We get a one-time scheme for set of size at least t, with just $O(\log(n/t))$ increase in share size. We now describe in detail the "exact t" scheme. The distribution phase is depicted in Fig. 2.

Distribution in Exactly t-out-of-n one-time scheme

Input: secret $s \in \mathrm{GF}(q)^t$.

Consider the secret as t secrets $s_1, \ldots, s_t \in \mathrm{GF}(q)$.
Share each secret s_i using Shamir's t-out-of-n secret sharing scheme.
Let $b = \min\{2t - 3, n - 2\}$.
Generate shares using $(2, b)$-key distribution scheme with key domain
$\mathrm{GF}(q)^4$ (which we consider as 4 keys from $\mathrm{GF}(q)$).

Share of P_j: the j-th share of each s_i,
    and the share of the key distribution scheme.

Figure 2: Exactly t-out-of-n one-time scheme.

The reconstruction is done exactly as in the unrestricted scheme. The security of one reconstruction of a set of exactly t parties follows from the property of $(2, 2t - 3)$ key distribution schemes discussed in Section 2: Given the shares of any disjoint coalition of at most $t-1$ parties, the keys held by any set of size t are distributed uniformly and independently. Thus, when used as one-time pads, the reconstruction is secure (using the same arguments as in the unrestricted case). This scheme uses t shares of Shamir's t-out-of-n secret sharing scheme with secrets taken from $\mathrm{GF}(q)$. In addition, each party gets a share of a $(2, 2t - 3)$ key distribution scheme with keys taken from $\mathrm{GF}(q)^4$ and with shares taken from $\mathrm{GF}(q)^{4(2t-2)}$. Overall, each share contains $(9t - 8)$ elements from $\mathrm{GF}(q)$ (if $2t > n + 1$, then the shares are even shorter). Recall that the secret is taken from $\mathrm{GF}(q)^t$, and therefore the size of the share is less than 9 times the size of the secrets.

In this previous scheme the domain of secrets has to be $\mathrm{GF}(q)^t$ (for some prime-power q). Restricting the domain of the secret to such cardinality can cause problems when we employ simultaneously many schemes with the same secret but with different thresholds. To overcome this, given any domain of secrets we consider a slightly bigger domain whose size (which can depend on the threshold) is of the desired form. That is, given a secret of size m which is at least $t \log n$, we choose a prime power q such that $m \leq t \log q$, and use the previous scheme with secrets of size $m' = t \log q$. Choosing the smallest prime-power satisfying these conditions, we have $m' \leq m + t \leq 2m$. Thus,

Theorem 4. *Let m be a natural number such that $m > 9t \log n$. There exists a one-time sharing scheme with public reconstruction for exactly t-out-of-n, in which the size of the secret equals m, and the size of the share of each party is less than 10 times the size of the secrets.*

We next describe the one time scheme in which every set of *at least* t parties can securely reconstruct the secret.

One-time Secret Sharing Scheme

Distribution stage:
Input: secret s of size m
Share the secret s using two independent copies of a
  one time exactly t-out-of-n secret sharing schemes.
For every i, $1 \le i < \log(n/t)$:
  Share s with an exactly $2^i t$-out-of-n one time secret sharing scheme.

Reconstruction stage:
A set $G = \{P_{i_1}, \ldots, P_{i_\ell}\}$ that wants to reconstruct the secret ($\ell \ge t$).
Cover the set G by (possibly intersecting) sets of size $2^i t$
  (at most one set for every $i > 1$, and at most 2 sets of size t).
Each set of size $2^i t$ independently reconstructs the secret using
  the shares of the exactly $2^i t$-out-of-n secret sharing scheme.

Figure 3: t-out-of-n one-time scheme for every set.

Theorem 5. *The scheme of Figure 3 is a one-time t-out-of-n secret sharing scheme with public reconstruction in which every set of parties of size at least t can securely reconstruct the secret. If the size of the secrets m is larger than $9n \log n$, then the size of the shares of every party is less than $10(\log(n/t) + 1)$ times the size of the secrets.*

Remark If we require that the size of the secret m is greater than $n^2 \log n$, then we can construct a scheme in which the size of the shares is only $2 \log(n/t) + O(1)$ times the size of the secret, i.e a smaller leading constant.

5 Unrestricted Non-Reactive Schemes

A secret sharing scheme with public reconstruction is called *non-reactive* if the messages sent by each party depend only on his share (and not on messages received during the reconstruction). Non-reactive schemes are simpler to implement, as they require less synchronization. Therefore, they are desirable from practical point of view. In this section we present non-reactive, unrestricted t-out-of-n schemes. The size of the shares in these schemes is n times the size of the secret. This represents a slight improvements (by a factor of 2) over the reactive scheme of Section 3 for $t = 2$, but is strictly less efficient (in terms of share size) for $t \ge 5$. We extend these schemes to general access structures. The size of the share in our public reconstruction schemes is n times the size of the share in the original scheme. This is typically not a significant increase, as the best schemes for most access structures to date require shares whose size is exponential in n.

We first present a simple, non-reactive, 2-out-of-n secret sharing scheme. Let $s \in Z_m$ be the secret which the dealer wants to share. The dealer chooses

n independent random elements from $\mathcal{Z}_m$, denoted $r_1, \ldots, r_n$. The share of P_i is $r_1, \ldots, r_{i-1}, r_i + s, r_{i+1}, \ldots, r_n$. Each share is uniformly distributed in $\mathcal{Z}_m^n$, regardless of the secret. Hence, prior to any reconstruction every party has no information on the secret (as defined in Definition 1). To reconstruct the secret, P_i sends the message r_j, and P_j sends the message r_i. Now, P_i, who holds $r_i + s$, hears the message r_i, so he can reconstruct the secret. Every third party hears messages that he already knows, and gains no information on the secret. That is, the reconstruction is secure. The size of the shares in this scheme is n times the size of the secret. During the reconstruction in this scheme every party is deterministic and sends only one message that depends only on its share.

In general secret sharing schemes scenario, first suggested by [14], we are given a collection $\mathcal{A}$ of sets of parties called an *access structure*. We require that every set in $\mathcal{A}$ can reconstruct the secret, while every set not in $\mathcal{A}$ does not know anything about the secret. It follows that secret sharing schemes can exist only for monotone collections. Indeed, it is known that for every monotone collection there exists a traditional secret-sharing scheme [14, 5, 21]). However, the size of the shares in these schemes is typically exponential in the number of parties (i.e., of size $m2^{\Theta(n)}$ where n is the number of parties in the system and m is the size of the secret). Let $\mathcal{A}$ be any monotone access structures. The unrestricted, non-reactive, 2-out-of-n scheme can be generalized to a unrestricted, non-reactive scheme realizing the access structure $\mathcal{A}$, with the following properties:

Theorem 6. *Assume there exists a (traditional) secret sharing scheme realizing $\mathcal{A}$ with domain of secrets S and domain of shares U. Then there exists a unrestricted, non-reactive secret sharing scheme realizing $\mathcal{A}$ with public reconstruction for secrets taken from S. The share of each party is an element of $S \times U^{n-1}$. So the size of each share is at most n times the size of the shares in the original scheme.*

Corollary 7. *Let q be a prime-power such that $q > n$. There exists an unrestricted, non-reactive t-out-of-n secret sharing scheme with public reconstruction for secrets taken from $GF(q)$. The share of each party is an element of $GF(q)^n$. So the size of each share is n times the size of the secret.*

For most known schemes [5, 21], it is possible to design unrestricted *reactive* schemes with just an *additive* factor of n times the *secret* size (in these schemes it suffices for a party to send a message of the size of the secret, instead of his entire share). This is typically much better, as the shares tend to be much larger than the secret for general access structures. Additional details will be given in the final version.

6 Lower Bounds for Unrestricted Schemes

In this section we prove a $\Omega(n/t)$ lower bound on the increase in the shares' size for unrestricted t-out-of-n schemes. For $t = 2$ this lower bound is tight by the non-reactive scheme of Section 5. For $t > 2$ this lower bound is tight up to a constant factor (by the reactive scheme of Section 3). We first prove a $\Omega(n)$ lower bound on increase in size of shares for 2-out-of-n schemes. Then, we show that this lower bounds translates into $\Omega(n/t)$ increase for t-out-of-n schemes.

We start with the lower bound for $t = 2$. The proof uses entropy and mutual information. For definitions of these information theoretic terms, the reader can refer to [11]. We assume an arbitrary probability distribution on the secret. The intuition behind the proof is that P_i has to expose $H(S)$ extra bits of his share in each reconstruction. Finally, after all reconstructions, the uncertainty of S_1 has to remain at least $H(S)$, as an outsider who listened to all reconstructions still has $H(S)$ uncertainty on the secret. Since, P_1 participates in $n - 1$ reconstructions, the original entropy of the share has to be at least $n \cdot H(S)$.

Without loss of generality, we prove the claim for P_1. To prove the lower on P_1's share, we only use the requirement that P_1 can reconstruct the secret together with every other P_j (we do not care if other pairs can or cannot reconstruct the secret). We start with some notation. Denote by S_i the share given to P_i in the initial distribution phase, and by $C_{i,j}$ the messages exchanged between P_i and P_j (all these are random variables). We denote $C = C_{1,3} \ldots C_{1,n}$, the concatenation of all messages exchanged between P_1 and $P_3, \ldots, P_n$. Recall that the communication $C_{1,2}$, together with P_2's share S_2, enable P_2 to reconstruct the secret S. On the other hand, the communication C and S_2 give no information (to P_2) about the secret. These facts will imply the next claim.

Claim 8. $H(C_{1,2}|S_2C) \geq H(S)$.

Proof. Since P_2 can reconstruct the secret S, given his share S_2 and the messages $C_{1,2}$ exchanged between P_1 and P_2, the conditional entropy $H(S|C_{1,2}S_2)$ equals 0. On the other hand, P_2 gets no information about the secret S from his own share S_2 and all messages C exchanged between P_1 and the other $n - 2$ parties. Therefore the conditional entropy $H(S|S_2C)$ equals $H(S)$. Now, consider the conditional mutual information $I(C_{1,2}; S|S_2C)$ of the message $C_{1,2}$ and the secret S, given the share S_2 and C. We have

$$H(C_{1,2}|S_2C) - H(C_{1,2}|SS_2C) = I(C_{1,2}; S|S_2C)$$
$$= H(S|S_2C) - H(S|C_{1,2}S_2C) = H(S)$$

which implies $H(C_{1,2}|S_2C) \geq H(S)$. $\qquad\qquad\square$

The next claim is the heart of the proof of the lower bound. It states that the mutual information between S_1 and $C_{1,2}$ given the "other" communication C is at least $H(S)$. Intuitively, since P_2 does not know the secret prior to the reconstruction, and knows it after the reconstruction, P_2 has to receive $H(S)$ bits of information which could only originate in S_1 and passed through the communication $C_{1,2}$. Hence, $C_{1,2}$ must contain $H(S)$ bits of information originating from the share S_1. Claim 9 is stated for deterministic parties – the outgoing messages are determined by the given share and previous incoming messages. An analogous statement, for randomized protocols, will be included in the final version of this paper. In randomized, outgoing messages can also depend on random local inputs.

Claim 9. *For deterministic reconstruction protocols we have*

$$I(C_{1,2}; S_1|C) = H(S_1|C) - H(S_1|C_{1,2}C) \geq H(S) \ .$$

Proof. Since P_1 and P_2 are deterministic, and their domain of shares is finite, there is a bound k on the maximum number of communication rounds which take place during the reconstruction of the secret. Denote by A_i the i-th message sent by P_1 to P_2, and similarly, let B_i be the i-th message sent by P_2 to P_1. Then, without loss of generality, $C_{1,2} = A_1 B_1 \ldots A_k B_k$. The message A_i is determined by the share S_1 and previous messages, that is, $H(A_i|S_1 A_1 B_1 \ldots A_{i-1} B_{i-1}) = 0$. The following inequality holds for any deterministic communication protocol:

$$
\begin{aligned}
H(\,C_{1,2}\,|S_1 C) &= H(A_1 B_1 \ldots A_k B_k|S_1 C) \\
&= \sum_{i=1}^{k}(H(A_i|S_1 C A_1 B_1 \ldots A_{i-1} B_{i-1}) + H(B_i|S_1 C A_1 B_1 \ldots A_{i-1} B_{i-1} A_i)) \\
&= \sum_{i=1}^{k} H(B_i|S_1 C A_1 B_1 \ldots A_{i-1} B_{i-1} A_i) \\
&\leq \sum_{i=1}^{k} H(B_i|C A_1 B_1 \ldots A_{i-1} B_{i-1} A_i) \ .
\end{aligned}
$$

Similarly, $H(C_{1,2}|S_2 C) \leq \sum_{i=1}^{k} H(A_i|C A_1 B_1 \ldots A_{i-1} B_{i-1})$. Combing the two inequalities

$$
\begin{aligned}
H(C_{1,2}|S_1 C) + H(C_{1,2}|S_2 C) &\leq \sum_{i=1}^{k} H(B_i|C A_1 B_1 \ldots A_{i-1} B_{i-1} A_i) \\
&\quad + \sum_{i=1}^{k} H(A_i|C A_1 B_1 \ldots A_{i-1} B_{i-1}) \\
&= H(A_1 B_1 \ldots A_k B_k|C) = H(C_{1,2}|C) \ .
\end{aligned}
$$

This inequality, together with Claim 8, implies

$$I(C_{1,2}; S_1|C) \ = \ H(C_{1,2}|C) - H(C_{1,2}|S_1 C) \ \geq \ H(C_{1,2}|S_2 C) \geq H(S) \quad \square$$

Claim 10. *In any unrestricted 2-out-of-n secret sharing scheme with public reconstruction, the share of each participant, S_i, satisfies $H(S_i) \geq n \cdot H(S)$.*

Proof. We first note that by Definition 2 a listener, who overhears all communication involving P_1, gets no information on the secret. Therefore,

$$H(S|C_{1,2} C_{1,3} \ldots C_{1,n}) = H(S) \ .$$

On the other hand, given P_1's share, this communication determines the secret, so $H(S|S_1 C_{1,2} C_{1,3} \ldots C_{1,n}) = 0$. Therefore,

$$
\begin{aligned}
H(S) &= H(S|C_{1,2} C_{1,3} \ldots C_{1,n}) - H(S|S_1 C_{1,2} C_{1,3} \ldots C_{1,n}) \\
&= I(S; S_1|C_{1,2} C_{1,3} \ldots C_{1,n}) \\
&= H(S_1|C_{1,2} C_{1,3} \ldots C_{1,n}) - H(S_1|S C_{1,2} C_{1,3} \ldots C_{1,n}) \ ,
\end{aligned}
$$

and in particular $H(S_1|C_{1,2} C_{1,3} \ldots C_{1,n}) \geq H(S)$. Claim 9 (or analog claim for the case of randomized protocols, which will appear in the final version) states that

$$
H(S_1|C_{1,3} \ldots C_{1,n}) - H(S_1|C_{1,2} C_{1,3} \ldots C_{1,n}) \geq H(S) \ .
$$

Similarly it holds that

$$
H(S_1|C_{1,4} \ldots C_{1,n}) - H(S_1|C_{1,3} C_{1,4} \ldots C_{1,n}) \geq H(S) \ .
$$

$$
\vdots
$$

$$
H(S_1) - H(S_1|C_{1,n}) \geq H(S) \ .
$$

Summing these n inequalities, we conclude that $H(S_1) \geq n \cdot H(S)$. $\quad\square$

We next show that this lower bounds on increase in size of shares for 2-out-of-n schemes translates into $\Omega(n/t)$ increase for t-out-of-n schemes.

Theorem 11. *In every unrestricted t-out-of-n secret sharing scheme with public reconstruction the size of the shares of every party is at least $\lfloor 1 + (n - 1)/(t - 1) \rfloor$ times the size of the secrets.*

Proof. Consider any t-out-of-n scheme. Denote the party whose share is shortest by P_1. We construct an unrestricted 2-out-of-($\lfloor 1 + (n - 1)/(t - 1) \rfloor$) scheme in which the entropy of S_1 – the share of P_1 – is the same. Hence, by Claim 10 its entropy is at least $(\lfloor 1 + (n - 1)/(t - 1) \rfloor) H(S)$. Since the scheme is secure whatever the distribution on the secrets is, we can assume uniform distribution on the secrets. In this case $H(S) = \log |S|$, which is the size of the secret. Since $H(S_1) \leq \log |S_1|$, the size of the share of P_1 is at least $\lfloor 1 + (n - 1)/(t - 1) \rfloor$ times the size of the secrets.

The construction is simple: the dealer gives P_1 the share of P_1 in the original scheme, and every other party gets shares of $t - 1$ disjoint parties. Since every party has at most $t - 1$ shares, he does not gain any information on the secret even after hearing communications. On the other hand, every 2 parties have at least t shares, therefore they can communicate on a public channel, and securely reconstruct the secret. $\quad\square$

Acknowledgments We would like to thank Carlo Blundo and Hugo Krawczyk for helpful discussions, and Eyal Kushilevitz for helping us spell Hugo's name.

References

1. R. Ahlswede and I. Csiszar. Common Randomness in Information Theory and Cryptography – Part I: Secret Sharing. *IEEE IT*, 39(4):1121–1132, July 1993.
2. A. Beimel and B. Chor. Interaction in Key Distribution Schemes. To appear in *IEEE IT*. An extended abstract appears in *Crypto '93*, Springer-Verlag, *LNCS* 773, D.R. Stinson, ed. pp. 444–455.
3. M. Ben-Or and T. Rabin. Verifiable Secret sharing and Multiparty Protocols with Honest Majority. In *Proceeding 21th STOC*, pages 73–85. 1989.
4. J. Benaloh. Secret Sharing Homomorphisms: Keeping Shares of a Secret Secret. In A. M. Odlyzko, editor, *CRYPTO '86*, volume 263 of *LNCS*, pages 251–260.
5. J. Benaloh and J. Leichter. Generalized Secret Sharing and Monotone Functions. In S. Goldwasser, ed., *CRYPTO '88*, volume 403 of *LNCS*, pages 27–35.
6. B. Blakley, G. R. Blakley, A. H. Chan, and J. Massey. Threshold Schemes with Disenrollment. In E. F. Brickell, ed., *CRYPTO '92*, vol. 740 of *LNCS*, pp. 540–548.
7. G. R. Blakley. Safeguarding Cryptographic Keys. In *Proc. AFIPS 1979 NCC, vol. 48*, pages 313–317, June 1979.
8. R. Blom. An Optimal Class of Symmetric Key Generation Systems. In T. Beth, N. Cot, and I. Ingemarsson, editors, *Eurocrypt 84*, volume 209 of *LNCS*, pages 335–338. Springer-Verlag.
9. C. Blundo, A. Cresti, A. De Santis, and U. Vaccaro. Fully Dynamic Secret Sharing Schemes. In D. R. Stinson, ed., *CRYPTO '93*, vol. 773 of *LNCS*, pages 110–125.
10. C. Blundo, A. De Santis, A. Herzberg, S. Kutten, U. Vaccaro, and M. Yung. Perfectly-Secure Key Distribution for Dynamic Conferences. In E. F. Brickell, editor, *CRYPTO '92*, volume 740 of *LNCS*, pages 471–486. Springer-Verlag, 1993.
11. T. M. Cover and J. A. Thomas. *Elements of Information Theory*. John Wiley & Sons, 1991.
12. W. Diffie and M. E. Hellman. New Directions in Cryptography. *IEEE IT*, 22(6):644–654, 1976.
13. S. Goldwasser and S. Micali. Probabilistic Encryption. *JCSS*, 28(21):270–299, 1984.
14. M. Ito, A. Saito, and T. Nishizeki. Secret Sharing Schemes Realizing General Access Structure. In *Proc. IEEE Globecom 87*, pages 99–102, 1987.
15. E. D. Karnin, J. W. Greene, and M. E. Hellman. On Secret Sharing Systems. *IEEE IT*, 29(1):35–41, 1983.
16. S. C. Kothari. Generalized Linear Threshold Scheme. In G. R. Blakley and D. Chaum, editors, *CRYPTO '84*, volume 196 of *LNCS*, pages 231–241. Springer-Verlag, 1985.
17. U. M. Maurer. Secret Key Agreement by Public Discussion from Common Information. *IEEE IT*, 39(3):733–742, May 1993.
18. R. J. McEliece and D. V. Sarwate. On Sharing Secrets and Reed-Solomon Codes. *Communications of the ACM*, 24:583–584, September 1981.
19. A. Shamir. How to Share a Secret. *Communications of the ACM*, 22:612–613, 1979.
20. G. J. Simmons. An Introduction to Shared Secret and/or Shared Control and their Application. In G. J. Simmons, editor, *Contemporary Cryptology*, pages 441–497. IEEE Press, 1991.
21. G. J. Simmons, W. Jackson, and K. M. Martin. The Geometry of Shared Secret Schemes. *Bulletin of the ICA*, 1:71–88, 1991.

On General Perfect Secret Sharing Schemes

G. R. Blakley[1] and G. A. Kabatianski[2]

[1] Department of Mathematics, Texas A&M University, USA
blakley@math.tamu.edu
[2] IPPI, Moscow, Russia, on leave at Department of Mathematics, Texas A&M
University
kaba@ippi.ac.msk.su

Abstract. The purpose of this paper is to provide a general information-theoretic framework extensible to arbitrary access structures and to establish the correspondence between ideal SSS and matroids without invoking the more restrictive combinatorial definition of ideal scheme.

1 Introduction

The history of secret sharing schemes (SSS) began in 1979 when this problem was introduced and partially solved for the case of (n, τ)-threshold schemes (see [1], [2]). R. McEliece and D. Sarwate pointed out [3] a relationship between threshold schemes and codes in 1981. In 1983, E. Karnin, J. Greene, and M. Hellman [4] gave an information-theoretic (IT) approach to SSS. Later this approach was developed in [5]. In our opinion the most important step in the classification of ideal perfect SSS was taken by E. F. Brickell and D. M. Davenport [6] by establishing the relationship between *combinatorial ideal* (*i.e.* such that secret and shadows belong to the same alphabet) schemes and matroids. To do this they used not only combinatorial techniques but also combinatorial definitions. Recently K. Kurosawa *et al.* [11], following the ideas of [6], have shown that this relationship is also true for the case of IT ideal perfect SSS, but only under the restrictive assumption of uniform distribution of secrets.

In this paper we treat SSS by information-theoretic tools, in continuation of the approach of [4], [5]. We prove a new bound on the "cardinality" of a perfect SSS which shows, in particular, that such a scheme should have properties similar to the properties of a well-known combinatorial object—an orthogonal array. For ideal SSS we generalize the result of Brickell and Davenport under a general definition of ideal scheme based on information-theoretic notions. Our paper is self-contained and it seems to us that our proof is not only simpler than [6], [11], but sheds a different light on this problem.

2 Definitions

The problem of SSS can be formulated in the following way. There is a secret s_0 chosen from the set S_0 of all possible secrets with probability $p(s_0)$. And there is a dealer who provides information to n participants in a such a way that some

D. Coppersmith (Ed.): Advances in Cryptology - CRYPTO '95, LNCS 963, pp. 367-371, 1995.
© Springer-Verlag Berlin Heidelberg 1995

sets of participants, called allowed coalitions, can recover the secret exactly, but participants forming any other set cannot get additional information beyond their *a priori* information about the value of the secret. To do this, the dealer uses some (finite) alphabets $S_1, \ldots, S_n$, whose elements are called "shares" (or "shadows"). For a given s_0, the dealer distributes shares $s_1, \ldots, s_n$ (the i-th participant receives share s_i, and has no information about the values of other shares) chosen by him with probability $P_{s_0}(s_1, \ldots, s_n)$. We define the probability distribution P on a set $S = S_0 \times \cdots \times S_n$, where $P(s) = P(s_0, s_1, \ldots, s_n) = p(s_0)P_{s_0}(s_1, \ldots, s_n)$. Equivalently one can start from some distribution P on a set $S = S_0 \times \cdots \times S_n$. Any such a pair (P, S) can be considered to be an SSS. We call a point $(s_0, s_1, \ldots, s_n)$ a "sharing rule". P is called the distribution of sharing rules. We regard the share values s_i as random variables with joint distribution P, and denote them S_i.

Let Γ be some access structure, *i.e.* let Γ be a set of subsets of $\{1, \ldots, n\}$ with the monotonic property $(A \in \Gamma, A \subset B$ imply $B \in \Gamma)$. W.l.o.g. one can restrict consideration to Γ containing neither "negligible" participants (*i.e.* j such that $j \in A \in \Gamma$ always implies $A \backslash \{j\} \in \Gamma$) or "super" participants (*i.e.* j such that $\{j\} \in \Gamma$). We call a pair (P, S) a perfect SSS, realizing the access structure Γ, if

1. $P(S_0 = c_0 \mid S_i = c_i, i \in A) \in \{0, 1\}$ if $A \in \Gamma$
2. $P(S_0 = c_0 \mid S_i = c_i, i \in A) = P(S_0 = c_o)$ if $A \notin \Gamma$

Following [4] we reformulate the above definition in the language of entropy, *i.e.*

1. $H(S_0 \mid S_i, i \in A) = 0$ if $A \in \Gamma$
2. $H(S_0 \mid S_i, i \in A) = H(S_0)$ if $A \notin \Gamma$

Define a set $V = \{s \in S \mid P(s) > 0\}$ and call it the "array" (or the "code") of the SSS (P, S). Roughly speaking, combinatorial treatments of SSS deal with an array V whose rows are uniformly distributed, *i.e.* *only* with a uniform distribution of sharing rules (or, if one allows repetitions of row, probabilities have to be only rational numbers). Then the definition of the perfect SSS can be reformulated in "cardinality" language (see [8]). E. F. Brickell and D. M. Davenport gave [6] another definition of "perfect scheme" which is weaker than the usual one(s) given above. In fact, they replaced Property 2 by the following property. If the set of rows of V having given entries c_i in positions belonging to a set A, $A \notin \Gamma$, is not empty then any value of s_0 (*i.e.* "0"-entry) occurs among these rows. The usual combinatorial definition demands that all values of s_0 occur equally often. It is easy to see that the array of any perfect SSS is perfect according to the definition of [6], but there are examples of arrays which generate perfect SSS in the sense defined in [6] and which do not give rise to any perfect SSS in the usual sense of "perfect". Luckily, for the case of *combinatorial ideal* SSS, these notions coincide [6].

As we mentioned above, the combinatorial(C) definition of *ideal* is that $|S_i| = |S_0|$ for all $i = 1, \ldots, n$. The information-theoretic (IT) definition is based on the amount of information which the dealer should send through a channel to the i-th participant. It equals $H(S_i)$. It is known (see [5]) that for any perfect SSS, $H(S_i) \geq H(S_0)$ for all $i = 1, \ldots, n$. Therefore we call a perfect SSS IT-ideal (or C-ideal) if $H(S_i) = H(S_0)$ (or, $|S_i| = |S_0|$, correspondingly) for all $i = 1, \ldots, n$.

3 New bound for the cardinality of general perfect SSS

Instead of considering a pair (P, S) we will consider only a pair (P, V), because $P(s) = 0$ for $s \notin V$. We call $|V|$ the cardinality of the SSS. We do not know to whom to attribute the following simple result.

Lemma 1 *If a pair (P, V) is a perfect SSS for some access structure Γ and some probability distribution p on the set of secrets S_0 then the pair (P', V) perfectly realizes the same Γ and a distribution p', where $P'(s_0, \ldots, s_n) = \frac{p'(s_0)}{p(s_0)} P(s_0, \ldots, s_n)$.*

Denote by $\Gamma_{\min}$ the set of minimal subsets of Γ. The following lemma (see [5]) is very useful.

Lemma 2 $H(S_j \mid S_i, i \in A \backslash \{j\}) \geq H(S_0)$ *for any $A \in \Gamma_{\min}$ and any $j \in A$.*

Corollary 1 $H(S_i, i \in A) \geq |A| \cdot H(S_0)$ *for any $A \in \Gamma_{\min}$.*

Denote by V_A the minor of the array V whose columns belong to the set $A \subseteq \{1, \ldots, n\}$ and by $\|V_A\|$ the number of different rows of this minor. Consider any perfect SSS (P, V) with corresponding access structure Γ. According to Lemma 1, there is a perfect realization of the same access structure Γ with uniform distribution of secrets and with the same array V. Hence, for the new SSS, $H(S_0) = \log q$, where $q = |S_0|$ is the number of secrets. On the other hand, $H(S_i, i \in A) \leq \log \|V_A\|$, with equality if and only if different rows of V_A occur equally often. Therefore the following result is true:

Theorem 1. $\|V_A\| \geq q^{|A|}$ *for any perfect SSS (P, V) and any $A \in \Gamma_{\min}$ with equality only if different rows of V_A occur equally often.*

Corollary 2 *For any perfect SSS, the cardinality of its array satisfies the inequality $|V| \geq q^\gamma$, where $\gamma = \max\{|A|, A \in \Gamma_{\min}\}$.*

4 Ideal schemes and matroids

We will distinguish between two definitions of an ideal SSS. For the combinatorial definition, Theorem 1 guarantees that the array of any C-ideal perfect SSS has the property that all possible rows occur equally often within the "subarray" V_A. Such a property (similar to a corresponding property of an orthogonal array) already provides enough power to give a new proof of the result of [6] for the particular case of perfect C-ideal SSS. But we will do more. We will prove it for the more general information-theoretic definition of an ideal SSS.

Let us recall the definition of matroid (see [10]). A matroid is a finite set X and a collection $\mathcal{I}$ of subsets of X (called independent sets) such that the following properties hold.

1. $\oslash \in \mathcal{I}$
2. If $A \in \mathcal{I}$ and $B \subseteq A$ then $B \in \mathcal{I}$
3. If $A, B \in \mathcal{I}$ and $|A| = |B| + 1$ then there exists $a \in A \setminus B$ such that $a \cup B \in \mathcal{I}$

There are other equivalent definitions of matroid. One is based on the *rank* function, and another on minimal dependent sets, called *circuits*. Let (P, V) be an IT-ideal SSS for the access structure Γ. Define a function h on subsets A of the set $\{0, \ldots, n\}$ in the following way: $h(A) = H(S_i, i \in A)$. W.l.o.g. let $h(0) = 1$. Then well known properties of entropy assure us that the following properties of a rank function of matroid hold (see [10]):

1. $h(\oslash) = 0$
2. $h(A) \le h(b \cup A) \le h(A) + 1$
3. If $h(A \cup b) = h(A \cup c) = h(A)$ then $h(A \cup b \cup c) = h(A)$

The entropy is not always an integer-valued function, unfortunately. Otherwise we could immediately conclude that our definition produces a matroid. The main point of the proof of [6] was to prove that $\log_q |V_A|$ is an integer-valued function. It is clear that, if one assigns uniform probability distribution to rows of V, then $\log_q |V_A|$ is exactly the above-defined function $h(A)$. K. Kurosawa *et al.* [11] applied this approach to IT-ideal SSS and proved that, under the assumption of uniformly distributed secrets, $h(A)$ is an integer-valued function, and hence serves as the rank function of the matroid. We use another, simpler, way to provide a proof without appealing to a uniform distribution on secrets. We replace the desired "integer-valued" property by a weaker one, which is much simpler to prove for general IT-ideal perfect SSS.

Lemma 3 *If $h(A) = |A|$, then $h(A \cup b)$ equals either $|A|$ or $|A| + 1$ for any b.*

Based on this lemma, the following generalization of result of E. F. Brickell and D. M. Davenport [6] can be proved.

Theorem 2. *For any IT-ideal perfect SSS the independent sets A such that $h(A) = |A|$ define a matroid. All circuits of this matroid which contain the point 0 are of the form $0 \cup A, A \in \Gamma_{\min}$.*

The second part of the statement of Theorem 2 is very important (see[9]), because all circuits of a matroid can be uniquely (and rather simply, see [10]) determined by all its circuits containing a given point. Hence, a matroid's structure can be derived directly from the access structure Γ. This fact provides a tool for proving that some access structures cannot be realized by IT-ideal SSS.

5 Acknowledgment

Shortly after submitting this paper, we received a preprint of the forthcoming [7] by W.-A. Jackson and K. M. Martin. Among its numerous results, it already provided a different sort of proof of Theorem 2, very much in the spirit of [6]. We are indebted to Keith Martin for thoughtful commentary and criticism in the ensuing discussions.

References

1. G. R. Blakley, Safeguarding cryptographic keys, *Proceedings of AFIPS 1979 National Computer Conference*, vol.48, N. Y., 1979,pp. 313–317.

2. A. Shamir, How to share a secret, *Communications of the ACM*, vol.22,no.1, 1979, pp. 612–613.

3. R. J. McEliece and D. V. Sarwate, On secret sharing and Reed-Solomon codes, *Communications of the ACM*, vol. 24, 1981, pp. 583–584.

4. E. D. Karnin, J. W. Greene, and M. E. Hellman, On secret sharing systems, *IEEE Transactions on Informatiom Theory*, vol. 29,no. 1, 1983, pp. 231–241.

5. R. M. Capocelli, A. De Santis, L. Gargano, and U. Vaccaro, On the size of shares for secret sharing schemes, *Journal of Cryptology*, vol. 6, 1993, pp. 157–167.

6. E. F. Brickell and D. M. Davenport, On the classification of ideal secret sharing schemes, *Journal of Cryptology*, vol. 4, 1991, pp. 123–134.

7. W.-A. Jackson and K. M. Martin, Combinatorial models for perfect secret sharing schemes, Submitted to *Journal of Combinatorial Mathematics and Combinatorial Computing*

8. D. R. Stinson, An explication of secret sharing schemes, *Designs, Codes and Cryptography*, vol.2, 1992, pp. 357–390.

9. K. M. Martin, Discrete structures in the theory of secret sharing, Ph. D. thesis, University of London, 1991.

10. D. J. A. Welsh, *Matroid Theory*, Academic Press, 1976.

11. K. Kurosawa, K. Okada, K. Sakano, W. Ogata and S. Tsujii, Nonperfect secret sharing schemes and matroids, Advances in Cryptology— EUROCRYPT'93, Lect. Notes in Comput. Sci. vol. 765, 1993, pp. 126–141.

NFS with Four Large Primes: An Explosive Experiment

Bruce Dodson[1], Arjen K. Lenstra[2]

[1] Department of Mathematics, Lehigh University, Bethlehem, PA 18015-3174, U. S. A
E-mail: bad0@Lehigh.EDU
[2] MRE-2Q330, Bellcore, 445 South Street, Morristown, NJ 07960, U. S. A
E-mail: lenstra@bellcore.com

Abstract. The purpose of this paper is to report the unexpected results that we obtained while experimenting with the multi-large prime variation of the general number field sieve integer factoring algorithm (NFS, cf. [8]). For traditional factoring algorithms that make use of at most two large primes, the completion time can quite accurately be predicted by extrapolating an almost quartic and entirely 'smooth' function that counts the number of useful combinations among the large primes [1]. For NFS such extrapolations seem to be impossible—the number of useful combinations suddenly 'explodes' in an as yet unpredictable way, that we have not yet been able to understand completely. The consequence of this explosion is that NFS is substantially faster than expected, which implies that factoring is somewhat easier than we thought.

1 Introduction

For the last ten years all 'general purpose factoring records', i.e., those that are relevant for cryptographical applications of factoring, have been obtained by the quadratic sieve factoring algorithm (QS, cf. [13, 15]). The most recent of these records was the factorization of the 129-digit RSA Challenge number, which was published in 1977 and factored in 1994 using the double large prime multiple polynomial variation of QS [1]. The authors of [1], however, suspected that their factorization would be the last factoring record obtained by QS, and that future records will be set by another, faster method, the general number field sieve (NFS, [8]).

It has been known since 1989 that NFS is asymptotically superior to any of the variants of QS: for $n \to \infty$ it can be expected, on loose heuristic grounds, that it takes time

$$\exp((1.923 + o(1))(\log n)^{1/3}(\log \log n)^{2/3})$$

to factor a composite number n using NFS, as opposed to a (slower) heuristic expected run time

$$\exp((1 + o(1))(\log n)^{1/2}(\log \log n)^{1/2}),$$

D. Coppersmith (Ed.): Advances in Cryptology - CRYPTO '95, LNCS 963, pp. 372-385, 1995.
© Springer-Verlag Berlin Heidelberg 1995

also for $n \to \infty$, for QS. These heuristic run time estimates do not imply that NFS is also faster than QS in practice. Indeed, it has for some time been suspected that NFS would *never* be practical at all, and that, even if we would be able to get it to work, the crossover point with QS would be far beyond our current range of interest.

In this paper we present some evidence that NFS is actually more practical than expected, and that the crossover point with QS is easily within reach of our current computational resources. Our results indicate that NFS is already substantially faster than QS for numbers in the 115 digit range. Since the 'gap' between the factorization times for these methods only widens for larger numbers, our results imply that the 129-digit number factored in [1] could be factored by NFS in about a quarter of the time spent by QS. The consequences for the strength of 512-bit composites, as sometimes used in cryptographic applications, will be commented on in future work.

One of the major reasons that NFS is performing so much better than expected, is that NFS has a certain advantage over QS that has almost no relation to the theoretical advantage of NFS over QS. Roughly speaking, QS only allows efficient usage of two 'large primes', whereas in NFS it should be possible to use four large primes. Practical experiments that exploit this large prime advantage have so far been limited to three large primes [3]. These experiments did not indicate a distinct advantage of three over two large primes, possibly because the numbers that were factored were rather small. The NFS implementation from [6] allowed us to carry out some large scale experiments with four large primes, which, for the first time, unequivocally proved the advantage of more than two large primes.

In Section 2 we describe why it is easier for NFS to take advantage of large primes than it is for QS. Our experiments and results are presented in Section 3, followed by some of the new methods that were used to obtain our results: an alternative 'cycle' counting method in Section 4, and a discussion of the matrix step in Section 5.

2 Large primes in QS and NFS

Let n be an odd number that is not a prime power.

Large primes in QS. To factor n with QS, one begins by selecting a 'factor base' P, consisting of -1 and certain primes up to some bound B. One then employs a 'sieving' process to efficiently collect a set V of more than $\#P$ 'relations', which are identities modulo n of the form

$$v^2 \equiv \prod_{p \in P} p^{e(v,p)} \bmod n,$$

with $v, e(v,p) \in \mathbf{Z}$. Since $\#V > \#P$, the vectors $e(v) = e(v,p)_{p \in P}$ are linearly dependent. This implies that $\#V - \#P$ subsets W of V can be found (using linear algebra) for which there are linearly independent dependencies of the form $\sum_{v \in W} e(v) = 2(w(p))_{p \in P}$ with $w(p) \in \mathbf{Z}$. Each W therefore leads to an identity

$$\left(\prod_{v \in W} v \right)^2 \equiv \left(\prod_{p \in P} p^{w(p)} \right)^2 \bmod n$$

of the form $x^2 \equiv y^2 \bmod n$ with $x, y \in \mathbf{Z}$. For each such identity there is a chance of at least $1/2$ that $\gcd(x - y, n)$ will be a nontrivial factor of n.

In [9] is was shown that it is advantageous to collect identities of a slightly more general form, namely

$$v^2 \equiv q_1(v) q_2(v) \prod_{p \in P} p^{e(v,p)} \bmod n$$

with $v, e(v,p) \in \mathbf{Z}$ and $q_1(v)$, $q_2(v)$ either equal to 1 or to some prime not in P satisfying $q_1(v) \le q_2(v) \le B_2 < B^2$, for some bound B_2 with $B_2^2 < B^3$. If $q_1(v) = q_2(v) = 1$ these are the same as the earlier relations, which will be called 'full' relations from now on; otherwise a relation is called 'partial'. The q_i's are referred to as the 'large primes'. Partial relations are potentially useful because it might be possible that they can be combined into 'cycles': collections of partial relations where all occurring large primes can be combined into squares, thus making the combination 'look like' a full relation. As an example, if v and w are two different partial relations for which $q_1(v) = q_1(w) = 1$ and $q_2(v) = q_2(w)$, then

$$(vw/q_2(v))^2 \equiv \prod_{p \in P} p^{e(p,v)+e(p,w)} \bmod n$$

is just as useful as a full relation, unless $\gcd(q_2(v), n) \ne 1$. This implies that the condition that the number of full relations is larger than $\#P$ can be replaced by the condition that the number of full relations plus the number of independent cycles is larger than $\#P$. The large primes that occur in the cycles can be thought of as cheap factor base extenders—cheap because they are found almost for free, without having to sieve or to trial divide with them. The cycles are simply linear combinations of exponent vectors where the coordinates corresponding to the factor base extenders, the large primes, are even.

We explain how partial relations can efficiently be collected during the search for full relations. During the sieving, candidate v's are identified in an efficient manner. For each candidate v the least absolute residue $v^2 \bmod n$ is trial divided, to see if it factors using the elements of P. If so, a full relation has been found. If not, and the remaining cofactor t after trial division with the primes $< B$ is $< B^2$, a partial relation with $q_1(v) = 1$, $q_2(v) = t$ has been found. How many of such partial relations will be found depends on how easily candidates are accepted after the sieving—if many near misses for fulls are accepted, many partials will be found. If we accept even more candidates, we might also find near misses for the partials with $q_1(v) = 1$: if $t > B^2$, $t < B_2^2$, and t is composite we find a partial relation if both of the prime factors of t are $< B_2$ (note that t can have at most two prime factors because $B_2^2 < B^3$). Since compositeness tests are cheap, because composites of this form are fairly easy to factor, and because relatively many t's will have their factors in the right range, it follows that these partials can also be found at relatively small cost.[3]

Of course, it only makes sense to spend this extra effort if the partial relations are useful in practice, i.e., if cycles among the partials indeed turn up. In [9] it is shown that if only partials with $q_1(v) = 1$ are allowed, the total number of independent combinations (of the type as shown in the example above) can be expected to behave as $c \cdot m^2$, where m is the number of partial relations (with $q_1(v) = 1$), and c is some very small constant depending on the bounds (cf. [9], and the 'birthday paradox'). This quadratic behavior can indeed be observed in practice. Using these restricted partials leads to a speed-up of about a factor 2.5 compared to using only full relations.

Using *all* partials, i.e., also those with $q_1(v) \neq 1$ leads to another speed-up of about a factor 2.5, for sufficiently large numbers. A theoretical analysis of the expected number of cycles has not been given yet, but according to the data from [1] the number of cycles seems to grow almost as $c' \cdot m^4$, where c' is another small constant, and m is now the total number of partials. In any case, the number of independent cycles as a function of the number of partials seems to behave as a very smooth curve, at least over the intervals where it has been observed so far. Thus, reliable estimates of the expected completion time of the relation collection stage can easily be derived from this curve.

Large primes in NFS. To factor n with NFS, one begins by selecting two bivariate polynomials $f_1(X, Y)$, $f_2(X, Y) \in \mathbf{Z}[X, Y]$ that satisfy certain properties that are not relevant for this paper (cf. [8]). Given f_1 and f_2 one selects two factor bases P_1 and P_2, consisting of the primes $\leq B_1$ and $\leq B_2$, respectively. Relations are given by coprime integers a, b, with $b > 1$, such that

$$|f_1(a, b)| = \prod_{p \in P_1} p^{e_1(a,b,p)} \quad \text{and} \quad |f_2(a, b)| = \prod_{p \in P_2} p^{e_2(a,b,p)},$$

with $e_1(a, b, p)$, $e_2(a, b, p) \in \mathbf{Z}$. If more than (approximately) $\#P_1 + \#P_2$ relations have been found, a factorization of n can, with high probability, be derived from linear dependencies modulo two among the $(\#P_1 + \#P_2)$-dimensional vectors consisting of the concatenation of $(e_1(a, b, p))_{p \in P_1}$ and $(e_2(a, b, p))_{p \in P_2}$. How this is done is beyond the scope of this paper (but see [8]; [10]), and neither will we discuss the influence of the (small) amount of 'free' relations.

As in QS, we can allow large primes in the factorizations of the $|f_i(a, b)|$, where those that occur in cycles among the partial relations can be regarded as cheap factor base extenders (where it should be noted that a large prime dividing $f_1(a, b)$ cannot be combined with the same prime dividing $f_2(a, b)$, and

---

[3] If we relax the conditions on the candidates even more, we might be able to allow three large primes in the factorization of $v^2 \bmod n$. So far, however, this does not seem to lead to a speed-up because it leads to a huge amount of numbers to be trial divided, the vast majority of which will lead to cofactors that do not have the right factoring pattern. Also the factorization of the composite cofactors is substantially more expensive.

vice versa—even more restrictive, a large prime q dividing $f_i(a, b)$ can only be combined with the same large prime q dividing $f_i(a', b')$ if $ab' \equiv a'b \bmod q$). In the NFS implementations described in [8] at most one large prime per $|f_i(a, b)|$, for a total of at most two per relation, was used. Candidate relations are identified using a sieve, similar to QS. For each candidate, $|f_1(a, b)|$ is trial divided with the primes $\leq B_1$, and upon success $|f_2(a, b)|$ is trial divided with the primes $\leq B_2$. This implies that, in principle and as explained above, two large primes per $|f_i(a, b)|$ can quite easily be recognized. In [8] it was reported, however, that actually finding these relations with up to $2 + 2$ large primes was prohibitively expensive. Fortunately, it was shown in [6] that they can efficiently be found with better sieving and trial division methods. These methods do not seem to apply to QS to efficiently generate three or more large primes per relation in QS. The important difference is that in QS we have one composite cofactor that has to be factored into three or more factors in the right range, whereas in NFS we are dealing with two composite cofactors that each have to factor in the right way — the latter both occurs with higher probability and is easier to decide.

The experiments from [8] indicated that the number of cycles in the $1 + 1$ large prime variation of NFS is consistently lower than the number of cycles found in the (two) large prime variation of QS, i.e., for NFS substantially more relations are needed than for QS to get the same number of cycles. This is due to the fact that in QS we have a single set of large primes, whereas in NFS we have two 'incompatible' sets of large primes, one for f_1 and one for f_2. Thus, for NFS it takes longer for the 'birthday paradox' to take effect. On the other hand, $1 + 1$ large primes in NFS behaves markedly better than the single large prime variation of QS, i.e., if only partials with $q_1(v) = 1$ are considered in QS.

Based on these observations, the work from [3], and the 'almost quartic' behavior observed in [1], we hoped that $2 + 2$ large primes in NFS would produce somewhat better than quartic cycle growth. We also expected that the number of independent cycles as a function of the number of partial relations would, as usual, behave as a nice and smooth curve that would allow easy extrapolation to predict the completion time of the relation collection stage. These expectations turned out to be wrong, as we will see in the next section.

In the sequel, partials with i large primes in $|f_1(a, b)|$ and j large primes in $|f_2(a, b)|$ will be called 'i, j-partials'. Partials for which $i + j = k$ will be called 'k-partials', if $i + j \leq k$ they will be called '$\leq k$-partials', and similarly for '$\geq$'.

3 Experiments and results

In this section we describe the details of three NFS factoring experiments in chronological order: a 116, a 119, and a 107-digit number.

Factoring a 116-digit integer. Let n be the following 116-digit composite factor of the 11887th partition number:

$$n = 1\ 50802\ 87457\ 98463\ 07441\ 49612\ 94413\ 35408\ 90110\ 76626\ 79218\ 10826\ 04486$$

$$78500\ 16206\ 10665\ 65455\ 29820\ 06606\ 21307\ 78648\ 81680\ 71410\ 39443.$$

To factor n using NFS we first spent a few workstation days to find 5 reasonable candidate polynomials f_1 and f_2. Next we sieved a while with each candidate pair, using reasonably sized factor bases. This yielded one pair that stood out from the others with a more than 10% better yield than the next best one:
$f_1(X, Y) = X - 49999\ 99918\ 54766\ 46567\,Y$ and

$$f_2(X, Y) = 48\ 25692\ 37961\ 89830\,X^5 + 35\ 68080\ 39372\ 65531\,X^4 Y$$
$$- 4\ 65605\ 61818\ 75120\,X^3 Y^2 - 59\ 69883\ 14526\ 21728\,X^2 Y^3$$
$$- 13\ 44285\ 55250\ 45260\,X Y^4 + 29\ 65432\ 72740\ 38354\,Y^5,$$

with common root $X/Y = 49999\ 99918\ 54766\ 46567$ modulo n. We could not observe any correlation with properties that were thought to be relevant, like coefficient sizes or number of roots modulo small primes. The issue of polynomial selection in NFS needs to be understood better; we have not pursued this yet as our current trial-and-error approach seems to work satisfactorily, for the moment.

Having thus decided on f_1 and f_2, we selected $\#P_1 = 100,001$, $\#P_2 = 400,001$, and $B_2 = 2^{30}$. From our sieving experiments we derived that this choice could lead to approximately 50 million partials in about 250 mips years, using [6]. Given our experience with QS and NFS with fewer large primes, and the expected counts of the various types of relations (i.e., with a total of 0, 1, 2, 3, or 4 large primes), we expected that this choice would be enough to produce more than $500,000$ fulls plus cycles, even without relying on any ≥ 3-partials. Furthermore, since 250 mips years is less than QS would need for this number (about 400 mips years), our choice of $\#P_1$, $\#P_2$ and B_2 seemed not too bad.

Initial results were not surprising. The relations were found at the expected rates and the cycle-yield followed our worst-case scenario. The $\leq$ 2-partials combined more or less at their expected rate, and the other partials hardly contributed to the cycles: at $7,336,602$ partials (and $12,607$ fulls) there were only five 3-partials involved in the $1,474$ cycles, and no 2,2-partials.

The results were mixed after we had completed more than half our anticipated sieving, at $28,243,830$ partials (and $43,555$ fulls). Even though the curve of the cycle-yield as a function of the number of partials resembled a quartic function, extrapolation suggested that we would not even be close to what we needed by the time we would be finished sieving. On the other hand, of the 2,2-partials there were $10,495,464$ of which $5,645$ (0.054%) that occurred in cycles (of which there were $41,366$), which was a marked improvement.

It all looked different by the next time we attempted to count the cycles; attempted, because the counting program failed to work properly, a first indication that something had changed. After replacing the failing counting approach by a new one (cf. Section 4), we found $317,862$ cycles among $33,264,762$ partials (and $49,680$ fulls), with the 2,2-partials at $12,460,866$ with $672,773$ (5.4%) participating in the cycles. Although this was still not enough, we did not fail to notice that our nice smooth curve had effectively been cut short by an almost vertical line, implying that the sieving was almost complete.

A side effect associated with this explosive cycle behavior was that the average length of the cycles (i.e., the number of partials that together form a cycle)

Table 1

total fulls	total partials	useful partials	%	cycles
12607	7336602	3368	0.04	1474
19456	11521334	8913	0.07	3842
25782	15773678	18627	0.11	7491
29744	18368031	28982	0.15	10934
33215	20736047	42299	0.20	14895
37724	23902815	73039	0.30	22329
40466	25972266	111358	0.42	29415
43555	28243830	224217	0.79	41365
49680	33264762	3486661	10.48	317862
58325	42202890	9369843	22.20	1746609
61849	45876382	13970578	30.45	2605463

seemed to grow rapidly (we failed to keep the number, but see Table 6). Because longer cycles could lead to problems in the matrix step, and because we were curious to see if and how the explosion would continue, we decided to keep sieving for a while—hoping that if we had many cycles, we would also be able to find enough short ones, and restrict ourselves to those short ones in the matrix.

This led to two additional counts. At $42,202,890$ partials, there were $1,746,609$ cycles of average length 93. Of the $16,079,778$ 2,2-partials 15.4% were useful, i.e., occurred in cycles. At $45,876,382$ partials, there were $2,605,463$ cycles (and $61,849$ fulls). The average length of these cycles had dropped to 74, and the proportion of useful 2,2-partials rose to 23.4% with $4,111,077$ useful 2,2-partials out of a total of $17,572,446$. Further, we note that 87.3% of the cycles used a 2,2-partial. More details can be found in Tables 1, 2 and 3.

So the explosion indeed continued. Further, the cycle length went down sufficiently that there were $459,922$ cycles of length ≤ 23, which was acceptable for our matrix processing method back then. Note that $459,922 + 61,849 > 500,002$. At this point we stopped sieving, after about 220 mips years. We could have sieved more, and attempted to find the 50 million partials that we hoped to find in 250 mips years, to see if indeed the ≤ 2-partials would have sufficed, as expected. This is unlikely: where we stopped those partials led to only $93,328$ cycles (with only $56,328$ cycles among the i,j-partials with $i,j \leq 1$ as used in [8]. The ≤ 3-partials would have had a better chance, with $330,485$ cycles where we stopped.

We have not yet been able to understand why this very sudden growth in the number of cycles occurs. Counting arguments failed to prove anything, and probabilistic arguments are complicated by the inhomogeneous nature of the data. Connections with well-known 'crystalization' behavior of random graphs have been suggested, but have so far not given any insights that could be used to prove or predict cycle explosions. Obviously this would be useful for a better selection of factor base sizes and minimization of the total sieving time: if no explosion had occurred we would have needed larger factor bases, but if we had

Table 2

cycles	total 1-partials	useful 1-partials	%	total 2-partials	useful 2-partials	%
10934	515439	21906	4.24	3157511	6224	0.19
14895	579057	29991	5.27	3556296	10529	0.29
22329	660552	45965	6.95	4076913	21808	0.53
29415	709775	62463	8.80	4395522	37346	0.84
41365	768018	97711	12.72	4772352	85653	1.79
317862	884118	371434	42.01	5551149	1063330	19.15
1746609	1065782	600653	56.35	6856874	2319722	33.83
2605463	1135723	720965	63.48	7385166	3143071	42.55

cycles	total 3-partials	useful 3-partials	%	total 2,2-partials	useful 2,2-partials	%
10934	7946983	808	0.01	6748098	44	0.00
14895	8970893	1666	0.01	7629801	113	0.00
22329	10338014	4823	0.04	8827336	443	0.00
29415	11194689	10389	0.09	9591557	1160	0.01
41365	12207997	35208	0.28	10495464	5645	0.05
317862	14368629	1379124	9.59	12460866	672773	5.39
1746609	18200456	3968763	21.80	16079778	2480705	15.42
2605463	19783047	5995465	30.30	17572446	4111077	23.39

Table 3

	fulls	%	0,1-partials	%	0,2-partials	%
total	61849		668995		1670521	
useful			417840	62.45	753575	45.11
wrt total useful				2.99		5.39
	1,0-partials		1,1-partials		1,2-partials	
total	466728		5038434		12521329	
useful	303125	64.94	2043376	40.55	3670595	29.31
wrt total useful		2.16		14.62		26.27
	2,0-partials		2,1-partials		2,2-partials	
total	676211		7261718		17572446	
useful	346120	51.18	2324870	32.01	4111077	23.39
wrt total useful		2.47		16.64		29.42

known about it we could have settled for smaller ones, so that the explosion would occur exactly when the sieving is done. This would lead to either denser or much larger matrices than we were used to, and thus would require better matrix techniques than used at the time we sieved this 116-digit number.

For the present number the matrix did not pose a big problem, because of all the extra sieving we had done. After inclusion of 256 columns for quadratic signatures, we had a $521,771 \times 500,258$ bit-matrix, with on average 277 one-bits per row. This matrix was reduced to an almost 9 gigabyte dense $274,696 \times 274,496$ bit-matrix using 'structured Gaussian elimination' [7; 12; 14]. It took 6

CPU-days on a 16K MasPar MP-1 to find the dependencies in the dense matrix using plain Gaussian elimination.

The first dependency was processed by Peter Montgomery at the CWI in Amsterdam, using the method from [10], in less than 2.5 CPU-days on a MIPS R4400 processor. This resulted in the following factorization:

$$n = 3\ 73787\ 18590\ 84719\ 25152\ 67256\ 20648\ 89920\ 58833\ 29019 \times 40344\ 58115$$

$$87486\ 91262\ 33674\ 44522\ 38913\ 92270\ 04005\ 21248\ 10720\ 24860\ 30673\ 43497.$$

More data points can be obtained by pruning the data that we have for smaller values of $B_2 = 2^{30}$, and checking if and where the resulting sets of relations lead to cycle-explosions. We have not done this for the present number, but we did for the 119-digit number discussed below. From these and other experiments of this sort that we carried out in [5] (for a QS factorization) it follows that using a higher value of B_2 indeed saves sieving time, at the cost of substantial amounts of disk space.

Factoring a 119-digit integer. To convince ourselves that the cycle-explosion described above was not an isolated incident, we tried factoring the following 119-digit composite factor of the 13171th partition number:

$$n = 1472\ 39730\ 37795\ 02230\ 11857\ 21506\ 65046\ 38946\ 42104\ 16013\ 39117\ 46791\ 27360$$

$$74474\ 37214\ 92509\ 46318\ 17633\ 03651\ 67483\ 02069\ 42164\ 60898\ 11241.$$

We again found one pair of polynomials that stood out from the rest, for no reasons that we could find: $f_1(X, Y) = X - 1\ 44999\ 99959\ 86876\ 68083\,Y$ and

$$
\begin{aligned}
f_2(X, Y) =\ & 229\ 71270\ 09947\ 70930\,X^5 - 75\ 24490\ 95044\ 76954\,X^4 Y \\
& - 349\ 19223\ 42428\ 31010\,X^3 Y^2 + 213\ 34303\ 57653\ 48142\,X^2 Y^3 \\
& - 133\ 73262\ 31271\ 45009\,X Y^4 - 83\ 88784\ 35301\ 30136\,Y^5,
\end{aligned}
$$

with common root $X/Y = 1\ 44999\ 99959\ 86876\ 68083$ modulo n. Anticipating the explosion this time, we chose relatively small factor bases ($\#P_1 = 100,001$ and $\#P_2 = 320,001$). This turned out to be too small: after 195 mips years sieving was complete with only about $24,000$ cycles among 30 million partials (and $30,000$ fulls). Extension of $\#P_2$ to $360,000$ and about 45 additional mips years of sieving led to $470,000$ cycles among 35.8 million partials (and $39,000$ fulls), occupying almost 2 gigabytes of storage. Details can be found in Tables 4 and 5.

As expected this led to an unusually large matrix problem. Structured Gaussian elimination would have required in excess of 15 gigabytes of storage and more than two CPU-weeks on a 16K MasPar MP-1, which is hardly feasible. Instead we used an experimental MasPar implementation of the blocked Lanczos method from [11] (cf. Section 5, and [4]). This took 2.5-CPU-days. The dependencies were again processed by Peter Montgomery, this time in one CPU-day per dependency—the factorization was found on the third one:

Table 4

(with $\#P_2 = 320,001$)

total fulls	total partials	useful partials	%	cycles	total ≥ 3- partials	useful ≥ 3- partials	%
5607	3683540	588	0.01	280	2898623	2	0.00
10426	7262853	2525	0.03	1157	5771533	6	0.00
13157	9387071	4451	0.04	1997	7492921	24	0.00
16363	12152617	7840	0.06	3371	9755753	69	0.00
18297	14017542	10881	0.07	4549	11301609	138	0.00
21100	16950688	16743	0.09	6621	13765642	365	0.00
25233	21930116	31702	0.14	11118	18018822	1339	0.00
28194	26465817	56158	0.21	16716	21950471	4146	0.01
30289	30107586	110348	0.36	23886	25129694	15258	0.06

(after recounting with $\#P_2 = 360,001$)

35953	30101922	219051	0.72	34728	24735646	44746	0.18
38199	34567876	3980288	11.51	337351	28636652	2548939	8.90
38741	35763524	4725203	13.21	472426	29680006	3116297	10.49

Table 5

	fulls	%	0,1-partials	%	0,2-partials	%
total	38741		459082		1254636	
useful			207979	45.30	314563	25.07
wrt total useful				4.40		6.66
	1,0-partials		1,1-partials		1,2-partials	
total	303127		3603958		9746884	
useful	148271	48.91	794788	22.05	1197510	12.29
wrt total useful		3.14		16.82		25.34
	2,0-partials		2,1-partials		2,2-partials	
total	462715		5476700		14456422	
useful	143305	30.97	770361	14.07	1148426	7.94
wrt total useful		3.04		16.30		24.30

n = 54 67135 70838 33359 03092 89109 24162 82702 67063 91975 73477 × 26 93178

62646 37991 75226 13452 71731 62372 36050 27370 88674 71432 35361 24533.

With less than 250 mips years spent on sieving, this factorization was completed about 2.5 times faster than it would with QS. Had we known how well our blocked Lanczos implementation would work, then we would have used larger factor bases, but sieved shorter per 'special-q' (cf. [6]), which would have reduced the sieving time considerably. Combining this observation with many other possible improvements, the 250 mips year figure should not be taken too seriously.

As indicated above, we obtained more data points by pruning the final data sets for smaller values of B_2. Some results are given in Table 6, including results for the data sets as they were at 88%, 90%, and 95% of the sieving. Note the sharp increase of the cycle length (of the cycles as found by our method, cf. Section 4) 'early' in the explosion, and the subsequent decrease of the cycle length as the explosion continues. Apparently, a first indication of an upcoming

Table 6

B_2	total partials	% useful	cycles	longest cycle	average length	total ≥ 3- partials	% useful
	using the data sets after 88% of the sieving:						
2^{26}	3716735	2.12	22920	66	4	2435866	0.28
2^{27}	7141411	1.49	26652	106	5	5172320	0.27
2^{28}	12566475	1.14	29867	564	6	9708135	0.22
2^{29}	20536267	5.37	32464	1072	8	16560014	3.17
2^{30}	31471940	5.92	61402	$\geq 10^6$	≥ 388360	26118418	3.95
	using the data sets after 90% of the sieving:						
2^{26}	3801311	2.28	23477	65	4	2491247	0.34
2^{27}	7303838	1.69	27794	134	5	5289915	0.33
2^{28}	12852293	1.42	31665	997	8	9928854	0.35
2^{29}	21003354	8.26	59148	$\geq 10^6$	≥ 214885	16936513	5.56
2^{30}	32187175	7.35	98433	$\geq 10^6$	≥ 57252	26712006	5.18
	using the data sets after 95% of the sieving:						
2^{26}	4012410	3.93	33026	230	6	2629594	0.95
2^{27}	7709526	12.30	55897	127536	742	5583741	7.52
2^{28}	13566166	15.49	137720	50136	948	10480347	11.30
2^{29}	22169973	13.75	237435	34279	835	17877242	10.44
2^{30}	33975344	11.02	319882	32783	788	28195708	8.46
	using the final data sets:						
2^{26}	4223679	4.86	37133	619	8	2768053	1.41
2^{27}	8115376	17.25	86015	≥ 29761	≥ 664	5877684	11.87
2^{28}	14280326	19.21	216173	21728	663	11032061	14.77
2^{29}	23337060	16.62	357792	17235	612	18818348	13.13
2^{30}	35763524	13.21	472426	16524	590	29680006	10.49

cycle-explosion is the sudden growth of the number of useful relations and of the average cycle length. Note also that for larger B_2 there are relatively more partials with more large primes.

Factoring a 107-digit integer. As a third experiment we factored a 107-digit number, with assistance from Magnus Alvestad and Paul Leyland (using our program described in [6]), Peter Montgomery (using his implementation at the CWI in Amsterdam), and Jörg Zayer (using his implementation from [3]). Zayer's and our program both use 'special q's' (cf. [6]), so that non-overlapping sieving tasks could easily be distributed among Zayer and the users of our program. To avoid overlap with Zayer's and our results, Montgomery used his more traditional siever with 'large prime special q's', i.e., special q's that would be considered as large primes by the other programs. As a consequence, Montgomery's program also produced i, j-partials with $i, j > 2$ (but with $i \leq 4$ and $j \leq 6$).

Although smaller numbers lead to smaller factor bases, and therefore to smaller large primes with a higher cycle-yield, the cycle-yield for the 107-digit number was initially even smaller than the cycle-yield for the 116 and 119-digit numbers reported above: at 12,921 fulls and 24,660,318 partials, there were only 7,552 cycles among 26,070 useful partials, compared to 22,239 cycles among 73,039 usefuls in 23,902,815 partials (for the 116-digit number) and 11,118 cycles among 31,702 usefuls in 21,930,116 partials (for the 119-digit number). This lower than expected cycle-yield cannot be explained by the $\geq$ 5-partials, because at this point we had hardly received any partials from Montgomery yet.

At 29,061,638 partials (and 13,918 fulls) on the other hand, the cycle-yield was better than for the 119-digit number: 202,387 cycles among 2,856,606 usefuls, compared to 34,728 cycles among 219,051 usefuls in 30,101,922 partials for the 119-digit number (good comparison data with the 116-digit number were not available). When we stopped sieving, we had 34,135,923 partials, 5,531,053 of which were useful (we did not count the cycles).

We conclude that also for this smaller number we got a cycle-explosion, even a bit more dramatic than before. The behavior of the cycle-yield is however sufficiently erratic that we have not been able to derive a reasonable 'rule-of-thumb' that could be useful to predict the explosion for future NFS factorizations.

4 Counting

In [9] some elementary methods were given to count and build the useful combinations among relations with at most two large primes per relation. A generalization of these methods to the case of at most three large primes per relation was presented in [2]. This generalization does not extend to our case where relations can have four (or more) large primes. We give an outline of our methods.

From the data presented in the previous section it should be clear that we are dealing with large amounts of data: it takes at least 24 bytes to store one a, b pair with four large primes, which already implies several hundred megabytes for the 2,2-partials (which actually take more space than that). The first concern while counting cycles therefore is to quickly weed out partials that are useless, i.e., that do not occur in any cycle (of course they will be kept, because they might be useful in later counts). Note that a partial is useless if it contains a large prime that does not occur in any other partial.

We hashed each large prime, without collision resolution, to a 2-bit location where the hits were counted (not counting further than 2). We kept only those relations for which all large primes had hashed to a location with count 2. This process was repeated on the resulting collection, until the resulting relations could be handled by another version of the same program that did use collision resolution. The latter version was repeated until no relations were deleted. The number of cycles among the partials in the resulting collection can then easily be estimated by subtracting the total number of distinct primes from the number of partials. The latter can also be done on 'earlier' collections of partials as long as there is enough disk space for the sorting and uniqueing.

To do an exact count of the cycles, or to build them, we always first computed the collection of useful partials, as sketched above, and next applied the following 'greedy elimination'. First we process the single large prime partials, storing each large prime when it is encountered for the first time, and counting (and building) a cycle each time it is encountered after that. Next, we remove all those stored primes from all other partials (counting, and building, cycles for partials where all large primes get removed), separating the resulting partials that still have at least one large prime into those with one, and those with at least two large primes. The entire process is repeated until no new partials with one remaining large prime are kept. Usually, at this point, no other partials remain; if there are, the cycles among them can easily be found using a similar strategy. Note that for the 'building' this greedy elimination requires some additional administration to account for the 'history' of the large prime deletions; so far this has not taken serious amounts of disk space.

5 The matrix step

In all previous QS and NFS factorizations, including the 116-digit one from Section 3, we used the following strategy for the matrix step: first build all cycles to get a, roughly, $\#P \times \#P$ (or $(\#P_1 + \#P_2) \times (\#P_1 + \#P_2)$) bit-matrix, next apply structured Gaussian elimination [7; 12; 14], and finally apply ordinary Gaussian elimination. The extent to which the first step should be carried out before applying the last two steps is debatable. That is, we may remove all of the large primes before starting elimination on the matrix; some of the large primes may be removed in the first step, with the others included in the matrix; or none of the large primes need to be removed—without a clear advantage among these possible strategies. For the 119-digit number from Section 3, only the large primes that occurred at most three times in the useful partials were removed by constructing cycles. This resulted in a sparse $1,475,898 \times 1,472,607$ bit-matrix, which could have been 'reduced' to a dense $362,597 \times 362,397$ bit-matrix by structured Gauss. In our experience about the same would have happened if we had removed all large primes and processed the (unusually dense) 'sparse' $461,001 \times 460,001$ matrix with structured Gauss.

We did, however, not actually build this dense $362,597 \times 362,397$ bit-matrix. Instead we used the blocked Lanczos method from [11] to process the sparse $1,475,898 \times 1,472,607$ bit-matrix for the 119-digit number. The reason that we removed only the large primes that occur ≤ 3 times in the useful partials before using blocked Lanczos, is the following. The expected run time of blocked Lanczos applied to a matrix of (approximately) m rows and m columns with on average w non-zero entries per row, is proportional to m times the total weight (i.e., mw) of the matrix. Initially, we may assume that all rows have about equal weight. Therefore, removal of a large prime that occurs in k different rows results in $m - 1$ rows, $m - k$ of the same average weight w as before, and $k - 1$ of average weight at most $2w - 2$, and thus expected run time for blocked Lanczos proportional to $m - 1$ times $(m - k)w + (k - 1)(2w - 2)$. In realistic circumstances, the latter is less than $m^2 w$ as long as $k \leq 3$.

Evidently, removal of large primes destroys the even distribution of the non-zeros over the rows, so the same argument cannot be used to analyse the effect of removing more large primes. Nevertheless, similar arguments imply that the run time can reasonably be expected to decrease if large primes that occur in at most 3 rows are removed, but that an increase can be expected if large primes that occur more often are removed. This explains the choice that we made for the 119-digit number. For a detailed description of the blocked Lanczos algorithm we refer to [11], and to [4] for a description of the implementation that we used.

Acknowledgments. Acknowledgments are due to M. Alvestad, S. Contini, P. Leyland, P. Montgomery, and J. Zayer for their assistance.

References

1. D. Atkins, M. Graff, A.K. Lenstra, and P.C. Leyland, *THE MAGIC WORDS ARE SQUEAMISH OSSIFRAGE*, Asiacrypt'94, to appear.
2. J. Buchmann, J. Loho, and J. Zayer, *Triple-large-prime variation*, manuscript, 1993.
3. J. Buchmann, J. Loho, and J. Zayer, An implementation of the general number field sieve, Advances in Cryptology, Crypto'93, Lecture Notes in Comput. Sci. **773** (1994), 159–165.
4. S. Contini and A. K. Lenstra, *Implementations of blocked Lanczos and Wiedemann algorithms*, in preparation.
5. T. Denny, B. Dodson, A. K. Lenstra, and M. S. Manasse, *On the factorization of RSA-120*, Advances in Cryptology, Crypto'93, Lecture Notes in Comput. Sci. **773** (1994), 166–174.
6. R. Golliver, A.K. Lenstra, and K. McCurley, *Lattice sieving and trial division*, ANTS'94, Lecture Notes in Comput. Sci. **877** (1994), 18–27.
7. B. A. LaMacchia and A. M. Odlyzko, *Solving Large Sparse Linear Systems over Finite Fields*, Advances in Cryptology, Crypto'90, Lecture Notes in Comput. Sci. **537** (1991), 109–133.
8. A. K. Lenstra and H. W. Lenstra, Jr. (eds), *The development of the number field sieve*, Lecture Notes in Math. **1554**, Springer-Verlag, Berlin, 1993.
9. A. K. Lenstra and M. S. Manasse, *Factoring with two large primes*, Math. Comp **63** (1994), 785–798.
10. P. L. Montgomery, *Square roots of products of algebraic numbers*, Proceedings of Symposia in Applied Mathematics, Mathematics of Computation 1943-1993, Vancouver, 1993, Walter Gautschi, ed.
11. P. L. Montgomery, *A block Lanczos algorithm for finding dependencies over GF(2)*, Advances in Cryptology, Eurocrypt'95, Lecture Notes in Comput. Sci. **921** (1995), 106–120.
12. A. M. Odlyzko, *Discrete Logarithms in Finite Fields and their Cryptographic Significance*, Advances in Cryptology, Eurocrypt'84, Lecture Notes in Comput. Sci. **209**, 224–314.
13. C. Pomerance, *The quadratic sieve factoring algorithm*, Advances in Cryptology, Eurocrypt'84, Springer, Lecture Notes in Comput. Sci. **209**, 169–182.
14. C. Pomerance and J. W. Smith, *Reduction of huge, sparse matrices over finite fields via created catastrophes*, Experiment. Math. **1** (1992) 89–94.
15. B. Silverman, *The multiple polynomial quadratic sieve*, Math. Comp. **48** (1987), 329–339.

Some Remarks on Lucas-Based Cryptosystems

Daniel Bleichenbacher[1], Wieb Bosma[2], Arjen K. Lenstra[3]

[1] Institut für Theoretische Informatik, ETH Zentrum, 8092 Zürich, Switzerland
E-mail: bleichen@inf.ethz.ch
[2] School of Mathematics and Statistics, University of Sydney,
Sydney, NSW 2006, Australia
E-mail: wieb@maths.su.oz.au
[3] MRE-2Q330, Bellcore, 445 South Street, Morristown, NJ 07960, U. S. A.
E-mail: lenstra@bellcore.com

Abstract. We review the well-known relation between Lucas sequences and exponentiation. This leads to the observation that certain public-key cryptosystems that are based on the use of Lucas sequences have some elementary properties their re-inventors were apparently not aware of. In particular, we present a chosen-message forgery for 'LUC' (cf. [21; 25]), and we show that 'LUCELG' and 'LUCDIF' (cf. [22; 26]) are vulnerable to subexponential time attacks. This proves that various claims that were made about Lucas-based cryptosystems are incorrect.

1 Introduction

The application of Lucas sequences in various branches of number theory is well known (cf. [18]), and their properties have been studied extensively. Applications of Lucas sequences to public-key cryptography, phrased in terms of the equivalent Dickson-polynomials, were proposed and analysed by a series of authors [13; 14; 12; 16; 17; 11]. More recently, the system from [13] re-emerged, by a different author and in slightly altered form, as 'LUC' (cf. [21], and later [25]), and was subsequently extended to 'LUCDIF', 'LUCELG PK', and 'LUCELG DS' (cf. [22; 26]). The difference between [13] and [21; 25] is that the latter introduce 'message-dependent' keys.

The main selling point of the Lucas-based cryptosystems as presented in these later publications (cf. [21; 22; 25; 26]) is that they are not formulated in terms of exponentiation. This would make them unsusceptible to various well-known attacks that threaten the security of more traditional exponentiation-based cryptosystems like 'RSA' (cf. [19]) and 'Diffie-Hellman' (cf. [4]). This is illustrated by the following quotes from [21]:

> This opens RSA to a cryptographic attack known as *adaptive chosen-message forgery*. ... LUC is not multiplicative and therefore not susceptible to this attack.

and from [22]:

D. Coppersmith (Ed.): Advances in Cryptology - CRYPTO '95, LNCS 963, pp. 386-396, 1995.
© Springer-Verlag Berlin Heidelberg 1995

This problem has the advantage that the subexponential algorithms do not appear to generalize to it, so breaking these ciphers is much more expensive.

Concerning the first quote, it was shown independently in [2] and [6] that LUC is susceptible to 'existential forgeries', a restricted variant of chosen-message forgeries. LUC seemed to avoid a true chosen-message forgery, however, which is, according to the response to [6] in [23], 'the most important advance of LUC over RSA'.

Concerning the second quote, LUCDIF and LUCELG would require far shorter key sizes than traditional systems to provide the same level of security. Or, alternatively, with the same key sizes they would provide security far superior to the older systems.

In this paper we address these two quotes. We review the relation between Lucas sequences and exponentiation, and derive some properties of the Lucas-based cryptosystems that the authors of [21; 22; 25; 26] might not have been aware of. As a result, we present a chosen-message forgery for LUC that is more general than the 'existential forgery' referred to above, thus undermining LUC's main advantage over RSA.

Furthermore, we show that LUCDIF and LUCELG are vulnerable to subexponential time attacks[4]. We do not claim that the security of LUCDIF and LUCELG is threatened by these subexponential attacks to the same extent as RSA or standard ElGamal cryptosystems are threatened by subexponential time attacks. In the latter systems one typically works in groups of order $\approx m$, for some integer m. They can be broken in time $L_m[1/3; (64/9)^{1/3} + o(1)]$, for $m \to \infty$, where

$$L_m[u, v] = \exp(v(\log m)^u (\log \log m)^{1-u}),$$

either by factoring m (cf. [10]) or by computing a discrete logarithm in a group of order $\approx m$ (cf. [1; 5; 7; 20]).

The situation for LUCDIF and LUCELG is reminiscent of the Schnorr variation of ElGamal as used in the US government Digital Signature Algorithm ('DSA', cf. [15]). In DSA one works in a subgroup of order q of a group of order $\approx p$, with q substantially smaller than p. As above, DSA can be broken in time $L_p[1/3; (64/9)^{1/3} + o(1)]$, which is subexponential in p, but an attack that is subexponential in the subgroup order q seems to be infeasible. So, in a subexponential attack on DSA nobody knows how to take advantage of the small subgroup size. As we will see below, in LUCDIF and LUCELG one works in a subgroup of order $\approx p$ of a group of order $\approx p^2$. A subexponential attack would require time $L_{p^2}[1/3; (64/9)^{1/3} + o(1)] = L_p[1/3; (128/9)^{1/3} + o(1)]$. Although this is subexponential in p, it is much slower than time $L_p[1/3; (64/9)^{1/3} + o(1)]$ which one would want to take full advantage of the small subgroup size.

---

[4] This fact was independently noted by Burt Kaliski, Scott Vanstone, and the authors of [9]. We are grateful to an anonymous member of the Crypto'95 program committee for bringing the latter paper to our attention.

This greater resistance against subexponential attacks, however, might be offset by possible greater speed of the more traditional systems, like RSA, if comparable parameter sizes are used. It is conceivable that one could use substantially larger parameters in RSA, and still attain the same speed as a Lucas-based system with smaller parameters. Naturally, this would affect the relative security of the two systems. Because these considerations depend heavily on implementation details, we do not elaborate. In any case, we conclude that the situation is not as bright for LUCDIF and LUCELG as suggested in [26], where it is assumed that the best attacks 'may take time proportional to $p^{1/5}$'.

The paper is organized as follows. First we review some properties of Lucas sequences. Next we present LUC and a chosen-message forgery for LUC, and then we discuss the relative strengths of LUC and RSA. Finally, we present LUCELG PK and a subexponential time attack against it. Similar attacks on LUCELG DS and LUCDIF follow immediately.

2 Lucas sequences

Let P, Q be integers, and let α be a root of $x^2 - Px + Q = 0$ in the field $\mathbf{Q}(\sqrt{\Delta})$, where $\Delta = P^2 - 4Q \in \mathbf{Z}$ is assumed to be a non-square (but not necessarily squarefree). Then α is an element of the ring of integers $\mathcal{O}_\Delta$ of the quadratic field $\mathbf{Q}(\sqrt{\Delta})$, and there exist integers $v = v(\alpha)$ and $u = u(\alpha)$ such that $\alpha = \frac{v+u\sqrt{\Delta}}{2}$. In fact, for every $k \geq 1$ it holds that $2\alpha^k \in \mathbf{Z}[\sqrt{\Delta}]$, and we can write $\alpha^k = \frac{v_k+u_k\sqrt{\Delta}}{2}$, for certain integers $v_k = v(\alpha^k) = v_k(\alpha)$ and $u_k = u(\alpha^k) = u_k(\alpha)$.

Choosing $\alpha = \frac{P+\sqrt{\Delta}}{2}$ and its conjugate $\beta = \bar{\alpha} = \frac{P-\sqrt{\Delta}}{2}$, we find that $v_1(\alpha) = v(\alpha) = P$ and $u_1(\alpha) = u(\alpha) = 1$ and it is easy to see by induction that the v_k and u_k are given by the recurrence relations

$$u_{k+2} = u_{k+2}(P,Q) = Pu_{k+1} - Qu_k, \quad u_1 = 1, \quad u_0 = 0,$$
$$v_{k+2} = v_{k+2}(P,Q) = Pv_{k+1} - Qv_k, \quad v_1 = P, \quad v_0 = 2.$$

Remarks. Thus the v_k, u_k may be seen as the 'coefficients' of the powers of α that may be computed by the above recurrence relations. Knowing v_k and u_k implies knowledge of α^k, which immediately ties the problem of determining k from v_k and u_k to the discrete logarithm of α^k with respect to the base α.

Depending on which view we like to stress we will write $v_k(\alpha)$ or $v_k(P,Q)$, and these are related via $\alpha = \frac{P+\sqrt{P^2-4Q}}{2}$.

Of the many relations between the u_k, v_k we derive a few that are relevant for what is to follow. The first lemma deals with the u and v of conjugates, traces and norms of powers.

Lemma 1. With notation as above, for every α and every $k \geq 0$:
(i) $v_k(\beta) = v_k(\alpha)$ and $u_k(\beta) = -u_k(\alpha)$.

(ii) $\alpha^k + \beta^k = v_k(\alpha) = v_k(\beta)$.

(iii) $\alpha^k \beta^k = Q^k = \dfrac{v_k^2(\alpha) - \Delta u_k^2(\alpha)}{4}$.

Proof. The first and second assertions are immediate from the fact that exponentiation and conjugation commute:

$$\beta^k = (\bar{\alpha})^k = \overline{\alpha^k} = \frac{v_k(\alpha) - u_k(\alpha)\sqrt{\Delta}}{2}.$$

Multiplying this by α^k yields (iii).

Lemma 2. *For all* $k \geq \ell \geq 0$:

$$v_{k+\ell} = v_k v_\ell - Q^\ell v_{k-\ell}.$$

Proof. Use Lemma 1(ii) and (iii).

This shows that v_k for large k can be easily computed since exponentiation can be done by repeated squaring and multiplication. Alternatively, if both sequences are needed, the following lemma can be used.

Lemma 3. *For* $k \geq 1$:

(i) $u_{2k} = u_k v_k$, $\quad v_{2k} = v_k^2 - 2Q^k$;

(ii) $u_{2k+1} = (Pu_{2k} + v_{2k})/2$, and $v_{2k+1} = (\Delta u_{2k} + Pv_{2k})/2$.

Proof. Write out the coefficients of $(\alpha^k)^2$ and of $\alpha(\alpha^{2k})$ respectively.

The other relevant relation is most easily formulated in terms of recurrent sequences. It expresses the fact that the coefficients of the powers of a fixed power α^m can be found from a recursion with parameters depending on α^m in a simple fashion.

Lemma 4. *For every* P *and* Q:

$$v_{km}(P,Q) = v_k(v_m(P,Q), Q^m),$$

and

$$u_{km}(P,Q) = u_k(v_m(P,Q), Q^m)u_m(P,Q).$$

Proof. Let α be as before; then

$$\alpha^m = \frac{v_m + u_m\sqrt{\Delta}}{2} = \frac{v_m + \sqrt{u_m^2 \Delta}}{2} = \frac{v_m + \sqrt{v_m^2 - 4Q^m}}{2},$$

by Lemma 1, so

$$\alpha^m = \frac{P' + \sqrt{P'^2 - 4Q'}}{2},$$

where $P' = v_m(P,Q)$ and $Q' = Q^m$, and thus

$$(\alpha^m)^k = \frac{v_k(P',Q') + u_k(P',Q')\sqrt{P'^2 - 4Q'}}{2}$$

$$= \frac{v_k(P',Q') + u_k(P',Q')u_m(P,Q)\sqrt{\Delta}}{2}.$$

Now write

$$\alpha^{km} = \frac{v_{km}(P,Q) + u_{km}(P,Q)\sqrt{\Delta}}{2}$$

and compare the coefficients.

In the applications, Lucas sequences are often considered modulo a fixed modulus. If we choose a prime $p \neq 2$ for which the Legendre symbol $\left(\frac{\Delta}{p}\right) = -1$ then $\mathcal{O}_\Delta/p \cong \mathbf{F}_{p^2}$, the finite field of p^2 elements, via an isomorphism that we will denote by ϕ_p. The following lemma gives information about the order of α in $\mathcal{O}_\Delta/p$, and hence of $\phi_p(\alpha)$ in $\mathbf{F}_{p^2}$, which we will refer to in section 6.

Lemma 5. Let $\alpha = \frac{P \pm \sqrt{P^2 - 4Q}}{2}$ and let p be an odd prime, with $\left(\frac{P^2 - 4Q}{p}\right) = -1$. Then:

$$\alpha^{p+1} \equiv Q \bmod p.$$

Proof. In O_Δ/p:

$$\alpha^{p+1} = \alpha\left(\frac{v_1 + u_1\sqrt{\Delta}}{2}\right)^p = \alpha\left(\frac{v_1^p + u_1^p\sqrt{\Delta}^p}{2^p}\right) = \alpha\left(\frac{v_1 + \left(\frac{\Delta}{p}\right)u_1\sqrt{\Delta}}{2}\right) = \alpha\beta = Q$$

because

$$\Delta^{\frac{p-1}{2}} \equiv \left(\frac{\Delta}{p}\right) \equiv -1 \bmod p$$

by Euler's criterion.

3   LUC

In [21] the following cryptographic application of Lucas sequences was proposed, apparently independent of earlier publication in [13] and [14]. See also [25].

Public Key System (LUC). Each user publishes the product n of two large primes p and q, and an index e with $\gcd(e, (p^2-1)(q^2-1)) = 1$. The corresponding d such that $de \equiv 1 \bmod (p^2-1)(q^2-1)$ is kept secret (cf. [25: page 115]).

A message m is an integer satisfying $1 \leq m \leq n-1$ with $\gcd(m,n) = 1$. To encrypt a message m meant for some user, one looks up the user's n and e, and computes the encrypted message $y = v_e(m,1) \bmod n$ — i.e., P is equal to the message, and $Q = 1$. This computation can be carried out using the recurrence given in Lemma 2 in $O(\log e)$ elementary operations on integers modulo n. To decrypt the message, the user calculates

$$v_d(y, 1) \equiv v_d(v_e(m, 1), 1) \equiv v_{de}(m, 1) \equiv m \bmod n$$

(cf. Lemma 4). The final identity holds because $\alpha^{de} \equiv \alpha$ modulo both p and q.

Alternatively, to use LUC as a signature scheme, the user's signature on a message m equals $v_d(m, 1) \bmod n$, which can be verified by checking that $v_e(v_d(m, 1), 1) \equiv m \bmod n$.

Remarks. Our description of the choice of e and d is more general than the message-dependent choices from [21] or [25]; we refer to [21] and [25] for details. We would like to stress that the Lucas function using these message-dependent secret keys and the Lucas function using our choice of d are the same functions, since in both cases the inverse of $e \mapsto v_e(m, Q)$ is computed. In practical circumstances one would probably prefer to use message-dependent secret keys for efficiency reasons [24], for instance as follows.

Note that $v_{d(p)}(m, 1) \equiv v_d(m, 1) \bmod p$ if $d(p) \equiv d \bmod (p - \left(\frac{m^2-4}{p}\right))$ (use Lemma 5 if $\left(\frac{m^2-4}{p}\right) = -1$; otherwise use that $\alpha \in \mathbf{F}_p$). Signatures can therefore be generated substantially faster than computing $v_d(m, 1) \bmod n$ by computing $v_{d(p)}(m, 1) \bmod p$ and $v_{d(q)}(m, 1) \bmod q$, followed by an application of the Chinese remainder theorem. However, no message-dependent d will be used in the sequel, because a message-independent d simplifies the analysis of LUC, and because in this paper we are not concerned with efficiency issues of LUC.

The choice $Q = 1$ is not essential for LUC as a public-key system: y could have been defined as $y = v_e(m, Q) \bmod n$, for some Q depending on the intended recipient, who can calculate $v_d(y, Q^e) \equiv v_d(v_e(m, Q), Q^e) \equiv v_{de}(m, Q) \equiv m$ modulo n. This would be slightly less efficient and offers no additional security. To use LUC as a signature system, however, either Q has to be equal to 1, or $Q^d \bmod n$ has to be included in the user's public key. Otherwise a verifier of the signature $v_d(m, Q)$ on message m would not be able to verify that $v_e(v_d(m, Q), Q^d)$ is indeed equivalent to m modulo n.

The signature $v_d(m, 1)$ on message m can be used to generate signatures $v_d(v_k(m, 1), 1) \equiv v_{dk}(m, 1) \equiv v_k(v_d(m, 1), 1) \bmod n$ on message $v_k(m, 1)$ for any $k \geq 0$. This 'existential forgery' was mentioned in [2] and [6].

4 A chosen-message forgery for LUC

Let $n = pq$, e, and d be as above the public and secret data of some user, and let $Q = 1$. To forge the signature of this user on message m, an adversary could proceed as follows. First, integers a, b, c, s, and t are selected such that

$$bs - ct = 1, \qquad bs + ct = ae.$$

This can for instance be done by selecting c, h, and t such that $ct = (e-1)/2 + eh$ (note that e is odd), and selecting b and s such that $bs = 1 + ct$. It follows that $bs - ct = 1$, and that $bs + ct = 1 + 2ct = e + 2eh$, so that $a = 2h + 1$.

Next, the adversary calculates the messages $m_s = v_s(m, 1) \bmod n$ and $m_t = v_t(m, 1) \bmod n$ and obtains the user's signatures $v_d(m_s, 1) \bmod n$ and $v_d(m_t, 1) \bmod n$ on these messages. Finally, $v_d(m, 1)$ is computed as

$$v_d(m, 1) = v_b(v_d(m_s, 1), 1)v_c(v_d(m_t, 1), 1) - v_a(m, 1) \bmod n.$$

The correctness follows from Lemma 4, the choice of a, b, c, s, t, m_s, and m_t, and from Lemma 2 with $k = dbs$, $\ell = dct$, and $Q = 1$:

$$\begin{aligned}
v_b(v_d(m_s, 1), 1)v_c(v_d(m_t, 1), 1) &\equiv v_{dbs}(m, 1)v_{dct}(m, 1) \\
&\equiv v_{d(bs+ct)}(m, 1) + v_{d(bs-ct)}(m, 1) \\
&\equiv v_{dae}(m, 1) + v_d(m, 1) \\
&\equiv v_a(m, 1) + v_d(m, 1) \bmod n.
\end{aligned}$$

Remarks. The mapping sending k to $v_k(m, 1) \bmod n$ is not generally a random map into the message space (since it need not be surjective). As a consequence, the messages $v_s(m, 1) \bmod n$ and $v_t(m, 1) \bmod n$ that are to be signed are not always completely 'blind'.

If m, s, t and the signatures for $v_s(m, 1) \bmod n$ and $v_t(m, 1) \bmod n$ are given and if s, t, and e are pairwise relatively prime, then b, c satisfying $bs - ct = 1$ and $bs + ct \equiv 0 \bmod e$ can be found. Thus the signatures for m and $v_k(m, 1)$ can be computed.

The choice of a, b, c, s, t is supposed to make it difficult for the user to find out which past signatures were used to make the forgery. The latter would be easy if we would have chosen $a = b = c = 1$, $s = (e + 1)/2$, $t = (e - 1)/2$, and

$$v_d(m, 1) = v_d(m_s, 1)v_d(m_t, 1) - m \bmod n.$$

5 LUC and RSA

In the abstract and the introduction of [25] the authors of [25] announce a proof that LUC is cryptographically stronger than RSA. We have not been able to locate this proof in [25][5], and neither have we been able to derive such a proof ourselves. Here we offer some observations that might be pertinent to this matter.

Because $\alpha^{de} \equiv \alpha \bmod n$ and $u_1 = 1$, it follows from the second identity in Lemma 4 that

$$u_d(P, 1) \equiv u_e(v_d(P, 1), 1)^{-1} \bmod n.$$

Thus $u_d(P, 1)$ can be computed whenever $v_d(P, 1)$ is known. Moreover, the following equation can be shown to hold by induction on k, using the recurrence relations for v_k and u_k:

---

[5] In [25: 3.4], however, the authors 'say, with confidence, that LUC is cryptographically stronger than RSA'.

$$2P^k \equiv v_k(P + P^{-1}, 1) + (P - P^{-1})u_k(P + P^{-1}, 1) \bmod n.$$

In particular, the above relations show that $P^d \bmod n$ can be derived once $v_d(P + P^{-1}, 1)$ is known.

To break an RSA-cryptogram $E(m)$, where $E(m) \equiv m^e \bmod n$ for some message m, it suffices to compute $E(m)^d \bmod n$, where e and d are as in the description of LUC, since $de \equiv 1 \bmod (p^2 - 1)(q^2 - 1)$ implies that $de \equiv 1 \bmod (p - 1)(q - 1)$. According to the above, this can be achieved if $v_d(E(m) + E(m)^{-1}, 1) \bmod n$ can be computed. Thus the RSA-cryptogram $E(m)$ can be broken if LUC can be broken for the message $E(m) + E(m)^{-1} \bmod n$.

This does not imply, however, that LUC is stronger than RSA. It is conceivable that LUC can only be broken for some particular set of messages, whereas RSA is secure. For instance, it might be the case that $v_d(X, 1)$ can only efficiently be derived from X, e, and n for X for which $\left(\frac{X^2 - 4}{p}\right) = \left(\frac{X^2 - 4}{q}\right) = -1$, where p and q are the prime factors of n. This would allow us to break 25% of all LUC-cryptograms, but since $\left(\frac{(X + X^{-1})^2 - 4}{p}\right) = \left(\frac{(X - X^{-1})^2}{p}\right) = 1$, the method cannot be used in the above manner to break RSA.

We are not aware of any further results in this direction.

6   LUCELG

In [26] the following cryptographic application of Lucas sequences was proposed.

Public Key System (LUCELG PK). A prime p and the start values P and $Q = 1$ are published, chosen such that $P^2 - 4Q \bmod p$ is a quadratic non-residue, and such that $v_\ell(P, Q) \not\equiv 2 \bmod p$ for any ℓ less than and dividing $p + 1$. Every user also chooses a private key x, and publishes the public key $y \equiv v_x(P, Q) \bmod p$ (cf. Lemma 2).

A message m is an integer satisfying $1 \le m \le p - 1$. To encrypt a message meant for some user, one looks up the user's y, chooses a secret k, which will also be an integer satisfying $1 \le k \le p - 1$, computes $G \equiv v_k(y, Q) \bmod p$, as well as $d_1 \equiv v_k(P, Q) \bmod p$ and $d_2 \equiv Gm \bmod p$. The encrypted message consists of the pair (d_1, d_2).

To decrypt the message, the user calculates

$$v_x(d_1, Q) \equiv v_x(v_k(P, Q), Q^k) \equiv v_{kx}(P, Q) \equiv G \bmod p,$$

inverts the result modulo p and recovers $m \equiv d_2 G^{-1} \bmod p$.

Remarks. Note that it seems essential in this scheme that $Q \equiv 1 \bmod p$: the recipient needs to know $Q^k \bmod p$ for the secret value k in order to be able to compute $v_{kx}(P, Q)$ from $v_k(P, Q)$ using the fourth lemma above. This can be achieved by taking $Q \equiv 1 \bmod p$; in [21; 22; 25; 26] it is assumed that $Q = 1$.

Let $\alpha = \frac{P + \sqrt{P^2 - 4Q}}{2}$; the condition that $v_\ell(P, Q) \not\equiv 2 \bmod p$ for proper divisors ℓ of $p + 1$ ensures that the multiplicative order of the image $\phi_p(\alpha) \in \mathbf{F}_{p^2}$

equals $p+1$. Namely, if $\phi_p(\alpha^n) = 1$ then $v_n(\alpha) \equiv 2 \bmod p$ and $u_n(\alpha) \equiv 0 \bmod p$, which does not happen for any proper divisor of $p+1$ by this condition. On the other hand, by Lemma 5 (with $Q \equiv 1 \bmod p$) the order divides $p+1$.

The condition that $\left(\frac{P^2-4Q}{p}\right) = -1$ (which is nowhere explicitly stated in [26]) guarantees that one is working in the the finite field $\mathbf{F}_{p^2}$ rather than $\mathbf{F}_p$; the latter contains a square root of $P^2 - 4Q$ if the Legendre symbol equals 1 instead. In that case the attack described in the next section merely requires a discrete logarithm computation in $\mathbf{F}_p$. The recursive relations are still valid, but the order of α in O_Δ/p will be a divisor of $p - 1$.

7 A subexponential time attack on LUCELG

Unfortunately, choosing $Q \equiv 1 \bmod p$ also provides the key to an attack on the proposed system: noting that

$$v_k(\alpha) = \alpha^k + \beta^k = \alpha^k + \left(\frac{Q}{\alpha}\right)^k \equiv \alpha^k + \alpha^{-k} \bmod p$$

in this case, enables an adversary to obtain $\alpha^{\pm k}$ from v_k, since it is a root in $\mathbf{F}_{p^2}$ of the equation $z^2 - v_k z + 1 = 0$ (this is equivalent to deriving $\pm u_k(\alpha)$ and therefore $\alpha^{\pm k}$ from $v_k(\alpha)$ using Lemma 1(iii)). Then retrieving $\pm k$ from $\alpha^{\pm k}$ is a discrete logarithm problem in $\mathbf{F}_{p^2}$, which with the currently best available methods can be done in subexponential time $L_{p^2}[1/3; (64/9)^{1/3} + o(1)]$, for $p \to \infty$ (cf. [20]). Note that the sign of k does not matter, since $v_k = v_{-k}$ for all k when $Q = 1$, and that roots in $\mathbf{F}_{p^2}$ can be computed in expected polynomial time (cf. [3]). Other subexponential time methods to compute discrete logarithms in $\mathbf{F}_{p^2}$ can be found in [1; 5].

This implies that an adversary can derive x from y in subexponential time for any user, and decrypt all intercepted messages sent to that user. Alternatively, an adversary can decide only to derive k from the intercepted d_1, in subexponential time, after which G and thus m follow trivially from y and d_2.

Remarks. In [22; 26] an ElGamal-type signature scheme based on Lucas sequences was proposed (LUCELG DS). Since in this system both v_k and u_k are explicitly given, a direct analogue of the discrete logarithm attack on ElGamal (but here in $\mathbf{F}_{p^2}$) applies. Note that the 'double key size' problems of LUCELG DS as mentioned in [26] can be avoided if one uses Lemma 1(iii) to derive $\pm u_k$ from v_k. This would also avoid the serious weakness in LUCELG DS that is pointed out in [8]. Another variant of ElGamal based Lucas functions is discussed in [8]. The security of that system relies on the difficulty of computing discrete logarithms in $\mathbf{F}_p$.

In [22] a Diffie-Hellman-type key agreement scheme based on Lucas sequences was proposed (LUCDIF). Since LUCDIF again uses $Q = 1$, a subexponential attack similar to the one described above applies to it.

Acknowledgments. The authors are grateful to Eric Bach, Burt Kaliski, and Scott Vanstone, for their support of this article, which parallels similar remarks they sent or intended to send to the developers of Lucas-based cryptosystems. Christopher Skinner kindly communicated the effectiveness of our chosen-message attack in his 'message-dependent' implementation of LUC.

References

1. L. M. Adleman and J. DeMarrais, *A subexponential algorithm for discrete logarithms over all finite fields*, Proceedings Crypto'93, Lecture Notes in Comp. Sci. **773** (1994), 147–158.

2. E. Bach, *Comments on Peter Smith's LUC public-key encryption system*, manuscript, March 1993.

3. E. R. Berlekamp, *Factoring polynomials over large finite fields*, Math. Comp. **24** (1970), 713–735.

4. W. Diffie and M. E. Hellman, *New directions in cryptography*, IEEE Trans. Info. Theory, vol IT-33 (1976), 644–654.

5. T. ElGamal, *A subexponential-time algorithm for computing discrete logarithms over GF(p^2)*, IEEE Trans. Info. Theory, vol IT-32 (1985), 469–472.

6. T. ElGamal and B. Kaliski, *Letter to the editor*, Dr. Dobb's Journal (May 1993), 10.

7. D. Gordon, *Discrete logarithms in GF(p) using the number field sieve*, SIAM J. Disc. Math. **6** (1993), 124–138.

8. P. Horster, H. Petersen, and M. Michels, *Digital signature schemes based on Lucas functions*, University of Technology Chemnitz-Zwickau, Technical Report TR-95-1; to appear in: Communications and Multimedia Security, IT-Sicherheit '95, Joint working conference IFIP TC-6 TR-11 and Austrian Computer Society, Graz, Sept. 20–21, 1995.

9. C.-S. Laih, F.-K. Tu, and W.-C. Tai, *On the security of the Lucas function*, Information Processing Letters **53** (1995), 243–247.

10. A. K. Lenstra and H. W. Lenstra, Jr. (eds), *The development of the number field sieve*, Lecture Notes in Math. **1554**, Springer-Verlag, Berlin, 1993.

11. R. Lidl and W. B. Müller, *Permutation polynomials in RSA-cryptosystems*, Proceedings of Crypto'83, Plenum Press (1984), 293–301.

12. W. B. Müller, *Polynomial functions in modern cryptology*, Contributions to general Algebra 3, Proceedings of the Vienna conference (1985), 7–32.

13. W. B. Müller and W. Nöbauer, *Some remarks on public-key cryptosystems*, Studia Sci. Math. Hungar. **16** (1981), 71–76.

14. W. B. Müller and W. Nöbauer, *Cryptanalysis of the Dickson-scheme*, Proceedings of Eurocrypt'85, Springer (1985), 50–61.

15. NIST, *A proposed federal information processing standard for digital signature standard (DSS)*, Federal Register **56** (1991), 42980–42982.

16. W. Nöbauer, *Cryptanalysis of the Rédei-scheme*, Contributions to general Algebra 3, Proceedings of the Vienna conference (1985), 255–264.

17. W. Nöbauer, *Cryptanalysis of a public-key cryptosystem based on Dickson-polynomials*, Mathematica Slovaca **38** (1989), 309–323.

18. H. Riesel, *Prime numbers and computer methods for factorization*, Progr. Math. **57**, Boston: Birkhauser, 1985.

19. R. L. Rivest, A. Shamir, L. Adleman, *A method for obtaining digital signatures and public-key cryptosystems*, Comm. ACM **21** (1978), 120–126.

20. O. Schirokauer, *Using number fields to compute general discrete logarithms*, in preparation.

21. P. Smith, *LUC public-key encryption*, Dr. Dobb's Journal (January 1993), 44–49.

22. P. Smith, *Cryptography without exponentiation*, Dr. Dobb's Journal (April 1994), 26–30.

23. P. Smith, *Response to [6]*, Dr. Dobb's Journal (May 1993), 10–11.

24. P. Smith, Personal communication, February 1995.

25. P. J. Smith and M. J. J. Lennon, *LUC: a new public key system*, Proceedings of the Ninth IFIP Int. Symp. on Computer Security (1993), 103–117.

26. P. Smith and C. Skinner, *A public-key cryptosystem and a digital signature system based on the Lucas function analogue to discrete logarithms*, Pre-proceedings Asiacrypt'94, 298–306.

Threshold DSS Signatures without a Trusted Party

Susan K. Langford*

Department of Electrical Engineering
Stanford University
Stanford CA 94305-4055

Abstract. A t-out-of-l threshold signature scheme allows l members of a group to own shares of a private key such that any t of them can create a signature, while fewer than t cannot. Most of these schemes require a single trusted party to create the secret key and calculate the l shares. Harn [10] and Li, Hwang, and Lee [13] have devised threshold schemes based on the difficulty of solving the discrete logarithm problem which do not require such a trusted party. This paper extends that property to 2-out-of-l threshold signatures based on the Digital Signature Standard and describes two possible generalizations to a t-out-of-l scheme.

1 Introduction

A digital signature, like a handwritten signature, is used to verify the sender of a particular message. The signer is generally a single person. However, when the message is on behalf of an organization, a valid message may require the approval of several people. A common example of this policy is a large bank transaction, which requires signatures from two people. Such a policy could be implemented by having a separate digital signature for every required signer, but this solution increases the effort to verify the message linearly with the number of signers.

An alternative method is a threshold signature scheme. In a t-out-of-l threshold signature scheme, there is one public key for the group while the private key is shared among the l members of the group. Any t members can cooperate to create a digital signature without revealing their shares of the private key. Fewer than t members cannot create a valid signature.

Threshold signatures are closely related to the concept of threshold cryptography, first introduced by Desmedt [5]. A number of group oriented systems have been proposed and, for an overview of the field, the reader is referred to [6].

Threshold signature schemes have been proposed using RSA signatures [7, 11, 9]. These schemes require the use of a trusted center to generate the private key and calculate the share values. While the existence of such a center is not an unreasonable assumption, there are two potential problems. First, for many

---

* This work was supported by the National Semiconductor Corporation under the Fellow/Mentor/Advisor [FMA] program of Stanford University Center for Integrated Systems.

D. Coppersmith (Ed.): Advances in Cryptology - CRYPTO '95, LNCS 963, pp. 397-409, 1995.

applications, there is no one person or device which can be completely trusted by all members of the group. Second, the use of a key center creates a single point failure. Any security lapse at the key center can reveal the private key. Once the shares are distributed, careless handling of any one share reveals no information about the private key. At least t shares must be compromised before an attacker has any useful information.

To avoid these problems, Harn [10] introduced a scheme based on a modified ElGamal signature which does not require a trusted key center. Li, Hwang, and Lee used Harn's scheme to allow the receiver to verify which members of the group signed a given message [13]. These schemes show that it is possible to generate threshold signatures without a trusted party, but leave open the question of how to create such schemes for other signature algorithms.

This paper extends the idea of distributed generation of the private key by proposing three threshold schemes based on the Digital Signature Standard (DSS). The first is a 2-out-of-l threshold scheme which requires a small amount of interaction between the co-signers, but does not require the use of a trusted center to generate the shares. This scheme does need a third person as combiner, but the combiner need not be any more trusted than other members of the group. In fact, the combiner may be a member of the group who is not a co-signer. The second scheme uses precomputed values to create t-out-of-l threshold signatures. The key generation center for this scheme can be distributed into three centers, any two of which must be compromised to expose the private key. The third scheme does not require the use of a semi-trusted combiner or precomputed lists. However, the scheme requires $t^2 - t + 1$ signers for t-out-of-l security. These three schemes are the first threshold schemes to be proposed using DSS.

2 Digital Signature Standard

The Digital Signature Standard (DSS) [1] is based on the difficulty of computing discrete logs. For this algorithm, we need the following parameters.

1. p = a prime modulus, L bits long, where $512 \leq L \leq 1024$, and L is a multiple of 64.
2. q = a 160-bit prime divisor of $p - 1$.
3. $g = h^{(p-1)/q} \bmod p$, where h is any integer with $0 < h < p$ such that $g \bmod p > 1$ (g is an element of order q in $\mathrm{GF}(p)$).
4. x = an integer with $0 < x < q$.
5. $y = g^x \bmod p$.
6. M = a one-way hash of the message to be signed.
7. k = a random integer with $0 < k < q$.

The system parameters p, q, and g are public. The user's public key and private key are y and x, respectively. To create a signature, the user chooses a random k, computes $k^{-1} \bmod q$, and calculates

$$r = (g^{k^{-1}} \bmod p) \bmod q, \tag{1}$$

and

$$s = k(M + xr) \bmod q. \tag{2}$$

The pair (r, s) is the desired signature. A new value for k must be generated for each new signature.

Note: For this paper, k and k^{-1} are switched from the standard DSS notation. This is merely to simplify the following discussion. Since k is random, this interchange is immaterial.

To verify a signature, the recipient calculates $g^{Ms^{-1}} y^{rs^{-1}} \bmod p$ and checks that the result equals r. Since $y = g^x$, when we multiply the two values together, the exponent becomes $Ms^{-1} + xrs^{-1} = (M + xr)s^{-1} \bmod q$. This expression reduces to k^{-1} in the exponent, yielding the desired result r as in equation (1).

3 Secret Sharing of Sums and Products

Shamir [15] developed a secret sharing scheme based on polynomials over a finite field. A polynomial f of degree $t - 1$ is chosen such that $f(0)$ is the secret. Every participant i, $(i \neq 0)$, in the scheme is given $f(i)$, a share of the secret. Any t of these shares can be used to calculate $f(0)$. Further, if any $t - 1$ of these shares are combined, then no information is revealed about $f(0)$.

In general, from any t distinct points of a degree $t - 1$ polynomial, we can calculate the value of any other point of the polynomial by interpolation. For a set π of t shareholders with shares $f(x_i)$, over the field $GF(q)$, the formula for this interpolation is

$$f(x) = \sum_{i \in \pi} f(x_i) \prod_{j \in \pi, j \neq i} \frac{(x - x_j)}{(x_i - x_j)} \bmod q. \tag{3}$$

Therefore the formula for calculating the secret $f(0)$ is the linear combination of the shares,

$$f(0) = \sum_{i \in \pi} c_{i,\pi} f(x_i) \bmod q, \tag{4}$$

where $c_{i,\pi}$ is a constant which can be publicly determined because its value does not depend on any secret information.

Although generally defined over a field, this scheme can be defined mod n, where n is not prime, if the quantities $x_i - x_j$ have inverses for all i and j. If the shareholders are numbered from 1 to l, this is equivalent to requiring that n has no prime factors less than l.

It is natural to think of secret sharing in terms of an existing secret owned by an individual or device. The owner of the secret then creates and distributes the shares. However, we can also implement secret sharing so that the secret is created as part of the process of creating the shares. The value of the secret is determined by the value of the shares, but the secret itself is not explicitly calculated at the time the shares are created. This process will be useful for creating threshold signatures, where we want to create shares of a private key.

The private key itself is never used directly, so there is no reason that it ever needs to be explicitly calculated.

Shamir's scheme can be used to implement this idea of implicitly creating the secret. Suppose that a group wishes to create shares in a secret which is the sum of two numbers, without any one person ever knowing that sum. Person A creates a secret sharing polynomial $f_A(x)$ with secret S_A (i.e., $f_A(0) = S_A$), and distributes the shares. Similarly, person B creates $f_B(x)$ with secret S_B and distributes the resulting shares. Then $f(x) = f_A(x) + f_B(x)$ is a new polynomial with secret $f(0) = f_A(0) + f_B(0) = S_A + S_B$. If shareholder i has received $f_A(i)$ and $f_B(i)$, he can create the share of the new polynomial by adding the two shares, $f(i) = f_A(i) + f_B(i)$. Clearly, this scheme is not limited to two equations, the only requirement for additional equations is that the numbering of the shareholders remains consistent. Shareholder i must receive the polynomial evaluated at point i for all polynomials involved.

If we use these polynomials in the exponent, then we create shares which multiply together, rather than add. Given a prime q and an element β of order n in $GF(q)$, we can create a polynomial $f(x) \bmod n$ and define the secret as $\beta^{f(0)} \bmod q$. The secret can be recovered by exponentiating the normal linear interpolation formula. As explained above, n must not have any small prime factors or this interpolation is not possible. We will use this multiplicative sharing technique to create the shares of the private key in our threshold scheme.

The methods for creating additive or multiplicative shares without a trusted center are well known. However, generating DSS threshold signatures leads to a new problem—creating shares of an arbitrary product. We would like to be able to have every member of a group generate a number and calculate shares which combine linearly to form the product of those numbers. We have not solved this problem in general. The remainder of this section explains the solution for the case of the product of two numbers.

Suppose Alice knows secret A and Bob knows secret B. Together, they would like to create shares which can produce the secret $AB \bmod q$. However, they would like to create these shares without having any one person know the secret AB. To do this, they need the help of a third person, Charles.

1. Alice generates a random number R_A and Bob generates R_B, where $0 \leq R_A, R_B < q$.

2. Alice privately sends Charles $A - R_A \bmod q$ and privately sends Bob R_A. Bob privately sends Charles $B - R_B \bmod q$ and privately sends Alice R_B.

3. Alice calculates $S_A = AR_B - \frac{1}{2}R_AR_B \bmod q$, and Bob calculates $S_B = BR_A - \frac{1}{2}R_AR_B \bmod q$. Charles calculates $S_C = (A - R_A)(B - R_B) \bmod q$.

The sum of the three secrets, S_A, S_B, and S_C, is the desired product, AB. Notice that no single person has sufficient information to calculate AB, but that any two of the three can work together to recover the secret. Thus, Charles must be trusted exactly the same amount as Alice or Bob.

4   2-out-of-l Threshold Signature Algorithm

4.1   DSS Parameters

The 2-out-of-l threshold signature scheme places an additional requirement on
the parameters of DSS. We require that $(q-1)/2$ has no prime divisors smaller
than l, the number of shareholders.[2] This requirement will slow the generation
of the system parameters slightly. We will also need an additional parameter, β,
where β is the square of a primitive element mod q. (β has order $(q-1)/2$.)

4.2   Creating the Shares

To create the 2-out-of-l threshold signature algorithm, every shareholder needs
a multiplicative share of the secret key, x. We will define $x = \beta^{f(0)} \bmod q$, where
f is a secret sharing polynomial. Since β is the square of a primitive element,
arithmetic in the exponent is done mod $(q-1)/2$. Therefore, we will define f
mod $(q-1)/2$. All necessary inverses exist since q has been chosen such that
$(q-1)/2$ has no prime factors less than l.

As described in Section 3, the generation of the secret sharing polynomial
f can be accomplished cooperatively, rather than relying on a trusted center.
Each member of the group generates a degree one polynomial mod $(q-1)/2$.
For instance, if Alice is member number 1, she generates $f_1(z) = \gamma_1 z + \delta_1 \bmod$
$(q-1)/2$, where γ_1 and δ_1 are random. She now generates shares of f_1 and
distributes them privately to each member of the group, taking care to give each
member i the share $f_1(i)$.

When every member of the group has generated and distributed shares of
their polynomial, Alice has $f_1(1), f_2(1), \ldots f_l(1)$. She now adds them together.
Since the polynomials are linear, the result gives her $f(1)$, where $f(z) = f_1(z) +$
$f_2(z) + \ldots + f_l(z)$. Now everyone has a share of the polynomial f, yet no single
person knows $f(0)$.

To calculate the public key, member a computes $x_a = \beta^{c_a, \pi f(a)} \bmod q$ and
member b computes $x_b = \beta^{c_b, \pi f(b)} \bmod q$. The product of these two numbers is
the secret key since $x_a x_b = \beta^{c_a, \pi f(a) + c_b, \pi f(b)} = \beta^{f(0)} = x \bmod q$. Members a and
b can use their values of x_a and x_b in a Diffie-Hellman exchange to calculate
$y = g^{x_a x_b} \bmod p$. This process can be repeated for any pair, allowing group
members to check their shares without revealing them.

4.3   Creating the Signature

The signature is created using the method for creating shares of a product described in Section 3. We will be creating shares of the values k and kx. Creating
a signature requires two shareholders, Alice and Bob, and a trusted combiner,
Charles. Any two of these three can cooperate to recover the secret, so Charles
must be trusted the same amount as any shareholder.

---

[2] For simplicity, we could require that $(q-1)/2$ be prime. However, this stronger
requirement is not necessary.

To create a signature, Alice generates three independent random numbers, k_A, R_{A1}, and R_{A2}, uniformly between 0 and $q-1$. She then sends $g^{k_A^{-1}}$ mod p, R_{A1}, and R_{A2} privately to Bob. Similarly, Bob sends $g^{k_B^{-1}}$ mod p, R_{B1}, and R_{B2} privately to Alice. Alice and Bob can avoid the need for a secure channel by generating R_{A1}, R_{A2}, R_{B1}, and R_{B2} by a Diffie-Hellman exchange.

Alice and Bob can now calculate $r = (g^{k_A^{-1}k_B^{-1}} \bmod p) \bmod q$, and send the result publicly to Charles. Alice calculates $x_A = \beta^{c_{A,x}f(A)} \bmod q$ for her private use. Note that $x_A x_B = x \bmod q$ and $k_A k_B = k \bmod q$. So Alice can use the number k_A and the method of Section 3 to create a sharing scheme for k. Similarly, she can create a sharing scheme for kx. Let S_{A1} be Alice's share of k and S_{A2} be Alice's share of kx. Alice computes

$$s_A = S_{A1}M + S_{A2}r = (k_A - \frac{1}{2}R_{A1})R_{B1}M + (k_A x_A - \frac{1}{2}R_{A2})R_{B2}r \bmod q. \quad (5)$$

She then sends s_A, $k_A - R_{A1}$, and $k_A x_A - R_{A2}$ privately to Charles. Similarly, Bob computes

$$s_B = S_{B1}M + S_{B2}r = (k_B - \frac{1}{2}R_{B1})R_{A1}M + (k_B x_B - \frac{1}{2}R_{B2})R_{A2}r \bmod q, \quad (6)$$

and sends s_B, $k_B - R_{B1}$, and $k_B x_B - R_{B2}$ privately to Charles.

Charles now computes his shares of k and kx,

$$S_{C1} = (k_A - R_{A1})(k_B - R_{B1}) \bmod q \quad (7)$$

and

$$S_{C2} = (k_A x_A - R_{A2})(k_B x_B - R_{B2}) \bmod q. \quad (8)$$

He can now calculate $s = s_A + s_B + (S_{C1}M + S_{C2}r) \bmod q$. The result (r, s) is a valid DSS signature, since

$$s = (S_{A1} + S_{B1} + S_{C1})M + (S_{A2} + S_{B2} + S_{C2})r \quad (9)$$
$$= kM + kxr \quad (10)$$
$$= k(M + xr) \bmod q. \quad (11)$$

4.4   Maintenance of Shares

One of the benefits of a threshold scheme is that if a single share is compromised, the private key is not revealed. If the shareholders know that a share has been compromised, for security they should generate a new public key and shares of the private key. However, the shareholders may not realize that a share has been compromised, or the cost of changing the public key may be quite high. To help preserve the security of the system in these cases, we can change the shares in such a way that although the public and private keys do not change, the old shares and the new shares are incompatible. This process is similar to the disenrollment capability described in [2] and [3].

To create the new shares, one of the shareholders should distribute shares of a polynomial with a "secret" of zero (ie., $g_Z(x) = Rx + 0 \bmod (q-1)/2$, where R is random). Each shareholder then adds the share of g_Z to the share of the private key. Adding the two polynomials does not change the secret, but does change the value of all shares. Visualizing the secret sharing polynomial as a line, we have changed the slope of the line, but not the intercept. Once the shareholders confirm that their shares are still valid, they destroy their shares of g_Z. Since the old shares and the new shares now correspond to two different polynomials, they cannot be combined to calculate the private key. The private key is only compromised if someone has access to two shares which are valid at the same time.

Any two shareholders can cooperate to replace a lost or damaged share without revealing their shares or calculating the secret. To create share i, the two shareholders use the interpolation formula (3) to calculate the value of $f(i)$. Each calculates one of the terms of this sum and adds a share of a polynomial with secret zero as described in the preceding paragraph. The result is transmitted secretly to shareholder i. Share i is the sum of the these two results. The two shareholders generating these pieces must use independent polynomials. Otherwise, shareholder i would have sufficient information to calculate the secret. Once share i is created, shares of the two polynomials with secret zero should be distributed to the remaining shareholders.

4.5 Security of 2-out-of-l Scheme

Because the partial signer, Alice (or by symmetry Bob), knows only a share of x, a share of k, and four random numbers, she has no information that could not be simulated by an external attacker. She therefore cannot break the system faster than an attacker who sees only the final DSS signature which is a result of this scheme.

The only participant with information which cannot be simulated by an attacker is the combiner. Coppersmith [4] has observed that the combiner Charles can compute quadratic residues involving x_A and x_B, but that this does not seem to lead to an attack. Essentially, this "attack" manipulates the equations Charles knows to generate a quantity which is a function of x_A and x_B and known information. Charles does not know the quantity itself, merely that it is a quadratic residue. This information does not appear to help Charles calculate x_A and x_B. For a description of the method for calculating quadratic residues, see [12].

5 t-out-of-l Threshold Signatures

To create general t-out-of-l threshold signatures, we will precompute and store information for each signature. While this approach raises non-trivial implementation problems, it may be suitable for certain applications.

5.1 With a Trusted Center

For simplicity, we will first assume the existence of a trusted center to create the shares. The next section will show how to remove this requirement. The center creates the private key x, and the public key y. For a single signature, the center generates a random k and creates two independent secret sharing schemes, one with secret k and the other with secret kx. Then the center privately gives Alice her share of k, her share of kx, and the quantity $r = (g^{k^{-1}} \bmod p) \bmod q$. Since the sharing schemes for k and kx are both mod q, and q is 160 bits long, the entire share is 480 bits. The center repeats the process for the other shareholders. Notice that, unlike the 2-out-of-l scheme, this scheme does not place any additional restrictions on the parameters of the system.

In order to sign a message, any t of the shareholders must create a partial signature. Let U_A be Alice's share of k and V_A be Alice's share of kx. Then her partial signature would be $s_A = U_A M + V_A r$. She sends this partial signature to a combiner. Once the combiner has accumulated t partial signatures for a message, he can combine them into a single signature using equation (3),

$$s = \sum_{i \in \pi} c_{i,\pi} s_i \tag{12}$$

$$= \sum_{i \in \pi} (c_{i,\pi} U_i) M + (c_{i,\pi} V_i) r \tag{13}$$

$$= kM + kxr = k(M + xr). \tag{14}$$

The interpolation constant $c_{i,\pi}$ can be calculated by the combiner, so the signers do not need to know who else is signing. Notice that unlike the method from Section 4, this method does not require trusting the combiner or keeping the partial signatures secret.

Once the share has been used to create a signature, it must never be used to create another signature. Therefore, the center will create a list of shares, with each share to be used exactly once. Since about 20,000 shares would fit on a floppy disk, the storage requirements for this are not unreasonable. However, using these lists creates a new problem. The shareholders need some system to insure that two messages are never signed using the same k. One of the simplest methods would be to have one person in charge of sending out message signing requests which include the number of the share to be used to create the signature. Each signing request would be signed by the individual. The shareholders could then keep track of what requests had been sent. If the requester cheats and requests that two different messages be signed with the same k, the shareholders' records will show this, and he will be eventually caught.

Other protocols may be possible depending on the individual application. For example, the following protocol might be acceptable for a small group of shareholders which creates signatures infrequently. Whenever a message to be signed is sent to the group, each member signs with her personal key an acknowledgement giving the message and the current k value. Group members wait to receive acknowledgements from all members before partially signing the message.

5.2 Distributing the Share Generation

To avoid the single point failure of Section 5.1, we can replace the single trusted center with three share creation centers. The three centers use the method of sharing a product described in section 3 to create additive shares of k and kx, then create t-out-of-l secret sharing schemes and distribute these shares to the shareholders. This method requires a great deal of interaction between the share creation centers at the time when they create all of the shares.

If $t > 2$ then this system puts more trust in the centers than in the shareholders since any two of the centers can compromise the private key, while two shareholders cannot reveal any information about it.

As in Section 4, we need some method for checking the validity of the shares. The shareholders can check their shares by choosing random shares within the list and computing

$$y^k = \prod_{i \in \pi} y^{c_{i,\pi} U_i} \bmod p, \tag{15}$$

a function of their shares of k, and

$$g^{xk} = \prod_{i \in \pi} g^{c_{i,\pi} V_i} \bmod p, \tag{16}$$

a function of their shares of kx. Since $y = g^x \bmod p$, if equation (15) equals equation (16), then the shares are legitimate.

5.3 Security of t-out-of-l Scheme

The security of this scheme depends critically on preventing shareholders from signing more than one message with the same k. If the protocol for determining the current k value is not secure, then two messages may be signed using the same k, revealing the private key. While the reliance on lists is clearly a disadvantage of this system, the t-out-of-l scheme does have the nice property that the partial signatures are zero-knowledge. Therefore, we can prove the following theorem.

Theorem 1. *If the shareholders never partially sign different messages with the same k value, then breaking the t-out-of-l signature scheme is equivalent to breaking DSS.*

To prove this theorem, we need to show that an external attacker, Eve, can simulate the threshold signature scheme given a standard DSS signature. Without loss of generality, we show that Eve can simulate a threshold signature scheme in which she knows the $t-1$ shares $f(1), \ldots, f(t-1)$. Eve generates $t-1$ random numbers, $U_1, \ldots, U_{t-1}$, which are shares of k. For every possible value of k, there is exactly one degree $t-1$ polynomial, $f(z)$, such that $f(i) = U_i$ and $f(0) = k$. Eve's "shares" are therefore perfectly legitimate shares of the secret k. Similarly, Eve generates $t-1$ random numbers, $V_1, \ldots, V_{t-1}$, which are shares of kx. Let $s_j = f(j)M + g(j)r \bmod q$, where f is the polynomial determined by Eve's choice of U_i and the value of k, and g is the polynomial determined by Eve's choice of V_i and the value of kx. Eve creates $t - 1$ partial signatures, $s_i = U_i M + V_i r \bmod q$, for $i = 1, \ldots, t - 1$. Eve does not know $f(j)$ or $g(j)$ for $t \leq j \leq l$, but she does know that

$$s = c_{j,\pi} s_j + \sum_{i=1}^{t-1} c_{i,\pi} s_i \bmod q. \tag{17}$$

Therefore, using s, she can solve for s_j, for any j of interest.

Eve can thus simulate a threshold signature scheme in which she knows $t-1$ of the shares and sees partial signatures created by all shareholders. Therefore, she can use any attack on the threshold scheme to attack standard DSS signatures.

6 $(t^2 - t + 1)$-out-of-l Scheme

This section presents an alternate way to generalize the 2-out-of-l signature scheme. This method avoids the precomputed lists of the previous section, but requires $t^2 - t + 1$ signers to participate in a signature which has t-out-of-l security. The additional signers are required because we are multiplying secret sharing polynomials. Let $h(z) = f(z) \times g(z)$, then $h(0) = f(0) \times g(0)$, and the degree of h is equal to the degree of f plus the degree of g. For the signature algorithm, t people generate shares of k, and multiply the secret sharing polynomials. While this method is clearly impractical for large t, it may be usable for small values. For example, a scheme with 2-out-of-l security requires only three signers.

For this algorithm, we will place the same requirements on q as those of Section 4 and create the private key as described in that section. An individual signature is created by the following steps:

1. Identify t of the signers to act as dealers for k. Let π represent this group.
2. Each dealer, i, generates k_i and $k_i^{-1} \bmod q$.
3. The dealers calculate $r = (g^{k_1^{-1} k_2^{-1} \cdots k_t^{-1}} \bmod p) \bmod q$.
4. Each dealer generates and distributes shares of the following polynomials mod q:
   $\kappa_i(z)$, degree $t - 1$ polynomial, $\kappa_i(0) = k_i$.
   $\lambda_i(z)$, degree $t - 1$ polynomial, $\lambda_i(0) = k_i x_i$, where $x_i = \beta^{c_{i,\pi} f(0)}$.
   $\gamma_i(z)$, degree $t^2 - t$ polynomial, $\gamma_i(0) = 0$.
   $\delta_i(z)$, degree $t^2 - t$ polynomial, $\delta_i(0) = 0$.
5. Signer j creates the shares

$$U_j = [\prod_{i \in \pi} \kappa_i(j)] + [\sum_{i \in \pi} \gamma_i(j)] \bmod q, \tag{18}$$

and

$$V_j = [\prod_{i \in \pi} \lambda_i(j)] + [\sum_{i \in \pi} \delta_i(j)] \bmod q. \tag{19}$$

6. Signer j creates the partial signature

$$s_j = U_j M + V_j r, \qquad (20)$$

and sends the result to the combiner. (Note: the combiner is not trusted as in Section 4.)

7. After receiving $t^2 - t + 1$ partial signatures, the combiner calculates

$$s = \sum_{j \in \rho} c_{j,\rho} s_j \bmod q, \qquad (21)$$

where ρ is the set of $t^2 - t + 1$ signers and the $c_{j,\rho}$'s are the interpolation constants for a $t^2 - t$ degree polynomial.

6.1 Security of the $(t^2 - t + 1)$-out-of-l Scheme

Like the algorithm of Section 5, the partial signatures of this algorithm are zero-knowledge. Therefore, we can prove the following theorem.

Theorem 2. *If $t - 1$ or fewer of the shareholders collude, then breaking the $(t^2 - t + 1)$-out-of-l signature scheme is no more than twice as difficult as breaking a standard DSS signature with the same p, q, and g parameters.*

Again, to prove this theorem, we need to show that an external attacker, Eve, can simulate the threshold signature scheme given a standard DSS signature. Recall that the method for creating the private key, x, caused x to be a quadratic residue mod q. Since x may not be a quadratic residue for a standard DSS signature, Eve first conducts her simulation using the signature (r, s), then if the attack fails she repeats the attack using the value $(r, \alpha s)$. This value corresponds to a signature with message αM and private key αx. Let α be a primitive element mod q, so either x or αx must be a quadratic residue, and one of Eve's simulations should succeed.

Now Eve can simulate the knowledge of $t-1$ signers of the threshold scheme.[3] For $1 \leq j < t$ and $1 \in \pi$, Eve generates $\kappa_i(j)$, $\lambda_i(j)$, $\gamma_i(j)$, and $\delta_i(j)$ randomly. For every value of k_i and $k_i x_i$, and any $t - 1$ values for $\kappa_i(j)$ and $\lambda_i(j)$, there is exactly one polynomial κ_i such that $\kappa_i(0) = k_i$ and one polynomial λ_i such that $\lambda_i(0) = k_i x_i$. Eve then computes U_j and V_j according to equations (18) and (19) and calculates the partial signatures s_j.

Now she generates $t^2 - 2t + 1$ additional independent, random s_j, where $j = t, \ldots, t^2 - t$. We need to show that these are valid partial signatures for those $t^2 - 2t + 1$ signers. Let $\mathcal{X}(z)$ be a $t^2 - t$ degree polynomial with $\mathcal{X}(0) = 0$ and $\mathcal{X}(j) = \sum_{i=1}^{t} \gamma_i(j)$. Similarly, let $\mathcal{Y}(z)$ be a $t^2 - t$ degree polynomial with $\mathcal{Y}(0) = 0$ and $\mathcal{Y}(j) = \sum_{i=1}^{t} \delta_i(j)$. Rewriting the formula for the partial signature, equation (20), we have

$$s_j = ([\prod_{i \in \pi} \kappa_i(j)] + \mathcal{X}(j))M + ([\prod_{i \in \pi} \lambda_i(j)] + \mathcal{Y}(j))r. \qquad (22)$$

---

[3] The proof that Eve can simulate $t - 1$ dealers is a straightforward extension of this, but is more complicated to follow.

The values of $\kappa_i(j)$ and $\lambda_i(j)$ were fixed for all j by the random numbers Eve generated. The polynomials $\mathcal{X}(z)$ and $\mathcal{Y}(z)$ each have degree $t^2 - t$, and $t^2 - t$ unknown terms. (The zeroth order term is fixed at zero.) For arbitrary s_j, $j = t, \ldots, t^2 - t$, there are $t^2 - 2t + 1$ linear equations in these unknowns created by equation (22). There are an additional $2(t - 1)$ equations from our above definitions of $\mathcal{X}(z)$ and $\mathcal{Y}(z)$. Therefore, there are a total of $t^2 - 1$ linear equations in $2t^2 - 2t$ unknowns. Since $2t^2 - 2t > t^2 - 1$ for all $t > 1$, the number of unknowns is greater than the number of equations, and there must be a solution for the system. Therefore, the random s_j's are legitimate partial signatures.

Eve now has $t^2 - t$ valid partial signatures. She can use the equation

$$s = c_{i,\rho} s_i + \sum_{j=1}^{t^2 - t} c_{j,\rho} s_j, \tag{23}$$

to find any other partial signature, s_i. Therefore, she can perfectly simulate the threshold signature scheme using a standard DSS signature.

7 Conclusion

This paper describes three threshold signature schemes for the Digital Signature Standard. The 2-out-of-l scheme does not require a trusted party to create the shares or to recreate lost shares. The more general t-out-of-l scheme is less practical, requiring precomputation and storage of shares for each individual signature. The $(t^2 - t + 1)$-out-of-l scheme does not require precomputed lists, but approximately t^2 signers are required for t-out-of-l security. Both generalized schemes have provable levels of security.

Acknowledgements

The author would like to thank Jimmy Upton for suggesting this problem and Don Coppersmith and Rajeev Motwani for their helpful comments.

References

1. "A proposed federal information processing standard for digital signature standard (DSS)," *Federal Register*, August 1991.
2. B. Blakley, G. Blakley, A. Chan, and J. Massey. "Threshold schemes with disenrollment. *Advances in Cryptology–Crypto '92 proceedings*, Springer-Verlag, 1993. p. 540-8.
3. C. Charnes, J. Pieprzyk, and R. Safavi-Naini, "Conditionally secure secret sharing schemes with disenrollment capability," 2nd ACM conference on Computer and Communications Security, Fairfax, Virginia, Nov. 2-4, 1994, pp. 89-95.
4. D. Coppersmith, personal communication.
5. Y. Desmedt, "Society and group oriented cryptography," *Advances in Cryptology–Crypto '87 proceedings*, Springer-Verlag, 1988, pp. 120-127.

6. Y. Desmedt, "Threshold Cryptography," *European Transactions on Telecommunications and Related Technologies*, Vol. 5, No. 4, July-August 1994, pp. 35-43.

7. Y. Desmedt and Y. Frankel, "Shared generation of authenticators and signatures," *Advances in Cryptology–Crypto '91 proceedings*, Springer-Verlag, 1992, pp. 457-269.

8. T. ElGamal, "A public key cryptosystem and a signature scheme based on discrete logarithms," *IEEE Trans. Inform. Theory*, Vol. 31, 1985, pp. 521-539.

9. Y. Frankel and Y. Desmedt, "Parallel reliable threshold multisignature," *Tech. Report TR-92-04-02*, Dep. of EE & CS, Univ. of Wisconsin-Milwaukee, April 1992.

10. L. Harn, "Group-oriented (t, n) threshold digital signature scheme and digital multisignature," *IEE Proc.-Comput. Digit. Tech.*, Vol. 141, No. 5, September 1994, pp. 307-313.

11. C. Laih and L. Harn, "Generalized threshold cyrptosystems," *Advances in Cryptology – Asiacrypt '91 proceedings*, Springer-Verlag, 1993, pp. 159-166.

12. S. Langford, *Differential-Linear Cryptanalysis and Threshold Signatures*, Ph.D. Thesis, Stanford University, June 1995.

13. C. Li, T. Hwang, N. Lee, "(t, n)-threshold signature scheme based on discrete logarithm," *Advances in Cryptology – Eurocrypt '94 proceedings*, to appear.

14. R. Rivest, A. Shamir, L. Adleman, "A method for obtaining digital signatures and public key cryptosystems," *Communications of the ACM*, Vol. 21, April 1978, pp. 294-299.

15. A. Shamir, "How to share a secret," *Commun. ACM*, 22:612-613, November 1979.

t-Cheater Identifiable (k, n) Threshold Secret Sharing Schemes

Kaoru KUROSAWA[1]   Satoshi OBANA[1]   *and*   Wakaha OGATA[2]

[1] Department of Electrical and Electronic Engineering,
Faculty of Engineering, Tokyo Institute of Technology
2-12-1 O-okayama, Meguro-ku, Tokyo 152, Japan
E-mail kkurosaw@ss.titech.ac.jp
[2] Department of Computer Engineering,
Faculty of Engineering, Himeji Institute of Technology
2167 shosha, Himeji-shi, Hyougo 671-22, Japan

Abstract. In this paper, we show that there exists a t-cheater identifiable (k, n) threshold secret sharing scheme such as follows for cheating probability $\varepsilon > 0$. If $k \geq 3t + 1$, then
1. Just k participants are enough to identify who are cheaters.
2. $|V_i|$ is independent of n. That is, $|V_i| = |S|(1/\varepsilon)^{(t+2)}$, where S denotes the set of secrets and V_i denotes the set of shares of a participant P_i, respectively.

(Previously, no schemes were known which satisfy both requirements.)
Further, we present a lower bound on $|V_i|$ for our model and for the model of Tompa and Woll. Our bound for the TW model is much more tight than the previous bound.

1 Introduction

In a (k, n) threshold secret sharing scheme [1, 2], a secret s is distributed by the dealer to n participants, $P_1, \ldots, P_n$, in such a way that k or more participants can recover s and $k - 1$ or less participants have no information on s. A piece of information held by P_i is called a share and it is denoted by v_i.

Various researches considered the problem of cheaters in threshold schemes. Some participants may attempt to cheat, that is, to deceive other participants by lying about shares they hold. A threshold scheme is said to be unconditionally secure against cheating if the probability of successful cheating is limited to a specified probability even if the cheaters are infinitely powerful. Assume that the dealer is honest. Then, several constructions have been given such as follows. (If the cheaters are polynomially time bounded, this problem is easily solved by using a digital signature scheme.)

McEliece and Sarwate [3] showed that Shamir's scheme itself has a cheater detection capability. Any set of $k + 2t$ participants containing at most t cheaters can detect who are cheating. They can reveal the correct secret, as well. This scheme, however, requires more than k participants to detect who are cheaters. A secret sharing scheme of [4] also requires more than k participants to detect cheaters.

D. Coppersmith (Ed.): Advances in Cryptology - CRYPTO '95, LNCS 963, pp. 410-423, 1995.

T.Rabin and Ben-Or [5] showed a scheme such that each participant can always detect who are cheating with high probability. Therefore, any set of participants containing at least k honest participants can reveal the correct secret with high probability. In this scheme, however, $|V_i|$ is very large, where V_i denotes the set of shares v_i. That is, $|V_i| = |S|^{(3n-2)}$, where S denotes the set of secrets.

At the same time, Brickell and Stinson [6] showed such a nonperfect scheme in which $|V_i| = |S|^{(n+2t-3)}$. In this scheme, $k - 1$ participants can have a small amount of information on the secret. In this scheme, $|V_i|$ is also an exponential function on n.

On the other hand, Tompa and Woll [7] showed a scheme such that an honest participant can detect only the fact of cheating with high probability. Honest participants, however, cannot detect who are cheating nor reveal the correct secret. Carpentieri, De Santis and Vaccaor [8] showed a lower bound on $|V_i|$ for this model.

Here, we note that cheater identifiable schemes proposed so far are either

1. $k + 1$ or more participants are necessary to detect who are cheaters, or
2. $|V_i|$ is an exponential function on n.

Further, no lower bound on $|V_i|$ is known.

In this paper, we consider a model in which there are at most t cheaters. After formulation, we show that there exists a t-cheater identifiable (k, n) threshold scheme such as follows for cheating probability $\varepsilon > 0$. If $k \geq 3t + 1$, then

1. Just k participants are enough to identify who are cheating.
2. $|V_i|$ is independent of n. That is, $|V_i| = |S|(1/\varepsilon)^{(t+2)}$.

The proposed scheme uses an orthogonal array $OA(t+1, n|S|, \frac{1}{\varepsilon})$. It is interesting to compare with a t error correcting BCH codes in which a generator polynomial $G(x)$ has $2t$ consecutive zeros. Further, we present a lower bound on $|V_i|$ for our model and for the model of Tompa and Woll [7]. Our bound for the TW model is much more tight than the previous bound [8].

Our scheme and bound are closely related to unconditionally secure authentication codes [12]~[16]. Especially, our bound on $|V_i|$ is derived from a bound for splitting authentication codes shown in [16].

In section 3, we give a formulation of our problem. The proposed scheme is shown in section 4. A lower bound on $|V_i|$ for our model is presented in section 5. A lower bound on $|V_i|$ for the TW model is presented in section 6.

2 Preliminaries

2.1 (k, n) threshold scheme

In a secret sharing scheme, a dealer D randomly produces $(v_1, \ldots, v_n)$ on input s, where s is a secret and v_i is called a share of the secret s. v_i is given to a participant P_i, where $i = 1, 2, \cdots, n$. Let S be the random variable induced by s and V_i be the random variable induced by v_i, respectively. We also use S to denote the set of s and V_i to denote the set of v_i, respectively. In a (k, n) threshold scheme [1, 2], any k or more participants can recover s but no subset of less than k participants can determine any partial information on s.

Shamir's (k, n) threshold scheme [1] has error correcting capability such as follows [3].

Proposition 1. *[3] Let $i_1, \ldots, i_m$ be fixed distinct values of $GF(p)$. Let*

$$C \triangleq \{(f(i_1), \ldots, f(i_m)) | f(x) \text{ is a degree } t \text{ polynomial over } GF(p)\}$$

where $m \geq t$. Then, C is a linear code with the minimum Hamming distance $m - t$.

Proof. It is clear that C is a linear code. Since the degree of $f(x)$ is t, the number of zeros of $f(x)$ is at most t. Therefore, the Hamming weight of $(f(i_1), \ldots, f(i_m))$ is greater than or equal to $m - t$. Further, there exists a $f(x)$ which has t zeros. Hence, the minimum Hamming weight of C is $m - t$. In a linear code, the minimum Hamming distance is equal to the minimum Hamming weight. $\square$

[9, 10] showed that $\log_2 |V_i| \geq H(S)$ for (k, n) threshold schemes. This bound was improved by Kurosawa and Okada as follows [11]

Proposition 2. *[11] In a (k, n) threshold scheme, $|V_i| \geq |S|$ for any probability distribution on S.*

2.2 Authentication code

In the model of unconditionally secure authentication codes, there are three participants, a transmitter T, a receiver R and an opponent O. T and R share a common encoding rule e. On input a source state u, T sends a message $m = e(u)$ to R. R accepts or rejects m based on e. O tries to cheat R by an impersonation attack or a substitution attack. We assume independent probability distributions on source states and on encoding rules. In the impersonation attack, O sends m to R before T sends. O succeeds if R accepts m. This cheating probability P_I is defined by

$$P_I = \max_m \Pr[\text{R accepts } m]$$

In the substitution attack, O observes a message m transmitted by T and substitutes it with another message $\hat{m}$. This cheating probability P_S is defined by

$$P_S = \sum_m \Pr(m) \max_{\hat{m}} \Pr[\text{R accepts } \hat{m} | \text{O observed } m] \tag{2.1}$$

where the maximum is taken over $\hat{m}$ such that the source state of $\hat{m}$ is different from that of m.

Definition 3. An authentication code is called no splitting if $|\{m \mid e(u) = m\}| = 1$ for $\forall u$ and $\forall e$. Otherwise, it is called splitting.

Let U be the set of source states and *Message* be the set of messages respectively. Further, let

$$Message(e, u) \stackrel{\triangle}{=} \{m \mid e(u) = m\}, \qquad Message(e) \stackrel{\triangle}{=} \bigcup_u Message(e, u).$$

Proposition 4. *In an authentication code,*

(i) [14, 15] if it is no splitting,

$$P_S \geq (|U| - 1)/(|Message| - 1)$$

(ii) [16] if it is splitting,

$$P_S \geq \min_e \frac{|Message(e)| - \max_{u \in U} |Message(e, u)|}{|Message| - \min_{u \in U} |Message(e, u)|}$$

3　Formulation

In this section, we formulate our model of t-cheater identifiable (k, n) threshold scheme. In section 3~5, we assume the following assumption.

Assumption 5. *The dealer is honest. There are at most t cheaters in n participants. (Cheaters may collude.)*

Informally, our model is defined as follows.

(T1) Completeness Any set of participants containing at least k honest participants can reveal the original secret s with high probability.

(T2) Soundness No subset of less than k participant can determine any partial information on the secret s.

(T3) Detectability There exists a Turing machine M which detects who are cheating with high probability if k or more participants open their shares.

First, we define (k, n) threshold schemes with cheaters. In what follows, let $A = \{i_1, \ldots, i_m\}$.

Definition 6. $(d_1, \ldots, d_m) \in V_{i_1} \times \cdots \times V_{i_m}$ is honest on A if

$$\Pr[V_{i_1} = d_1, \ldots, V_{i_m} = d_m] > 0$$

Definition 7. A (k, n) threshold scheme is a secret sharing scheme such as follows.

(i) If $m \geq k$, for any honest $(d_1, \ldots, d_m)$,

$$\exists s, \Pr[S = s | V_{i_1} = d_1, \ldots, V_{i_m} = d_m] = 1 \tag{3.1}$$

(ii) If $m < k$, for any honest $(d_1, \ldots, d_m)$,

$$\forall s, \Pr[S = s | V_{i_1} = d_1, \ldots, V_{i_m} = d_m] = \Pr[S = s]$$

Definition 8. If $m \geq k$ and $(d_1, \ldots, d_m)$ is honest, s is uniquely determined from eq.(3.1). Denote such s by $Secret(d_1, \ldots, d_m) = s$.

Next, we divide $V_{i_1} \times \cdots \times V_{i_n}$ into three subsets.

$$Honest(A) \triangleq \{(d_1, \ldots, d_m) \mid (d_1, \ldots, d_m) \text{ is honest on } A\}$$

Let M be a deterministic Turing machine. For M, define

$$Dishonest_M(A) \triangleq \{(d_1, \ldots, d_m) \mid (d_1, \ldots, d_m) \notin Honest(A),$$
$$M \text{ detects who are cheaters from } (d_1, \ldots, d_m) \text{ correctly}\}$$

$$Semihonest_M(A) \triangleq \{(d_1, \ldots, d_m) \mid (d_1, \ldots, d_m) \notin Honest(A),$$
$$(d_1, \ldots, d_m) \notin Dishonest_M(A)\}$$

Suppose that $P_{i_1}, \ldots, P_{i_t}$ are cheaters and they open $\hat{d}_1, \ldots, \hat{d}_t$ while the dealer distributed $(d_1, \ldots, d_m)$ to A. If $(\hat{d}_1, \ldots, \hat{d}_t, d_{t+1}, \ldots, d_m) \in Dishonest_M(A)$, M can detect who are cheaters. Successful cheating occurs if case 1 or case 2 below occurs.

(case 1) $(\hat{d}_1, \ldots, \hat{d}_t, d_{t+1}, \ldots, d_m) \in Semihonest_M(A)$. In this case, M can detect only the fact of cheating.

(case 2) $(\hat{d}_1, \ldots, \hat{d}_t, d_{t+1}, \ldots, d_m) \in Honest(A)$ and
$Secret(\hat{d}_1, \ldots, \hat{d}_t, d_{t+1}, \ldots, d_m) \neq Secret(d_1, \ldots, d_m)$ $(=$ original secret$)$.
In this case, cheaters succeed completely.

If case 1 or case 2 occurs, M cannot identify who are cheaters. This probability $Cheat_M(A)$ is formulated as follows. Let

$$Fool_{(M,A)}(\hat{d}_1, \ldots, \hat{d}_t \mid d_1, \ldots, d_t)$$
$$\triangleq \{(d_{t+1}, \ldots, d_m) \mid (d_1, \ldots, d_t, d_{t+1}, \ldots, d_m) \in Honest(A),$$
$$\text{case 1 or 2 occurs for } (\hat{d}_1, \ldots, \hat{d}_t, d_{t+1}, \ldots, d_m)\}$$

$$Cheat_M(A \mid i_1, \ldots, i_t)$$
$$\triangleq \sum_{(d_1, \ldots, d_t)} \Pr[d_1, \ldots, d_t]$$
$$\times \max_{(\hat{d}_1, \ldots, \hat{d}_t)} \sum_{Fool_{(M,A)}(\hat{d}_1, \ldots, \hat{d}_t \mid d_1, \ldots, d_t)} \Pr[d_{t+1}, \ldots, d_m \mid d_1, \ldots, d_t]$$

Definition 9.

$$Cheat_M(A) \triangleq \max\{Cheat_M(A \mid i_1), Cheat_M(A \mid i_1, i_2), \ldots, Cheat_M(A \mid i_1, \ldots, i_t)\}$$

Definition 10. We say that a (k, n) threshold scheme is a (t, ε) cheater identifiable (k, n) threshold scheme if there exists a deterministic Turing machine M such as follows.

$Cheat_M(A) \leq \varepsilon$ for $\forall A$ such that $|A| \geq k$.

4 Proposed scheme

In this section, we show a (t, ε) cheater identifiable (k, n) threshold scheme for $k \geq 3t + 1$ such that $|V_i|$ is independent of n. To obtain our scheme, we use an orthogonal array of strength $t + 1$ as an unconditionally secure authentication code and combine it with Shamir's (k, n) threshold scheme and the linear code of proposition 1. Each share v_i of the proposed scheme has a form of $(\alpha_i, \beta_i, \gamma_i)$. α_i is a share of the secret generated by Shamir's (k, n) threshold scheme. β_i is an authenticator for α_i of our authentication code. The key of the authentication code is encoded as a codeword $(\gamma_1, \ldots, \gamma_n)$ by the code of proposition 1. In the reconstruction phase, the key is reconstructed first. For any $i_1, \ldots, i_m$ such that $m \geq k$, $(\gamma_{i_1}, \ldots, \gamma_{i_m})$ has t-error correcting capability. Therefore, cheaters cannot forge the key. Once the key is reconstructed, a forged $(\hat{\alpha}_i, \hat{\beta}_i)$ is detected with high probability by the property of the authentication code. Thus, our scheme is (t, ε) cheater identifiable.

Definition 11. An orthogonal array $OA(t + 1, np, q)$ is a $q^{t+1} \times np$ array of q symbols such that, in any $t + 1$ columns of the array, every one of the possible q^{t+1} ordered tuples of symbols occurs in exactly one row.

In what follows, let $S = \{0, 1, \ldots, p - 1\}$, where p is a prime power.
[Proposed Scheme]
Dealer D produces a share $v_i = (\alpha_i, \beta_i, \gamma_i)$ such as follows for $i = 1, 2, \ldots, n$.

(C1) As in Shamir's scheme, D chooses a $(k - 1)$-th order random polynomial over $GF(p)$ such that,

$$f(x) = s + a_1 x + \cdots + a_{k-1} x^{k-1}$$

Let $\alpha_i = f(i)$ for $i = 1, 2, \ldots, n$.
(C2) Let $OA(t + 1, np, q)$ be an orthogonal array such that q is a prime power. D chooses a random number e such that $1 \leq e \leq q^{t+1}$. Let β_i be the e-th row and the
$(i - 1)p + \alpha_i$ th column element of $OA(t + 1, np, q)$.
(C3) D chooses a t-th order random polynomial over $GF(q^{t+1})$ such that

$$g(x) = e + b_1 x + \cdots + b_t x^t$$

Let $\gamma_i = g(i)$ for $i = 1, 2, \ldots, n$.

Remark. We assume that $OA(t + 1, np, q)$ is publicly known.

Theorem 12. *The above scheme is a (t, ε) cheater identifiable (k, n) threshold scheme if $k \geq 3t + 1$, where $\varepsilon = 1/q$.*

Proof. (1) Def.7 is satisfied because α_i is a share of Shamir's (k, n) threshold scheme.

(2) Suppose that $A = \{i_1, \ldots, i_m\}$ and $m \geq k$. Let $d_j = (\alpha_{i_j}, \beta_{i_j}, \gamma_{i_j})$ for $j = 1, 2, \ldots, m$.

Clearly,

$$Honest(A) = \{(d_1, \ldots, d_m) \mid (d_1, \ldots, d_m) \text{ satisfies (C1)} \sim \text{(C3).}\}$$

We show that there exists a deterministic Turing machine M such that

$$Dishonest_M(A) = \{(d_1, \ldots, d_m) \mid (d_1, \ldots, d_m) \text{ doesn't satisfy (C2) or (C3).}\} \tag{4.1}$$

Suppose that $(d_1, \ldots, d_m)$ doesn't satisfy (C3). That is,

$$(\gamma_{i1}, \ldots, \gamma_{im}) \neq (g(i_1), \ldots, g(i_m)),$$

where $g(x)$ is a degree t polynomial chosen by D. From Proposition 1, $\{(g(i_1), \ldots, g(i_m))\}$ is a linear code with the Hamming distance $d = m - t$. In our case,

$$m \geq k \geq 3t + 1$$

Hence

$$d \geq 3t + 1 - t = 2t + 1$$

Therefore, there is a deterministic algorithm which can identify t errors in $(\gamma_{i1}, \ldots, \gamma_{im})$.

Also, there is a deterministic algorithm which can recover $g(x)$. Then, e is reconstructed. Now, we see that there is a deterministic algorithm which finds e in any case. Once e is found, it is easy to detect which β_i violates (C2).

Thus, there exists a deterministic Turing machine which detects who are cheaters if $(d_1, \ldots, d_m)$ doesn't satisfy (C2) or (C3). Let M be a deterministic Turing machine which detects who are cheaters if $(d_1, \ldots, d_m)$ doesn't satisfy (C2) or (C3). Then, eq.(4.1) holds.

(3) Finally, we prove that $Cheat_M(A) \leq 1/q$ for the above M. Suppose that $P_{i_1}, \ldots, P_{i_t}$ are cheaters and they open $\hat{d}_1, \ldots, \hat{d}_t$ while the dealer distributed $(d_1, \ldots, d_m)$ to A. Let

$$F_{(M,A)}(\hat{d}_1, \ldots, \hat{d}_t \mid d_1, \ldots, d_t)$$
$$\stackrel{\triangle}{=} \{(d_{t+1}, \ldots, d_m) \mid (d_1, \ldots, d_m) \in Honest(A),$$
$$(\hat{d}_1, \ldots, \hat{d}_t, d_{t+1}, \ldots, d_m) \notin Dishonest_M(A)\}$$

Then, it is easy to see that

$$F_{(M,A)}(\hat{d}_1, \ldots, \hat{d}_t \mid d_1, \ldots, d_t) \supseteq Fool_{(M,A)}(\hat{d}_1, \ldots, \hat{d}_t \mid d_1, \ldots, d_t)$$

Let

$$F1_A(d_1, \ldots, d_t) \stackrel{\triangle}{=} \{(d_{t+1}, \ldots, d_m) \mid (d_1, \ldots, d_m) \in Honest(A)\}$$
$$F2_A(\hat{d}_1, \ldots, \hat{d}_t) \stackrel{\triangle}{=} \{(d_{t+1}, \ldots, d_m) \mid (\hat{d}_1, \ldots, \hat{d}_t, d_{t+1}, \ldots, d_m) \notin Dishonest_M(A)\}$$

Then

$$F_{(M,A)}(\dot{d}_1,\ldots,\hat{d}_t|d_1,\ldots,d_t) = F1_A(d_1,\ldots,d_t) \cap F2_A(\hat{d}_1,\ldots,\hat{d}_t)$$

Let's compute $|F1_A(d_1,\ldots,d_t)|$. Fix an honest $d_1,\ldots,d_t$ arbitrarily. From the definition of $OA(t+1,np,q)$ and since $\deg g(x) = t$, there are q rows which matches $(d_1,\ldots,d_t)$. That is,

$$|\{e \mid \beta_{i_j} = \text{ the } (e,(i_j-1)p+\alpha_{i_j}) \text{ element of } OA(t+1,np,q), 1 \le j \le t\}| = q$$

For each e, $g(x)$ is uniquely determined. Then, γ_{i_j} is uniquely determined for $t+1 \le j \le m$. On the other hand, there are p^{k-t} possible $f(x)$ which matches $(d_1,\ldots,d_t)$. For each $f(x)$ and e, $(\alpha_{i_j},\beta_{i_j})$ is uniquely determined for $t+1 \le j \le m$. Therefore, $|F1_A(d_1,\ldots,d_t)| = qp^{k-t}$.
Similarly, we see that

$$|F_{(M,A)}(\hat{d}_1,\ldots,\hat{d}_t|d_1,\ldots,d_t)| = 0 \text{ or } p^{k-t}$$

Then,

$$\max_{(\hat{d}_1,\ldots,\hat{d}_t)} \sum_{Fool_{(M,A)}(\hat{d}_1,\ldots,\hat{d}_t|d_1,\ldots,d_t)} \Pr[d_{t+1},\ldots,d_m|d_1,\ldots,d_t]$$

$$\le \max_{(\hat{d}_1,\ldots,\hat{d}_t)} \sum_{F_{(M,A)}(\hat{d}_1,\ldots,\hat{d}_t|d_1,\ldots,d_t)} \Pr[d_{t+1},\ldots,d_m|d_1,\ldots,d_t]$$

$$= \max_{(\hat{d}_1,\ldots,\hat{d}_t)} \frac{|F_{(M,A)}(\hat{d}_1,\ldots,\hat{d}_t|d_1,\ldots,d_t)|}{|F1_A(d_1,\ldots,d_t)|}$$

$$= \frac{p^{k-t}}{qp^{k-t}} = \frac{1}{q}$$

Therefore,

$$Cheat_M(A|i_1,\ldots,i_t) \le \sum_{(d_1,\ldots,d_t)} \Pr[d_1,\ldots,d_t] \times (1/q) = 1/q$$

Similarly,

$$Cheat_M(A|i_1,\ldots,i_j) \le 1/q, \quad 1 \le j \le t-1$$

Therefore,

$$Cheat_M(A) \le 1/q$$

$\square$

In this scheme,

$$\log_2 |V_i| = \log_2 |\alpha_i| + \log_2 |\beta_i| + \log_2 |\gamma_i|$$
$$= \log_2 |p| + \log_2 |q| + \log_2 |q^{t+1}|$$
$$= \log_2 |S| + (t+2)\log_2(1/\varepsilon)$$

Equivalently,

$$|V_i| = |S|(1/\varepsilon)^{t+2}$$

Thus, $|V_i|$ is independent of n.

5 Lower bound on $|V_i|$

From proposition 2, $|V_i| \geq |S|$ in any (k, n) threshold scheme. In general, a perfect secret sharing scheme is called ideal if $|V_i| = |S|$ for $\forall i$. First, we show a refinement of this bound and the concept. Let $A = \{i_1, \ldots, i_k\}$ ($|A| = k$). Let d_j be the share of P_{i_j}, where $j = 1, \ldots, k$.

Theorem 13. *In a* (k, n) *threshold scheme,*

$$|\{d_1 \mid \Pr(s, d_1, d_2, \ldots, d_k) > 0\}| \geq 1$$

for any secret s and any honest $(d_2, \ldots, d_k)$.

Proof. From Def.7 (ii), for any honest $(d_2, \ldots, d_k)$, s can take any value of S. From Def.7 (i), each s of S must be determined by some d_1 together with this $(d_2, \ldots, d_k)$. This means that Theorem 13 holds. $\square$

Definition 14. A (k, n) threshold scheme is c-compact on (A, i_1) if

$$|\{d_1 \mid \Pr(s, d_1, d_2, \ldots, d_k) > 0\}| = c$$

for any secret s and any honest $(d_2, \ldots, d_k)$.

Next, for simplicity, suppose that only P_{i_1} is a cheater. Let $Cheat(A|i_1)$ be the cheating probability that the case 2 occurs for $(\hat{d}_1, d_2, \ldots, d_k)$ for a forged $\hat{d}_1$. Formally,

$$Fool_A(\hat{d}_1|d_1) \triangleq \{(d_2, \ldots, d_k) \mid (d_1, d_2, \ldots, d_k) \in Honest(A),$$
$$\text{case 2 occurs for } (\hat{d}_1, d_2, \ldots, d_k)\}$$
$$Cheat(A|i_1) \triangleq \sum_{d_1} \Pr[d_1] \times \max_{\hat{d}_1} \sum_{Fool_A(\hat{d}_1|d_1)} \Pr[d_2, \ldots, d_k|d_1]$$

Note that the subscript M is dropped in the above definitions because case 2 is independent of M.

Theorem 15. *Suppose that a* (k, n) *threshold scheme is c-compact on* (A, i_1) *for some* $A \ni i_1$. *If* $Cheat(A|i_1) \leq \varepsilon$,

$$|V_{i_1}| \geq c \left(\frac{|S| - 1}{\varepsilon} + 1 \right) \tag{5.1}$$

Proof. Consider a splitting authentication code such as follows (see subsection 2.2). The receiver R has an encoding rule $e = (d_2, \ldots, d_k)$ such that $(d_2, \ldots, d_k)$ is honest. R accepts a message $m = d_1$ if $(d_1, d_2, \ldots, d_k)$ is honest. The source state u conveyed by d_1 is a secret s such that $s = Secret(d_1, d_2, \ldots, d_k)$. (see Def.8.) P_{i_1} is the opponent. He observes d_1 and substitutes d_1 with another $\hat{d}_1$ (substitution attack). Then

$$Cheat(A|i_1) = \sum_{d_1} \Pr[d_1] \max_{\hat{d_1}} \sum_{Fool_A(\hat{d_1}|d_1)} \Pr[d_2, \ldots, d_k|d_1]$$

$$= \sum_{d_1} \Pr[d_1] \max_{\hat{d_1}} \Pr[\text{R accepts } \hat{d_1}|P_{i_1} \text{ observed } d_1] \quad (5.2)$$

where the maximum is taken over $\hat{d_1}$ such that the source state (secret) determined by $\hat{d_1}$ is different from that of d_1. Now, we see that $Cheat(A|i_1)$ is equal to the substitution attack probability P_S (compare eq.(5.2) with eq.(2.1)). In this authentication code,

$$|Message| = |\{d_1\}| = |V_{i_1}|$$
$$|Message((d_2, \ldots, d_k), s)| = |\{d_1 \mid \Pr(s, d_1, d_2, \ldots, d_k) > 0\}| = c$$
$$|Message((d_2, \ldots, d_k))| = |\bigcup_s Message((d_2, \ldots, d_k), s)| = c|S|$$

Then from proposition 4 (ii),

$$\varepsilon \geq Cheat(A|i_1) = P_S \geq \min_{(d_2, \ldots, d_k)} \frac{c|S| - c}{|V_{i_1}| - c} = \frac{c(|S| - 1)}{|V_{i_1}| - c}$$

Therefore, eq.(5.1) is obtained □

Finally, we show a lower bound on $|V_i|$ of our model.

Corollary 16. *If a (t, ε) cheater identifiable (k, n) threshold scheme is c-compact on (A, i_1) for some $A \ni i_1$,*

$$|V_{i_1}| \geq c\left(\frac{|S| - 1}{\varepsilon} + 1\right) \quad (5.3)$$

Proof. There exists a deterministic Turing machine M such that $Cheat_M(A|i_1) \leq Cheat_M(A) \leq \varepsilon$. Clearly, $Cheat(A|i_1) \leq Cheat_M(A|i_1)$. Therefore, $Cheat(A|i_1) \leq \varepsilon$. Then, from Theorem 15, we obtain this corollary. □

6 Lower bound for TW model

Tompa and Woll [7] showed a (k, n) threshold scheme such that an honest participant can detect only the fact of cheating with high probability. Honest participants, however, cannot detect who are cheating nor reveal the correct secret. Carpentieri, De Santis and Vaccaor [8] showed a lower bound on $|V_i|$ for this model.

In this section, we show a much more tight lower bound on $|V_i|$ for the model of Tompa and Woll. (We don't assume Assumption 5 in this section while we use the same notation as before.)

In the model of Tompa and Woll [7], the cheating probability, P_{TW}, is defined as the probability that from $k - 1$ forged shares $d'_1, \ldots, d'_{k-1}$ and any d_k, the secret s' reconstructed is legal but not a correct one. Formally, it should be defined as follows.

$$cheated \triangleq P_k \text{ has } d_k \text{ such that } (d'_1, \ldots, d'_{k-1}, d_k) \text{ is honest and}$$
$$Secret(d_1, \ldots, d_k) \neq Secret(d'_1, \ldots, d'_{k-1}, d_k).$$

$$P_{TW} \triangleq \sum_{d_1, \ldots, d_{k-1}} \Pr(d_1, \ldots, d_{k-1}) \max_{d'_1, \ldots, d'_{k-1}} \Pr(cheated | d_1, \ldots, d_{k-1})$$

For a technical reason, [8] defined the following probability.

$$P_{over} \triangleq \sum_{d_1, \ldots, d_{k-1}, s} \Pr(d_1, \ldots, d_{k-1}, s) \max_{d'_1, \ldots, d'_{k-1}} \Pr(cheated | d_1, \ldots, d_{k-1}, s)$$

Lemma 17. *For any function $f(x_1, x_2)$,*

$$\sum_{x_2} \max_{x_1} f(x_1, x_2) \geq \max_{x_1} \sum_{x_2} f(x_1, x_2)$$

Proof. For $\forall \hat{x}_1$,

$$\sum_{x_2} \max_{x_1} f(x_1, x_2) \geq \sum_{x_2} f(\hat{x}_1, x_2)$$

□

Lemma 18. $P_{over} \geq P_{TW}$

Proof. From Def.7 (ii), $\Pr(d_1, \ldots, d_{k-1}, s) = \Pr(d_1, \ldots, d_{k-1}) \Pr(s)$. Then,

$$P_{over} = \sum_{d_1, \ldots, d_{k-1}, s} \Pr(d_1, \ldots, d_{k-1}) \Pr(s) \max_{d'_1, \ldots, d'_{k-1}} \Pr(cheated | d_1, \ldots, d_{k-1}, s)$$

$$= \sum_{d_1, \ldots, d_{k-1}, s} \Pr(d_1, \ldots, d_{k-1})$$
$$\times \max_{d'_1, \ldots, d'_{k-1}} \Pr(S = s \text{ and } cheated | d_1, \ldots, d_{k-1})$$

$$= \sum_{d_1, \ldots, d_{k-1}} \Pr(d_1, \ldots, d_{k-1})$$
$$\times \sum_s \max_{d'_1, \ldots, d'_{k-1}} \Pr(S = s \text{ and } cheated | d_1, \ldots, d_{k-1})$$

From lemma 17,

$$P_{over} \geq \sum_{d_1, \ldots, d_{k-1}} \Pr(d_1, \ldots, d_{k-1})$$
$$\times \max_{d'_1, \ldots, d'_{k-1}} \sum_s \Pr(S = s \text{ and } cheated | d_1, \ldots, d_{k-1})$$
$$= P_{TW}$$

□

[7] showed a scheme such that $P_{over} \leq \varepsilon$ and

$$|V_i| \geq \{(|S|-1)(k-1)/\varepsilon + k\}^2$$

[8] showed that, if $P_{over} \leq \varepsilon$ and S is uniformly distributed, then

$$|V_i| \geq |S|/\varepsilon$$

From lemma 18, these results [7, 8] can be restated as follows.

1. There exists a (k, n) threshold scheme such that $P_{TW} \leq \varepsilon$ and

$$|V_i| \geq \{(|S|-1)(k-1)/\varepsilon + k\}^2 \tag{6.1}$$

2. If $P_{TW} \geq \varepsilon$ and S is uniformly distributed, then

$$|V_i| \geq |S|/\varepsilon \tag{6.2}$$

However, there is a big gap between eq.(6.1) and eq.(6.2). In what follows, we show a much more tight lower bound on $|V_i|$ than eq.(6.2).

Lemma 19. $P_{TW} \geq Cheat(A|i_1)$.

Proof.

P_{TW}
$$= \sum_{d_1} \Pr(d_1) \sum_{d_2,\ldots,d_{k-1}} \Pr(d_2,\ldots,d_{k-1}|d_1) \max_{d_1',\ldots,d_{k-1}'} \Pr(cheated|d_1,\ldots,d_{k-1})$$
$$= \sum_{d_1} \Pr(d_1) \sum_{d_2,\ldots,d_{k-1}} \max_{d_1',\ldots,d_{k-1}'} \Pr(cheated \text{ and } P_2,\ldots,P_k \text{ has } d_2,\ldots,d_{k-1}|d_1)$$

Let

$$B \triangleq P_2,\ldots,P_{k-1} \text{ has } d_2,\ldots,d_{k-1}.$$
$$C \triangleq P_k \text{ has } d_k \text{ such that } (d_1'd_2,\ldots,d_k) \text{ is honest}$$
$$\text{and } Secret(d_1,\ldots,d_k) \neq Secret(d_1',d_2,\ldots,d_k).$$

Then,

$$\max_{d_1',\ldots,d_{k-1}'} \Pr(cheated \text{ and } B|d_1) \geq \max_{d_1'} \Pr(C \text{ and } B|d_1)$$

Therefore, from lemma 17,

$$P_{TW} \geq \sum_{d_1} \Pr(d_1) \sum_{d_2,\ldots,d_{k-1}} \max_{d_1'} \Pr(C \text{ and } B|d_1)$$
$$\geq \sum_{d_1} \Pr(d_1) \max_{d_1'} \sum_{d_2,\ldots,d_{k-1}} \Pr(C \text{ and } B|d_1)$$
$$= \sum_{d_1} \Pr(d_1) \max_{d_1'} \sum_{Fool_A(d_1'|d_1)} \Pr[d_2,\ldots,d_k|d_1]$$
$$= Cheat(A|i_1)$$

□

Corollary 20. *Suppose that a (k, n) threshold scheme is c-compact on (A, i_1) for some $A \ni i_1$. If $P_{TW} \leq \varepsilon$, then*

$$|V_{i_1}| \geq c \left(\frac{|S| - 1}{\varepsilon} + 1 \right) \tag{6.3}$$

Proof. From lemma 19, if $P_{TW} \leq \varepsilon$, then $Cheat(A|i_1) \leq \varepsilon$. Then, from Theorem 15, we obtain this corollary. $\square$

Eq.(6.3) is more tight than eq.(6.2) if $c \geq 2$. The (k, n) threshold scheme of Tompa and Woll [7] is c-compact on $\forall (A, i)$ such that $c \geq (|S| - 1)/(k - 1)/\varepsilon$. Then, our bound becomes

$$|V_i| \geq \{(|S| - 1)/\varepsilon + 1\}(|S| - 1)(k - 1)/\varepsilon$$

This bound is much closer to eq.(6.1) than eq.(6.2).

Remark. In Theorem 15, Corollary 16 and Corollary 20, we can eliminate "c-compact" to obtain a more general bound on $|V_i|$, which is as general as Proposition 4 (ii). The details will be given in the final paper.

References

1. A.Shamir, How to share a secret, Comm.ACM, 22(1979), pp.612–613.
2. G.R.Blakley, Safeguarding cryptographic keys, Proc. National Computer Conference, AFIPS Conference Proceedings, 48(1979), pp.313–317.
3. R.J.McEliece and D.V.Sarwate, On sharing secrets and Reed-Solomon codes, Comm.ACM, 24(1981), pp.583–584.
4. G.Simmons, Robust shared secret schemes or "how to be sure you have the right answer even though you don't know the question," Congr.Numer., 68(1989), pp.215–248.
5. T.Rabin and M.Ben-Or, Verifiable secret sharing and multiparty protocols with honest majority, Proc. 21st ACM Symposium on Theory of Computing (1989), pp.73–85.
6. E.F.Brickell and D.R.Stinson, The Detection of Cheaters in Threshold Schemes, SIAM J. DISC. MATH, Vol.4, No.4, Nov.1991, pp.502–510.
7. M.Tompa and H.Woll, How to share a secret with cheaters, Journal of Cryptology, vol.1(1988), pp.133–138.
8. Marco Carpentieri, Alfredo De Santis and Ugo Vaccaro, Size of Shares and Probability of Cheating in Threshold Schemes, Proceedings of Eurocrypt'93, Lecture Notes in Computer Science, LNCS 765, Springer Verlag (1993), pp.118–125.
9. E.D.Karnin, J.W.Greene, and M.E.Hellman, On Secret Sharing Systems, IEEE Trans. on Inform. Theory, Vol.IT-29 (1983), pp.35–41.
10. R.M.Capocelli, A.De Santis, L.Gargano and U.Vaccaro, On the size of shares for secret sharing schemes, Proceedings of Crypto'91, Lecture Notes in Computer Science, LNCS 576, Springer Verlag (1991), pp.101-113.
11. K.Kurosawa and K.Okada, Combinatorial interpretation of secret sharing schemes, In Pre-Proceedings of Asiacrypt'94 (1994), pp.38–48.

12. G.J.Simmons, A survey of Information Authentication, in Contemporary Cryptology, The science of information integrity, ed. G.J.Simmons, IEEE Press, New York (1992).

13. G.J.Simmons, Message authentication: a game on hypergraphs, Congr. Numer. 45 (1984), pp.161–192.

14. D.R.Stinson, Some constructions and bounds for authentication codes, Journal of Cryptology, vol.1 (1988), pp.37–51.

15. J.L.Massey, Cryptography – a selective survey, in Digital Communications, North-Holland (pub.) (1986), pp.3–21.

16. M.De Soete, New Bounds and Constructions for Authentication/Secrecy Codes with Splitting, Journal of Cryptology, vol.3, no.3 (1991), pp.173–186.

Quantum Cryptanalysis of Hidden Linear Functions

(Extended Abstract)

Dan Boneh
dabo@cs.princeton.edu

Richard J. Lipton*
rjl@cs.princeton.edu

Department of Computer Science
Princeton University
Princeton, NJ 08544

Abstract. Recently there has been a great deal of interest in the power of "Quantum Computers" [4, 15, 18]. The driving force is the recent beautiful result of Shor that shows that discrete log and factoring are solvable in random quantum polynomial time [15]. We use a method similar to Shor's to obtain a general theorem about quantum polynomial time. We show that any cryptosystem based on what we refer to as a 'hidden linear form' can be broken in quantum polynomial time. Our results imply that the discrete log problem is doable in quantum polynomial time over any group including Galois fields and elliptic curves. Finally, we introduce the notion of 'junk bits' which are helpful when performing classical computations that are not injective.

1 Introduction

The general discrete log problem can be phrased as follows: Let G be a finite group for which the group operation can be computed efficiently(given $x, y \in G$ we can find $x + y$). Let $h : \mathbb{Z} \to G$ be a homomorphism from the integers to G which can also be computed efficiently. Given $\beta = h(\alpha)$ the general discrete log problem is to find the smallest positive integer x such that $h(x) = \beta$. For example, in the standard discrete log problem over $\mathbb{Z}_p^*$ the homomorphism h is defined by $h(\alpha) = g^\alpha \pmod{p}$ for some generator g of $\mathbb{Z}_p^*$. Here $\mathbb{Z}_p^*$ is the multiplicative group of residues modulo a prime p.

A large variety of cryptosystems are based on the discrete log problem for various groups G. Specific groups that are being used are the multiplicative groups of large Galois fields [6], the multiplicative group of residues modulo a composite number [9, 10], elliptic curves over finite fields [11, 7] and the class group of imaginary quadratic fields [17].

Recently Shor [15] showed that the discrete log problem where $G = \mathbb{Z}_p^*$ can be solved in polynomial time on a quantum machine. We generalize this result to show that any type of cryptosystem which is based on what we refer to as

* Supported in part by NSF CCR–9304718.

a "hidden linear form" can be broken in quantum polynomial time(QP). An immediate application of this result shows that the general discrete log problem for any finite group G can be solved in QP. Thus, QP can break any of the cryptosystems discussed above.

Simon [14] observed that in QP it is possible to find a period of a function defined over $\mathbb{Z}_2^n$. We show that it is possible to detect the period of any function defined over $\mathbb{Z}$, even when the function is not one to one in its fundamental domain. Our method is similar to Shor's factoring algorithm and is crucial for solving the general discrete log problem.

These results raise a natural question of trying to detect periods over arbitrary groups G. The problem can be stated as follows: given a function $f : G \to D$ for some range D, find an element $g \in G$ such that $f(x+g) = f(x)$ for all $x \in G$. For instance, the problem of detecting periods of functions over S_n is of significant importance since the problem of graph isomorphism can be reduced to it. Fourier analysis is a natural tool to use when trying to detect a period of a function. It is well known that one can define a Fourier transform over any group G ([13]). Now, suppose that for a given group G, the Fourier transform of G can be computed in QP (in time polynomial in $\log |G|$). Does this imply that a period of the function $f : G \to D$ can be found in QP? We have so far been unable to resolve this general problem. However, our results can be generalized to solve this problem for any finite Abelian group.

We assume that the reader is familiar with the general model of quantum computations. See [4, 15, 18] for further details.

2 Main Results

In this section we will state our main results. We begin by introducing some terminology. A function $h : \mathbb{Z} \to S$ has *period* q if for any integer x we have $h(x + q) = h(x)$. Such a function h can be regarded as a function from $\mathbb{Z}_q$ to S. Here $\mathbb{Z}_q$ is the group of residues modulo q. We say that the function h has *order at most* m provided that h does not map more than m elements of $\mathbb{Z}_q$ to one, i.e. all $z \in S$ satisfy $|h^{-1}(z) \pmod{q}| \leq m$.

Let $f(x_1, ..., x_k)$ be a function from the integers $\mathbb{Z}^k$ to some arbitrary range S. Say that f has *hidden linear structure* over q provided there are integers $\alpha_2, ..., \alpha_k$ and some function h with period q so that

$$f(x_1, ..., x_k) = h(x_1 + \alpha_2 x_2 + ... + \alpha_k x_k)$$

for all integers $x_1, ..., x_k$. We say that f has order at most m if h has order at most m.

Theorem 1. *Suppose that $f(x_1, ..., x_k)$ is a function which has a hidden linear structure over q of order at most m. We impose two technical conditions:*

1. *Let $n = \log q$ then m and k are at most $n^{O(1)}$.*
2. *Let p be the smallest prime divisor of q; then $m < p$.*

For such a function f, in random quantum polynomial time in n we can recover the values of all the $\alpha_2, ..., \alpha_n$ (mod q) from an oracle for f.

The point of this theorem is that random quantum polynomial time is able to solve a kind of cryptanalysis problem. With just the ability to evaluate the function f we can find the "secret" linear structure of f. The two restrictions on the function f are critical. The first one restricts m, the order of h. This is crucial since for example, if h is a constant function then trivially it is impossible to recover the values of the α's.

The second restriction on m ensures that the $\alpha_2, \ldots, \alpha_n$ are unique modulo q. In fact, as we shall see in Section 6, this condition enables us to test if a proposed solution $\alpha'_2, \ldots, \alpha'_n$ is the correct one. Note that when q has no small factors the second restriction is subsumed by the first.

Another important problem which can be solved in quantum polynomial time is that of determining the period of a function.

Theorem 2. *Suppose the function $h : \mathbb{Z} \to S$ is periodic. Let q be the smallest positive period of h and assume h has order at most m. We impose two conditions:*

1. Let $n = \log q$ then m is at most $n^{O(1)}$.
2. Let p be the smallest prime divisor of q; then $m < p$.

For such a function h, in random quantum polynomial time in n it is possible to recover the period q of h.

The two technical conditions are required so that we will be able to test that the output of the algorithm is correct. Theorem 2 shows that the value of q need not be known for Theorem 1 to hold. Indeed, as we shall see, in many important applications the value of q is not known.

3   Applications

There are several applications of these theorems. First, we generalize the original results of Shor [15] to show how to compute discrete log over an arbitrary group. To achieve this we show how to phrase the general discrete log problem as a hidden linear form.

Let $h : \mathbb{Z} \to G$ be a homomorphism and let $\beta = h(\alpha)$. Given β we wish to find the smallest positive integer x such that $\beta = h(x)$. Let d be the order of $h(1)$ in the group G. Clearly, the homomorphism h has period d. Note that in general d in unknown, e.g. when $G = \mathbb{Z}_n^*$ for some composite n or when G is the class group of a quadratic field.

Define the function $f : \mathbb{Z}^2 \to G$ as $f(x, y) = h(x + \alpha y)$. By the remarks above, the function f has a hidden linear form over d of order 1. An important observation is that the function f can be efficiently evaluated as follows:

$$f(x, y) = h(x)h(\alpha y) = h(x)h(\alpha)^y = h(x)\beta^y \ .$$

To solve the general discrete log problem we apply the following two steps: first use Theorem 2 to find d, the period of the homomorphism h. The theorem can be applied since the function h has order 1, i.e. m=1. Then apply Theorem 1 to find an integer $\alpha' < d$ such that $\alpha' \equiv \alpha \pmod{d}$. Since α' is the smallest positive integer such that $h(\alpha) = h(\alpha')$, it is the required solution to the general discrete log problem. We have proved the following corollary to Theorems 1 and 2.

Corollary 3. *The general Discrete Log problem can be solved in random quantum polynomial time.*

This shows that we can find Discrete Log over composite modulus, Galois fields, and elliptic curves. An immediate corollary of Theorem 2 is the following.

Corollary 4. *Factoring can be solved in random quantum polynomial time.*

Proof. Suppose we wish to factor an n bit odd integer q. For an element $g \in \mathbb{Z}_q^*$, define the function $h : \mathbb{Z} \to \mathbb{Z}_q^*$ by $h(x) = g^\alpha \pmod{q}$. Let d be the order of g in $\mathbb{Z}_q^*$ then the function h has period d and oder 1, i.e. m=1. Theorem 2 can be used to find the period of h and hence the order of g. The ability to find the order of an element in $\mathbb{Z}_q^*$ enables us to factor as is described in [15]. $\square$

Another application of Theorem 1 concerns what are sometimes called "garbled" linear equations. Consider the following family of linear equations over $\mathbb{Z}_q$:

$$\alpha_1 x_{11} + \ldots \alpha_n x_{1n} = y_1 + e_1$$
$$\vdots$$
$$\alpha_1 x_{m1} + \ldots \alpha_n x_{mn} = y_m + e_m$$

where $e_1, \ldots, e_m$ are unknown "errors" and the x's are known values. The general garbled linear equation problem is to find the value of the α's given $m \gg n$ large enough and given that most of the errors are equal to 0. This is a known difficult problem. However, suppose that the errors are determined by some polynomial time rule, i.e. some polynomial time function $e()$ satisfies $e(y_i) = e_i$. Then the function

$$f(x_1, \ldots, x_n) = h(\alpha_1 x_1 + \ldots + \alpha_n x_n) \quad \text{where} \quad h(y) = y + e(y)$$

has a hidden linear structure. By Theorem 1 we can, in random quantum polynomial time, find the α's provided h does not collapse too much. Note, that we assume that we have an oracle that given $x_1, \ldots, x_n$ supplies us with the value of $y + e(y)$. Of course we do not assume we know when $e(y) = 0$ or not.

4 Basic Lemmas

Before we can prove Theorem 1 we need several lemmas. The following lemma is the main lemma which enables us to handle the fact that h may not be one-to-one in Theorems 1 and 2.

Lemma 5. *Let W be some integer and let $R < W$. Then for any integers $b_1, \ldots, b_m$ there are at least R/m^2 integers $0 \leq x \leq R$ satisfying*

$$\left| \sum_{k=1}^{m} \exp(\frac{2\pi i x b_k}{W}) \right| > \frac{1}{2} .$$

Lemma 5 relies on the following lemma.

Lemma 6. *Let $\lambda_1, \ldots, \lambda_m$ be m complex numbers each of norm 1. Let $S_k = \sum_{j=1}^{m} \lambda_j^k$ then there exists a $1 \leq k \leq m$ such that $|S_k| > \frac{1}{2}$.*

Proof. Assume that for all $k = 1, \ldots, m-1$ we have $|S_k| \leq \frac{1}{2}$. We show that this implies that $|S_m| > m/2$ proving the lemma. Let C_k be the m'th symmetric polynomial in $\lambda_1, \ldots, \lambda_m$, i.e.

$$C_k = \sum_{1 \leq j_1 < j_2 < \ldots < j_k \leq m} \lambda_{j_1} \cdot \lambda_{j_k} .$$

First we prove by induction on k that $|C_k| \leq \frac{1}{2}$ for $k = 1, \ldots, m-1$. For $k = 1$ this is clear since $|C_1| = |S_1| \leq \frac{1}{2}$. Now, assume that $|C_j| \leq \frac{1}{2}$ for $j = 1, \ldots, k-1 < m-1$. We show that $|C_k| \leq \frac{1}{2}$. For $k > 1$ define

$$A_k = C_1 S_{k-1} - C_2 S_{k-2} + \ldots + (-1)^k C_{k-1} S_1 .$$

The Newton relations (see [8]) state that $S_k - A_k + (-1)^k k C_k = 0$ for $k \leq m$. The induction hypothesis implies that $|A_k| < \frac{k-1}{2}$ since the norm of each term in the sum is less than $1/2$. Hence,

$$|C_k| = \frac{1}{k}|S_k - A_k| \leq \frac{1}{k}(|S_k| + |A_k|) \leq \frac{1}{2} .$$

To conclude the proof of the lemma we show that $|S_m| > m/2$. The fact that for $k = 1, \ldots, m-1$ we have $|S_k| \leq \frac{1}{2}$ and $|C_k| \leq \frac{1}{2}$ implies that $|A_m| \leq m/2$. Furthermore, Since $C_m = \prod_{k=1}^{m} \lambda_k$ we know that $|C_m| = 1$. Hence, by Newton's relations

$$|S_m| = |A_m - (-1)^m m C_m| \geq |m C_m| - |A_m| \geq m - \frac{m}{2} = m/2 .$$

$\square\square$

Proof of Lemma 5. Define

$$\beta(x) = \left| \sum_{k=1}^{m} \exp(\frac{2\pi i x b_k}{W}) \right| \ .$$

By Lemma 6, for any x, one of $\beta(x), \beta(2x), \ldots, \beta(mx)$ must be bigger than $\frac{1}{2}$. Observe that the integers $\{0, \ldots, R\}$ can be partitioned into R/m^2 distinct sequences of the form $\{x, 2x, \ldots, mx\}$. Hence, the lemma follows. □

The following lemma provides a lower bound on the sum of roots of unity which are close to 1.

Lemma 7. *Suppose that* $|\theta_k| \le \epsilon < 1$ *are real numbers for* $k = 1, \ldots, m$. *Then,*

$$\left| \sum_{k=1}^{m} \exp(i\,\theta_k) \right| \ge (1 - \epsilon^2)m \ .$$

Proof. This follows directly from the fact that the real part of $\exp(i\theta_i)$ is at least $\cos(\epsilon) > 1 - \epsilon^2$. □

5 An Overview of the Proofs

Before we present the proofs of Theorems 1 and 2 we will outline a general paradigm for proving that a problem of size n can be solved in quantum polynomial time. We will describe a certain quantum experiment $\mathcal{E}$. Each time we perform this experiment we will get some observable value. Let $\mathcal{V}$ be some subset of all the possible observable values. We will arrange things so that the following are true:

1. Given *any* value from $\mathcal{V}$ we can in polynomial time (on a conventional computer) solve the given problem.
2. The probability of observing a specific element of $\mathcal{V}$ is at least $1/Wn^c$ for some integer W and constant c.
3. The cardinality of the set $\mathcal{V}$ is at least $W/n^{c'}$ for some constant c'.

We refer to the observables in $\mathcal{V}$ as the "good" observables. By 2 and 3 above, The probability of sampling an observable from $\mathcal{V}$ is at least $1/n^{O(1)}$. Once such an observable is found it will be used to solve the given problem. Hence, in expected polynomial time the problem will be solved.

An important point is that we do not know which observables lie in the set $\mathcal{V}$. When an observable is observed, we try to use it to solve the hidden linear problem as if it is in $\mathcal{V}$. Then, we check that the computed result works correctly. If it does we are done; otherwise, we try again.

6   The Proof of Theorem 1

We now turn to the proof of Theorem 1. We will prove the theorem for a hidden linear form with two variables $f(x, y) = h(x + \alpha y)$. This is enough to prove the general theorem, since we can find all the α's one by one by setting all the irrelevant variables to zero.

Let $f(x, y) = h(x + \alpha y)$ be a hidden linear form over q, an n-bit number. The assumptions of Theorem 1 state that h has order at most $m = n^d$ for some constant d and if p is the smallest prime divisor of q, then $m < p$. Our objective is to find α.

We first show that given an α' it is easy to test if $\alpha \equiv \alpha'$ (mod q). This is the only place where we use the fact that $m < p$. Let $A_{\alpha'}$ be the set of pairs $\{(-k\alpha', k)\}$ for $k = 0, \ldots, m$.

Lemma 8. *If for all $(x, y) \in A_{\alpha'}$ we have $f(x, y) = h(x + \alpha' y)$ then $\alpha \equiv \alpha'$ (mod q).*

Proof. Observe that for all $(x, y) \in A_{\alpha'}$ we have $x + \alpha' y = 0$. Hence, all $(x, y) \in A_{\alpha'}$ satisfy $h(x + \alpha y) = f(x, y) = h(0)$. Now, suppose $\alpha \not\equiv \alpha'$ (mod q). For two distinct pairs (x, y) and (x', y') in $A_{\alpha'}$ we have that $x + \alpha y \not\equiv x' + \alpha y'$. This follows from the fact that

$$\alpha \not\equiv \alpha' \equiv \frac{x - x'}{y' - y} \pmod{q} .$$

The division by $y - y'$ is valid since $|y - y'| \leq m < p$ where p is the smallest prime divisor of q. Hence, $y - y'$ is relatively prime to q and hence invertible. This shows that h maps the $m + 1$ pairs in $A_{\alpha'}$ to the same value, $h(0)$. However, by assumption h had order at most m. This contradiction proves the lemma.   □

6.1   The Quantum Experiment

Let $W_1 < W_2 < \ldots$ be the first primes that are relatively prime to q. Define $W = \prod_{i=1}^{k} W_i$ as the first product that exceeds $\max\{2q, mq\}$. Note that W and q are relatively prime. Since $m < n^{O(1)}$ we have $W < qn^{O(1)}$.

Let F_W be the Fourier transform unitary matrix:

$$(F_W)_{x,y} = \frac{1}{\sqrt{W}} e^{2\pi i x y / W} .$$

Shor shows that for the W constructed above the transformation F_W can be carried out by a quantum machine in polynomial time. In general this holds whenever W is smooth, i.e. contains no large prime factors,

The quantum experiment $\mathcal{E}$ is as follows: First, the quantum machine writes two random numbers r_1, r_2 from $\mathbb{Z}_q$ on its tape. So the state after this first step is

$$\frac{1}{q} \sum_{r_1, r_2} |r_1, r_2 > .$$

The algorithm next computes the function f in a reversible manner so that the machine is in state

$$\frac{1}{q} \sum_{r_1, r_2} |r_1, r_2, f(r_1, r_2) > .$$

We now use the mapping $(F_W)_{x,y} = e^{2\pi i x y/W}$ to send each r_i to s_i for $i = 1, 2$ with amplitude $\frac{1}{\sqrt{W}} \exp(2\pi i r_i s_i/W)$. This places the machine in the state

$$\frac{1}{qW} \sum \exp(2\pi i (r_1 s_1 + r_2 s_2)/W) |s_1, s_2, f(r_1, r_2) >$$

where the sum is over all r_1, r_2 and s_1, s_2. Thus, the machine will end up in state $|s_1, s_2, b >$ with probability

$$\left| \frac{1}{qW} \sum \exp(2\pi i (r_1 s_1 + r_2 s_2)/W) \right|^2$$

where the sum is over all r_1, r_2 such that $f(r_1, r_2) = b$.

We now describe the special set of observables $\mathcal{V}$. We denote the residue of x modulo W by $\{x\}_W$. The observable (s_1, s_2, b) is in $\mathcal{V}$ provided the following properties are satisfied:

1. $s_1 q \geq W$;
2. $\{s_1 q\}_W \leq W/m$;
3. Let $C = s_2 - s_1 \alpha + \frac{\alpha}{q} \{s_1 q\}_W$. Then $C = tW + \delta$ for some integer t and $|\delta| < 1$.
4. $\left| \sum_{k=1}^{m} \exp(2\pi i b_k s_1/W) \right| \geq 1/2$ where $b_1, \dots, b_m$ are distinct elements so that $h(b_k) = b$ for $k = 1, \dots, m$. Recall that m is the order of the function h.

In what follows we will refer to these conditions as $(1),(2),(3)$ and (4). It remains to prove that the set $\mathcal{V}$ satisfies the three properties specified in Section 5.

6.2   Using a "Good" Observable

Let (s_1, s_2, b) be an observable from the set $\mathcal{V}$. We show how this observable can be used to find α. Condition (3) implies that

$$s_2 - \frac{\alpha}{q}(s_1 q - \{s_1 q\}_W) = tW + \delta .$$

Write $s_1 q = vW + u$ with $0 \leq u < W$. Observe that $v = \frac{s_1 q - \{s_1 q\}_W}{W}$. Since t is an integer, and $|\delta| < 1$, dividing the above equality by W leads to

$$\left\| \frac{s_2}{W} - \alpha \frac{v}{q} \right\| < \frac{1}{W}$$

where $\|x\|$ is the fractional part of x, i.e. $\min|x + i|$ over all integers i.

Let s be the integer which makes the values of $\frac{s}{q}$ the closet to $\frac{sv}{W}$. That is, $\frac{s}{q}$ is the fraction we get when we round $\frac{sv}{W}$ to the closest rational with denominator q. Since $W > 2q$ it is not difficult to see that for the above inequality to hold we must have

$$\left\| \frac{s}{q} - \alpha \frac{v}{q} \right\| = 0 \ .$$

This means that $s - \alpha v \equiv 0 \pmod{q}$. By condition (1) we know that $v \geq 1$. Hence, when q and v are relatively prime we can easily recover α.

When q and v are not relatively prime we proceed as follows: let $z = q/\gcd(q, v)$. Observe that v is invertible modulo z and let $\alpha' = s/v \pmod{z}$. Clearly $\alpha' \equiv \alpha \pmod{z}$. For $0 \leq \alpha' < z$ we have that $\alpha' \equiv \alpha \pmod{z}$ if and only if $\alpha' \frac{q}{z} \equiv \alpha \frac{q}{z} \pmod{q}$. Hence, it is easy to check that the resulting α' satisfies $\alpha' \equiv \alpha \pmod{z}$ by using Lemma 8 on the function $f'(x, y) = f(x, \frac{q}{z}y)$.

Once a pair α', z satisfying $\alpha' \equiv \alpha \pmod{z}$ is found, write $\alpha = \alpha' + zk$. Define a new function $f''(x, y) = f(zx - \alpha'y, y)$. Then

$$f''(x, y) = h(zx - \alpha'y + \alpha y) = h(z(x + ky)) \ .$$

Hence, $f''(x, y)$ has a hidden linear structure over q/z. We can now recursively apply the algorithm to f'' to find k and thus find $\alpha \pmod{q}$.

6.3 The Amplitude of a "Good" Observable

For an observable (s_1, s_2, b), we denote by $\sigma(s_1, s_2, b)$ the probability of observing (s_1, s_2, b) at the end of the quantum experiment. To simplify the exposition in this section we assume that the order of the function f satisfies $m \geq 10$. This is not a restriction since a function which has order less than 10 may be regarded as a function with order 10.

Let (s_1, s_2, b) be an observable from the set $\mathcal{V}$. Recall that the probability of this observation is

$$\sigma(s_1, s_2, b) = \left| \frac{1}{qW} \sum \exp\left(\frac{2\pi i}{W}(r_1 s_1 + r_2 s_2) \right) \right|^2$$

where the sum is over all r_1, r_2 such that $f(r_1, r_2) = b$. The key is that f has a hidden linear structure, i.e. $f(r_1, r_2) = b$ if and only if $h(r_1 + \alpha r_2) = b$. Since h need not be one to one there are *distinct* $b_1, .., b_{m'}$ so that $h(b_k) = b$ for $k = 1, \ldots, m'$ and $m' \leq m$. WLOG we assume $m = m'$. Thus, $\sigma(s_1, s_2, b)$ is equal to

$$\frac{1}{q^2 W^2} \left| \sum_{k=1}^{m} \sum \exp(2\pi i(r_1 s_1 + r_2 s_2)/W) \right|^2$$

where the inner sum is over all r_1, r_2 so that $r_1 \equiv b_k - \alpha r_2 \bmod q$. Since $1 \leq r_1 < q$, given an r_2 the value of r_1 is equal to $b_k - \alpha r_2 - q\lfloor (b_k - \alpha r_2)/q \rfloor$. Thus, the key is to bound the absolute value of the following double summation,

$$\sum_{k=1}^{m} \exp(2\pi i b_k s_1/W) \sum_{r_2=0}^{q-1} \exp\left[\frac{2\pi i}{W}(r_2 s_2 - \alpha s_1 r_2 - s_1 q\lfloor (b_k - \alpha r_2)/q \rfloor) \right] \ .$$

First we bound the inner sums. For a given k, rewrite the inner sum as

$$\sum_{r_2=0}^{q-1} \exp\left[\frac{2\pi i}{W}r_2(s_2 - \alpha s_1 + \frac{\alpha}{q}\{s_1 q\}_W)\right] \exp\left[-\frac{2\pi i}{W}\left(\frac{\alpha r_2}{q} + \left\lfloor\frac{b_k - \alpha r_2}{q}\right\rfloor\right)\{s_1 q\}_W\right] .$$

By condition (3), and the fact that $r_2/W < q/W < 1/m$, the argument of the first exponent is always less than $2\pi i/m$. For the second exponent we know $b_k < q$. The fact that all reals $A, B > 0$ satisfy $|B + \lfloor A - B\rfloor| \leq \lfloor A\rfloor + 1$ implies that

$$\left|\frac{\alpha r_2}{q} + \left\lfloor\frac{b_k - \alpha r_2}{q}\right\rfloor\right| \leq \left\lfloor\frac{b_k}{q}\right\rfloor + 1 \leq 1 .$$

Combining this with condition (2) we see that the argument of the second exponent is always less, in absolute value, than $2\pi i/m$. Hence, the total exponent is less than $4\pi i/m$. Using Lemma 7, we get that the inner sum is always bigger than $[1 - O(\frac{1}{m^2})]\, q$. On the other hand the inner sum is clearly less than q. It follows that $\sigma(s_1, s_2, b)$ is equal to

$$\sigma(s_1, s_2, b) = \frac{1}{W^2}\left|\sum_{k=1}^{m}(1 - \epsilon_k)\exp(2\pi i b_k s_1/W)\right|^2$$

where $0 \leq \epsilon_k \leq O(\frac{1}{m^2})$ for all $k = 1, \ldots, m$. Now, since the ϵ_k are small it is not difficult to see that condition (4) implies that $\sigma(s_1, s_2, b) > \Omega(\frac{1}{W^2})$. Hence, a "good" observable (s_1, s_2, b) has the required probability.

6.4 Cardinality of Set of "Good" Observables

The last step is to show that $\mathcal{V}$ has the required cardinality. First, observe that for any s_1 there exists an s_2 satisfying condition (3). This follows by setting s_2 to the integer closest to $\alpha s_1 + \frac{\alpha}{q}\{s_1 q\}_W$. We only need to lower bound the number of s_1 satisfying

1. $s_1 q \geq W$;
2. $\{s_1 q\}_W \leq W/m$;
3. $\left|\sum_{k=1}^{m} \exp(2\pi i b_k s_1/W)\right| \geq 1/2$

We will show that the number of s_1 satisfying conditions (2) and (3) is at least W/m^3. The number of s_1 violating condition (1) is at most W/q which is negligible in comparison. Hence, throwing away the s_1 that violate condition (1) will make no difference.

Let $x = q s_1 \pmod{W}$ and $c_k = b_k q^{-1} \pmod{W}$. Since q and W are relatively prime by construction, q^{-1} exists modulo W. Conditions (2) and (3) can now be rewritten as

1. $0 \leq x \leq W/m$
2. $\left|\sum_{k=1}^{m} \exp(2\pi i c_k x/W)\right| \geq 1/2$

By Lemma 5, the number of x that satisfy these two conditions is at least W/m^3. Since $m < n^{O(1)}$, the number of such x is at least $W/n^{O(1)}$.

Hence, the total number of pairs s_1, s_2 satisfying conditions (1),(2),(3) and (4) in the definition of $\mathcal{V}$ is $W/n^{O(1)}$. Putting this together with the fact that there are q possible value for b, we get that the number of triplets (s_1, s_2, b) in $\mathcal{V}$ is $qW/n^{O(1)}$. By definition of W we know that $W = qn^{O(1)}$. Hence, $|\mathcal{V}| > W^2/n^{O(1)}$, which is what we had to show.

7 The Proof of Theorem 2

Say we are given a function $h : \mathbb{Z} \to S$ which is periodic. We wish to find the smallest period q of h. Let $n = \log q$. We assume that h is of order at most m where $m = n^{O(1)}$.

Without loss of generality we can assume that we are given an upper bound q' on q such that $q' < 2q$. This upper bound can be found by guessing some initial q' and running the algorithm. If the algorithm fails to find the period, double q' and rerun the algorithm. After at most n steps q' will be the required upper bound.

Let p be the smallest prime factor of q. As in the previous section, the assumption of Theorem 2 that $m < p$ implies that when the algorithms outputs q' as the period, we can test that $q = q'$.

7.1 The Quantum Experiment

Let W be a smooth number constructed as in the previous section such that $W > \max\{q'^2, mq'^2\}$ and $W < q'^2 n^O(1)$. The quantum experiment $\mathcal{E}$ is as follows: First, the quantum machine writes a random numbers r from $\mathbb{Z}_W$ on its tape. So the state after this first step is

$$\frac{1}{\sqrt{W}} \sum_r |r> \quad .$$

The algorithm next computes the function h in a reversible manner so that the state of the machine is now

$$\frac{1}{\sqrt{W}} \sum_r |r, h(r)> \quad .$$

We now use the Fourier unitary transformation F_W to send r to s with amplitude $\frac{1}{\sqrt{W}} \exp(2\pi i r s/W)$. It places the machine in the state

$$\frac{1}{W} \sum_{r,s} \exp(2\pi i r s/W)|s, h(r)> \quad .$$

The probability that the machine ends in the state $|s, b>$ is

$$\left| \frac{1}{W} \sum \exp(2\pi i r s/W) \right|^2$$

where the sum is over all r such that $h(r) = b$.

As before, we now describe the special set of observables $\mathcal{V}$. An observable (s, b) is in $\mathcal{V}$ provided the following properties are satisfied:

1. $\{sq\}_W < q/m$;
2. $\left| \sum_{k=1}^{m} \exp(2\pi i b_k s/W) \right| \geq 1/2$ where $b_1, \ldots, b_m$ are distinct elements so that $h(b_k) = b$ for $k = 1, \ldots, m$. Recall that m is the order of the function h.

It remains to prove that the set $\mathcal{V}$ satisfies the three properties specified in Section 5:

1. Given an observable (s, b) in $\mathcal{V}$ Condition (1) implies that we can find a non trivial factor z of q using a method similar to Shor's [15]. We can then define a new function $h'(x) = h(zx)$ which will have period q/z. The algorithm can be applied recursively on h' to recover q/z. This shows that given a "good" observable we can find the period q.
2. Using condition (2) and an argument similar to the one in the previous section we can show that the amplitude of a "good" observable is $\Omega(\frac{1}{q^2})$.
3. Using Lemma 5 we can show that the cardinality of $\mathcal{V}$ is at least $q^2/n^{O(1)}$.

8   Junk Bits

In both algorithms described in the previous sections the first step was to pick a random number between 1 and $q - 1$ for some integer q. This means that the machine should be in state

$$\frac{1}{\sqrt{q}} \sum_{r=0}^{q-1} |r> .$$

However, when q is a large prime, this state can not be easily constructed using a quantum circuit.

An easy method for generating a random number between 0 and $q - 1$ is to pick an integer W which is the closest power of 2 to q. Then generate a random number $x \pmod{W}$. If $x < q$ then use x, otherwise generate a new x and repeat this until a number in the required range is generated. This will clearly generate a number uniformly distributed on $0, \ldots, q-1$. The problem is that this procedure can not be carried out on a quantum machine since all the "bad" samples (the ones larger than q) can not be erased from the tape. Erasure is not a reversible operation. Clearly the bad samples can not be left on the tape since they would prevent the interference effects which are so useful in quantum computing.

Another approach is to pick some large integer $W > q^2$ which is a power of 2. Then generate a random number $x \pmod{W}$ and compute $x \pmod{q}$. The resulting value will be exponentially close to being uniformly distributed between 0 and $q - 1$ which is good enough. However, as before, we run into the problem that the map sending x to $x \pmod{q}$ is not reversible. As before keeping extra information on the tape to make this map reversible is risky since it may prevent interference effects.

The solution is to keep just enough extra information on the tape so that the computation is reversible, however the extra information on the tape should be independent of the computation taking place. We call this extra information *Junk bits*.

Definition 9. Let $f : \{0,1\}^n \to Y$ be some polynomial time computable function which is not one to one. A function $J : \{0,1\}^n \to Y'$ will be called a "junk" function for f if the following are satisfied:

1. The map $x \to (f(x), J(x))$ is one to one and polynomial time computable. Furthermore, the inverse map is in QP;
2. $\big| \Pr[f(x) = y \mid J(x) = j] - \Pr[f(x) = y] \big| < 2^{-\Omega(n)}$.

Thus, the value of $J(x)$ and $f(x)$ should be almost independent of one another. Condition (1) implies that the map sending x to $(f(x), J(x))$ can be computed in QP using a result due to Bennett [2]. It should be clear that once we have computed $(f(x), J(x))$, the computation can proceed to use the value of $f(x)$ as if $J(x)$ was not written on the tape. The independence property will guarantee that the interference effects will change by an exponentially small amount. The full details of this method will be given in the final version of the paper.

To generate a random number between 0 and $q - 1$ we follow the second method. Let $W > q^2$ be a large power of 2. Generate a random number between 0 and $W - 1$. We now wish to compute the function $f(x) = x \bmod q$. A possible junk function for f is $J(x) = \lfloor x/q \rfloor$. It is not difficult to see that $J(x)$ is indeed a junk function for $f(x)$. Using similar methods we can show that it is possible to generate random permutations and other random objects.

9 Conclusions and Open Problems

We have shown that QP can solve two types of problems: recovering the hidden linear structure of a function and detecting periods over $\mathbb{Z}$. Our results hold even when the function h used is not one to one. Using both theorems we were able to show that the discrete log problem can be solved in quantum polynomial time over any group.

The problem of recovering the hidden linear structure can be generalized to any ring. Similarly, the problem of detecting periods can be generalized to any group. As was mentioned in the introduction, graph isomorphism is reducible to the problem of detecting periods of functions defined over the symmetric group S_n. This example shows the importance of these generalizations. We hope that Fourier methods analogous to the ones used in this paper can be used to detect periods over S_n. This will show that the graph isomorphism problem can be solved in random quantum polynomial time. We mention that Beals [1] has shown that the Fourier transform over the group S_n can be carried out in quantum polynomial time.

We have also introduced the concept of Junk bits which enables quantum machine to carry out certain non invertible functions in a way that does not effect the interference patterns. A natural problem is to try and understand which deterministic computations can be done using junk bits.

Acknowledgments

We wish to thank Robert Beals and Merrick Furst for helpful discussions about this work.

References

1. R. Beals, *Computing Fourier Transform over S_n in QP*, unpublished manuscript.
2. C. Bennett, *Logical reversibility of computation*, IBM J. Res. Develop. vol. 17, 1973, pp. 525-532.
3. C. Bennett, E. Bernstein, G. Brassard, U. Vazirani, *Strengths and Weaknesses of Quantum Computing*, to appear.
4. E. Bernstein and U. Vazirani, *Quantum Complexity Theory*, Proc. 25th ACM Symp. on Theory of Computation, 1993.
5. D. Coppersmith, *An Approximate Fourier Transform Useful in Quantum Factoring*, IBM Research Report 19642, 1994.
6. W. Diffie and M. Hellman, *New Directions in Cryptography*, IEEE transactions on Information Theory, vol. 22, no. 6, pp. 644-654, 1976.
7. N. Koblitz, *Elliptic Curve Cryptosystems*, Mathematics of Computations 48, 1987, pp. 203-209.
8. S. Lang, *Algebra*.
9. U. Maurer and Y. Yacobi, *Non-interactive public-key cryptography*, EuroCrypt-91, pp.498-507, 1991.
10. K. McCurley, *A Key Distribution System Equivalent to Factoring*, Journal of Cryptology, vol. 1, no. 2, pp. 95-105.
11. V. Miller, *Uses of Elliptic Curves in Cryptography*, In Proceedings of Crypto 1985, pp. 417-426.
12. B. Preneel, R. Govaerts, J. Vandewalle, *Hash Functions Based on Block Ciphers: A Synthetic Approach*, in Proc. of Advances in Cryptology, CRYPTO '93.
13. J. P. Serre, *Linear Representations of Finite Groups*, Springer-Verlag, 1977.
14. D. Simon, *On the Power of Quantum Computation*, Proc. FOCS, 1994, pp. 116-123.
15. P. Shor, *Algorithms for Quantum Computation*, Proc. FOCS, 1994, pp. 124-134.
16. L. Washington, *Introduction to Cyclotomic Fields*, Springer-Verlag, 1982.
17. J. Buchmann and H. Williams, *A Key Exchange System Based on Imaginary Quadratic Fields*, Journal of Cryptology, vol. 1, no. 2, pp. 107-118, 1988.
18. A. Yao, *Quantum Circuit Complexity*, Proc. 34th IEEE Symp. on Foundations of Computer Science, 1993, pp. 352-360.

An Efficient Divisible Electronic Cash Scheme

Tatsuaki Okamoto

NTT Laboratories
1-2356 Take, Yokosuka-shi, 238-03 Japan
Email: okamoto@sucaba.isl.ntt.jp

Abstract. Recently, several "divisible" untraceable off-line electronic cash schemes have been presented [8, 11, 19, 20]. This paper presents the first practical "divisible" untraceable[1] off-line cash scheme that is "single-term"[2] in which every procedure can be executed in the order of $\log \mathcal{N}$, where $\mathcal{N}$ is the precision of divisibility, i.e., $\mathcal{N} = $ (the total coin value)/(minimum divisible unit value). Therefore, our "divisible" off-line cash scheme is more efficient and practical than the previous schemes. For example, when $\mathcal{N} = 2^{17}$ (e.g., the total value is about \$ 1000, and the minimum divisible unit is 1 cent), our scheme requires only about 1 Kbyte of data be transfered from a customer to a shop for one payment and about 20 modular exponentiations for one payment, while all previous divisible cash schemes require more than several Kbytes of transfered data and more than 200 modular exponentiations for one payment.
In addition, we prove the security of the proposed cash scheme under some cryptographic assumptions. Our scheme is the first "practical divisible" untraceable off-line cash scheme whose cryptographic security assumptions are theoretically clarified.

1 Introduction

Recently, much research has been performed in the area of off-line electronic currency [2, 5, 8, 9, 11, 12, 13, 16, 18, 19, 20, 25]. Protocols have been proposed enabling consumers to withdraw "electronic coins" from a bank, and later spend these coins at a shop in an "off-line" manner. Here, off-line refers to the property that communication with a bank or authorized center is unnecessary during the payment protocol. In addition, electronic coins should be anonymous, i.e., "untraceable".

A "divisible" coin worth some amount of money, say \$x, is a coin that can be spent many times as long as the sum total of all its the transactions does not exceed \$x. This property, divisibility, is very useful and convenient for a

---

[1] Note that coins divided from the same coin can be linked each other in the proposed scheme, although they are anonymous, i.e., "untraceable" from the customer's identity. In other words, the unlinkability among divided coins is not satisfied, although the untraceability is satisfied.

[2] In the first generation of the practical off-line cash schemes [5, 16, 18, 19, 20], the cut-and-choose method is used, in which cash consists of many terms of the same form (e.g., 40 terms). A "single-term" cash scheme [2, 11, 12] means a practical cash scheme in which the cut-and-choose method is not used and cash consists of a single term. The basic idea of "single-term" is from [13], but the technique to realize the "single-term" property is specific to each scheme [2, 11, 12].

D. Coppersmith (Ed.): Advances in Cryptology - CRYPTO '95, LNCS 963, pp. 438-451, 1995.

customer. If a coin is not divisible, the customer must withdraw a coin whenever he spends it, or withdraw many coins of various values and store them in his electronic wallet (e.g., smart card). Since real cash does not satisfy this property, we must use many various bills and coins in daily life. On the other hand, prepaid cards[3] satisfy this property, and this is the major merit of such cards over real cash. Therefore, in practice, the divisibility is a very important requirement for electronic cash systems.

So far, several "divisible" untraceable off-line electronic cash schemes have been presented [8, 11, 19, 20]. The scheme in [8] is far from practical, since their scheme utilizes a non-interactive zero-knowledge proof for a general NP predicate. Pailles' divisible coin construction [20] is also inefficient. The size of data transfered from a customer to a shop during payment is linear in $\mathcal{N}$ where $\mathcal{N}$ is the divisibility precision, i.e., $\mathcal{N} = $ (the total coin value)/(minimum divisible unit value). A system in which a coin worth \$5367 consists of 5367 \$1 coins is a rather unwieldy and inefficient divisible cash system. [19] has also shortcomings in efficiency: their scheme utilizes a cut-and-choose method, a paid/deposited coin consists of many terms (e.g., 40 terms), and hence, the resulting complexities (the transfered data size and computation amount for a payment) can be quite large.

[11] partially solved the efficiency problems by constructing the first "single-term" divisible cash scheme. The cut-and-choose method is not used, so the transfered data size of [11] is less than that of [19]. However, the major shortcoming of [11] is that the required computation amount for a payment is of the order of the divisibility precision, $\mathcal{N}$. Hence, the amount of computation required for a payment in [11] is almost as large as that of [19]. Another shortcoming of [11] is that the size of the divisible coin depends on the selection of routes in the binary tree. If the selection of routes is almost optimal, the transfered data can be much smaller than that of [19]. However, if no such selection exists (such cases often occur), the transfered data size becomes almost as large as that of [19].

This paper presents a divisible untraceable off-line electronic cash scheme which solves these shortcomings, and is much more efficient than all previous schemes. Our scheme is the first practical "single-term divisible" cash scheme in which every procedure can be executed in the order of $\log \mathcal{N}$, and every transfered data sizes are of the order of $\log \mathcal{N}$. Therefore, the amount of the required computation and communication for a payment is, with our scheme, much less than those of the previous schemes.

For example, when $\mathcal{N} = 2^{17}$ (e.g., the total value is about \$ 1000, and the minimum divisible unit is 1 cent), our scheme requires only about 1 Kbyte of data be transfered from a customer to a shop for one payment and about 20 modular exponentiations (for a customer and a shop respectively) for one payment. All previous practical divisible cash schemes require more than several Kbytes of transfered data and more than 200 modular exponentiations for one payment. [4]

---

[3] Prepaid cards such as telephone cards are a kind of electronic cash, but their security heavily depends on physical tricks. Here, electronic cash means a cash system whose security depends only on mathematical techniques. In this sense, electronic cash can be considered to be an ideal version of prepaid cards and so is suitable for smart cards.

[4] Note that the evaluation here is based on the degree of security of factoring a 512 bit composite (and a 512 bit modulus discrete logarithm). For example, if the modulus size is 1024 bits, then the data sizes of the cash schemes should be twice. On the other hand, the number of the modular exponentiations can only depend on an arbitrary

The withdrawal procedure of our scheme is very efficient; it requires around 0.1 Kbyte of data be exchanged between a customer to a bank for one withdrawal and just one modular exponentiation (for a customer and a bank respectively) for one withdrawal, regardless of the total value and the divisibility precision. Moreover, the amount of data (electronic license and coin) stored in an electronic wallet (e.g., a smart card) is around 0.2 Kbytes. All previous practical divisible cash schemes require more than several Kbytes be stored in a wallet.

In addition, we prove that the proposed cash scheme satisfies four security requirements (no forging, no tracing, no overspending and no swindling) under some assumptions. Our scheme is the first "practical divisible" untraceable cash scheme whose cryptographic security assumptions, which are relatively primitive, are theoretically clarified. Even from a practical viewpoint, such a security proof is very important especially for a complicated cryptographic protocol, which consists of many primitive protocols. Although it is unclear whether these four security requirements are sufficient, our security proof guarantees that if there exists an attack on our scheme, then it should reside outside these four security requirements, unless our security assumptions are broken.

Note that the unlinkability among coins divided from the same coin cannot be satisfied in the proposed scheme as well as the previous practical divisible untraceable off-line cash schemes, although these divided coins are anonymous, i.e., "untraceable". In addition, since the procedure for interrupting the linking chain among the withdrawn coins (i.e., the opening protocol) is less efficient than the other procedures of our scheme, the procedure cannot be executed so often.

2 Number Theoretic Conventions

Since our scheme is constructed using some number theoretic techniques developed by [19], this paper follows the notations and propositions of the number theoretic techniques in [19]. However, Lemma 1 and Lemma 2 are new in this paper (A similar technique is used in [3]). They constitute a new technique to prevent a customer from double-spending a coin (or a node of a tree).

Lemma 1. *Let $N = PQ$ be the Williams integer, and t be an integer which is greater than 1. Then, for any $x \in QR_N$, and for any $e \in Z_{2^t}$, there exists a unique solution y such that $y^{2^t} = 2^{2e}x \bmod N$ and $y \in QR_N$.*

Lemma 2. *Let $N = PQ$ be the Williams integer, and t be an integer which is greater than 1 and $t = O(|N|)$. Then, there exits a deterministic poly-time (i.e., $O(|N|^3)$) algorithm to factor N, given N, t, $x \in QR_N$, $e_1 \in Z_{2^t}$, $e_2 \in Z_{2^t}$, $(e_1 \neq e_2)$, y_1, and y_2 such that*

$$y_i^{2^t} \equiv 2^{2e_i}x \bmod N \quad and$$

$$y_i \in QR_N \ (i = 1, 2).$$

security parameter and $\mathcal{N}$ practically, although, theoretically, it also depends on the modulus size. In our evaluation, this security parameter, the probability that a dishonest Customer is accepted in the payment protocol, is supposed to be at most $1/2^{40}$.

3   Binary Tree Approach

We will adopt the binary tree approach as do all previous divisible cash schemes [8, 11, 19, 20]. Each coin of worth $w = 2^l$ is associated with a tree of $(1 + l)$ levels and w leaves.

Each node of the tree represents a certain denomination. The root node, n_0, is assigned a monetary value of w, and the value of all other nodes, $n_{j_1 \cdots j_i}$, is found by halving the value of the node's parent, $n_{j_1 \cdots j_{i-1}}$ ($j_1 = 0$, $j_i \in \{0, 1\}$ for $i = 2, \ldots, l$).

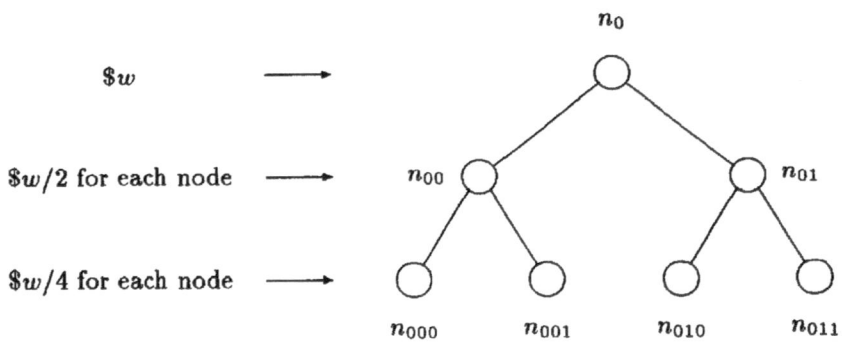

With this tree, we will show that for a single coin of worth w, it will be possible for a consumer to engage in several transactions, such that the sum total of the amounts of each transaction is less than or equal to w.

Divisibility can be implemented under the following two rules:

1. **(Route node rule:)** When a node is used, all descendant nodes and all ancestor nodes of this node cannot be used.
2. **(Same node rule:)** No node can be used more than once.

Preserving both rules implies that the set of past transactions involving the coin is legitimate and vice versa. Spending more than w, the value of a coin, will result in violation of at least one of these rules.

Moreover, in our concrete cash scheme, which will be shown in the following sections, two values are used for each node in the tree (Γ value and Λ value); Γ values are used to realize the route node rule, and Λ values to realize the same node rule. $\Gamma_{j_1 \cdots j_i}$ and $\Lambda_{j_1 \cdots j_i}$ denote Γ value and Λ value for node $n_{j_1 \cdots j_i}$ respectively. In addition, Ω values are introduced to calculate Γ values.

4   Bit Commitment Schemes

A bit commitment scheme is used in the opening stage of the proposed scheme, in place of the cut-and-choose method of the previous schemes [19, 20]. That is, the commitment scheme plays an essential role in realizing the "single-term" property of the proposed cash scheme.

One type of commitment scheme was used in the gradually releasing protocol [7]. Our paper, however, uses another type of commitment scheme, since the commitment scheme in [7] is not appropriate in our scheme. The new commitment scheme is based on the discrete logarithm problem [5], while the scheme in [7] is based on the factoring problem. However, almost all techniques developed in [7] can be used with slight modification in our scheme.

4.1 Bit Commitments

Assume that B sets up the commitment scheme and U commits to a number. Finally U proves to B that a value is correctly generated without revealing committed information, by using some protocols to be described later.

To set up the commitment scheme, B generates prime $\mathcal{P}$ satisfying $\mathcal{P} - 1 = 2 \cdot Prime$ ($Prime$ is a prime number), G and g whose orders in the multiplicative group $Z_{\mathcal{P}}^*$ are $Prime$. B sends $\mathcal{P}$, G and g. U checks whether $Prime = (\mathcal{P} - 1)/2$ is a prime by a probabilistic primality (or composite) test, and whether the orders of G and g are $Prime$ by checking that they are not 1 and $G^{Prime} \equiv 1 \pmod{\mathcal{P}}$ and $g^{Prime} \equiv 1 \pmod{\mathcal{P}}$.

U can commit to any integer $s \in Z_{Prime}$ by choosing R uniformly at random in Z_{Prime} and computing the commitment

$$BC_g(R, s) = G^R g^s \bmod \mathcal{P}.$$

This is called a base-g commitment. A commitment is opened by revealing R and s.

4.2 Protocols of Checking the Contents of Bit Commitments

This subsection introduces some useful protocols in which U can prove to B in a zero-knowledge manner that a committed value is in an interval, and that two committed values are equivalent.

Let the interval be $I = [a, b]$ $(= \{x | a \le x \le b\})$, $e = b - a$, and $I \pm e = [a - e, b + e]$.

Protocol: CHECK COMMITMENT
  Common input: x and $(\mathcal{P}, G, g, I)$.
  What to prove: U knows (R, s) such that $x = BC_g(R, s)$ and $s \in I \pm e$.
  Execute the following k times:

1. U chooses t_1 uniformly in $[0, e]$, and sets $t_2 = t_1 - e$. U sends to B the unordered pair of commitments $T_1 = BC_g(S_1, t_1)$, $T_2 = BC_g(S_2, t_2)$.
2. B selects a bit $\beta \in \{0, 1\}$ and sends it to U.
3. U sends to B one of the following:
   (a) if β is 0, opening of both T_1 and T_2
   (b) if β is 1, opening of $x \cdot T_i \bmod N$ $(i \in \{1, 2\})$, such that $s + t_i \in I$.
4. B checks the correctness of U's messages.

Protocol: COMPARE COMMITMENTS
  Common input: x, x' and $(\mathcal{P}, G, g, I)$.
  What to prove: U knows (R, R', s) such that $x = BC_g(R, s)$, $x' = BC_g(R', s)$ and $s \in I \pm e$.
  Execute the following k times:

---

[5] The underlying technique has been well known (e.g., [6, 21]).

1. U chooses t_1 uniformly in $[0, e]$, and sets $t_2 = t_1 - e$. U sends to B the unordered pair of commitments $(T_1, T_1'), (T_2, T_2')$, where each component of the pair is ordered and $(T_i, T_i') = (BC_g(S_i, t_i), BC_h(S_i', t_i))$.
2. B selects a bit $\beta \in \{0, 1\}$ and sends it to U.
3. U sends to B one of the following:
   (a) if β is 0, opening of both (T_1, T_1') and (T_2, T_2')
   (b) if β is 1, opening of $x \cdot T_i \bmod N$ and $x' \cdot T_i' \bmod N$ ($i \in \{1, 2\}$), such that $s + t_i \in I$.
4. B checks the correctness of U's messages.

Protocol: CHECK MOD-MULT
   Common input: x, y, z, n and $(\mathcal{P}, G, g, I = [n, 2n])$. $(|(\mathcal{P}-1)/2| \geq 2|n|+6)$
   What to prove: U knows $(R, R', R'', s, t, \alpha)$ such that $x = BC_g(R, s)$, $y = BC_g(R', t)$, $z = BC_g(R'', \alpha)$, $\alpha \equiv st \pmod{n}$, and $s, t, \alpha \in [0, 3n](= I \pm n)$.

1. U uses the CHECK COMMITMENT protocol with $I = [n, 2n]$ to prove that U knows how to open x to reveal a value in $[0, 3n]$ ($= I \pm n$).
2. U sends $v = BC_x(R''', t) = BC_g(R''' + tR, st)$, and uses the COMPARE COMMITMENTS protocol with I to prove that U knows how to open $y = BC_g(R, t)$ and $v = BC_x(R''', t)$.
3. U sends $u = BC_g(R'''', d)$, where d is defined by $st = \alpha + dn$, $\alpha \equiv st \pmod{n}$, and $s, t, \alpha \in I$.
4. U uses the CHECK COMMITMENT protocol with $[n-1, 4n-1]$ to prove that U knows how to open u to reveal a value in $[-2n-1, 7n-1]$ ($= [n-1, 4n-1] \pm 3n$). U uses the CHECK COMMITMENT protocol with I to prove that U knows how to open z to reveal a value in $I \pm n$.
5. U opens (as a base-g commitment) the product $zu^n v^{-1} \bmod \mathcal{P}$ to reveal a 0 (i.e., reveals R^* such that $BC_g(R^*, 0) = zu^n v^{-1} \bmod \mathcal{P}$).

5 Efficient Divisible Cash Scheme

This section outlines the various protocols in our divisible electronic cash scheme.

As in [18, 19], the electronic cash in our proposed scheme consists of an electronic license and electronic coins.

The electronic license is issued by the bank to a customer during an "opening protocol". This protocol is done once per customer, typically when a customer opens an account. If, however, a customer prefers to change the license at some later time, or desires several licenses, this protocol is run again for each additional license. As mentioned in [18], the frequency of license changes should be determined after considering the trade-offs between the degree of unlinkability and efficiency desired[6].

5.1 The Opening Protocol

As a result of the one-time opening protocol, customer U obtains an *electronic license* $(N, L_1 = (N+a_1)^{1/K} \bmod n_1, L_2 = (N+a_2)^{1/K} \bmod n_2)$. This basically grants U permission to use the electronic cash of bank B. Here, (n_1, K) and

---

[6] Even if a customer does not change the license at all, the customer's privacy (no traceability) is mathematically preserved as shown in Theorem 4. Then, anonymous purchase histories with the same license are linkable.

(n_2, K) are B's RSA public keys and (a_1, a_2) is also B's public key. They are common to many (or all) customers. (a_i, n_i, K) $(i = 1, 2)$ are used to check the validity of U's license, (N, L_1, L_2), in the payment and deposit protocols. N and (L_1, L_2) are kept secret from B in this opening protocol, to keep the untraceability.

Roughly, in this protocol, U secretly generates N, which is the composite of two primes P and Q, and gives $x = g^P \bmod \mathcal{P}$ and $y = g^Q \bmod \mathcal{P}$ to B as U's identity, where $\mathcal{P}$ and g are B's public key, which can be common for many customers. Then, U asks B to sign N in a blind manner (through the RSA blind signature), after U proves B that N is honestly generated (in the relation with x and y) in a zero-knowledge manner. So, finally U gets B's RSA signature, (L_1, L_2), for N, while B has no information on N and (L_1, L_2).

The opening protocol is as follows:

1. Bank B chooses the parameters of the bit commitment scheme, $\mathcal{P}$, G and g, and sends them to Customer U. B also sends RSA public-keys, (n_1, K) and (n_2, K), and public key (a_1, a_2), as the public keys of B's blind signatures. Here, K is the form of $2^J + 1$ (J: integer). For simplicity of explanation here, for $i = 1, 2$, we assume that $n_i = p_i q_i$ (p_i, q_i: prime) and $(p_i - 1)/2$ and $(q_i - 1)/2$ are primes. (When n_i is a more general form, a slight modification is required for the value of K such that, for example, K must be prime.) We also assume $|n_1| = |n_2|$ and $|a_i| < |n_i|/2$. W.l.o.g, we assume $n_1 < n_2$.

2. Customer U checks whether $Prime = (\mathcal{P} - 1)/2$ is a prime, and whether the orders of G and g are $Prime$. U also checks whether $|(\mathcal{P} - 1)/2| \geq 2|n_i| + 6$. U generates two primes P and Q, and random numbers, $r_i \in Z_{n_i}$ $(i = 1, 2)$. Here, $|P| < K$ and $|Q| < K$, and $(1/4)n_1^{1/2} < P, Q < (1/2)n_1^{1/2}$. U calculates $x = g^P \bmod \mathcal{P}$, $y = g^Q \bmod \mathcal{P}$, and $s_i = (N + a_i)r_i{}^K \bmod n_i$ $(i = 1, 2)$. U sends (x, y, s_1, s_2) to B with U's signature, as U's identity.

3. U and B execute the following protocol: (Informally, U proves to B in a zero-knowledge manner that (s_1, s_2, x, y) is honestly generated.)

   (a) U generates random numbers, R_N, $R_{r_i}^{(0)}$, $R_{r_i}^{(1)}$, $\cdots$, $R_{r_i}^{(J)}$, $R_{r_i}^* \in Z_{Prime}$, $(i = 1, 2)$.

   U calculates $\alpha_i^{(1)}$, $\alpha_i^{(2)}$, $\cdots$, $\alpha_i^{(J)}$ such that $\alpha_i^{(j)} \equiv r_i^{2^j} \pmod{n_i}$ ($i = 1, 2; j = 1, \cdots, J$) ($\alpha_i^{(0)} = r_i$) and $\alpha_i^{(j)} \in [n_i, 2n_i]$ ($i = 1, 2; j = 0, \cdots, J$). U also calculates α_i^* such that $\alpha_i^* \equiv r_i^K \pmod{n_i}$ ($\equiv r_i^{2^J+1} \pmod{n_i}$) and $\alpha_i^* \in [n_i, 2n_i]$. U also sends B the following values: $z = BC_x(R_N, Q) = BC_g(R_N, N)$, $(N = PQ)$, $u_i^{(0)} = BC_g(R_{r_i}^{(0)}, \alpha_i^{(0)})$, ..., $u_i^{(J)} = BC_g(R_{r_i}^{(J)}, \alpha_i^{(J)})$, $u_i^* = BC_g(R_{r_i}^*, \alpha_i^*)$.

   (b) U uses the CHECK COMMITMENT protocol with interval $[(1/4)n_1^{1/2}, (1/2)n_1^{1/2}]$ to prove that U knows how to open x to reveal a value in $[0, (3/4)n_1^{1/2}]$. (Although x is a specific type of bit commitment: $x = BC_g(0, P)$, the CHECK COMMITMENT protocol can be used similarly.)

   (c) U uses the COMPARE COMMITMENTS protocol with $[(1/4)n_1^{1/2}, (1/2)n_1^{1/2}]$ to prove that U knows how to open $y = BC_g(0, Q)$ and $z = BC_x(R_N, Q)$ in $[0, (3/4)n_1^{1/2}]$.

   (d) The following procedure is repeated for $j = 0, \cdots, J - 1$ and $i = 1, 2$:

U uses the MOD-MULT protocol with $[n_i, 2n_i]$ for $(u_i^{(j)}, u_i^{(j)}, u_i^{(j+1)}, n_i, \mathcal{P}, G, g)$ to prove that $\alpha_i^{(j+1)}$ is committed to by $u_i^{(j+1)}$. (Duplicated procedures between j-th step and $(j+1)$-th step such as CHECK COMMITMENT protocol for $u_i^{(j)}$ can be omitted.)

(e) U uses the MOD-MULT protocol with $[n_i, 2n_i]$ for $(u_i^{(0)}, u_i^{(J)}, u_i^*, n_i, \mathcal{P}, G, g)$ to prove that α_i^* is committed to by u_i^*.
U uses the MOD-MULT protocol with $[n_i, 2n_i]$ for $(z_i, u_i^*, BC_g(0, s_i), n_i, \mathcal{P}, G, g)$ to prove that $s_i = (N + a_i)r_i^K \bmod n_i$ $(i = 1, 2)$, where $z_i = zg^{a_i} \bmod \mathcal{P} = BC_g(R_N, N + a_i)$.

4. B gives $\delta_i = s_i^{1/K} \bmod n_i$ $(i = 1, 2)$ to U.
5. U obtains $L_i = (N + a_i)^{1/K} \bmod n_i$ by $\delta_i / r_i \bmod n_i$ $(i = 1, 2)$.

5.2 The Withdrawal Protocol

When the customer wants to withdraw \$$w$ from the account, an electronic coin of worth \$$w$ is then obtained by executing the withdrawal protocol with the bank.

The withdrawal protocol itself is very simple: Bank B just issues a blind signature to user U. Assume the consumer wishes to withdraw a divisible coin worth $w = 2^l$ dollars from his account at bank B. (That is, assume U sends B U's signed message to request the withdrawal. Here, we assume that the key for U's signature is independently generated from the other parameters except the size.) Also, B has a public key of the RSA signatures, (e_w, n_w), which corresponds to $w = 2^l$ dollars. The following steps occur:

1. U chooses a random value b, then forms and sends Z to B.

$$Z = r^{e_w} H(N \parallel b) \bmod n_w,$$

where $r \in Z_{n_w}$ is a random integer and H is a one-way hash function.
2. B gives $Z^{1/e_w} \bmod n_w$ to U and charges U's account \$$w$.
3. U can then extract the electronic coin $C = (H(N \parallel b))^{1/e_w} \bmod n_w$.

5.3 Payment

Assume that customer U spends \$$y$ $(\leq w)$ at shop V through the payment protocol. The payment protocol consists of two stages: coin authentication and denomination revelation. During the coin authentication phase, the shop verifies that the coin bears the bank's signature. During the second phase, the customer reveals information about a certain set of nodes in the coin's binary tree representation depending on the denomination being spent. We assume that f_r, f_Ω, f_Λ and h are truly random (or pseudo random) functions[7]. These stages are described in more detail as follows:

---

[7] Here we omit the explicit description of the domains and codomains of these functions, since they are naturally determined by the input and output variables.

Coin Authentication Customer U supplies shop V with (L_1, L_2), N, (C, b) and w. The shop checks that $L_i{}^K \equiv N + a_i \pmod{n_i}$ $(i = 1, 2)$, $C^{e_w} \equiv H(N \parallel b) \pmod{n_w}$, $(-1/N) = 1$, and $(2/N) = -1$.

If the number, k, of the nodes corresponding to U's payment to V is less than 10 $(k' = 10 - k > 0)$[8], V gives $z_1, \cdots, z_{2k'}$ to U, U calculates $[(< f_\Gamma(z_1) >_{QR})^{1/2} \bmod N]_1, \cdots, [(< f_\Gamma(z_{2k'}) >_{QR})^{1/2} \bmod N]_1$, and send them to V. V checks the validity.

Denomination Revelation Let $[y_{l+1} y_l \cdots y_1]$ $(y_i \in \{0, 1\}, i = 1, \ldots, l + 1)$ be the binary representation of y. Here the length, $l + 1$, of the binary string is fixed by the coin value $(w = 2^l)$, and some most significant bits such as y_{l+1} can be 0. Then, if $y_{l+2-t} = 1$ $(t = 1, \ldots, l + 1)$, U selects a node $n_{j_1 \cdots j_t}$ [9] among the nodes in the t-th level that do not violate the two binary tree rules (see Section 3). Here, U has memorized the nodes already spent. The average number of nodes to be spent per a payment is at most $(l + 1)/2$.

We will show the payment protocol when U spends node $n_{0j_1 j_2 \cdots j_t}$ to V. When several nodes are spent per a payment, the following protocol of each node must be executed simultaneously.

1. U computes $\Gamma_{j_1 \cdots j_t}$,

$$\Gamma_{j_1 \cdots j_t} = [(< (\Omega_{j_1 \cdots j_{t-1}})^{2^{t-1} j_t} (\Omega_{j_1 \cdots j_{t-2}})^{2^{t-2} j_{t-1}} \cdots (\Omega_{j_1})^{2 j_2} \times$$

$$f_\Gamma(C \parallel 0 \parallel N) >_{QR})^{1/2^t} \bmod N]_{-1},$$

where $\Omega_{j_1 \cdots j_i} = < f_\Omega(C \parallel j_1 \parallel \cdots \parallel j_i \parallel N) >_1$ $(i = 1, \ldots, t - 1)$.

2. V computes $\Omega_{j_1 \cdots j_i}$ when $j_{i+1} = 1$ $(i = 1, \ldots, t - 1)$. Then V verifies the validity of $\Gamma_{j_1 \cdots j_t}$ such that

$$(\Gamma_{j_1 \cdots j_t} / N) = -1,$$

$$(\Gamma_{j_1 \cdots j_t})^{2^t} \equiv$$

$$d (\Omega_{j_1 \cdots j_{t-1}})^{2^{t-1} j_t} (\Omega_{j_1 \cdots j_{t-2}})^{2^{t-2} j_{t-1}} \cdots (\Omega_{j_1})^{2 j_2} f_\Gamma(C \parallel 0 \parallel N) \pmod{N},$$

where $d \in \{\pm 1, \pm 2\}$. If they are valid, V selects a random value $e' \in \{0, 1\}^u$, and sends V's identity ID_V, time T, and e' to U, where $u = O(m)$, $m = |P|(= |Q|)$. Otherwise V halts this protocol. V computes $e = h(ID_V \parallel T \parallel e')$, where $e \in \{0, 1\}^u$.

3. U computes $e = h(ID_V \parallel T \parallel e')$. U also computes $\Lambda_{j_1 \cdots j_t}$ such that

$$(\Lambda_{j_1 \cdots j_t})^{2^{u+1}} \equiv 2^{2e} < f_\Lambda(C \parallel j_1 \parallel \cdots \parallel j_t \parallel N) >_{QR} \pmod{N}.$$

4. V verifies that

$$(\Lambda_{j_1 \cdots j_t})^{2^{u+1}} \equiv d' 2^{2e} f_\Lambda(C \parallel j_1 \parallel \cdots \parallel j_t \parallel N) \pmod{N},$$

where $d' \in \{\pm 1, \pm 2\}$. If verification succeeds, V accepts U's messages as payment of the amount due.

---

[8] The value, 10, is theoretically $O(m)$, where $m = |P|(= |Q|)$. This is required for checking whether N is the Williams integer with high probability.

[9] If U selects a node randomly among the valid nodes, U can conceal some information about U's purchase history to V, since if nodes to be spent are selected by a published rule, V may get some information about the purchase history from the nodes.

Remarks

1. Notations $\Gamma_{j_1\cdots j_i}$ and $\Lambda_{j_1\cdots j_i}$ are differently used from those in [19]. ($\Gamma_{j_1\cdots j_i}$ and $\Lambda_{j_1\cdots j_i}$ in this paper correspond to $X_{i,j_1\cdots j_i}$ and $Y_{i,j_1\cdots j_i}$ in [19].)

2. Function h can be a collision intractable hash function in place of a truly random (or pseudo random) function.

5.4 Deposit

Deposit is as before; a transcript of payment is forwarded to the bank.

5.5 Detection of Overspending

Although, formally, the security including the detection of overspending is described in Section 6, in this subsection, we will describe the detection procedure of overspending.

Clearly, if U overspends a coin, U must violate one of the two rules of the binary tree approach (see Section 3).

First, we show that "Route node rule" of the binary tree approach is securely realized. Assume that nodes $n_{n_{0j_1j_2\cdots j_i}}$ and $n_{n_{0j_1j_2\cdots j_i\cdots j_t}}$ are used. (Clearly this assumption violates the the route node rule.) Then U sends $\Gamma_{j_1\cdots j_i}$ and $\Gamma_{j_1\cdots j_t}$ to shops, and these values are finally sent to B. B can firstly detect that the violation of the rule occurs, by checking the coin values (C, b) along with (L_1, L_2, N) and the consumed nodes, $n_{n_{0j_1j_2\cdots j_i}}$ and $n_{n_{0j_1j_2\cdots j_i\cdots j_t}}$, in B's data base. (To efficiently find the violation practically, B can use a short hashed values (e.g., 32 bytes) of the coin values as a search key in the data base.)

Then,

$$\Gamma_{j_1\cdots j_i} = [(< (\Omega_{j_1\cdots j_{i-1}})^{2^{i-1}j_i} \cdots (\Omega_{j_1})^{2j_2} f_\Gamma(C \parallel 0 \parallel N) >_{QR})^{1/2^i} \bmod N]_{-1},$$

On the other hand, from $\Gamma_{j_1\cdots j_t}$, B computes

$$(\Gamma_{j_1\cdots j_t})^{2^{t-i}} \equiv ((\Omega_{j_1\cdots j_{i-1}})^{2^{t-i-1}j_i} \cdots (\Omega_{j_1\cdots j_i})^{2^0 j_{i+1}}) \times$$

$$[(< (\Omega_{j_1\cdots j_{i-1}})^{2^{i-1}j_i} \cdots (\Omega_{j_1})^{2j_2} f_\Gamma(C \parallel 0 \parallel N) >_{QR})^{1/2^i} \bmod N]_1. \quad (\bmod N),$$

since $(\Gamma_{j_1\cdots j_t})^{2^{t-i}} \bmod N$ is the quadratic residue and $(\Omega_{j_1\cdots j_i}/N) = 1$. Therefore, B can compute

$$[(< (\Omega_{j_1\cdots j_{i-1}})^{2^{i-1}j_i} \cdots (\Omega_{j_1})^{2j_2} f_\Gamma(C \parallel 0 \parallel N) >_{QR})^{1/2^i} \bmod N]_1.$$

Using this value and $\Gamma_{j_1\cdots j_i}$, B can efficiently and deterministically factor N and obtains P and Q, from which B can trace U's identity, x and y. Here, (P, Q) is the witness of U's violating one of these rules.

Next, we show that "Same node rule" of the binary tree approach is also securely realized. Assume that a node $n_{n_{0j_1j_2\cdots j_i}}$ is used twice at different time or place. Then U's challenge messages (say, e_1 and e_2) of the double spending should be different with overwhelming probability from the property of a random function, h). Then, clearly from Lemma 2, B can efficiently factor N and obtains P and Q, from which B can trace U's identity, x and y.

6 Security

In this section, we show that the proposed cash scheme satisfies the four security requirements under some assumptions.

First, we introduce the security requirements for our cash system. They are a modification of those of [13].

Definition 3. Let m be $|P| = |Q|$. The proposed cash system is secure if the following conditions are satisfied:

- **No forging:** For all integers $k > 0$, for any poly-time nonuniform algorithm Adv (e.g., dishonest customer), after Adv's k times execution as a customer of the (resp., opening, withdrawal) protocol with bank B, the probability that Adv computes $k + 1$ (resp., licenses, coins) that pass the coin authentication by a shop is negligible in m.
- **No tracing:** For any poly-time nonuniform algorithm Adv (e.g., dishonest bank), after Adv's execution as a bank and shops of the opening, withdrawal and payment protocols with customers U_0 and U_1, and given a payment transcript by customer U_r ($r \in \{0, 1\}$), the probability that Adv outputs r correctly is less than $1/2 + 1/m^c$ for all constant c and for all sufficiently large m.
- **No overspending:** Suppose that customer U withdraws a coin, C, worth w dollars through the valid opening and withdrawal protocols with bank B. For any possible value of w, if customer U spends more than w dollars by C through payment protocol with shops, then there exists a probabilistic poly-time algorithm, $DETECT$ (e.g., bank), which, given all payment transcripts regarding C, can compute P such that $x = g^P \mod \mathcal{P}$ with overwhelming probability in m, where x is U's identity authorized in the opening protocol.
- **No swindling:** For any possible value of w, for any poly-time nonuniform algorithm Adv (e.g., dishonest shop), given all transcripts of opening and withdrawal protocols and after Adv's executions as a shop of the payment protocol with customers, in which the total payment value is w dollars, the probability that Adv can deposit more than w dollars at Bank B is negligible in m.

Remark: In *No overspending* condition, $DETECT$ can not only trace U's identity x from the overspending payment transcripts, but also gives the evidence, P, that U cannot deny U's overspending, since any poly-time algorithm is hard to calculate P unless U overspends.

Next, the following assumptions are required to prove the security of the proposed cash scheme.

Assumptions:

- **(RSA signatures)**
    - RSA signatures used with one-way hash functions are existentially unforgeable against adaptive chosen message attacks[10].
    - Let (n_1, K) and (n_2, K) be two RSA public keys such that n_1 and n_2 are independently selected, and $|n_1| = |n_2|$. Let $a_1, a_2 \in \{0, 1\}^{|n_1|/2}$ be independently selected and published. Let $L_1^K \equiv N + a_1 \pmod{n_1}$ and $L_2^K \equiv N + a_2 \pmod{n_2}$. Then, the signature (L_1, L_2) for message N is existentially unforgeable against adaptive chosen message attacks.

---

[10] For the definition of "existentially unforgeable" etc., see [15].

- There is no efficient algorithm, given an RSA public key, (n, K), to output P and Z with $P < n^{1/2}$, $P \equiv Z^K \pmod{n}$, and $K > |P| > 1$.
- (**Factoring and Diffie-Hellman**) Let $\mathcal{P}$, P_0, Q_0, P_1 and Q_1 be primes, $N_0 = P_0 Q_0$, $N_1 = P_1 Q_1$, and $m = |P_0| = |Q_0| = |P_1| = |Q_1| < |\mathcal{P}| < dm$ for a constant d. Let $(\mathcal{P} - 1)/2$ be a prime, and the order of g in the multiplicative group $Z_{\mathcal{P}}^*$ is $(\mathcal{P} - 1)/2$. Then, for any probabilistic poly-time (non-uniform) machine M, given $\mathcal{P}$, g, $(x_0 = g^{P_0} \bmod \mathcal{P}, y_0 = g^{Q_0} \bmod \mathcal{P})$, $(x_1 = g^{P_1} \bmod \mathcal{P}, y_1 = g^{Q_1} \bmod \mathcal{P})$, and (N_r, N_{1-r}) $(r \in_R \{0, 1\})$, M can compute r with probability less than $1/2 + 1/m^c$ for all constant c and for all sufficiently large m.
- (**Random functions**) Commonly available functions, f_Γ, f_Λ, f_Ω and h are truly random or pseudo-random[11].

Remarks
1. The first and second assumptions seem to be reasonable in practice, but the third assumption may be controversial, since a truly random function cannot be realized in the real world (as more than an exponential size of memory is required), and a commonly available pseudo-random function requires a tamper-free device. However, by using this assumption, the theoretical security requirements for our scheme can be clarified. We believe that a truly random or pseudo-random function f can be replaced by a practical one-way hash function family without sacrificing the security of our scheme, in practice.
2. The first assumption implies that factoring an RSA modulus n is hard, since if it is solved, the first assumption does not hold. The second assumption implies that both factoring N_i and the discrete logarithm $g^{P_i} \bmod \mathcal{P}$ are hard, since if at least one of them is solved, the second assumption does not hold.

Theorem 4. *The proposed cash scheme is secure under the above-mentioned assumptions.*

7 Efficiency

For this section, let us assume that $|P| = |Q| = 256$ bits, $|N| = 512$ bits, $|n_i| = 514$ bits $(i = 1, 2)$, $((1/4)n_1^{1/2} < P, Q < (1/2)n_1^{1/2})$, $|Prime| \geq 1030$ bits, $|b| = 64$ bits, and $|n_w| = 512$ bits. Then, $K = 2^8 + 1$ (i.e., $J = 8$). We also assume the binary tree has 18 levels, i.e., $\mathcal{N} = 2^{17}$, where $\mathcal{N} = $ (the total coin value)/(minimum divisible unit value). For example, the total value is about \$ 1000, and the minimum divisible unit is 1 cent. Here, we also assume that efficient random hash functions are used, which are usually much faster than modular exponentiations.

Then, customer U uses 576 bits (72 bytes) of data for the electronic coin (C, b) worth \$1000 and U's proper data (electronic license, L_1, L_2, P, Q) is 1536 bits (192 bytes). Thus the total amount of data (264 bytes) is small enough to be stored on typical smart cards.

In the withdrawal protocol, customer U and bank B send 512 bit (64 bytes) messages respectively, and U and B just compute one exponentiation $\bmod n_w$ respectively.

---

[11] For the definitions of truly random and pseudo-random functions, see [14].

In the payment protocol (denomination revelation), 9 nodes are used on average for each payment when the tree has 18 levels [12]. For each node, two 512 bit values ($\Gamma_{j_1 \ldots j_t}$ and $\Lambda_{j_1 \ldots j_t}$) are transfered to the shop. Hence, a total of 1152 bytes is transfered to a shop in a payment on average. The computation required from a customer U for a payment is, in total, 18 (2×9) times the computation of the 2^u-th (or 2^{u+1}-th) root $\bmod N$, using the factors of N. Each root computation $\bmod N$ is almost comparable to exponentiation $\bmod N$. The computation required from the shop is almost the same, 18 exponentiations $\bmod N$.

The most time consuming part of our cash system is the opening protocol, although the protocol is still practical, as the amount of computation and communication is $O((\log|P|)|P|^4)$. But, this protocol can be executed much less frequently than the other procedures such as withdrawal and payment protocols (see Subsection 5.1). When $J = 8$ and $k = 20$, around 4000 multi-exponentiations $\bmod \mathcal{P}$ are required (from U and B respectively) for the opening protocol. (A multi-exponentiation can be computed almost as efficiently as a exponentiation by the extended binary method [17].) The amount is fairly heavy, but all of U's computation can be pre-computed, and the opening protocol can be executed by U's terminal (Workstation or PC) instead of a smart card. (After the opening protocol, U can store the data (just 192 bytes) in a smart card.)

8 Conclusion

This paper has presented a practical "divisible" off-line electronic cash scheme that is more efficient than previous schemes. Our scheme is the first practical divisible cash scheme that is single-term and in which every procedure can be executed in the logarithmic order of the precision of divisibility.

In addition, we proved the security of the proposed cash scheme under some cryptographic assumptions. Our scheme is the first practical divisible coin scheme whose cryptographic security assumptions are theoretically clarified.

The remaining problems are:

- Improve the efficiency of the opening protocol.
- Realize the unlinkability among coins divided from the same coin.
- Prove the security under more primitive assumptions such as the hardness of factoring and discrete logarithm.
- Find requirements which are formally shown to be sufficient for the security of electronic cash schemes. (The four requirements shown in this paper are still ad hoc.)

Acknowledgments

I would like to thank Ivan Damgård, Torben Pedersen and Moti Yung for informing me some attacks on preliminary versions of this paper. I would also like to thank Tony Eng for many useful comments. This work was partially conducted while visiting AT&T Bell Laboratories, Murray Hill, NJ, USA. I wish to thank Andrew Odlyzko and Bell Laboratories for the hospitality.

---

[12] In the coin authentication phase of the payment protocol, the modular square root operations are needed with probability of $4/18$, and the average number of the root operations is 5. So, the average amount of the computation and communication is comparable to 2.5 ($= 10 \times (10/18) \times (1/2)$) node operations. Here, we neglect the operations of this phase, to simplify the evaluation, since it is relatively small.

References

1. Blum, M., "Coin flipping by telephone", IEEE, COMPCON, pp.133-137 (1982).
2. Brands, S., "Untraceable Off-line Cash in Wallet with Observers", Proceedings of Crypto 93, pp.302-318 (1994).
3. Bleumer, G., Pfitzmann, B. and Waidner, M., "A Remark on a Signature Scheme Where Forgery can be Proved", Proceedings of Eurorypt 90, pp.441-445 (1991).
4. Chaum, D., "Security without Identification: Transaction Systems to Make Big Brother Obsolete," Comm. of the ACM, 28, 10, pp.1030-1044 (1985).
5. Chaum, D., Fiat, A., and Naor, M., "Untraceable Electronic Cash," Proceedings of Crypto 88, pp.319-327 (1990).
6. Chaum, D., van Heijst, E., and Pfitzmann, B., "Cryptographically Strong Undeniable Signatures, Unconditionally Secure for the Signer," Proceedings of Crypto 91, pp.470-484 (1992).
7. Damgård, I., "Practical and Provably Secure Release of a Secret and Exchange of Signatures," Proceedings of Eurocrypt 93 (1993).
8. D'amingo, S. and Di Crescenzo, G., "Methodology for Digital Money based on General Cryptographic Tools", to appear in the Proceedings of Eurocrypt 94.
9. De Santis, A. and Persiano, G., "Communication Efficient Zero-Knowledge Proofs of Knowledge (with Applications to Electronic Cash)" Proceedings of STACS 92, pp. 449-460 (1992).
10. Even, S., Goldreich, O. and Yacobi, Y., "Electronic Wallet", Proceedings of Crypto 83, pp.383-386 (1983).
11. Eng, T. and Okamoto, T. "Single-Term Divisible Coins," to appear in the Proceedings of Eurocrypt 94.
12. Ferguson, N., "Single Term Off-line Coins", Proceedings of Eurocrypt 93, pp.318-328 (1994).
13. Franklin, M. and Yung, M., "Secure and Efficient Off-Line Digital Money", Proceedings of ICALP 93, pp. 449-460 (1993).
14. Goldreich, O., Goldwasser, S., and Micali, S., "How to Construct Random Functions," Journal of ACM, Vol.33, No.4 (1986).
15. Goldwasser, S., Micali, S. and Rivest, R., "A Digital Signature Scheme Secure Against Adaptive Chosen-Message Attacks," SIAM J. Comput., 17, 2, pp.281-308 (1988).
16. Hayes, B., "Anonymous One-Time Signatures and Flexible Untraceable Electronic Cash," Proceedings of Auscrypt 90, pp.294-305 (1990).
17. Knuth, D.E. The Art of Computer Programming, Vol.2, 2nd Ed. Addison-Wesley (1981).
18. Okamoto, T., and Ohta, K., "Disposable Zero-Knowledge Authentication and Their Applications to Untraceable Electronic Cash", Proceedings of Crypto 89, pp. 481-496 (1990).
19. Okamoto, T., and Ohta, K., "Universal Electronic Cash", Proceedings of Crypto 91, pp. 324-337 (1992).
20. Pailles, J.C., "New Protocols for Electronic Money", Proceedings of Auscrypt 92, pp. 263-274 (1993).
21. Pedersen, T. P., "Non-Interactive and Information-Theoretic Secure Verifiable Secret Sharing", Proceedings of Crypto 91, pp. 129-140 (1992).
22. Pfitzmann, B. and Waidner, M., "How to Break and Repair a "Provably Secure" Untraceable Payment System," Proceedings of Crypto 91 (1992).
23. Rabin, M.O., "Digitalized Signatures and Public-Key Functions as Intractable as Factorization," Tech. Rep., MIT/LCS/TR-212, MIT Lab. Comp. Sci., (1979).
24. Vaudenay, S., "One-Time Identification with Low Memory," Eurocodes 92 (1992).
25. Yacobi, Y., "Efficient electronic money", to appear in the Proceedings of Asiacrypt 94.

Collusion-Secure Fingerprinting for Digital Data

(Extended Abstract)

Dan Boneh*
dabo@cs.princeton.edu

James Shaw
jhs@cs.princeton.edu

Department of Computer Science
Princeton University
Princeton, NJ 08544

Abstract. This paper discusses methods for assigning codewords for the purpose of fingerprinting digital data (e.g., software, documents, and images). Fingerprinting consists of uniquely marking and registering each copy of the data. This marking allows a distributor to detect any unauthorized copy and trace it back to the user. This threat of detection will deter users from releasing unauthorized copies.

A problem arises when users collude: For digital data, two different fingerprinted objects can be compared and the differences between them detected. Hence, a set of users can collude to detect the location of the fingerprint. They can then alter the fingerprint to mask their identities. We present a general fingerprinting solution which is secure in the context of collusion. In addition, we discuss methods for distributing fingerprinted data.

1 Introduction

The human fingerprint has been used to distinguish individuals since the turn of the century. Wishing to distinguish other objects, people have applied fingerprint-like identifiers to a variety of objects from maps and mathematical tables to diamonds and explosives. With the increasing use of digital computers and data, there has been a strong desire to fingerprint these data as well. This paper discusses a general method to fingerprint digital data. We use the following terminology:

A *mark* is one bit of information which is encoded in an object. For instance, in a book, the use of one word versus another in some point in the text, can be used as a mark.

A *fingerprint* is a collection of marks. Thus, a fingerprint can be thought of as a binary word of length L, where L is the number of marks in the object.

A *distributor* is the sole supplier of fingerprinted objects.

A *user* is the registered owner of a fingerprinted object.

The core idea of fingerprinting is that each user receives a copy of the object containing a unique fingerprint. This fingerprint can be used to identify the user.

---

* Supported in part by NSF CCR–9304718.

Because a fingerprinted object can be traced, the users will be deterred from releasing an unauthorized copy. The users will wish to destroy the fingerprint, while the distributor will wish to prevent this from happening.

Unlike human fingerprints, digitally-stored data can be altered with no harm to the object. It is the goal of the distributor to make fingerprints which are unalterable and undetectable. However, for digital data, the fingerprints of two users can be easily compared and the differences between them detected. Two users can collude to produce a new fingerprint, potentially masking their identities, unless this is explicitly prevented. To the best of our knowledge, previous work does not discuss fingerprints which resist these effects of collusion. We discuss fingerprints which can be traced back to their users even if multiple users collude. Our fingerprinting method can be applied to any form of digital data. A related result by Chor, Fiat and Naor [5] discusses a similar problem in a different setting. Within the paper, we use as an example the specific problem of software protection.

The Software Publishers Association estimates that in one year over seven billion dollars was lost due to software piracy: the illegal copying and redistribution of computer software [12]. A variety of copy protection schemes exist in an attempt to slow this piracy. However, most of these schemes lack rigor and are eventually bypassed. A rigorous treatment was given by Goldreich and Ostrovsky. They discusses how to protect software using an oblivious RAM which prevents an opponent from gaining any information about a running program [8]. The main difference between such schemes and fingerprinting is that fingerprinting does not actively prevent copying. Instead it used as a deterrent. Fingerprinting does not require any special hardware and has negligible effect on the running time.

By definition, each fingerprinted copy of an object is unique; however, sending each user a different copy may not be practical. The distribution problem is the avoidance of giving each user a different copy. By using redundancy between copies, we show how to reduce the cost of distributing fingerprinted data.

Our fingerprint consists of a set of binary marks[2]. We denote the two states of a mark by $\{0, 1\}$. By colluding, users can detect a specific mark if it differs between their copies; otherwise a mark can not be detected. The main property the marks should satisfy is that users can not change the state of an undetected mark without rendering the object useless. A detected mark's state can be changed to any one of the three states $\{0, 1, ?\}$. The symbol '?' means that the mark is unreadable, i.e. is impossible to determine if the mark is 0 or 1.

We assume that marks satisfying these properties exist for the objects being fingerprinted. We refer to this as the Marking Assumption for which a precise definition is given in the next section. Note that if there is no collusion, by the Marking Assumption, fingerprinting is trivial: the fingerprint assigned to each user will be the user's serial number.

There has been much research investigating the Marking Assumption in a

---

[2] For simplicity, we use binary marks throughout the paper. Using a larger fixed size alphabet will change our results by constant factors.

variety of domains. Wagner [15] gives a taxonomy of fingerprints and suggests subtle marks for computer software. Marks have been embedded in digital images [13, 4], in documents [3], and in computer programs [7]. In all of these domains, our scheme allows these marks to be combined to form secure fingerprints. Thus our results are general, allowing a variety of digital data to be fingerprinted.

The remainder of the paper is organized into the following sections: Section 2 defines codes which prevent a group of users from framing another user; we call such codes c-frameproof. Section 3 defines more general codes which enable the distributor to trace a copy in the presence of c users colluding; we refer to these codes as c-secure codes. A construction of c-secure codes is presented in Section 4. Section 5 describes an efficient method for distributing fingerprinted data. Section 6 concludes with a summary and discussion of open problems.

Throughout the paper, the following notation will be used. Given an l-bit word $w \in \{0,1\}^l$ and a set $I = \{i_1, \ldots, i_r\} \subseteq \{1, \ldots, l\}$ we denote by $w|_I$ the word $w_{i_1} w_{i_2} \ldots w_{i_r}$ where w_i is the i'th bit of w. We refer to $w|_I$ as the *restriction* of w to the bit positions in I.

2 Protection Against Naive Redistribution

In order to protect its product, a distributor could uniquely mark each copy, using marks that are undetectable by individual users. The general strategy used by the distributor is to assign a unique codeword to each user and to ship each user his data marked with that codeword. If we assume that users can not change the marks, then if an unauthorized copy is found, the codeword embedded in it must belong to some user u. Unfortunately, the assumption that users can not change marks is unrealistic. A coalition of users can detect some of the marks, namely the ones where their copies differ. They can then change these marks arbitrarily. Henceforth, we assume that this is the only type of mark which users can detect. The following definitions formally capture this notion.

Definition 1. A set $\Gamma = \{w^{(1)}, \ldots, w^{(n)}\} \subseteq \{0,1\}^l$ will be called an (l, n)-**code**. The codeword $w^{(i)}$ will be assigned to user u_i, for $1 \leq i \leq n$. We refer to the set of words in Γ as the **codebook**.

Definition 2. Let $\Gamma = \{w^{(1)}, \ldots, w^{(n)}\}$ be an (l, n)-code and C be a coalition of users. For $i \in \{1, \ldots, l\}$ we say that bit position i is **undetectable** for C if the words assigned to users in C match in their i'th bit. Formally, suppose $C = \{u_1, \ldots, u_c\}$. Then bit i is undetectable if $w_i^{(u_1)} = w_i^{(u_2)} = \ldots = w_i^{(u_c)}$.

Definition 3. Let $\Gamma = \{w^{(1)}, \ldots, w^{(n)}\}$ be an (l, n)-code and C be a coalition of users. Let R be the set of undetectable bit positions for C. Define the **feasible set** of C as

$$F(C; \Gamma) = \left\{ w \in \{0,1\}^l \text{ s.t. } w|_R = w^{(u)}|_R \right\}$$

for some user u in C. Thus the feasible set contains all words which match the coalition's undetectable bits. Usually we omit the Γ and denote $F(C; \Gamma)$ by $F(C)$.

Formally, the Marking Assumption discussed in the introduction states that any coalition of c users is only capable of creating an object whose fingerprint lies in the feasible set of the coalition. If a coalition changes one of the detectable marks to be unreadable, then we arbitrarily say that the mark is set to 0.

In this section we consider the simple problem of *naive redistribution* which occurs when a user redistributes his copy of the object without altering it. If an unauthorized copy of the object is found containing user u's codeword we would like to say that user u is guilty. However, u could claim that he was framed by a coalition who created an object containing his codeword. Thus, we would like to construct codes that satisfy the following property: no coalition can collude to frame a user not in the coalition. We usually relax this condition by limiting the size of the coalition to c users. We call such codes c-frameproof codes.

If the code used to fingerprint the object is kept hidden from the users, then the construction of frameproof codes becomes trivial: to every one of the n users assign a unique codeword chosen at random. A coalition of users can not frame a user not in the coalition since they do not know his codeword. We would like to construct codes that are c-frameproof even if the codebook is known to the users. This requirement can be formally stated as follows:

Definition 4. A code Γ will be called a c-**frameproof** code if every set $W \subset \Gamma$, of size at most c, satisfies $F(W) \cap \Gamma = W$.

The definition states that in a c-frameproof code, the only codewords in the feasible set of a coalition of at most c users are the codewords of members of the coalition. Thus, no coalition of at most c users can frame a user who is not a member of the coalition. It is interesting to note that for random codes the length of the code must be exponential in c. Otherwise, a coalition of c users is likely to detect all the bits.

2.1 Construction of c-frameproof codes

We now show a construction for c-frameproof codes. A similar idea to the one presented below will be used in Section 4 for the construction of more powerful codes. We begin by introducing a simple (n, n)-code which is n-frameproof. Define the code $\Gamma_0(n)$ to be the (n, n)-code containing all n-bit binary words with exactly one 1. For example, the code $\Gamma_0(3)$ for three users is $\{100, 010, 001\}$. The following claim is immediate.

Claim 5. $\Gamma_0(n)$ *is a* (n, n)-*code which is* n-*frameproof.*

It is not difficult to see that any n-frameproof code must have length at least n. Hence, the code length of Γ_0 is optimal. The length of Γ_0 is linear in the number of users and is therefore impractical. We will use the code Γ_0 to construct shorter codes. We first recall some basic definitions from the theory of error correcting codes. See [14] for more details.

Definition 6. A set, $\mathcal{C}$, of N words of length L over an alphabet of p letters is said to be an $(L, N, D)_p$-*Error Correcting Code* or in short, an $(L, N, D)_p$-ECC, if the Hamming distance between every pair of words in $\mathcal{C}$ is at least D.

The idea of this construction is to compose the code $\Gamma_0(n)$ with an error correcting code. Let $\Gamma = \{w^{(1)}, \ldots, w^{(n)}\}$ be an (l, p)-code and let $\mathcal{C}$ be a $(L, N, D)_p$-ECC. We denote the composition of Γ and $\mathcal{C}$ by Γ'. The code Γ' is an (lL, N)-code defined as follows: for a codeword $v = v_1 v_2 \ldots v_L \in \mathcal{C}$ let

$$W_v = w^{(v_1)} \| w^{(v_2)} \| \ldots \| w^{(v_L)}$$

where $\|$ means concatenation of strings. The code Γ' is the set of all words W_v, i.e. $\Gamma' = \{W_v \mid v \in \mathcal{C}\}$.

Lemma 7. *Let Γ be a c-frameproof (l, p)-code and $\mathcal{C}$ be a $(L, N, D)_p$-ECC. Let Γ' be the composition of Γ and $\mathcal{C}$. Then Γ' is a c-frameproof code, provided $D > L \left(1 - \frac{1}{c}\right)$* .

Proof. Let C be a coalition of c users. We show that $F(C; \Gamma')$ contains no words from Γ' other than those of C. Let $v^1, \ldots, v^c \in \mathcal{C}$ be the codewords of $\mathcal{C}$ from which the codewords of the coalition C were derived.

Assume towards a contradiction that $F(C; \Gamma')$ contains a word $W \in \{0, 1\}^{lL}$ which belongs to a user $u \notin C$. Let $z \in \mathcal{C}$ be the codeword from which W was derived. For all $k = 1, \ldots, c$, the words z and v^k match in less than L/c positions. This follows since the minimal Hamming distance of $\mathcal{C}$ is bigger than $L(1 - 1/c)$. Hence, there must exist a position $1 \leq j \leq L$ for which $z_j \neq v_j^k$ for all $k = 1, \ldots, c$.

Let $C_j = \{w^{(v_j^1)}, \ldots, w^{(v_j^c)}\}$. Since Γ is a c-frameproof code we know that $w^{(z_j)}$ is not in $F(C_j; \Gamma)$. Since $w^{(z_j)}$ is a subword of W, this implies that $W \notin F(C; \Gamma')$. This contradiction proves the lemma. $\qquad\square$

We note that the condition that $\mathcal{C}$ has a large minimal distance can be relaxed. To make the proof work it suffices to require that no set of c words of $\mathcal{C}$ "cover" a word of $\mathcal{C}$ outside the set. This property has been studied in [6]. Using this relaxed requirement does not improve the constructions.

The error correcting codes we are using have large minimal distance and hence, low rate. By picking the codewords randomly it is possible to obtain a good low rate code. We state this in the following lemma, which is immediate from the Chernoff bound [2, Appendix A].

Lemma 8. *For any positive integers p, N let $L = 8p \log N$. Then there exists a $(L, N, D)_{2p}$-ECC where $D > L(1 - \frac{1}{p})$.*

The main theorem of this section now follows.

Theorem 9. *For any integers $n, c > 0$ let $l = 16c^2 \log n$. Then there exist an (l, n)-code which is c-frameproof.*

Proof. By Lemma 8 we know that there exists an error correcting code with parameters $(L, n, L(1 - 1/c))_{2c}$ for $L = 8c \log n$. Combining this with the code $\Gamma_0(2c)$ and Lemma 7 we get a c-frameproof code for n users whose length is $2cL = 16c^2 \log n$. □

To make this construction explicit we must use an explicit low rate error correcting code. Explicit constructions of such codes are described in [1]. The explicit construction are not as good as the bounds provided by Lemma 8. Using a simple explicit low rate code it is possible to obtain codes of length $l = c^2 \log^2(n)$.

3 c-secure Codes

Codes which are secure against naive redistribution do not provide enough information when users collude. Indeed, two users can collude to generate an object which is marked by a word which is not a codeword. In this section we address this more general problem.

Assume a distributor marks an object with a code Γ. Now, suppose a coalition of users, C, colludes to generate an unregistered object marked by some word x and then distributes this new object. When this object is found, the distributor would like to detect a subset of the coalition who created it.

We illustrate the problem with an example. Let Γ be some code. Let C_1 and C_2 be two coalitions of c users each such that $C_1 \cap C_2 = \emptyset$. Suppose an unregistered object is found which is marked by a codeword x which is feasible for both C_1 and C_2. Then both coalitions are suspect. Since their intersection is empty, we can not determine with certainty who created the unregistered object. This situation should be avoided and therefore we require that when the intersection of C_1 and C_2 is empty, the intersection of their feasible sets $F(C_1)$ and $F(C_2)$ should also be empty. This is captured in the following definition.

Definition 10. A code Γ will be called **totally c-secure** if

$$C_1 \cap \ldots \cap C_r = \emptyset \quad \Rightarrow \quad F(C_1) \cap \ldots \cap F(C_r) = \emptyset \ .$$

for all coalitions $C_1, \ldots, C_r$ of at most c users each.

It seems that totally secure codes provide a good solution to the problem of collusion. Unfortunately, when $c > 1$, totally c-secure codes do not exist.

Theorem 11. *For $c \geq 2$ and $n \geq 3$ there are no totally c-secure (l, n)-codes.*

Proof. Clearly, it is enough to show that there are no totally 2-secure codes. Let Γ be an arbitrary (l, n)-code. Let w_1, w_2, w_3 be three distinct codewords assigned to users u_1, u_2, u_3 respectively. Let w be the bitwise majority of w_1, w_2, w_3. The intersection of the coalitions $\{u_1, u_2\}, \{u_1, u_3\}, \{u_2, u_3\}$ is empty; however, the word w is feasible for all three coalitions. This proves that Γ is not totally 2-secure. □

Note that the above proof relies on the fact that Γ is made up of binary strings. The proof can be generalized to show that there are no totally c-secure codes even when larger alphabets are used. This generalization is based on the fact that users can make a detectable mark unreadable, e.g. by removing it. This is the main difference between our model and the model used in [5].

The negative result proved in Theorem 11 forces us to weaken our requirements for marking schemes. We will allow the distributor to make some random choices when embedding the codewords in the objects. The point is that the random choices will be kept hidden from the users. This will enable us to construct codes which will capture a member of the guilty coalition with high probability.

Suppose a coalition C of c users creates an illegal copy of an object. Codes that enable the capture of a member of the coalition C with probability at least $1 - \epsilon$ are called *c-secure codes with ϵ-error*. Here the probability is taken over the random choices made by the distributor and the random choices made by the coalition C.

To simplify the exposition we don't give the formal combinatorial definition of c-secure codes. Instead we present a construction of a good c-secure code.

4 Construction of c-secure Codes

The idea for the construction c-secure codes is similar to the one used in Section 2.1. We first construct an (l, n)-code which is n-secure. Thus, no matter how large the coalition is, we will be able to trace an illegal copy back to a member of the coalition with high probability. The length of this code is $n^{O(1)}$ and hence, too large to be practical. We then show how this code can be used to construct c-secure codes for n users whose length is $\log^{O(1)}(n)$ when $c = O(\log n)$.

We begin by presenting an (l, n)-code which is n-secure with ϵ-error for any $\epsilon > 0$. Let c_m be a column of height n in which the first m bits are 1 and the rest are 0. The code $\Gamma_0(n, d)$ will consist of all columns $c_1, \ldots, c_{m-1}$, each duplicated d times. The amount of duplication will determine the error probability ϵ. For example, the code $\Gamma_0(4, 3)$ for four users A, B, C, D is

$$A : 111111111$$
$$B : 000111111$$
$$C : 000000111$$
$$D : 000000000$$

Let $w^{(1)}, \ldots, w^{(n)}$ denote the codewords of $\Gamma_0(n, d)$. Before the distributor embeds the codewords of $\Gamma_0(n, d)$ in an object he makes the following random choice: the distributor randomly picks a permutation[3] $\pi \in S_l$. User u_i's copy of the object will be fingerprinted using the word $\pi w^{(i)}$. Note that the same permutation π is used for all users. The point is that π will be kept hidden from

---

[3] S_l is the full symmetric group of all permutations on l letters. For a word $x \in \{0,1\}^l$ and a permutation $\pi \in S_l$ we denote by πx the l-bit word $x_{\pi(1)} x_{\pi(2)} \cdots x_{\pi(l)}$.

the users. Keeping the permutation hidden from the users is equivalent to hiding the information of which mark in the object encodes which bit in the code. It is a bit surprising that this simple random action taken by the distributor is sufficient to overcome the barrier of Theorem 11 and enables us to prove the following theorem:

Theorem 12. *For $n \geq 3$ and $\epsilon > 0$ let $d = 2n^2 \log(2n/\epsilon)$. The code $\Gamma_0(n,d)$ is n-secure with ϵ-error.*

The length of this code is $d(n-1) = O(n^3 \log \frac{n}{\epsilon})$. To prove the theorem we must describe an algorithm, which given a word x generated[4] by some coalition C, outputs a member of C with probability $1 - \epsilon$. First we introduce some notation.

1. Let B_m be the set of all bit positions in which the users see columns of type c_m. That is, B_m is the set of all bit positions in which the first m users see a 1 and the rest see a 0. The number of elements in B_m is d.
2. For $2 \leq s \leq n-1$ define $R_s = B_{s-1} \cup B_s$.
3. For a binary string x, let weight(x) denote the number of 1's in x.

Before we describe the algorithm we give some intuition. Suppose user s is *not* a member of the coalition C which produced the word x. The hidden permutation π prevents the coalition from knowing which marks represent which bits in the code $\Gamma_0(n,d)$. The only information the coalition has is the value of the marks it can detect. Observe that without user s a coalition sees exactly the same values for all bit positions $i \in R_s$. For instance, in the code $\Gamma_0(4,3)$ above, the coalition A, C, D sees the exact same bit pattern for all bit positions in R_2. Hence, for a bit position $i \in R_s$, the coalition C can not tell if i lies in B_s or in B_{s-1}. This means that whichever strategy they use to set the bits of $x|_{R_s}$, the 1's in $x|_{R_s}$ will be roughly evenly distributed between $x|_{B_s}$ and $x|_{B_{s-1}}$ with high probability. Hence, if the 1's in $x|_{R_s}$ are not evenly distributed then, with high probability, user s is a member of the coalition that generated x.

Algorithm 1. Given a word $x \in \{0,1\}^l$, find a subset of the coalition that produced x.
1. If weight($x|_{B_1}$) > 0 then output "User 1 is guilty".
2. If weight($x|_{B_{n-1}}$) $< d$ then output "User n is guilty".
3. for all $s = 2$ to $n - 1$ do: Let $k = $ weight($x|_{R_s}$). If

$$\text{weight}(x|_{B_{s-1}}) < \frac{k}{2} - \sqrt{\frac{k}{2} \log \frac{2n}{\epsilon}}$$

then output "User s is guilty".

---

[4] When we say that a coalition C generated a word x, we mean that the bits of x have already been unscrambled using π^{-1}. For example, the first bit of x is the value of the mark which encodes the first bit of the codewords.

The correctness of the algorithm is proved in the next two lemmas.

Lemma 13. *Consider the code $\Gamma_0(n,d)$ where $d = 2n^2 \log(2n/\epsilon)$. Let S be the set of users which algorithm 1 pronounces as guilty on input x. Then with probability at least $1 - \epsilon$, the set S is a subset of the coalition C that produced x.*

Proof. Suppose user 1 was pronounced guilty, i.e. $1 \in S$. Then $\text{weight}(x|_{B_1}) > 0$. This implies that user 1 must be a member of C (otherwise the bits in B_1 would be undetectable for C which would imply that $\text{weight}(x|_{B_1}) = 0$). Similarly if $n \in S$ then $n \in C$.

Suppose the algorithm pronounces user $1 < s < n$ as guilty. We show that the probability that s is not guilty is at most ϵ/n. This will show that the probability that there exists a user in S which is not guilty is at most ϵ.

Let s be an innocent user, i.e. $s \notin C$. As was discussed above, this means that the the coalition C can not distinguish between the bit positions in R_s. Since the permutation π was chosen uniformly at random from the set of all permutations, the 1's in $x|_{R_s}$ may be regarded as being randomly placed in $x|_{R_s}$. Let $k = \text{weight}(x|_{R_s})$. Define Y to be a random variable which counts the number of 1's in $x|_{B_{s-1}}$ given that $x|_{R_s}$ contains k 1's. For any integer r in the appropriate range:

$$\Pr[Y = r] = \frac{\binom{M}{r}\binom{M}{k-r}}{\binom{2M}{k}}$$

where $M = 2n^2 \log(2n/\epsilon)$ is the size of B_{s-1}. Clearly, the expectation of Y is $k/2$. To bound the probability that s was pronounced guilty we need to bound

$$\Pr\left[Y < k/2 - \sqrt{\frac{k}{2}\log\frac{2n}{\epsilon}}\right]$$

from above. It can be shown that the variance of Y is $\text{var}(Y) = \frac{k}{4}\frac{(2M-k)}{(2M-1)}$.

Let X be a binomial random variable over k experiments with success probability $1/2$. The variance of X is $k/4$. We see that the variance of Y is smaller than the variance of X, implying that for any $a > 0$

$$\Pr\left[\left|Y - \frac{k}{2}\right| > a\right] \le \Pr\left[\left|X - \frac{k}{2}\right| > a\right] \le 2e^{-2a^2/k}$$

where the last inequality follows from the standard Chernoff bound [2, Appendix A]. Plugging in $a = \sqrt{\frac{k}{2}\log\frac{2n}{\epsilon}}$ leads to

$$\Pr\left[Y < \frac{k}{2} - \sqrt{\frac{k}{2}\log\frac{2n}{\epsilon}}\right] \le \Pr\left[\left|Y - \frac{k}{2}\right| > \sqrt{\frac{k}{2}\log\frac{2n}{\epsilon}}\right] \le 2e^{-\log(2n/\epsilon)} = \frac{\epsilon}{n}\;.$$

Thus, if user s is innocent then the probability of her being pronounced guilty by Algorithm 1 is at most ϵ/n. Therefore, the probability that some innocent user will be pronounced guilty is at most ϵ. This proves the lemma. □

Lemma 14. *Consider the code* $\Gamma_0(n, d)$ *where* $d = 2n^2 \log(2n/\epsilon)$. *Let* S *be the set of users which algorithm 1 pronounces as guilty on input* x. *Then the set* S *in not empty.*

The proof of the lemma relies on the following claim.

Claim 15. *Suppose the set* S *is empty. Then for all* s *we have*

$$weight(x|_{B_s}) \leq 2s^2 \log(2n/\epsilon) \ .$$

Proof. The proof is by induction on s. For $s = 1$, the claim is immediate since if user 1 is not guilty then $\text{weight}(x|_{B_1}) = 0$.

Now, we assume the claim holds for $s < n - 1$ and prove it for $s + 1$. Define

$$k = \text{weight}(x|_{B_s}) \quad ; \quad k' = \text{weight}(x|_{B_{s+1}}) \quad ; \quad t = \text{weight}(x|_{R_{s+1}}) \ .$$

Then the following three conditions are satisfied:

$$t = k + k' \quad ; \quad k \leq 2s^2 \log(2n/\epsilon) \quad ; \quad k \geq \frac{t}{2} - \sqrt{\frac{t}{2} \log \frac{2n}{\epsilon}}$$

The first condition follows from the fact that $R_{s+1} = B_s \cup B_{s+1}$. The second is the inductive hypothesis and the third follows from the fact that user s was not pronounced guilty, i.e. $s \notin S$. We show that these three conditions imply $k' \leq 2(s + 1)^2 \log(2n/\epsilon)$ which will prove the claim for $s + 1$.

$$k' = t - k \leq \frac{t}{2} + \sqrt{\frac{t}{2} \log \frac{2n}{\epsilon}} = \frac{k + k'}{2} + \sqrt{\frac{1}{2}(k + k') \log \frac{2n}{\epsilon}}$$

$$\leq \frac{2s^2 \log(2n/\epsilon) + k'}{2} + \sqrt{\frac{1}{2}(2s^2 \log \frac{2n}{\epsilon} + k') \log \frac{2n}{\epsilon}}$$

which leads to

$$k' \leq 2s^2 \log \frac{2n}{\epsilon} + \sqrt{2(2s^2 \log \frac{2n}{\epsilon} + k') \log \frac{2n}{\epsilon}} \ .$$

Suppose $k' = 2r^2 \log(2n/\epsilon)$ for some constant r. Substituting for k' and dividing by $2 \log(2n/\epsilon)$ we get

$$r^2 \leq s^2 + \sqrt{s^2 + r^2} \ .$$

It is not difficult to see that for this inequality to be satisfied when $s \geq 1$ we must have $r \leq s + 1$. Hence, $k' \leq 2(s + 1)^2 \log(2n/\epsilon)$. $\qquad\square$

Proof of Lemma 14. Suppose S is empty. Since user n was not pronounced guilty we know that $\text{weight}(x|_{B_{n-1}}) = d = 2n^2 \log(2n/\epsilon)$. On the other hand, for $s = n - 1$ Claim 15 shows that $\text{weight}(x|_{B_{n-1}}) \leq 2(n - 1)^2 \log(2n/\epsilon)$. This contradiction proves the lemma. $\qquad\square$

This completes the proof of Theorem 12. The following theorem provides a weak lower bound on the length of n-secure codes. We leave the proof for the final version of the paper.

Theorem 16. *Any* (l, n)-*code which is* n-*secure with* ϵ-*error must have length* $l \geq n - 1$, *for any* $\epsilon > 0$.

4.1 Logarithmic length c-secure codes

The n-secure code constructed in the previous section enables us to use the techniques of [5] to construct c-secure (n, l)-codes of length $l = c^{O(1)} \log n$. We thank Naor [10] for pointing out this relation. In this section we demonstrate how to apply the simplest technique from [5] to construct a short c-secure code from the n-secure code of the previous section. The basic idea is to use the n-secure code as the alphabet over which the techniques of [5] can be applied.

Let $\mathcal{C}$ be an (L, N)-code over an alphabet of size n where the codewords are chosen independently and uniformly at random[5]. The idea is to compose our n-secure code $\Gamma_0(n, d)$ with the code $\mathcal{C}$ as we did in the proof of Lemma 7. We call the resulting code $\Gamma'(L, N, n, d)$. Thus, the code $\Gamma'(L, N, n, d)$ contains N codewords and has length $Ld(n-1)$. It is made up of L copies of $\Gamma_0(n, d)$. We will refer to these copies as the *components* of $\Gamma'(L, N, n, d)$. The point is that the codewords of the code $\mathcal{C}$ will be kept hidden from the users. This is in addition to keeping hidden the L permutations used when embedding the L copies of $\Gamma_0(n, d)$ in the object.

Theorem 17. *Given integers N, c and $\epsilon > 0$ set $n = 2c$; $L = 2c \log(2N/\epsilon)$ and $d = 2n^2 \log(4nL/\epsilon)$. Then, $\Gamma'(L, N, n, d)$ is a code which is c-secure with ϵ-error. The code contains N words and has length $l = O(Ldn) = O(c^4 \log(N/\epsilon) \log(1/\epsilon))$.*

To prove the theorem we show an algorithm that finds a member of the guilty coalition and then prove it's correctness.

Algorithm 2. Given a word $x \in \{0, 1\}^l$, find a subset of the coalition that produced x.

1. Apply Algorithm 1 to each of the L components of x. For each component $i = 1, \ldots, L$ arbitrarily choose one of the outputs of Algorithm 1. Set y_i to be this chosen output. Note that y_i is a number between 1 and n. Next, form the word $y = y_1 \cdots y_L$.
2. Find the word $w \in \mathcal{C}$ which matches y in the most number of position (ties are broken arbitrarily).
3. Let u be the user whose codeword is derived from $w \in \mathcal{C}$. output "User u is guilty".

Lemma 18. *Let x be a word which was produced by a coalition C of at most c users. Then with parameters as in Theorem 17, Algorithm 2 will output a member of C with probability at least $1 - \epsilon$.*

Proof (sketch). Let W be the set of codewords in $\mathcal{C}$ that correspond to the users in the coalition C. For every $1 \leq i \leq L$ Algorithm 1 guarantees that y_i will match w_i for some $w \in W$ with probability $1 - \epsilon/2L$. This follows from the choice of d and the fact that in component i the users in C see words from

---

[5] In [5] the codewords of $\mathcal{C}$ are considered as random hash functions
$h : \{1, \ldots, L\} \to \{1, \ldots, n\}$.

$\Gamma_0(n, d)$. It follows that the above condition will be satisfied in *every* component with probability at least $1 - \epsilon/2$. We refer to this as event A.

Recall that the size of W is at most c. Therefore, when event A occurs there must exist a word $w \in W$ which matches y in L/c positions. However, since the words in $\mathcal{C}$ are random and hidden from the users, any word in $\mathcal{C}$ which is not in W is expected to match y in only $L/n = L/2c$ positions. Using the Chernoff bound it can be shown that the probability that a random word will match y in L/c positions is less than $\epsilon/2N$. Hence, the probability the some word in $\mathcal{C} \setminus W$ will match y in L/c positions is at most $\epsilon/2$. This shows that when event A occurs, the algorithm will output a member of C with probability at least $1 - \epsilon/2$. Combining this with the fact that event A occurs with probability at least $1 - \epsilon/2$ proves the lemma. $\qquad\qquad\square$

5 Distribution Scheme

Up until now, we have been ignoring distribution of the uniquely fingerprinted copies. This is fair, as at worst we can send each user an entire unique copy. However, this is impractical for products such as electronic books, software or CD-ROMs which are mass produced. We would like to come up with a scheme in which a user receives a bulk of data common to all users, and a small number of extra bits unique to him. We refer to the bulk of data as the *public data* and denote it by D. We refer to the extra bits given privately to user u as the *private data* and denote it by M_u. For the distribution scheme to be secure, given (D, M_u), user u should not be able to deduce any information about the fingerprints in copies given to other users.

Throughout this section, let Γ be an (l, n)-code which is used to fingerprint the objects. We denote the object to be distributed by P and let L be its length. Assume the object P can be partitioned into l pieces with exactly one mark in each piece. Let $p(r, s)$ be the rth piece which contains the rth mark in state $s \in \{0, 1\}$. For any r, the pieces $p(r, 0)$ and $p(r, 1)$ are interchangeable, that is a copy with one replaced with the other will behave identically. Given a codeword $x = x_1 x_2 \ldots x_l \in \Gamma$, let $P(x) = \{p(1, x_1), p(2, x_2), \ldots, p(l, x_l)\}$ be the partition set implied by x.

Theorem 19. *It is possible to solve the distribution problem using $O(L)$ size public data and $O(L^{1/m} l / \log L)$ private data for some fixed constant m.*

Proof (sketch). We encode each piece $p(r, s)$ with a secure private key cryptosystem. See [9] for the precise definitions; such systems are known to exist under some standard hardness assumptions. Let $f_k : \{0, 1\}^{L'} \to \{0, 1\}^{L'}$ be such a system, where $L' = L/l$ is the length of a single piece $p(r, s)$. The key k will have length $K = (L')^{1/m}$ for some fixed constant m. The fact that f_k is a secure private key cryptosystem implies that for a random key k, given $f_k(x)$, no polynomial time predicate can extract one bit of information about x with non negligible probability. This property is crucial for the security of our distribution scheme.

For each piece, $p(r,s)$, pick a random key $k(r,s)$. The public data will be

$$D = \big\{ f_{k(r,s)}\big(p(r,s)\big) \mid 1 \leq r \leq l,\ s \in \{0,1\} \big\} \ .$$

The size of D is $2L$, twice the size of the original object. Let $x \in \Gamma$ be the word associated with user u. The private data given to user u is the collection of keys necessary to decrypt his pieces:

$$M_u = \{ k(1, x_1), k(2, x_2), \ldots, k(l, x_l) \} \ .$$

Using this scheme, given (D, M_u), any user u can construct a usable copy of the distributed object. The size of the private data, M_u is $lK = O(l^{1-1/m} L^{1/m}) = O(lL^{1/m})$.

It is possible to further reduce the size of the private data by using double encryption: we group the keys $k(r,s)$ into blocks. Each block is encrypted and made public. Each user will receive keys that will enable him to decrypt his block of keys. This method requires public data of size $O(L)$ and private data of size $O(L^{1/m} \frac{l}{\log L})$ for some constant $m > 1$. □

It is worth noting that when implementing this distribution scheme, one can use a standard private key cryptosystem such as DES. Such systems use fixed length keys. This leads to private data of length $O(l/\log(L))$.

6 Discussion and Open Problems

The most significant contribution of this paper is to show how to overcome collusion when fingerprinting digital data. To summarize our results, we restrict the size of coalitions to be at most $\log n$ where n is the number of users. For the problem of naive redistribution considered in Section 2, we constructed codes of length $\log^3 n$. For the general redistribution problem considered in Section 3 we constructed codes of length $\log^6 n$. Furthermore, we demonstrated an efficient method for shipping fingerprinted data which requires only a small constant factor increase in the size of the data.

There are still many open problems which remain to be solved. Recall that throughout the paper, we assumed that secure marks can be embedded in the fingerprinted data. A mark encodes one bit of information and is secure if it can only be detected by collusion. To emphasize the fact that we will not be dealing with the implementation of secure marks, we referred to the assumption that they exist as the "Marking Assumption". In many domains, one can construct secure marks with the aid of problems that are believed to be hard. For instance, when fingerprinting movies, a single mark can be encoded by using one camera view point versus another. The choice of one view point versus another in a specific scene, encodes one bit of information in the film. Given an image, the problem of transforming the image to an image taken from a different view point is believed to be hard. As this method of marking can be used to fingerprint movies, we say that the Marking Assumption holds in the domain of movies.

Showing that the Marking Assumption is satisfied for software is much harder. As was stated in the introduction, there is a great deal of empirical evidence to support the existence of secure marks in software. However, to the best of our knowledge, no formal results have been obtained.

To overcome the fact that secure marks are hard to construct we could make our codes resilient to noise. Suppose a coalition of users is able to change a small fraction of its undetectable bits. Is it possible to construct codes which are still c-secure?

Acknowledgments

The authors wish to thank Richard Lipton for some discussions on this work. We are grateful to Moni Naor for many suggestions and comments on this work.

References

1. N. Alon, J. Bruck, J. Naor, M. Naor and R. Roth, *Construction of asymptotically good low-rate error-correcting codes through pseudo-random graphs*, IEEE Transactions on Information Theory, vol. 38, 1992, pp. 509–516.
2. N. Alon and J. Spencer, *The probabilistic method*, Wiley, 1992.
3. J. Brassil, S. Low, N. Maxemchuk and L. O'Gorman, *Electronic marking and identification techniques to discourage document copying*, Proceedings of Infocom '94, pp. 1278–1287, June 1994.
4. G. Caronni, *Assuring ownership rights for digital images*, H.H. Brueggemann and W. Gerhardt-Haeckl (Ed.) Proceedings of 'reliable IT systems' (verlaessliche IT-Systeme) VIS '95 Vieweg Publishing Company, Germany , 1995.
5. B. Chor, A. Fiat and M. Naor, *Tracing traitors*, Proceedings of Crypto, 1994, pp. 257–270.
6. P. Erdös, P. Frankl, Z. Füredi, *Families of finite sets in which no set is covered by the union of r others*, Israel J. of math. 51, 1985, pp. 79–89.
7. D. Glover, *The protection of computer software*, Cambridge University, 2nd ed., 1992.
8. O. Goldreich and R. Ostrovsky, *Software protection and simulation on oblivious RAMs*, to appear in JASM.
9. S. Goldwasser, *The search for provably secure cryptosystems*, AMS Lecture notes cryptology and computational number theory, 1990.
10. M. Naor, *Private communications*.
11. B. Schneier, *Applied cryptography*, Wiley, 1994.
12. Software Publishers Association, Press release, 1994.
13. K. Tanaka, Y. Nakamura and K. Matsui. *Embedding secret information into a dithered multi-level image*, Proceedings of the 1990 IEEE Military Communications Conference, pp. 216–220, September 1990.
14. van Lint, *Introduction to coding theory*, Springer-Verlag, 1982.
15. N. Wagner, *Fingerprinting*, Proceedings of the 1983 IEEE Symposium on Security and Privacy, April, 1983, pp. 18–22.

Author Index